Chemical Process Analysis

PRENTICE HALL INTERNATIONAL SERIES
IN THE PHYSICAL AND CHEMICAL ENGINEERING SCIENCES

NEAL R. AMUNDSON, SERIES EDITOR, *University of Houston*

AMUNDSON *Mathematical Methods in Chemical Engineering: Matrices and Their Applications*
BALZHIZER, SAMUELS, AND ELLIASSEN *Chemical Engineering Thermodynamics*
BRIAN *Staged Cascades in Chemical Processing*
BUTT *Reaction Kinetics and Reactor Design*
DENN *Process Fluid Mechanics*
FOGLER *The Elements of Chemical Kinetics and Reactor Calculations: A Self-Paced Approach*
FOGLER AND BROM *Elements of Chemical Reaction Engineering*
HIMMELBLAU *Basic Principles and Calculations in Chemical Engineering, 4th edition*
HINES AND MADDOX *Mass Transfer: Fundamentals and Applications*
HOLLAND *Fundamentals and Modeling of Separation Processes: Absorption, Distillation, Evaporation, and Extraction*
HOLLAND AND ANTHONY *Fundamentals of Chemical Reaction Engineering*
KUBICEK AND HLAVACEK *Numerical Solution of Nonlinear Boundary Value Problems with Applications*
KYLE *Chemical and Process Thermodynamics*
LEVICH *Physiochemical Hydrodynamics*
LUYBEN AND WENZEL *Chemical Process Analysis: Mass and Energy Balances*
MODELL AND REID *Thermodynamics and its Applications, 2nd edition*
MYERS AND SEIDER *Introduction to Chemical Engineering and Computer Calculations*
OLSON AND SHELSTAD *Introduction to Fluid Flow and the Transfer of Heat and Mass*
PRAUSNITZ, LICHTENTHALER, AND DE AZEVEDO *Thermodynamics of Fluid-Phase Equilibria, 2nd edition*
PRAUSNITZ ET AL *Computer Calculations for Multicomponent Vapor-Liquid and Liquid-Liquid Equilibria*
RAMKRISHNA AND AMUNDSON *Linear Operator Methods in Chemical Engineering with Applications for Transport and Chemical Reaction Systems*
RHEE ET AL *First-Order Partial Differential Equations: Theory and Applications of Single Equations*
RUDD ET AL *Process Synthesis*
SCHULTZ *Diffraction for Materials Scientists*
STEPHANOPOULOS *Chemical Process Control: An Introduction to Theory and Practice*
VILLADSEN AND MICHELSEN *Solution of Differential Equation Models by Polynomial Approximation*

Chemical Process Analysis: Mass and Energy Balances

WILLIAM L. LUYBEN
LEONARD A. WENZEL

Department of Chemical Engineering
Lehigh University
Bethlehem, Pennsylvania

PRENTICE HALL, Englewood Cliffs, New Jersey 07632

Library of Congress Cataloging-in-Publication Data

Luyben, William L. (date)
Chemical process analysis.

Includes index.
1. Chemical engineering. 2. Chemical engineering—
Data processing. I. Wenzel, Leonard A.
II. Title.
TP155.L88 1988 660.2 87-6977
ISBN 0-13-128588-2

Editorial/production supervision: WordCrafters Editorial Services, Inc.
Manufacturing buyer: Loraine Fumoso

Printed in the United States of America
10 9 8 7 6 5 4 3 2 1

ISBN 0-13-128588-2 025

PRENTICE-HALL INTERNATIONAL (UK) LIMITED, *London*
PRENTICE-HALL OF AUSTRALIA PTY. LIMITED, *Sydney*
PRENTICE-HALL CANADA INC., *Toronto*
PRENTICE-HALL HISPANOAMERICANA, S.A., *Mexico*
PRENTICE-HALL OF INDIA PRIVATE LIMITED, *New Delhi*
PRENTICE-HALL OF JAPAN, INC., *Tokyo*
SIMON & SCHUSTER ASIA PTE. LTD., *Singapore*
EDITORA PRENTICE-HALL DO BRASIL, LTDA., *Rio de Janeiro*

Contents

Preface

This book is intended for use in a first-level course in chemical engineering. Its basic objectives are: (1) to introduce beginning students to the field of chemical engineering, and (2) to teach the principles of mass and energy balances in the context of designing and operating a chemical process in the most profitable and safest manner.

The book represents a new approach to introduce students to chemical engineering. Its origins go back almost a decade to our dissatisfaction with the first chemical engineering course here at Lehigh. This course covered the topics normally presented in the traditional "stoichiometry" course.

We found, however, that many of our students were not getting a clear picture of what chemical engineering was all about until they took a junior-level "unit operations" course. Thus, some students found out much too late that they really didn't want to be chemical engineers. Some students were also poorly motivated during their sophomore year. Therefore, we searched for some way to move some of this "unit operations" material down into the first sophomore chemical engineering course.

The objective was to provide a course that would permit students to get a good "feel" for chemical engineering at an early stage. This would better motivate those students who liked the material to put in the time and effort it takes to make it successfully through the tough chemical engineering curriculum.

The philosophical basis for this approach was provided by the early book of Thibaut Brian, *Staged Cascades in Chemical Processing* (Englewood Cliffs,

NJ: Prentice-Hall, 1972). We found the thrust of this book to be what we were looking for. However, Brian's book was short on quantitative technical material and contained no discussion of solutions to problems by computer techniques.

Accordingly, we began to put together a set of notes that covered much more material. After several revisions, we found that we could effectively cover mass and energy balances applied to three typical and very important staged operations. We could go into enough depth for students to appreciate the challenges and opportunities that a career in chemical engineering has to offer. Quantitative approaches to solving problems, examination of the many alternative process configurations, safety considerations, economics, and engineering compromises were all naturally and effectively covered by examining the staged operations of binary distillation, liquid-liquid extraction, and evaporation.

This unique approach also meant that the number of staged operations topics left to be covered in the unit operations courses was much reduced, permitting more time in these courses for other important material.

We have tested and honed this approach for almost ten years here at Lehigh, revising and fine-tuning our notes almost yearly. Our experience has been very positive. We feel that students leave the course with a good idea of what chemical engineering is and with the quantitative ability to apply mass and energy balances to any new process. The students who stay in the curriculum are better motivated and have a better perspective on how their subsequent courses fit into the chemical engineering knowledge base.

We recognize that this book is a break with tradition. We feel it is an exciting and effective new approach and hope you will give it a try.

The contributions of many people to the development of the book are gratefully acknowledged. Certainly, first on the list should be the many bright-eyed and bushy-tailed sophomores who have suffered through many revisions. Their incisive and sometimes painfully candid comments were invaluable. We thank Mike Luyben for proofreading the final version of the manuscript. Our colleagues Hugo Caram, Cesar Silebi, and Bryce Andersen also deserve thanks for team-teaching the course with us on several occasions. Jane Lenner was a model of thoroughness and patience in typing the many revisions of the manuscripts. Kathy Turoski helped in the final typing.

Finally, we would like to thank our families for their support and encouragement over some fifty years of collective teaching careers. All the crack-of-dawn and late-evening writing periods would have been impossible without a lot of understanding and love. We officially apologize for all of the times we got home late for dinner!

This book is dedicated to the memories of two outstanding engineers and teachers: Alan S. Foust and Jack A. Gerster.

William L. Luyben
Leonard A. Wenzel

Chemical Process Analysis

INTRODUCTION TO CHEMICAL ENGINEERING **PART I**

The Chemical Engineering Profession **1**

1.1 WELCOME AND OVERVIEW

Welcome to chemical engineering! If you are the typical reader of this book, you are a sophomore taking your first course in chemical engineering. You may have some doubts and many questions about what you are getting yourself into by tackling this curriculum. You need to learn more about the profession quickly in order to decide whether or not you should "tough it out" in a course of study that is correctly reputed to be difficult and demanding.

What are you getting yourself into? Well, the purpose of this book and the course that you are taking is to provide answers to some of your questions. The book has two basic objectives:

1. *To introduce beginning students to the field of chemical engineering.* Many students know very little about what chemical engineers do, where they work, how much money they make, and the like. Chapters 1 through 3 attempt to describe some typical chemical processes and typical chemical engineering activities. We strongly recommend that this material be supplemented by plant trips (to see, feel, and smell a chemical plant or oil refinery) and by talks from practicing chemical engineers (alumni of your school are particularly effective since they demonstrate that it is possible to get through the curriculum). Both are invaluable to inexperienced students in gaining some concept of the size and complexity of a modern chemical plant and in providing a perspective of the wide variety of career potentials available to chemical engineers.

2. *To teach the principles of mass and energy balances in the context of designing and operating a chemical process in the most profitable and safest manner.* These basic principles are presented along with a large number of examples and problems that illustrate their use both for processes that include chemical reactions and for those that do not. We have chosen to illustrate the application of the principles by analyzing in a fair amount of detail three relatively simple but extremely important countercurrent staged operations: distillation, evaporation, and extraction. These staged processes are very widely used in industry and are reasonably easy for a beginning student to understand. Therefore, sufficiently detailed knowledge can be gained in a beginning course to permit introduction to (1) the design and operating trade-offs that are so typical in chemical engineering, (2) a consideration of safety and economics, and (3) comparisons of alternative process configurations.

We hope this book will give you a "feel" for the field of chemical engineering early in your college career. The material includes both the traditional graphical solution techniques (McCabe-Thiele and Ponchon diagrams) as well as the more modern computer solutions. Mass and energy balances for multistaged processes are ideally suited for digital computer solutions; the computer permits you to solve difficult problems without having to make too many simplifying assumptions, and it allows you to explore a large number of process cases (e.g., different quantities and compositions of feed streams, different desired products, and alternative process configurations).

Please heed a well-intended word of warning about the use of computers in engineering: they are very powerful tools, but they are only as smart as the engineers who use them. Students sometimes get so wrapped up in *computing* that they stop *thinking*! You have to understand the basic fundamentals and the engineering aspects of a problem before you can effectively use a computer. That is why the computer solutions are given after the basic concepts have been learned. The traditional graphical methods are presented first because they show a picture of what is going on as process parameters are changed. Once this insight has been gained, you are ready to go to the computer for more rigorous and extensive calculations.

The material in the book is basically very simple. By the time you are juniors and seniors, you will look back on this material (fondly, we hope!) as being straightforward common sense. However, you will not find it easy as you go through it for the first time. The typical beginning student finds the first course in chemical engineering difficult for a couple of reasons:

1. You have to learn a new language, including many new terms that you must understand and remember. For example, what is a "tube-in-shell heat exchanger"? What do terms like "relative volatility" and "reflux ratio" mean? All this chemical engineering jargon takes

time and effort to assimilate. In addition, you must learn English engineering units. You are familiar with metric units and the SI system from your chemistry and physics courses. But many people in industry still use English units. So you must learn the language of "pounds," "feet," degrees "Fahrenheit" and "Rankine," "psig," "psia," "gpm," etc. The newer "SI" units (e.g., pressures in "pascals" instead of "psig" and "atmospheres") must also be applied in engineering situations. We will talk more about this later.

2. You will find that most chemical engineering courses are quite different from the science courses you have taken. The objective is not to memorize formulas (most chemical engineering tests are "open book"); rather, the emphasis is on learning to think and to apply basic principles to new situations. You have to learn to derive your own formulas for the specific situation you are dealing with. You simply go back to fundamental mass balances and energy balances. Indeed, the only equation you will need to memorize for this course is

$$\text{IN} = \text{OUT}$$

Sound simple? Well, it is, but it takes a little practice and sweat to get the hang of it. Once you get the idea, though, you will wonder why it took you so long to learn.

1.2 HISTORY

The chemical industry developed in Europe, especially in Germany, toward the end of the nineteenth century. The United States imported what few chemicals were used here, especially dyes and simple drugs. There was a small U.S. chemical industry making explosives and some of the more basic inorganic chemicals, refining metals, tanning leather, and assisting in papermaking. The mounting tensions as World War I approached, and the war itself, cut off the U.S. from easy access to Germany. Suddenly, a chemical industry had to be built here.

In Germany, industrial chemical operations were built and managed by mechanical engineers working with industrial chemists. That pattern has been retained almost to the present. In the U.S., however, few chemists existed, and they did not relish a shift from their laboratories. Nor did mechanical engineers in the U.S. have the classical background in chemistry afforded by the rigorous secondary school system of Europe. Instead, the basic principles were being evolved by faculty in what soon became chemical engineering departments at MIT and Michigan. Such men as A. D. Little, W. K. Lewis, and A. H. White developed the idea that chemical processes could be broken up into (1) unit operations such as heat transfer, distillation, evaporation, and filtration, and (2) unit processes such as oxidation, sulfonation, and nitration. These units could then be studied as separate entities. Because there was a strong need at that time for professionals who were hybrids combining the skills of the chemist and the mechanical engineer, the chemical engineering discipline began to grow.

Through the years, chemical engineers have played a large and vital role in developing the U.S. chemical and petroleum industries. They have served as designers, construction consultants, and operators of chemical plants. They have also been important participants in research and development. The need to learn more about the processing operations themselves and the fundamental mechanisms involved was recognized and pursued. Finally, chemical engineers have assumed a role in the chemical business field in forecasting needs, consulting on product and equipment use, and marketing.

The discipline of chemical engineering has spread throughout the world. Curricula in chemical engineering, closely modeled after those in the U.S., are to be found on all the continents.

1.3 CURRICULUM

As chemical engineering knowledge developed, it was inserted into university courses and curricula. Before World War I, chemical engineering programs were distinguishable from chemistry programs in that they contained courses in engineering drawing, engineering thermodynamics, mechanics, and hydraulics taken from engineering departments. Shortly after World War I the first text in unit operations was published (W. H. Walker, W. K. Lewis, and W. H. McAdams, *Principles of Chemical Engineering*, New York: McGraw-Hill, 1923). Courses in this area became the core of chemical engineering teaching.

By the mid-1930s, chemical engineering programs included courses in (1) stoichiometry (using material and energy conservation ideas to analyze chemical process steps), (2) chemical processes or "unit operations," (3) chemical engineering laboratories (in which equipment was operated and tested), and (4) chemical plant design (in which cost factors were combined with technical elements to arrive at preliminary plant designs). The student was still asked to take the core chemistry courses, including general, analytical, organic, and physical chemistry. However, in addition, he or she took courses in mechanical drawing, engineering mechanics, electric circuits, metallurgy, and thermodynamics with other engineers.

Since World War II chemical engineering has developed rapidly. As new disciplines have proven useful, they have been added to the curriculum. Chemical engineering thermodynamics became generally formulated and taught by about 1945. By 1950, courses in applied chemical kinetics and chemical reactor design appeared. Process control appeared as an undergraduate course in about 1955, and digital computer use began to develop about 1960.

The idea that the various unit operations depended on common mechanisms of heat, mass, and momentum transfer developed about 1960. Consequently, courses in transport phenomena assumed an important position as an underlying, unifying basis for chemical engineering education. New general disciplines that have emerged in the last two decades include environmental and safety engineering, biotechnology, and electronics manufacturing processing. There has been an enormous amount of development in all fields, much of it arising out of more powerful computing and applied mathematics capabilities.

The new subjects forced some cuts in the non-chemical engineering part of the curriculum. Peripheral engineering courses were reduced, but the chemistry content of most curricula has remained strong. The reduction in the former areas was possible because of improved student preparation in high school science and mathematics. In fact, the number of courses in humanities and social sciences has actually increased, since it has been recognized that social, political, and human factors are vitally important in designing and operating a chemical process.

We suggest that you look at your college catalog and review the chemical engineering curriculum at your school. It will give you some perspective of where you are going and what courses you will be taking.

1.4 TYPICAL ACTIVITIES

The classical role of the chemical engineer is to take the discoveries made by the chemist in the laboratory and develop them into money-making, commercial-scale chemical processes. The chemist works in test tubes and Parr bombs with very small quantities of reactants and products (e.g., 100 ml), usually running "batch," constant-temperature experiments. Reactants are placed in a small container in a constant-temperature bath. A catalyst is added and the reactions proceed with time. Samples are taken at appropriate intervals to follow the consumption of the reactants and the production of products as time progresses.

By contrast, the chemical engineer typically works with much larger quantities of material and with very large (and expensive) equipment. Reactors can hold 1,000 gallons to 10,000 gallons or more. Distillation columns can be over 100 feet high and 10 to 30 feet in diameter. The capital investment for one process unit in a chemical plant may exceed $100 million!

Many commercial processes run in a "continuous" mode, as opposed to the "batch" mode that the chemist almost always employs. Feed streams and product streams are continuously fed and withdrawn from the process. The usual goal is to obtain "steady-state" operation in which all parameters in the continuous plant (temperatures, liquid levels, pressures, flow rates, compositions, etc.) are constant with time.

The chemical engineer is often involved in "scaling up" a chemist-developed small-scale reactor and separation system to a very large commercial plant. The chemical engineer must work closely with the chemist in order to understand thoroughly the chemistry involved in the process and to make sure that the chemist gets the reaction kinetic data and the physical property data needed to design, operate, and optimize the process. This is why the chemical engineering curriculum contains so many chemistry courses.

The chemical engineer must also work closely with mechanical, electrical, civil, and metallurgical engineers in order to design and operate the physical equipment in a plant—the reactors, tanks, distillation columns, heat exchangers, pumps, compressors, control and instrumentation devices, and so on. One big item that is always on such an equipment list is piping. One of the most impressive features of a typical chemical plant is the tremendous number of pipes running all over the site, literally hundreds of miles in many plants. These pipes transfer process materials (gases and liquids) into and out of the plant. They also carry utilities (steam, cooling water, air, nitrogen, and refrigerant) to the process units.

To commercialize the laboratory chemistry, the chemical engineer is involved in development, design, construction, operation, sales, and research. The terminology used to label these functions is by no means uniform from company to company, but a rose by any other name is still a rose. Let us describe each of these functions briefly. It should be emphasized that the jobs we shall discuss are "typical" and "classical," but are by no means the only things that chemical engineers do. The chemical engineer has a broad background in mathematics, chemistry, and physics. Therefore, he or she can, and does, fill a rich variety of jobs in industry, government, and academia.

1.4.1 Development

Development is the intermediate step required in passing from a laboratory-size process to a commercial-size process. The "pilot-plant" process involved in development might involve reactors that are five gallons in capacity and distillation columns that are three inches in diameter. Development is usually part of the commercialization of a chemical process because the scale-up problem is a very difficult one. Jumping directly from test tubes to 10,000-gallon reactors can be a tricky and sometimes dangerous endeavor. Some of the subtle problems involved which are not at all obvious to the uninitiated include mixing imperfections, increasing radial temperature gradients, and decreasing ratios of heat transfer areas to heat generation rates.

The chemical engineer works with the chemist and a team of other engineers to design, construct, and operate the pilot plant. The design aspect involves specifying equipment sizes, configuration, and materials of construction. Usually pilot plants are designed to be quite flexible, so that a wide variety of conditions and configurations can be evaluated.

Once the pilot plant is operational, performance and optimization data can be obtained in order to evaluate the process from an economic point of view. The profitability is assessed at each stage of the development of the process. If it appears that not enough money will be made to justify the capital investment, the project will be stopped.

The pilot plant offers the opportunity to evaluate materials of construction, measurement techniques, and process control strategies. The experimental findings in the pilot plant can be used to improve the design of the full-scale plant.

1.4.2 Design

Based on the experience and data obtained in the laboratory and the pilot plant, a team of engineers is assembled to design the commercial plant. The chemical engineer's job is to specify all process flow rates and conditions, equipment types and sizes, materials of construction, process configurations, control systems, safety systems, environmental protection systems, and other relevant specifications. It is an enormous responsibility.

The design stage is really where the big bucks are spent. One typical chemical process might require a capital investment of $50 to $100 million. That's a lot of bread! And the chemical engineer is the one who has to make many of the decisions. When you find yourself in that position, you will be glad that you studied as hard as you did (we hope) so that you can bring the best possible tools and minds to bear on the problems.

The product of the design stage is a lot of paper:

1. *Flow Sheets* are diagrams showing all the equipment schematically, with all streams labeled and their conditions specified (flow rate, temperature, pressure, composition, viscosity, density, etc.).
2. *P and I* (*Piping and Instrumentation*) *Drawings* are drawings showing all pieces of equipment (including sizes, nozzle locations, and materials), all piping (including sizes, materials, and valves), all instrumentation (including locations and types of sensors, control valves, and controllers), and all safety systems (including safety valve and rupture disk locations and sizes, flare lines, and safe operating conditions).
3. *Equipment Specification Sheets* are sheets of detailed information on all the equipment: precise dimensions, performance criteria, materials of construction, corrosion allowances, operating temperatures

and pressures, maximum and minimum flow rates, and the like. These "spec sheets" are sent to the equipment manufacturers for price bids and then for building the equipment.

Small-scale physical models of the plant are often built to help the engineers visualize and design the plant layout of equipment and piping. If you enjoy model railroading, you'll like to work with miniature plant models. Indeed, the design stage is where a lot of creative and innovative engineering can be translated into a real plant. The decisions made at this stage are long lived and very expensive to modify later in the field after the plant is built.

1.4.3 Construction

After the equipment manufacturers (vendors) have built the individual pieces of equipment, the pieces are shipped to the plant site (sometimes a challenging job of logistics, particularly for large vessels like distillation columns). The construction phase is the assembling of all the components into a complete plant. It starts with digging holes in the ground and pouring concrete for foundations for large equipment and buildings (e.g., the control room, process analytical laboratory, and maintenance shops). In some soils, pilings must be driven into the ground to support the structures. The technical aspects of these activities are usually handled by civil engineers.

After these initial activities, the major pieces of equipment and the steel superstructure are erected. Heat exchangers, pumps, compressors, piping, instrument sensors, and automatic control valves are installed. Control system wiring and tubing are run between the control room and the plant. Electrical wiring, switches, and transformers are installed for motors to drive pumps and compressors. As the process equipment is being installed, it is the chemical engineer's job to check that it is all hooked together properly and that each piece works correctly.

This is usually a very exciting and rewarding time for most engineers. You are seeing your ideas being translated from paper into reality. Steel and concrete replace sketches and diagrams. Construction is the culmination of years of work by many people. You are finally on the launch pad, and the plant is going to fly or fizzle! The moment of truth is at hand.

Once the check-out phase is complete, "startup" begins. Startup is the initial commissioning of the plant. It is a time of great excitement and round-the-clock activity. It is one of the best learning grounds for the chemical engineer. Now you find out how good your ideas and calculations really are. The engineers who have worked on the pilot plant and on the design are usually part of the startup team.

The startup period can require a few days or a few months, depending on the newness of the technology, the complexity of the process, and the quality of the engineering that has gone into the design. Problems are frequently encountered that require equipment modifications. This is time consuming and expensive: just the lost production from a plant can amount to thousands of dollars per day. Indeed, there have been some plants that have never operated, because of unexpected problems with control, corrosion, or impurities, or because of economic problems.

The engineers are usually on shift work during the startup period. There is a lot to learn in a short time period (just as there is for you in this course). Once the plant has been successfully operated at its rated performance, it is turned over to the operating or manufacturing department for routine production of products.

1.4.4 Manufacturing

Chemical engineers occupy a central position in manufacturing (or "operations" or "production," as it is called in some companies). Plant technical service groups are responsible for the technical aspects of running an efficient and safe plant. They run capacity and performance tests on the plant to determine where the bottlenecks are in the equipment, and then design modifications and additions to remove these bottlenecks.

Chemical engineers study ways to reduce operating costs by saving energy, cutting raw material consumption, and reducing production of off-specification products that require reprocessing. They study ways to improve product quality and reduce environmental pollution of both air and water. They are heavily involved in developing operating and control strategies that improve the dynamic performance and safety of the plant by:

1. Reducing the upsetting effects of disturbances (e.g., changes in cooling water temperature, feed stock quality, and feed rate).
2. Achieving smooth and rapid transition from one operating condition to another as desired products or feed compositions change.
3. Operating near constraints for maximum efficiency.
4. Detecting impending or existing unsafe conditions and taking appropriate emergency action to prevent disk ruptures, safety valve releases, or emergency venting.

In addition to serving in plant technical service, many engineers have jobs as operating supervisors. These supervisors are responsible for all aspects of the day-to-day operation of the plant, including supervising the plant operators who run the plant round the clock on a three-shift basis, meeting quality specifications, delivering products at agreed-upon times and in agreed-upon quantities, developing and maintaining inventories of equipment spare parts, keeping the plant well maintained, making sure safe practices are followed, avoiding excessive emissions into the local environment, and serving as spokespersons for the plant to the local community.

1.4.5 Technical Sales

Many chemical engineers find stimulating and profitable careers in technical sales. As with other sales positions, the work involves calling on customers, making recommendations on particular products to fill a customer's needs, and being sure that orders are handled smoothly. The sales engineer is the company's representative and must know the company's product line well. The sales engineer's ability to sell can greatly affect the progress and profitability of the company.

The marketing of many chemicals requires a considerable amount of interaction between engineers in the company producing the chemical and engineers in the company using the chemical. This interaction can take the form of advising on how to use a chemical or developing a new chemical in order to solve a specific problem of a customer.

When the sales engineer discovers problems that cannot be handled with confidence, he or she must be able to call on the expertise of specialists. The sales engineer may sometimes have to manage a joint effort among researchers from several companies who are working together to solve a problem.

1.4.6 Research

Chemical engineers are engaged in many types of research. They work with the chemist in developing new or improved products. They develop new and improved engineering methods (e.g., better computer programs to simulate chemical processes, better laboratory analysis methods for characterizing chemicals, and new types of reactors and separation systems). They work on improved sensors for on-line physical property measurements. They study alternative process configurations and equipment.

Research engineers are likely to be found in laboratories or at desks working on problems. They usually work as members of a team of scientists and engineers. Knowledge of the processes and common types of process equipment helps the chemical engineer make special contributions to the research effort. The chemical engineer's daily activities may sometimes closely resemble those of the chemist or physicist working on the same team.

1.5 SUPPLY OF AND DEMAND FOR CHEMICAL ENGINEERS

Both the supply and the demand for chemical engineers rose rather steadily for the 30-year period following World War II. There were a few brief periods of ups and downs, but the long-term steady growth of the chemical industry, both in the United States and abroad, required more and more chemical engineers. The pollution control activities of the late 1960s consumed the efforts of many chemical engineers.

The oil shortages of the 1970s triggered an explosion in the demand for chemical engineers. Almost all of the visionary planners were forecasting a permanent decrease in petroleum and natural gas supplies. The price of crude oil increased from $3 per barrel to $30 per barrel. The cost of energy in chemical plants jumped from $0.50 per million Btu to $5 per million Btu. This dramatic change resulted in an enormous demand for chemical engineers. They were needed to redesign plants in order to conserve energy and to develop, design, build, and operate the enormous new energy industries that were foreseen to arise on the basis of not foreign oil, but domestic coal, oil shale, and tar sands as raw materials.

The governmental and industrial planners envisioned massive plants located at the coal and shale fields, producing "syn-crude" (synthetic crude oil) which would be piped to new or revamped refineries to make gasoline, diesel fuel, heating oil, etc. The capital investment in these plants was to be in the billions of dollars. As a result, the projected demand for chemical engineers rose almost exponentially.

In response to these events, enrollments in chemical engineering doubled and tripled in universities around the U.S. Chemical engineering departments were stretched to, and sometimes beyond, the breaking point. The demand for additional faculty skyrocketed. At the same time, the supply of PhDs to fill these new positions was dropping. Few undergraduates were going on to graduate school, since it was financially very attractive to go into industry.

Despite all the rosy projections, the combination of reduced consumption and increased supplies of crude oil and natural gas in the early 1980s drove energy prices rapidly downward. Most of the syn-crude and shale-oil projects were scrapped. This drastic cutback in the energy companies, together with the recession of 1982–83 which adversely affected the chemical industry, resulted in a sharp weakening of the job market for chemical engineers. In most universities, many chemical engineering seniors found it quite difficult to find work as chemical engineers. Many companies laid off engineers or forced older engineers into early retirement. It was the worst period for chemical engineers since the depression of the 1930s!

The poor job market caused a roller-coaster drop in undergraduate enrollments in chemical engineering in 1983 and 1984. Then things stabilized somewhat. The number of graduating seniors dropped to about the point where supply was equal to demand. We all hope the experiences of the last decade are not repeated.

1.6 SALARIES AND MONEY

All of these ups and downs in supply and demand have had little effect on chemical engineering salaries. Starting B.S. engineers made about $5,000 per year back in 1955. In 1985 they made about $25,000 per year. An "average" engineer with twenty years of experience earns an annual salary in the $50,000 to $75,000 range. Of course some few engineers, the high fliers, earn much more than this.

While we are on the subject of finances, let's make one important point: don't expect to become filthy rich being an engineer! You can look forward to an interesting and stimulating career, and you will be well paid for your work. You should be able to live comfortably, support your family, send your children to college, and look forward to a secure retirement. But if you are really after the big bucks, get out of technical work. Start your own company, invest in real estate or the stock market, or start a rock music group. If money is your main motivation, engineering is not your profession. The hard work in engineering has its primary compensations in the satisfaction of a job well done and a contribution made to science and technology.

The preceding comments are by no means intended to discourage you from an engineering career, just to expose you to the reality of the situation. The authors personally would never have chosen any other careers. We've both had a lot of fun and gotten a lot of satisfaction out of the game for many years. So if you like solving problems and understanding how things work, hang in there in chemical engineering. Try it . . . you'll like it!

PROBLEMS

1–1. Find a family friend or neighbor who is a chemical engineer. Learn what he or she does at work, not just in terms of the job title but in terms of day-to-day activities. Ask what one project or job challenge was most rewarding during the past five years. What is it about the work that is discouraging or frustrating? Report your findings orally to your class.

1–2. Look up in your school catalog the details of your chemical engineering curriculum. Classify the various courses into the areas of basic science, mathematics, engineering science, engineering design, humanities, and social science. Calculate the fraction of the total credit hours devoted to each category.

The Chemical Process 2

2.1 INTRODUCTION

The people of today's world require a host of highly complex support systems: food, water, sanitary services, clothing, shelter, transportation systems, and health care. As the number and density of people increase, each of these support systems becomes more difficult to establish and maintain. We marvel today at the water supply system of ancient Rome, the sanitary and food supply systems of ancient Shanghai, and the agricultural methods by which Cuzco lived. Yet none of these cities exceeded a million people, which was thought of as the maximum city size before the modern era.

In maintaining our social structure, advanced technical and scientific capabilities are necessary. A major and vital factor among these is an understanding of the chemical nature of matter and of chemical changes. Upon this understanding and its practical application rest

1. Food supply systems, through fertilizer, food preservation, and nutrition technology
2. Fuel availability, through petroleum, natural gas, and coal production and conversion
3. Sanitation and health care systems, through pharmaceuticals and sterilization techniques
4. Clothing, housing, and transportation, through production of natural and synthetic materials, fibers, and polymers

Essentially all of the goods and services that modern society demands require components that are chemical in nature and that have been produced by the application of chemical knowledge. Let us consider, for example, a simple loaf of bread, traditionally considered the foundation of the human diet. The loaf in our kitchen lists on its label the following ingredients: flour, niacin, reduced iron, thiamine mononitrate, riboflavin, buttermilk, water, yeast, partially hydrogenated vegetable shortening (soybean, palm, and/or cottonseed oils), salt, sugar, calcium propionate, mono- and diglycerides (from hydrogenated vegetable oils), and lecithin.

Niacin, reduced iron, thiamine mononitrate, and riboflavin are vitamins and food minerals. They replace and extend the vitamins and minerals in wheat which have been lost to people since the earliest days of milling flour. The calcium propionate is a spoilage inhibitor that allows this loaf to be distributed through broad-scale food delivery systems. The diglycerides and lecithin are added basically for improvement in texture. Both occur naturally in the food chain, and lecithin is sometimes recommended medically to prevent cholesterol buildup. Thus, largely as a result of applied food chemistry, we have a loaf of bread that retains its texture, resists mold and other spoilage, and is more nutritious than the bread grandmothers are reputed to have made. Whether it tastes as good depends upon individual likes and dislikes and the tenor of one's nostalgia.

This example was chosen because of the general denunciation of "chemicals" in food. One can ask what makes lecithin or calcium propionate into "chemicals" while excluding yeast and salt as such. Perhaps it is the long name. Other examples could be chosen that raise different controversies—e.g., nylon, penicillin, automobile tires, house paint, vinyl tile flooring, synthetic detergents, lubricating oil, electric wire insulation, and aluminum. The fact is that human beings have acquired the knowledge necessary to rearrange atoms into molecules of modestly predictable properties. Of these, some are of special use and can be engineered into products that are indispensable to our existence.

2.2 THE CHEMICAL PROCESS INDUSTRY

The industry in which we apply the knowledge of chemical behavior is generally called the chemical process industry. It includes firms that alter the chemical composition of their raw materials. Chemical reactions and separation of compounds are used to produce products of desired properties. This industry is generally defined to include work with

1. Industrial chemicals such as alkalies, chlorine, ammonia, sulfuric acid, nitric acid, paint pigments, photographic chemicals, and industrial gases.

2. Organic chemicals such as dies and pigments, pesticides, plasticizers, and flavors and fragrances made primarily from hydrocarbon feedstocks such as petroleum and coal.
3. Polymers such as synthetic resins, plastics, fibers, and elastomers.
4. Other chemicals including bulk medicinals, surfactants, gum and wood chemicals, adhesives, explosives, carbon black, and fatty acids.

These groups of products represent about 5 percent of the U.S. GNP and would seem to be very broad indeed. However, they are overshadowed by a host of other products that are chemicals, but which enter our economy more directly and are thought of separately: refined metals and alloys, pulp and paper, detergents, paint and printing ink, medicinals, cosmetics and toiletries, etc. The list goes on nearly endlessly and represents in total the largest of the industry groupings in terms both of dollar value and the number of firms involved. Table 2–1 presents the scope of the chemical process industry. As

TABLE 2–1
THE CHEMICAL PROCESS INDUSTRY

1. Chemicals (including petrochemicals):
   - Alkalies and chlorine
   - Industrial gases
   - Cyclic intermediates and crudes
   - Industrial organic chemicals not elsewhere classified
   - Industrial inorganic chemicals not elsewhere classified
   - Synthetic rubber (vulcanizable elastomers)
   - Gum and wood chemicals
   - Chemicals and chemical preparations
2. Drugs and medicines:
   - Biological products
   - Medicinal chemicals and botanicals
   - Pharmaceutical preparations
   - Perfumes, cosmetics, and other toilet preparations
3. Explosives and fireworks:
   - Ammunition
   - Ordnance and accessories
   - Explosives
4. Fats and oils:
   - Cottonseed oil mills
   - Soybean oil mills
   - Vegetable oil mills,
   - Animal and marine fats and oils, including grease and tallow
   - Shortening, table oils, margarine, etc.
5. Fertilizers and agricultural chemicals:
   - Fertilizers
   - Agricultural pesticides, herbicides, growth regulators, etc.

TABLE 2–1 (*Con't*)
THE CHEMICAL PROCESS INDUSTRY

6. Food and beverages:
   - Condensed and evaporated milk
   - Wet corn milling
   - Cane sugar refining
   - Beet sugar
   - Malt liquors and malt
   - Wines, brandy, and brandy spirits
   - Distilled, rectified, and blended liquors
   - Flavoring extracts and syrups
   - Roasted coffee
   - Food preparations and mixes
7. Leather tanning and finishing
8. Lime and cement:
   - Hydraulic cement
   - Lime
9. Man-made fibers:
   - Cellulosic man-made fibers
   - Synthetic organic fibers (rayon, nylon, etc.)
10. Metallurgical and metal products:
    - Electrometallurgical products
    - Primary steel production
    - Primary smelting and refining of copper, lead, zinc, etc.
    - Primary production of aluminum
    - Primary smelting and refining of other nonferrous metals
    - Secondary smelting, refining, and alloying of nonferrous metals and alloys
    - Enameled iron and metal sanitary wear
    - Electroplating, plating, polishing, anodizing, and coloring
    - Coating, engraving, and allied services
11. Paints, varnishes, pigments, and allied products:
    - Inorganic pigments
    - Paints, varnishes, lacquers, enamels, and allied products
12. Petroleum refining and coal products:
    - Petroleum refining
    - Paving mixtures and blocks
    - Asphalt felts and coating
    - Lubricating oils and greases
    - Coke and coal chemicals
13. Plastic materials, synthetic resins, and nonvulcanizable elastomers
14. Rubber products:
    - Tires and inner tubes
    - Rubber footwear
    - Reclaimed rubber
    - Other fabricated rubber products
15. Soap, glycerin, and cleaning, polishing, and related products:
    - Soap and detergents
    - Specialty cleaning, polishing, and sanitation preparations
    - Surface-active agents, sulfonated oils, and assistants

TABLE 2–1 (*Con't*)
THE CHEMICAL PROCESS INDUSTRY

16. Stone, clay, glass, and ceramics:
    - Flat glass
    - Glass containers
    - Pressed and blown glass and glassware
    - Ceramic wall and floor tile
    - Brick and structural clay tile
    - Clay refractories
    - Structural clay products
    - Pottery and related products
    - Gypsum products
    - Abrasive products
    - Asbestos products
    - Insulation
    - Minerals and earth, ground or otherwise treated
    - Mineral wool
    - Nonclay refractories
    - Nonmetallic mineral products
17. Wood, pulp, paper, and board:
    - Wood preserving
    - Pulp mills
    - Paper mills, except building-paper mills
    - Paperboard mills
    - Paper coatings and glazes
    - Building-paper and building-board mills
18. Other chemically processed products:
    - Broad-woven-fabric mills
    - Dyeing and finishing textiles
    - Artificial leather, oilcloth, and other impregnated and coated fabrics
    - Glue and gelatin
    - Printing ink
    - Carbon black
    - Carbon and graphite products
    - Storage batteries
    - Primary batteries (dry and wet)
    - Some photographic equipment and supplies
    - Lead pencils, crayons, and artists' materials
    - Carbon paper and inked ribbons
    - Linoleum, asphalted felt-base, and other hard-surface floor coverings
    - Candles and matches

you can see, its breadth is staggering: the products listed affect us all in almost everything we do.

Table 2–2 lists the 50 largest-volume chemical products produced in the U.S. Statistics are given on the production in 1984, and on the production growth rate from 1974 to 1984. Four of the top five products are inorganics, as are nine of the top ten. However, the largest numbers of these products

TABLE 2–2
TOP 50 CHEMICALS PRODUCED IN THE U.S. IN 1984

Rank	*Chemical*	*Production (10^9 pound/yr)*	*Growth rate 1974–1984 (%/year)*	*Major uses*
1	Sulfuric Acid	79.37	1.6%	phosphate fertilizers, pigments, fibers, explosives, papermaking, petroleum products, steelmaking
2	Nitrogen	43.41	9.4%	inert blanket and purge gas, food freezing
3	Ammonia	32.41	0.3%	fertilizers, explosives, fibers, plastics
4	Lime	32.20	−2.3%	construction products, fertilizers, soil conditioning, steelmaking, wastewater treatment
5	Ethylene	31.18	2.7%	feedstock for petrochemical and polymers
6	Oxygen	31.04	0.4%	welding, life support, steelmaking, oxidizing agent for chemicals and for rocket fuel
7	Sodium Hydroxide	22.45	0.0%	reactant for organic and inorganic chemicals, petroleum and food processing
8	Phosphoric Acid	22.22	4.4%	fertilizers, phosphate salts, metal treatment, refractories, catalysts, food
9	Chlorine	21.45	0.0%	vinyl chlorides and other inorganic chlorides, wastewater treatment, inorganic chemicals and paper pulp
10	Sodium Carbonate	17.02	1.2%	glass, sodium silicates and phosphates, paper, detergents, and cleaners
11	Nitric Acid	16.05	0.1%	fertilizer (ammonium nitrate and urea), explosives, intermediates for polymers
12	Propylene	15.47	4.0%	fuels, petrochemicals, polymers
13	Urea	14.30	6.9%	fertilizer, animal feed supplement, polymers
14	Ammonium Nitrate	14.01	0.7%	fertilizer, industrial explosives, manufacture of nitrous oxide (NO)
15	Ethylene Dichloride	13.73	4.1%	vinyl chloride manufacture
16	Benzene	9.86	−1.0%	synthetic rubber, plastics, drugs, fuel additive, dyes, insecticides, solvent
17	Ethylbenzene	8.61	3.6%	plastics, synthetic rubber
18	Methanol	8.28	1.9%	formaldehyde, acetic acid, fuel additive
19	Carbon Dioxide	7.80	8.0%	beverage additive, dry ice, fire fighting, food preservation, oil-well tertiary recovery, greenhouses, mining operations

20	Styrene	7.71	2.6%	polystyrene and other plastics
21	Vinyl Chloride	7.51	2.9%	PVC and other plastics
22	Xylene	6.12	0.6%	gas additive, paint and coatings solvents, aromatics, organics production
23	Terephthalic Acid	6.05	3.6%	polymer especially for food packaging and beverage bottles
24	Ethylene Oxide	5.96	3.4%	fumigant and sterilizing agent, chemical intermediate
25	Hydrochloric Acid	5.72	1.5%	intermediate for organic synthesis, metal cleaning, hydrometallurgy
26	Formaldehyde	5.71	−0.1%	intermediary for organic synthesis, textile finish, preservative
27	Toluene	5.27	−2.4%	gasoline blending, benzene production, chemical intermediate
28	Ethylene Glycol	4.84	3.8%	antifreeze, polymer fibers and films, chemical intermediate
29	p-Xylene	4.27	4.7%	gasoline blending, solvent, intermediate for other aromatics
30	Ammonium Sulfate	4.13	−0.3%	fertilizer, fermentation process additive
31	Potash	3.53	−4.6%	fertilizer
32	Cumene	3.36	1.5%	polymerization initiator, acetone, phenol, aviation gasoline
33	Carbon Black	2.89	−1.5%	rubber compounding agent, brushes and electrodes, pigment, ore reduction and carburizing, high-temperature insulation
34	Phenol	2.85	2.0%	polymer resin formation
35	Acetic Acid	2.64	0.2%	general organic chemical intermediate, fermentation, solvent
36	Butadiene	2.53	−3.7%	synthetic elastomers, organic chemicals
37	Acrylonitrile	2.20	4.5%	polymer production
38	Aluminum Sulfate	2.16	−1.5%	pH control and water treatment, especially in papermaking, fire extinguishing, insulation, catalysis
39	Cyclohexane	2.13	−1.0%	nylon manufacture
40	Calcium Chloride	2.10	−0.5%	deicing, dust control, road stabilization
41	Vinyl Acetate	2.02	3.7%	polymerization to form adhesives
42	Propylene Oxide	1.90	0.8%	fumigant, cotton fiber treatment
43	Acetone	1.89	−0.5%	polymer intermediate, solvent
44	Sodium Sulfate	1.74	−4.3%	pulp and papermaking, detergents, glass
45	Titanium Dioxide	1.60	0.2%	white pigment
46	Sodium Silicate	1.50	−0.3%	soaps and detergents, water treatment, adhesives and binders, bleach stabilization
47	Methyl-tert-butyl Ethene	1.47	NA	high-octane gasoline
48	Adipic Acid	1.39	−0.6%	nylon, plasticizers, lubricants, films
49	Sodium Tripolyphosphate	1.33	−3.0%	detergent builder
50	Isopropyl Alcohol	1.15	−5.1%	manufacture of acetone, coatings, solvents, pharmaceuticals

Data reprinted with permission from *Chem. & Eng. News*, June 9, 1986.

are organic. Few of the chemicals are sold in large quantity to the ultimate consumer. Most are used as intermediates in making other products. In most cases the uses of these chemicals are diverse: they serve as chemical raw materials, solvents, and components of products ranging from cosmetics to synthetic detergents, and house paint to fertilizers.

Sulfuric acid is so widely used that its production figures are sometimes used to measure the level of development of a nation or the state of a regional economy. Urea (H_2N-CO-NH_2) was the first organic chemical to be synthesized in a laboratory (by Friedrich Wöhler in Germany in 1828). It is now used as a fertilizer, as a major raw material for polymer manufacture, and as an organic starting material in a wide range of specialty chemical products.

2.3 CHEMICAL COMPANIES

The 45 largest chemical companies in the U.S. are listed in Table 2–3 in order of the size of the chemical part of their sales. Thus, only 3.5 percent of the *total* sales of Exxon is included in the first column. In total sales of all products, Exxon would be over twice the size of DuPont. The table also shows the chemical operating profit of these companies and gives this profit as a percentage of the chemical assets of the company. The figures are similar to those of U.S. industry as a whole. Since profit margins fluctuate greatly from year to year, the figures for any given company should not be taken as typical for that company.

These companies are all producers of chemical products. In today's business and social climate, most chemical producers are large corporations. The development of a chemical product is a long and costly process, as will be described later. In general, the product must be made in large quantity in order to be sold cheaply enough to be accepted in the marketplace and still offer a profit to the producer. Widescale marketing is necessary, and the company must have the technical, legal, and financial resources to back up the product with services and to protect its workers and the public from any real or perceived hazards. Most chemical production facilities are capital intensive, and extensive resources must usually be applied to compete in the chemical market.

On the other hand, there are a host of small companies in the chemical industry. Generally, they make and sell chemical equipment, design and build chemical plants, or supply specialized services (e.g., in maintenance, plant startup, accounting, legal aid, or interaction with government agencies). Some small companies actually market chemical products, but usually these products tend to be specially tailored to the needs of a rather small group of customers. Typical products would include specially compounded detergents, catalysts, special-purpose paints, plastic products, and analytical standards.

TABLE 2–3
TOP 45 U.S. CHEMICAL PRODUCERS, RANKED BY CHEMICAL SALES

1985 rank	*Company*	*1985 chemical sales ($ millions)*	*Chemical operating profits as % of total operating profits*	*Identifiable chemical assets ($ millions)*	*Operating return on chemical assets*
1	Du Pont	$11,250	22.0%	$7,885	11.0%
2	Dow Chemical	9,508	64.5	8,047	5.1
3	Exxon	6,670	3.5	5,337	7.2
4	Monsanto	5,203	135.9	4,217	11.3
5	Union Carbide	3,961	50.0	4,330	9.5
6	Atlantic Richfield	3,804	4.9	2,296	3.2
7	Shell Oil	3,318	5.2	3,891	4.2
8	Celanese	3,046	100.0	2,809	4.5
9	Amoco	2,905	11.5	2,334	20.3
10	W. R. Grace	2,868	58.4	1,997	11.7
11	Chevron	2,611	def	3,010	def
12	BASF	2,600	NA	NA	NA
13	Eastman Kodak	2,348	32.6	2,136	8.6
14	General Electric	2,347	13.2	3,876	12.0
15	Phillips Petroleum	2,266	8.0	1,315	16.7
15	Mobil	2,266	1.3	2,045	4.0
17	Allied-Signal	2,055	21.8	1,979	9.9
18	Rohm & Haas	1,966	99.3	1,381	19.6
19	American Cyanamid	1,830	32.1	1,254	6.4
20	Hercules	1,743	35.9	1,409	3.3
21	Air Products	1,674	114.3	1,937	16.1
22	Occidental Petroleum	1,621	5.5	1,721	5.8
23	Mobay	1,599	100.0	1,215	12.4
24	Ciba-Geigy	1,540	NA	NA	NA
25	Borden	1,532	32.4	1,336	9.6
26	Ashland Oil	1,499	19.3	496	13.7
27	B. F. Goodrich	1,384	51.1	1,154	6.9
28	American Hoechst	1,361	NA	NA	NA
29	FMC	1,261	53.2	1,260	11.4
30	Ethyl	1,242	84.3	768	25.8
31	Texaco	1,220	0.3	1,010	1.2
32	U. S. Steel	1,217	def	587	def
32	Unocal	1,217	17.8	772	7.5
34	Olin	1,147	44.3	825	5.6
35	Chesebrough-Pond's	1,079	27.7	1,771	3.4
36	International Minerals	1,057	62.8	1,285	10.5
37	National Distillers	1,037	37.7	746	9.6
38	Borg-Warner	929	35.8	437	28.0
39	CF Industries	921	def	936	def
40	Dow Corning	901	100.0	NA	NA
41	National Starch	881	100.0	655	10.1
42	PPG Industries	841	17.9	1,129	7.1
43	Reichhold Chemicals	823	100.0	492	4.3
44	Lubrizol	804	97.5	603	15.5
45	Williams Cos.	742	def	553	def

NA—not available; def—deficit
Courtesy, American Chemical Society

2.3.1 An Example: Air Products and Chemicals, Inc.

It may help to understand the chemical industry if we examine a specific example. Air Products and Chemicals, Inc. was founded in 1940 by Leonard Pool in Detroit to supply on-site generation of industrial gases such as oxygen and nitrogen. At that time, such a concept of selling gases to another company was a new idea. The company moved to the Lehigh Valley in Pennsylvania after World War II and now has its headquarters, research laboratories, and engineering center at Trexlertown, just west of Allentown, Pennsylvania. By 1978, it had passed $1 billion in sales, employed 14,300 people in 11 countries, and had $1.2 billion in assets. It ranked 253rd in sales among U.S. corporations and about 25th among chemical companies. The 1985 figures are shown in Table 2–3 (21st entry). There are company offshoots (wholly owned or joint ventures) in England, Italy, Holland, Japan, South Africa, and Brazil.

Currently, Air Products is organized into the following divisions:

1. *Industrial Gases.* This division makes and markets oxygen, nitrogen, argon, helium, hydrogen, carbon dioxide, and specialty gases such as the nitrogen fluorides. Generally, these gases are distributed as liquids or gases in high-pressure cylinders or are produced in bulk on the site. They are used throughout industry all over the world.
2. *Equipment and Services.* Experience with industrial gases gave Air Products engineers special understanding of very low-temperature processes and problems. Hence, the firm offers engineering services and equipment in the cryogenic field. Examples include (a) processes and heat exchangers for liquefying natural gas, in which it leads the world, (b) air separation plants making oxygen and nitrogen, (c) plants for purifying radioactive gases, (d) ammonia processes, (e) light hydrocarbon separations, (f) high-efficiency wastewater purification using oxygen, (g) freezing of foods, (h) freezing and shredding tires, and (i) special refrigeration equipment for ultra-low temperatures.
3. *Chemicals.* This division manufactures a wide range of industrial chemicals, catalysts, and polymers. Included are the methyl and ethyl amines, alcohols, ammonia, toluene, diamine, dinitrotoluene, polymer emulsions, polyvinyl alcohols, polyvinylchloride, urethane catalysts, petroleum catalysts, acetylenic chemicals, automotive catalysts, and specialized surfactants. Use of these products is extremely diverse.
4. *Stearns Catalytic World Corp.* This subsidiary is primarily a plant engineering and construction group that designs, builds, starts up, and operates plants throughout the chemical industry. It also offers contract maintenance service to industry.

Obviously, it would be difficult to classify Air Products within any one of the industry groups mentioned, since it has branched very far from its early emphasis on industrial gases. However, it is fairly typical of companies in the chemical processing industry.

2.3.2 Another Example: Hercules, Inc.

Hercules was formed in 1913 when the U.S. Department of Justice forced the breakup of E. I. du Pont de Nemours, Inc. Three separate companies were formed: DuPont, Hercules, and Atlas (now ICI America). At that time, Hercules received part of the explosives manufacturing capacity of DuPont. Its headquarters remained in Wilmington, Delaware. A research center and engineering offices are also now located there.

By 1979, Hercules had net sales of $2.3 billion from over 1,000 industrial chemicals produced at 87 major plants located in 29 U.S. and overseas locations. There were 24,400 employees and 37,700 stockholders. In 1979 the wage costs were $600 million, and the return on the stockholders' equity was 19.6 percent.

Hercules products include (1) plastics such as polypropylene resins, films, and fibers, (2) organics, including resins, paper products, elastomers, and specialty chemicals, (3) water-soluble products, including pectin, oil drilling-mud additives, cellulose gum, and guar, (4) carboxy methylcellulose, (5) printing inks, (6) flavors, (7) water management chemicals, (8) explosives and rocket propellants, (9) magnetic recording tapes, (10) printing systems, and (11) pharmaceuticals.

The company is organized into three worldwide businesses: Specialty Chemicals, which produced about 45 percent of 1984 sales; Aerospace, with 29 percent of 1984 sales; and Engineered and Fabricated Products, with 14 percent of 1984 sales. Hercules also formed partnerships with other companies which accounted for the remaining 12 percent of sales. As with most chemical companies, Hercules depends heavily on petroleum and petrochemical raw materials. It has attempted to assure a long-term availability through large contracts. However, some 30 percent of the company's raw materials come from the southeastern part of the U.S. as waste from papermaking operations, particularly pine-tree stumps from which resins and wood chemicals are obtained.

2.4 PROCESS CHARACTERISTICS

The process by which a chemical product is made centers on the need to carry out a controlled chemical reaction. In some processes it may be sufficient to mix ingredients or to separate components of a naturally occurring mixture, as in air separation. Usually, however, a reaction must take place. In this case, reactants must be supplied at the required temperature, pressure, and purity, the reaction must be controlled at predetermined conditions, and the product mix must be removed from the reactor and the desired products separated from it.

2.4.1 Batch Processes

For small-volume products and in cases where the fundamental mechanisms of the reaction are not well known, *batch processing* is often used. The sequence of events copies that which would occur in the laboratory, but in larger-size vessels and batches. The raw materials are purified, perhaps by distillation or adsorption, and stored. Reactants are then pumped or poured into a reaction vessel. The agitation intensity, the rate of heating and/or cooling, and the rates of flow of other reactants or catalysts are controlled in such a manner that the reaction proceeds as planned. When the reaction is completed, the reactant mass is removed to a separation system. The desired products are separated from unreacted feed materials and undesired byproducts. The reactants are usually recycled for use in the next batch.

Soapmaking is a traditional batch process. Animal tallow and lye are mixed together in a large tank with a submerged steam inlet. The mass is then agitated by steam injection for perhaps four hours during which time the following reaction takes place:

$$\underset{\substack{\text{caustic}\\\text{soda}\\\text{(lye)}}}{3\,NaOH} + \underset{\substack{\text{glyceryl stearate}\\\text{(a typical fat)}}}{(C_{17}H_{35}COO)_3C_3H_5} \longrightarrow \underset{\substack{\text{sodium stearate}\\\text{(soap)}}}{3C_{17}H_{35}COONa} + \underset{\text{glycerin}}{C_3H_5(OH)_3}$$

Salt is added, and steam injection continues until two layers form. The bottom layer is salt water and glycerin; the upper layer is soap and fat. The bottom layer is withdrawn, water is added, and boiling continues. Glycerin caught in the soap separates into a new water layer and is removed. More lye is added to finish the saponification. It is later withdrawn from the bottom with more glycerin. Finally, another charge of water is added, boiled with the soap, and withdrawn. The exact sequence, timing, and ingredients used will vary depending on the grade of soap to be made. The total procedure takes about a week, during which time the progress of the batch is watched and controlled by a "soapmaker." Usually the soapmaker has spent years in this activity and may control the batch by the feel of the mass between the fingers or its motion as it slips off a trowel. The soapmaker may taste it and certainly observes its color, fluidity, and odor.

Eventually, the upper soap layer is transferred to a crutcher, a steam-jacketed tank with a spiral central agitator in a cylindrical shroud. Here, compounding agents such as builders (trisodium phosphate or borax, for example), silicates, perfumes, and color are added, and the batch is mixed until it is smooth. It is then dumped into wooden frames where it is allowed to cool, dry, and harden for several days until it is ready for cutting or forming into bars for wrapping and sale.

This process takes a long time, requires large pieces of equipment in relation to the production quantity, involves a large inventory, and requires a great amount of human supervision and interception. On the other hand, it is very flexible: one can make many kinds of soap with very modest process changes and can alter the process conditions if the batch seems to be somewhat off the specification at any time.

Small-scale processes and processes in which solids, gunks, and goos occur are likely to be batch. Examples are pharmaceuticals, polymers, most foods, and wine. Many of these continue to be made batchwise because quality is more important than price, the product has traditionally been so made, or the industry isn't large enough or technically sophisticated enough to operate successfully in a continuous mode.

2.4.2 Continuous Processes

For large production capacities where the process reaction mechanism is better known and reaction rates are not too slow, *continuous processing* is often possible. Here, the raw materials are prepared and fed to the reactor continuously. The reactor system is sized so that the materials reside in it long enough at the reaction conditions to achieve the desired extent of reaction. The reaction system may be a single vessel or a number of reactor vessels in series, each operating at different conditions. The product leaves the reaction zone continuously and passes to a sequence of separation steps where the desired products are obtained in continuous streams. Unreacted feed materials are obtained in other streams and continuously returned to the reactor. Any by-products are also removed.

Figure 2–1 shows a possible configuration for a continuous process plant. Two raw materials are fed to a reactor after each is purified. The reactor effluent is separated in three separation steps, and all waste streams are treated before release to the environment. Product X might be the main product desired and product Y a salable by-product.

Ideally, such a process operates under steady-state conditions, i.e., a stable operating condition where none of the process parameters (temperature, pressure, process stream composition, flow rate, etc.) vary with time. In any real process there will be a period of adjustment as the plant is started or stopped, and there will be disturbances as the process operates. The process control system attempts to minimize the effects of these process upsets.

In a plant operating under steady-state conditions, variables are different from point to point along the process path. But if you were to take a photograph of the control panel, where an array of instruments records the process conditions at all the crucial points in the process, you would see no differences between a picture taken at 8 a.m. and one taken at 8 p.m.

Some basically continuous processes have components—for example, driers, filters, and ion-exchange beds—that operate cyclically. These units must be

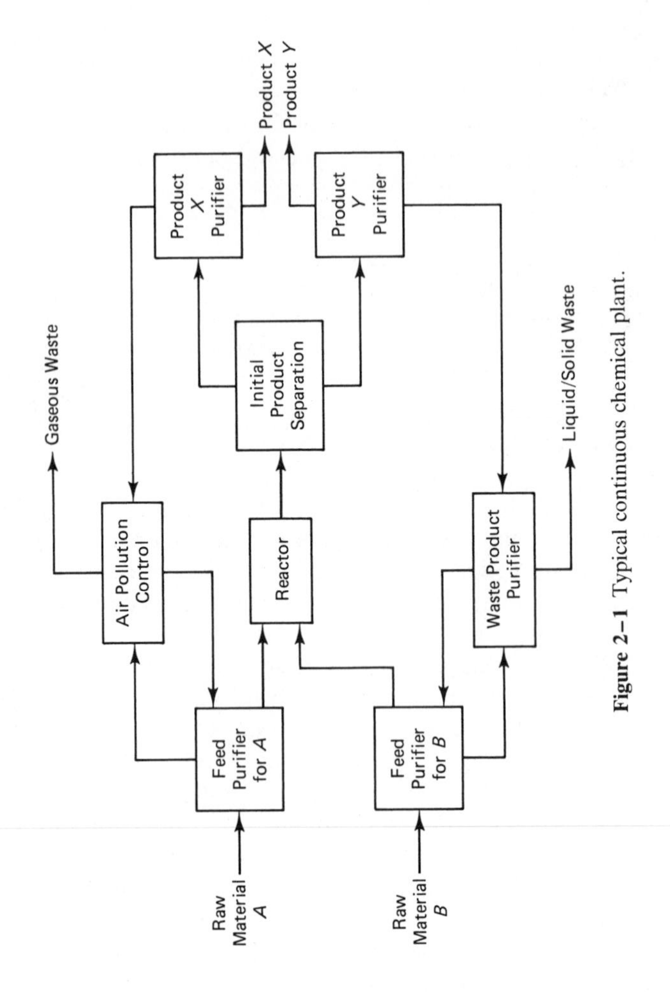

Figure 2–1 Typical continuous chemical plant.

taken off-stream periodically for regeneration. The period of on-stream operation may be several hours to several days.

Most large-scale processes operate continuously, especially ones in which gases and low-viscosity liquids are handled. Petroleum refining, the manufacture of bulk chemicals, and industrial gas manufacturing are typical examples.

2.5 PROCESS EQUIPMENT

Despite the very large number of chemical processes and products, there are only a modest number of different kinds of chemical process steps. This has allowed us to organize the field into manageable segments. When only physical changes occur, these process steps are called *unit operations*. Among such operations are distillation, evaporation, and liquid-liquid extraction, the topics we will be dealing with in this book. There are many others that you will learn about at other times, such as filtration, drying, heat exchange, classification, mixing, crystallization, and adsorption.

When a process step involves a chemical change, it is sometimes called a *unit process*, or, more appropriately a *chemical reaction step*. Each process step can be carried out in a variety of types of equipment. Usually the apparatus selected is chosen because it has some particular advantage in light of the properties of the materials being processed or the goal of the process step. The next subsection describes some of the more important equipment that is widely used in the chemical process industries.

2.5.1 Reactors

Figure 2–2 shows several common types of reactors. The agitated batch reactor shown in (a) is an extremely common device. The jacket can be used to heat or cool the reactor, typically with steam or cooling water. The vessel may be built with thick walls so that the reactions can take place under pressure. A mechanical or stuffing-box seal for the agitator shaft must be used to seal the vessel, and there may be various ports for feed addition and product withdrawal. There certainly will be thermometers, pressure gauges, and perhaps a "bulls eye" (a little round glass window) to look into the reactor. There will often be baffles on the inside to prevent vortexing of the liquid and to improve mixing. Normal construction materials include glass-lined steel, stainless steel, carbon steel, and various corrosion-resistant alloys. Normal sizes range from $\frac{1}{4}$ gallon to several thousand gallons. Internal cooling coils are sometimes used to provide additional heating or cooling capacity.

Sketch (b) shows a continuous stirred tank reactor. This is essentially identical to the agitated batch reactor but is operated continuously. Thus, feed and product are continuously added and removed. Obviously, it is not possible to operate this reactor so that every fluid molecule stays in for the

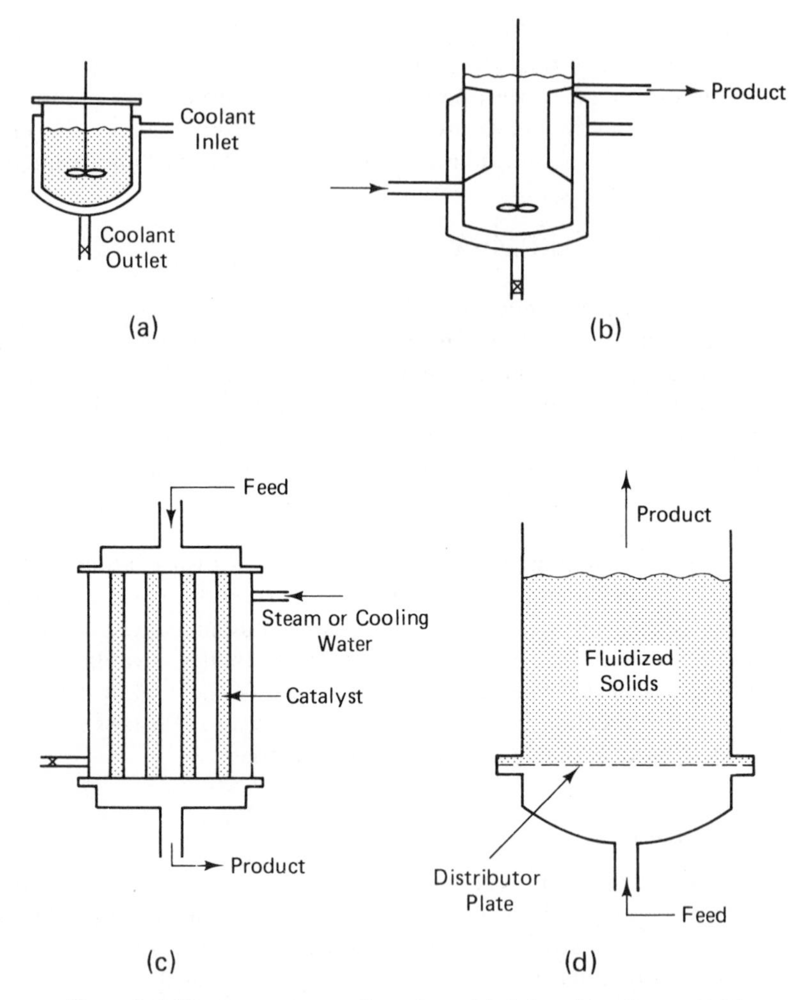

Figure 2–2 Various reactor configurations. (a) Agitated batch reactor (b) Continuous stirred tank reactor (c) Tubular packed-bed reactor (d) Fluidized-bed reactor (e) Packed-bed reactor (f) Screen as a reactor (g) Methanol reactor system. *Courtesy E.I. du Pont de Nemours, Inc.*

same length of time. Fairly intense agitation is usually employed to keep the reactor contents uniform. However, the degree of mixing decreases as the size of the reactor increases. The outlet stream contains molecules which have experienced residence times varying from almost zero to almost infinity. Of course, the average residence time can be calculated from feed and product flow rates and the reactor volume.

The tubular reactor shown in (c) is the most common type of reactor for continuous solid-catalyzed reactions. The fluid stream flows through many

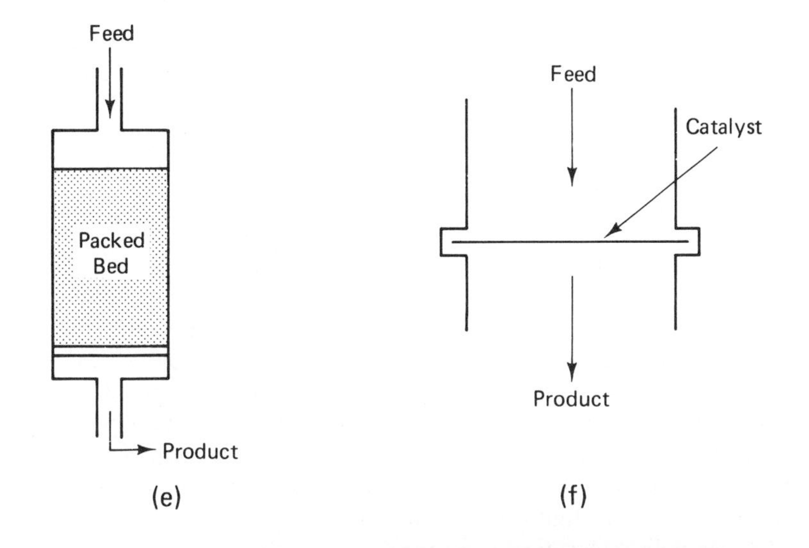

(g)

Figure 2–2 (*Con't.*)

parallel small metal tubes that are packed with solid catalyst particles. Heat can be removed from or added to a fluid flowing on the outside of the tubes. The main advantage of this type of reactor is that a large surface area is available for heat transfer. The tubular reactor is widely used for highly exothermic reactions that require a solid catalyst.

An alternative arrangement, shown in (e), would be a single packed-bed reactor operating adiabatically without addition or removal of heat. The length of the bed and the amount of diluent in the feed stream to the reactor must be adjusted to keep the temperature rise or drop through the bed within tolerable levels. Heating or cooling of the process stream between the beds is used as required.

The fluidized-bed reactor shown in (d) is similar to the packed-bed reactor except that the fluid flows up through the bed at a velocity high enough to dislodge the particles and suspend them in the gas or liquid stream. There is a large range of flow rates that will suspend the bed particles without carrying them out with the process stream. In these cases the bed behaves superficially like a pool of liquid: it is relatively cohesive, flows like a liquid, and is confined in the bottom of the available space. Light objects will float on the bed surface, and heavy objects will sink to the gas distributor plate at the base of the bed. The bed particles may be catalyst particles (as in petroleum fluid cracking) or a reactant (as in fluidized combustion).

Occasionally a reaction is so rapid that the reactor size can be reduced almost to zero. Such is the case of the ammonia oxidation reaction, which is the first step in nitric acid production. In this case a thin screen of catalyst wires is sufficient to produce the desired reaction. Such a reactor is shown in (f).

Figure 2–2(g) is a photograph of a large-scale commercial reactor system in a methanol plant. Note the size of the equipment compared to the two people pictured in the lower left part of the photograph.

2.5.2 Product Purification Units

The separation of the desired product or products from the material coming from the reactor is often a complex and expensive process. Many sequential separation steps may be needed. Although these may require additional reactors, more typically they involve distillation, extraction, absorption, adsorption, and crystallization. If solids are present, they may be removed by filtration or sedimentation.

Several fluid separation process units are shown in Figure 2–3. Sketch (a) shows a generalized distillation column. Distillation is by far the most commonly used separation process. It can be used when there are differences in the boiling points of the components to be separated. We will study distillation in some detail later on in the text.

In a distillation column, countercurrent contact between a liquid and a vapor on multiple stages produces a separation of chemical components. It is necessary to have a source of vapor at the bottom of the column (the reboiler) and one of liquid at the top (the condenser). The liquid and vapor streams have different concentrations of components. The easily vaporized components (the "light" components) concentrate in the vapor phase and

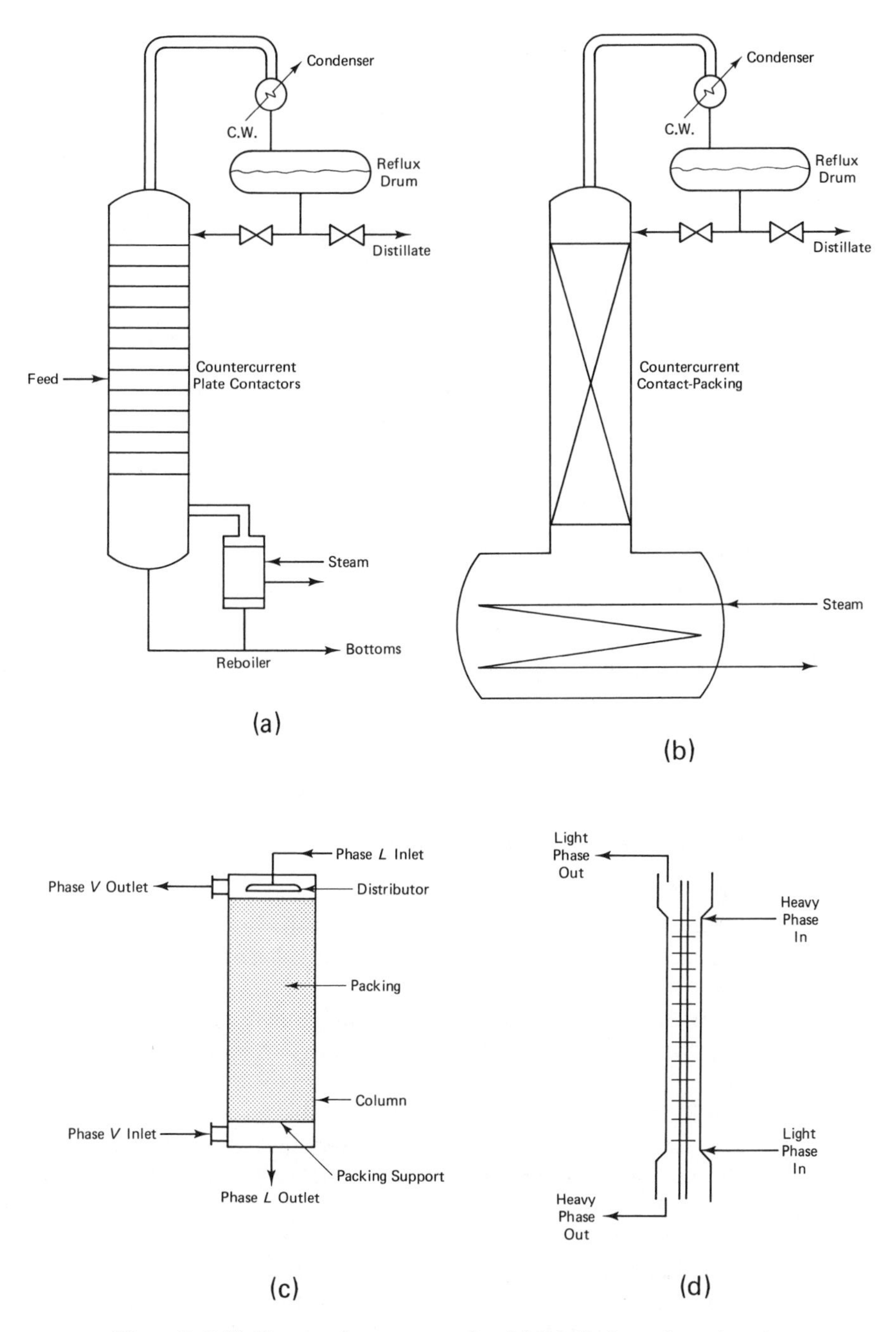

Figure 2–3 Fluid separation process units. (a) Distillation column (continuous) (b) Distillation column (batch wise) (c) Absorption column (d) Liquid-liquid extraction column (e) Adsorption column.

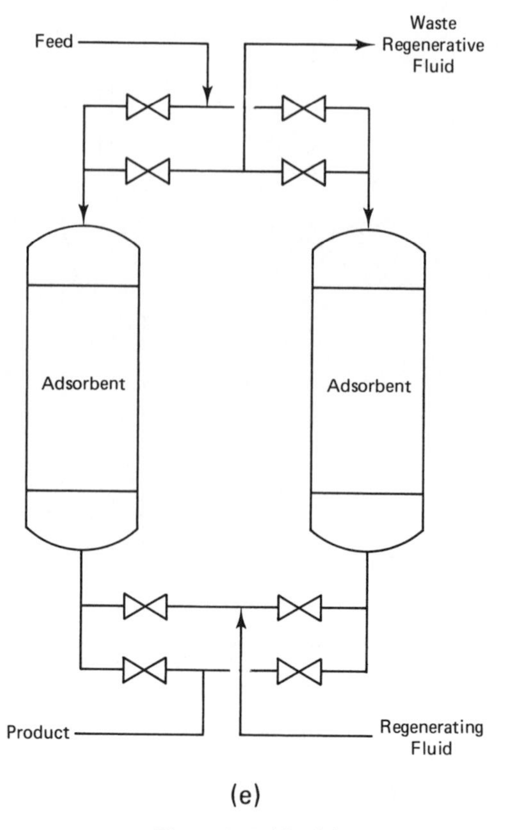

Figure 2–3 (*Con't.*)

ultimately at the top of the column. The less easily vaporized components (the "heavy" components) concentrate in the liquid phase and migrate toward the bottom of the column.

Sketch (b) also shows a distillation column, but here the column is arranged to operate in a batch mode rather than continuously. A charge of feed is first pumped into the still pot at the bottom of the column. Then heat is transferred into the liquid in the still pot, producing vapor which flows up the column. The vapor is condensed, and some of the condensed liquid is fed back into the top of the column, some removed as product. The lightest component comes off the batch distillation column first. When all of it has been removed from the material in the still pot, the next lightest component comes off. Thus, a single column can be used to separate a multicomponent mixture into more-or-less pure component streams. In batch distillation, the compositions of the distillate product stream and the remaining still pot liquid are both changing with time.

Separation in distillation is achieved by using thermal energy in the base of the column to produce vapor. However, this is the major disadvantage of distillation, because the cost of operation depends almost entirely on energy costs. Major efforts are under way to improve the energy efficiency of distillation or to use alternative, less energy-intensive separation steps.

Sketch (c) shows an absorption column. This unit repeatedly causes a falling liquid fed to the top of the column to come in contact with a rising vapor that enters the bottom of the column. Little of the liquid vaporizes, and little of the vapor condenses. Each of these streams contains a large fraction of nearly inert components with minor concentrations of one or more active components which distribute themselves between the two phases. The primary use of absorption is to remove minor components from a process gas stream. For example, SO_2 may be removed from stack gases by contacting the gas stream with an absorbing liquid which is basic in nature. The contacting liquid also collects particulates from the gas stream.

A liquid-liquid extraction column is shown in (d). Physically, this column is similar to the absorption column, but both phases are liquid. Since transfer across a liquid boundary is slow once such an interface forms, the column contains internal devices to break up liquid droplets and force them to reform. Sometimes this is done through a reciprocal movement of the sieve plates, as shown in the figure. In the extraction process, two liquid streams that have different densities (like oil and water) and limited solubilities pass countercurrently. A "solute" component distributes itself between these two liquid phases.

Extraction is used to transfer a solute component from a feed stream into a solvent stream. Then the solvent stream from the extractor is fed into another separation step (typically a distillation column), where the solute is separated from the solvent.

Liquid-liquid extraction is often used when the components in the feed cannot be separated directly by distillation because their boiling points are close together. A solvent must be found that both preferentially dissolves one of the components and is easily separated from that component.

Sketch (e) shows an adsorption system. Gas or liquid is fed into a vessel that is packed with solid particles. One or more components in the feed is attached by van der Waals forces to the surfaces and pores of a particulate solid. Other components are not affected and pass through the bed. When the bed becomes full of the adsorbed component, it is taken off-line. The adsorbed components are removed from the solid by heating, by pressure reduction, or by stripping with another gas. During this regeneration time, a second parallel adsorption vessel is used. The process is inherently cyclical.

Figure 2–4 shows several devices for effecting separation of a solid from a liquid or a vapor. Sketch (a) shows a rotary vacuum filter which is used to remove solids continuously from a liquid-solid stream. A filter medium on a rotating drum is passed through a trough containing the slurry to be filtered. A vacuum inside the drum sucks liquid through the cloth, depositing solid

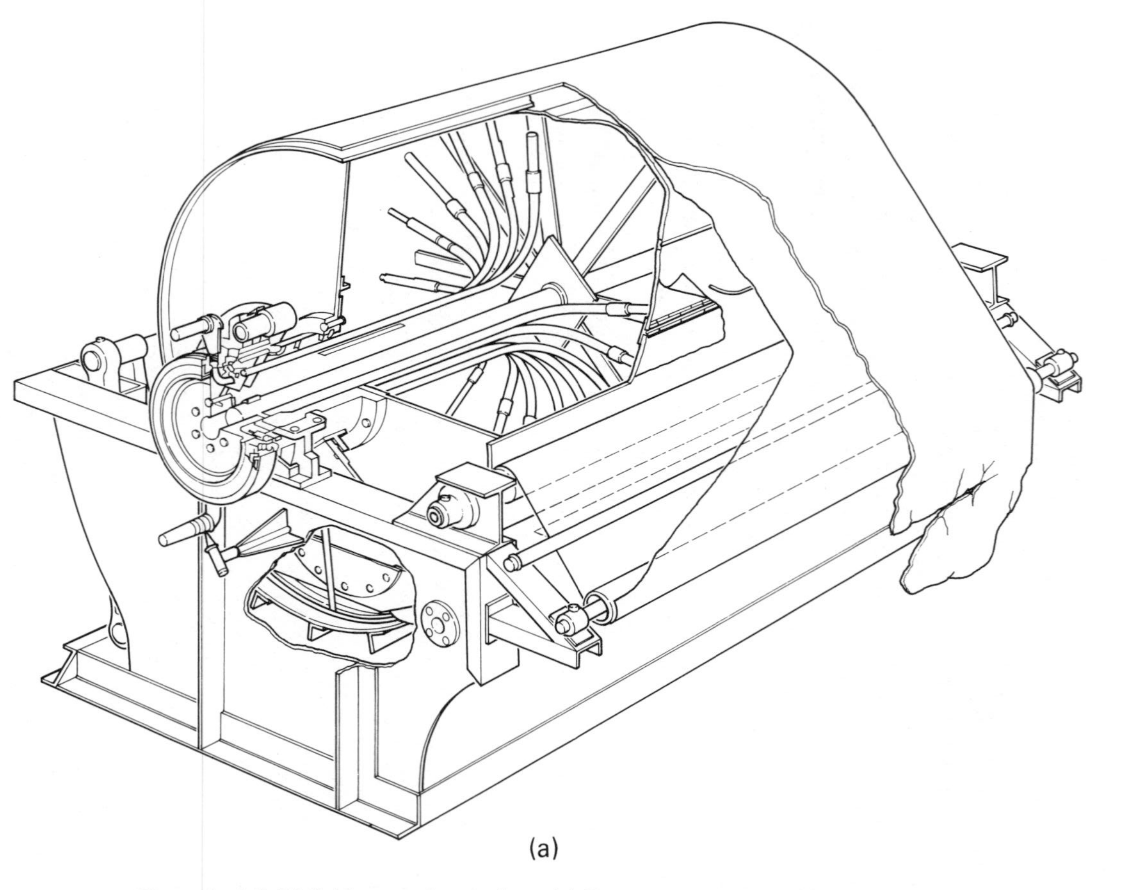

(a)

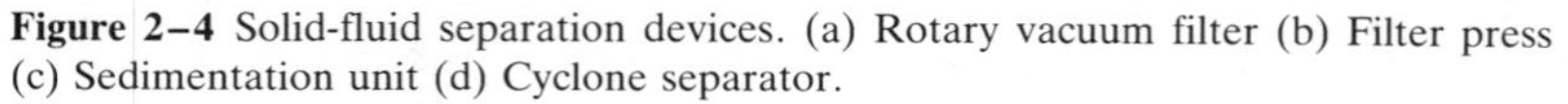

Figure 2–4 Solid-fluid separation devices. (a) Rotary vacuum filter (b) Filter press (c) Sedimentation unit (d) Cyclone separator.

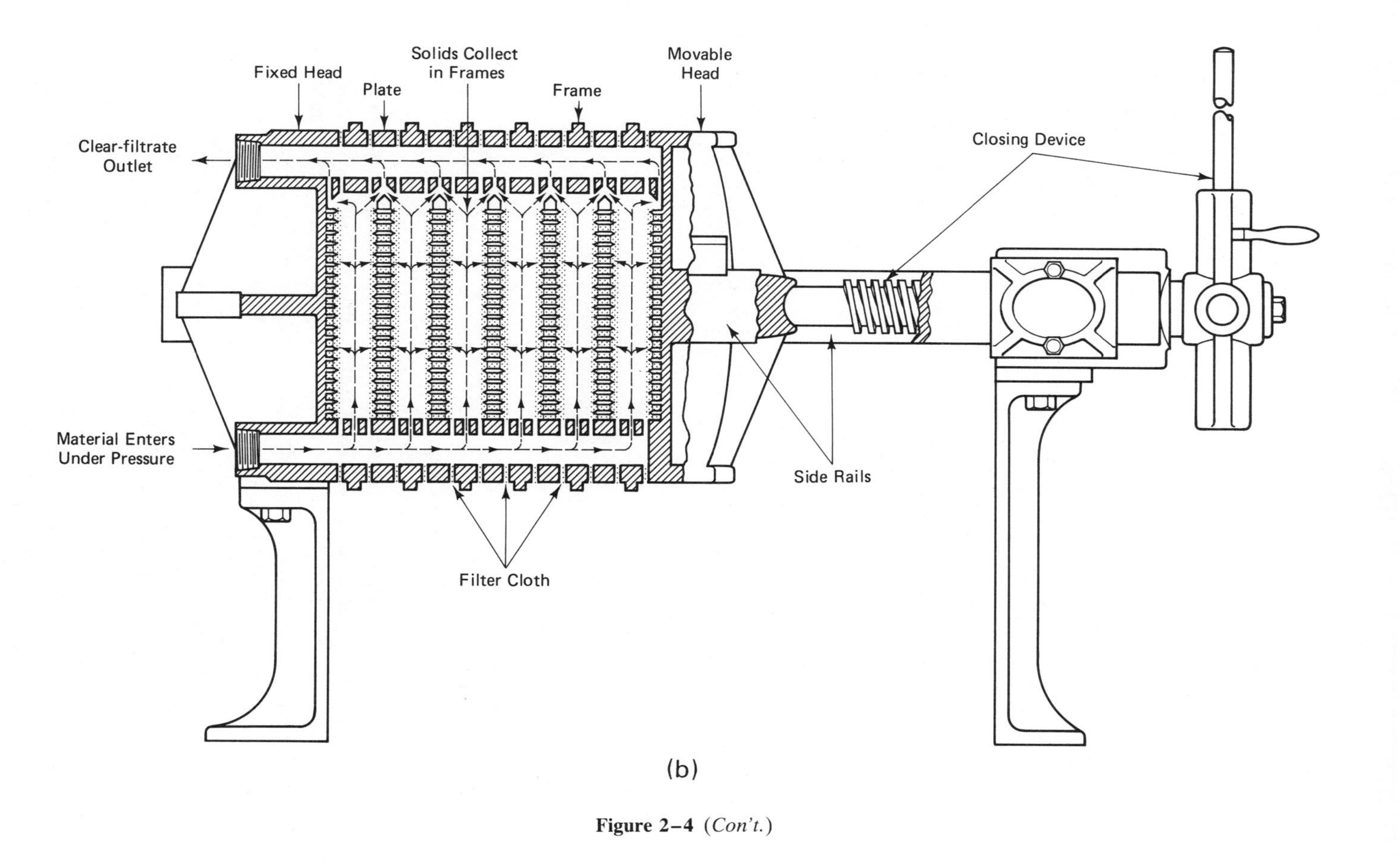

Figure 2–4 *(Con't.)*

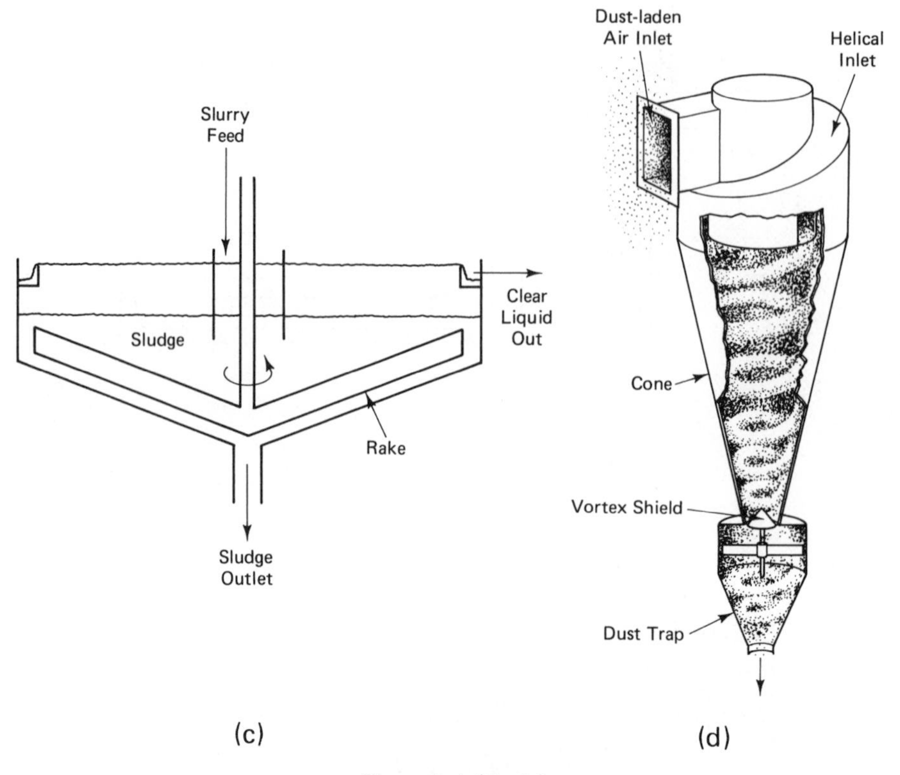

Figure 2–4 *(Con't.)*

on the cloth surface. As rotation continues, the cloth with its solid cake is washed by spraying liquid on the surface. The cake is dried by passing air through it and the cloth. Finally, the cake is scraped from the cloth with a "doctor" knife. A rotary vacuum filter is used for high-capacity streams.

A filter press, shown in (b), is used for batch operation. The filter cloth is suspended over a grill facing an open frame. Filter action is like that in a Buechner funnel. When the space fills with solids, the cake must be removed by disassembling the press. Although this is a simple and reliable unit, large amounts of manpower are consumed. Thus, it is primarily used for small-volume or valuable materials.

Sketch (c) shows a sedimentation unit, which is a large, shallow tank, often over 100 feet in diameter, into which a slurry is fed. As the solids separate by gravity to the bottom of the tank, a slow-moving rake pushes them toward a center bottom outlet. The liquid is decanted off the top continuously. Sedimentation tanks are used with dilute slurries. Sewage sludges, mine effluents, and phosphate tailings are typical sedimentation products.

Sketch (d) shows a cyclone separator. This unit is used for the continuous removal of solids from a gaseous stream. Separation is basically by centrifugal force as a result of the difference in density between the solids and the gas. The ease of separation is directly related to the particle size, so this unit is often used for a first, rough separation. Fine particles in the gas can be removed by using electrostatic precipitation, by scrubbing the gas with a liquid, or by passing the gas through a bag filter.

2.5.3 Other Important Process Units

In addition to purification and reaction, chemical processes usually include steps to heat or cool process streams and devices to move fluids. Heating and cooling devices include furnaces, air coolers, and heat exchangers. Furnaces and air coolers are familiar from your experiences with automobile radiators, household furnaces, refrigerators, and air conditioning units. Fluid-to-fluid heat exchangers are not so familiar, but they are very common in chemical processes.

The most common of these heat exchangers, the shell-and-tube heat exchanger, is shown in Figure 2–5. In this unit, one fluid flows through the shell over the outside surfaces of the tubes. Baffles are usually used to increase the velocity of the fluid passing through the shell. A second fluid enters the tubes, flowing through them in parallel in one or more passes. As it flows through the exchanger, the cold fluid is heated by heat transferred through the tube walls. At the same time, the hot fluid is cooled. The figure shows inlet and exit nozzles for two fluid streams. These streams could be gas or liquid and could change phase within the heat exchanger.

Fluid motion in chemical processing is generally produced by pumps or,

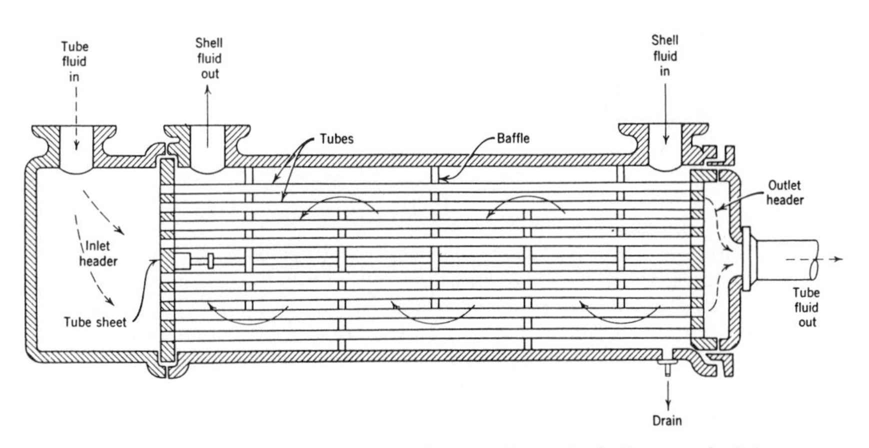

Figure 2–5 Shell-and-tube heat exchanger. *From A. S. Foust et al.,* Principles of Unit Operations, Second Edition. *Copyright © 1980 John Wiley & Sons, Inc. Reprinted by permission.*

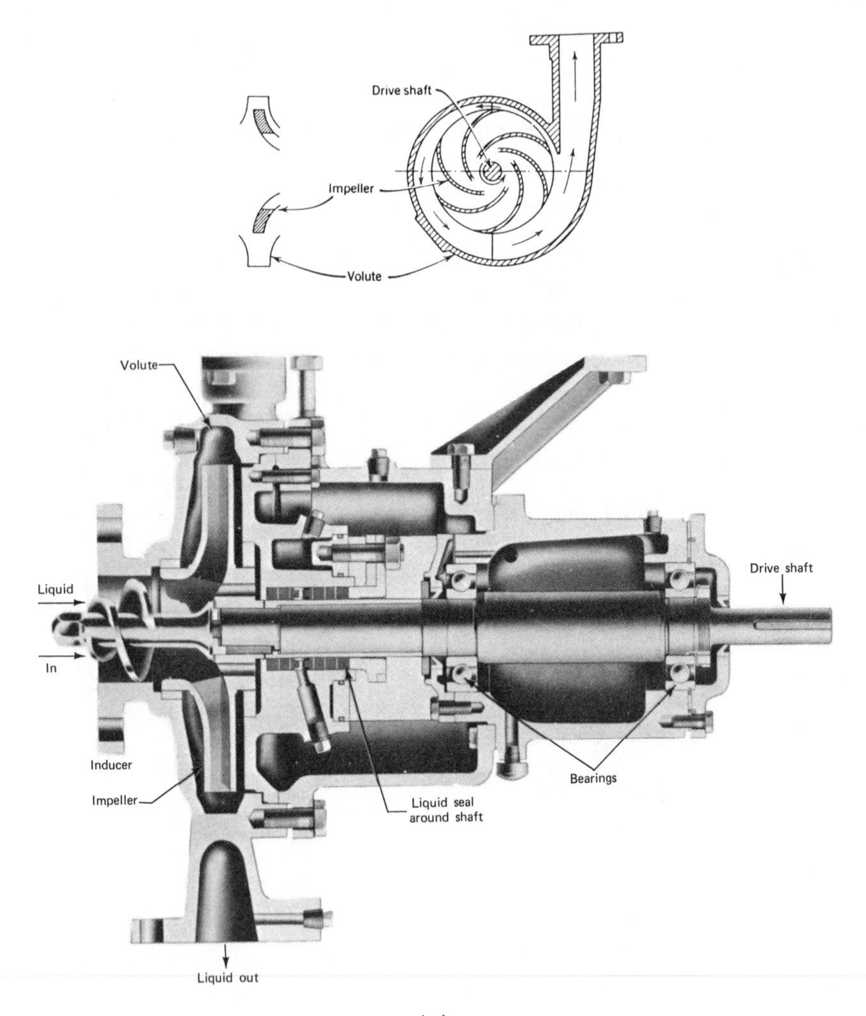

(a)

Figure 2–6 Pumps. (a) Centrifugal pump. *Courtesy Worthington Pump Div., Dresser Industries, Inc.* (b) Positive displacement pumps. The upper drawing is a cross section drawing of a plunger pump. *Courtesy Worthington Pump Div., Dresser Industries, Inc.* The lower drawing shows an internal gear pump. The four drawings show fluid entering the pump, being carried around between the gear teeth, and being forced out of the pump casing. *Courtesy Viking Pump-Houdaille, Inc.*

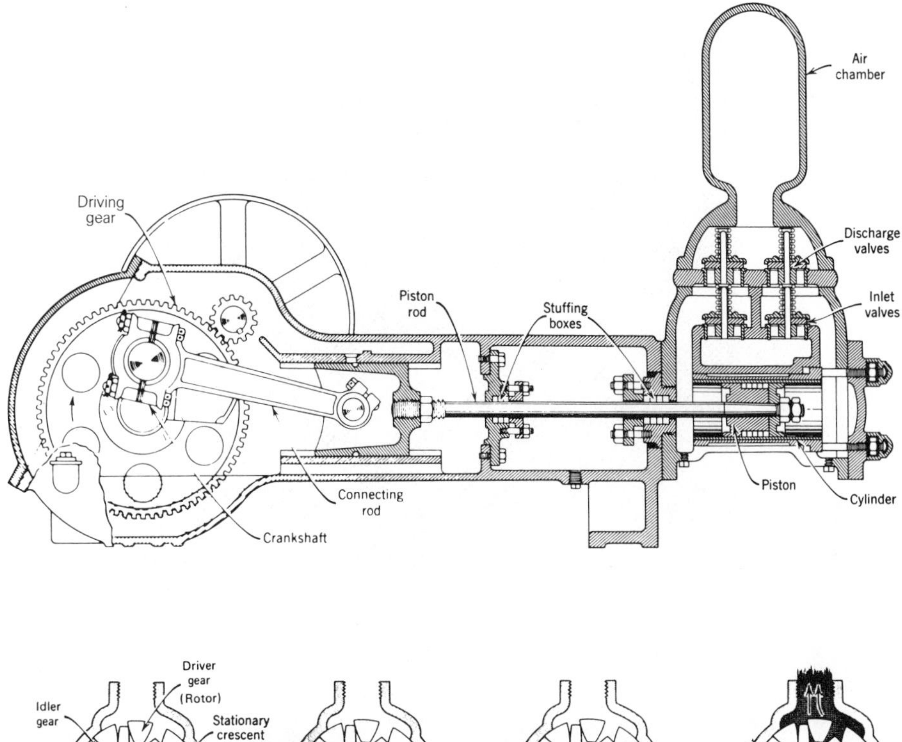

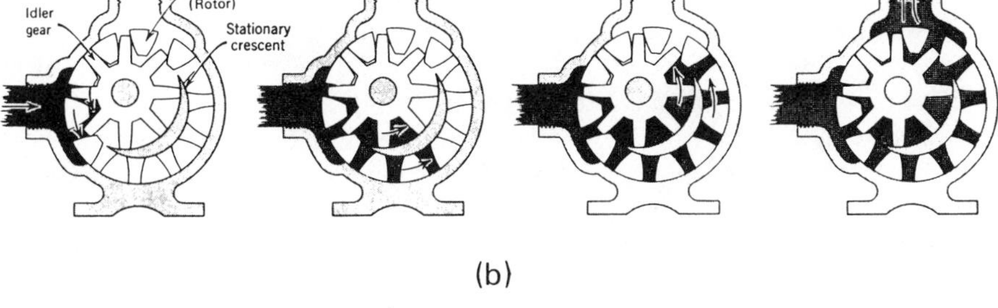

(b)

Figure 2–6 (*Con't.*)

if gases are being moved, by blowers and compressors. The most common type of pump is the centrifugal pump shown in Figure 2–6(a). Liquid flows into this pump and is accelerated into high-speed circular motion by the vanes of the pump impeller. The liquid exits into the pump volute, where it slows down, converting at least part of the kinetic energy into pressure. Because of the mechanism of its operation, the performance of this pump is sensitive to the pressure that it develops. The higher the pressure rise across the pump, the lower the pump capacity is. Thus, the pressure that the pump produces will decrease as the throughput through the pump is increased. Pump efficiency also depends on flow rate.

The other common type of liquid pump is a positive displacement pump.

Figure 2–7 Gas turbo-compressor. *Courtesy MAN GHH.*

(See Figure 2–6(b).) Capacity is directly related to the volume swept out by a piston operating in a cylinder or to the frequency with which a gear tooth enters the cavity between other like teeth. Pressure rise has much less of an effect on capacity in positive displacement pumps.

Gas compressors operate on the same principle as do liquid pumps but usually handle larger volumes and operate at higher speeds. Since large amounts of work are used to compress gases, cooling is usually needed during and after compression. Reciprocating, centrifugal, and axial-flow compressors are commonly used in the chemical process industries. Figure 2–7 shows the internals of a large gas compressor.

2.6 PROCESS SYNTHESIS

The development of a commercial product from a laboratory chemical involves the talents and efforts of many people. At the laboratory stage, the market potential of the product is estimated. The physical and chemical properties of the chemicals involved in the process are then determined, and the conditions under which the product can be produced in the laboratory are explored. But a long and complex effort still remains.

There are times when an engineer is called upon to devise a plausible process long before all of the development effort has taken place. Perhaps a decision whether to invest in the development effort is required. If a tentative process can be suggested, the cost and difficulty of the development of the process can be estimated. Also, a rough estimate of the cost of the product can be calculated, and the size of the market can be forecast. Like most economic forecasts, these "guesstimates" are subject to an enormous amount of uncertainty. Accordingly, many technically and economically sound processes and products have been shipwrecked on the rocks of poor marketing forecasts.

The development of a process scheme involves coming up with that configuration of processing steps that efficiently and safely produces the desired product. An enormous amount of art, skill, intuition, and innovation goes into developing the processing scheme. Literally millions of designs are possible. Somehow the chemical engineer must sort through the myriad of alternatives to come up with a good process. This process synthesis step is where vast amounts of money can be made or lost and where a good, innovative chemical engineer can be worth his or her weight in gold.

A series of questions must be answered as the engineer conceives and develops a process to make a marketable product:

1. What reaction steps are required to get the product?
2. What is the best type and size of reactor to use, and what are the optimum reactor operating conditions (temperature, pressure, agitation, catalyst concentration, etc.)?
3. What is the optimum reactor conversion, and how does it affect the design and operation of the downstream separation steps? The optimization must incorporate the entire plant.
4. Will the catalyst degrade or be poisoned?
5. What side reactions occur? What will the likely by-products be? What effect will they have on yield and performance?
6. What are the best raw materials? Air, water, petroleum, natural gas, coal, minerals, and agricultural products are the basic raw materials. However, a host of semiprocessed intermediates may also be used.
7. How can the raw materials be brought to conditions suitable for the reaction? Usually several purification steps are needed, as well as heating and compressing to the appropriate conditions.
8. How can the products and by-products be separated and purified to meet market specifications?
9. Should cooling water be used, or should air cooling be employed?
10. Are any of the materials used, produced, or present in the process toxic or carcinogenic? Are there other health or safety hazards in the process? Is there a potential for hot spots or explosions in the reactor? Should the reactor be shielded and remotely operated?

All of these questions have several possible answers. Each choice the engineer makes has technical, economic, social, and political repercussions.

The engineer is expected to make prudent and wise choices, and the impact on the environment must be considered.

Perhaps the best way to indicate some of the complex and interesting aspects of process synthesis is to give some examples. We will discuss two processes: the separation of a ternary mixture via distillation, a fairly narrow, focused process; and the conversion of coal to liquid diesel fuel, a broader, more complex process.

2.6.1 BTX Separation

The separation of a ternary mixture of benzene, toluene, and xylenes is a very widespread and important industrial process. Distillation is commonly used to perform the job because the boiling points of the components are reasonably different. There are several ways to implement the technique.

Method 1: Light Out First (Atmospheric Pressure) Consider the system shown in Figure 2–8. Two columns are used, operating in series. The lightest component, benzene, is removed out of the top of the first column. The mixture of the other two components is taken out of the bottom of the first column and is fed into a second column, where the toluene is taken out of the top and the xylenes are taken out of the bottom. Both columns run at atmospheric pressure.

This configuration, called the *light-out-first* (LOF) scheme, is very com-

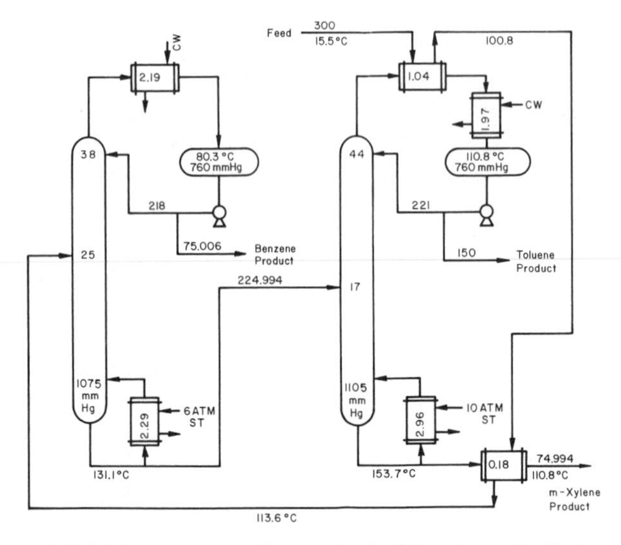

Figure 2–8 Light-product-out-first method of fractional distillation. *Reprinted with permission from H.C. Cheng and W.L. Luyben,* Ind. Eng. Chem. Process Des. & Devel. *24: 707. Copyright 1985 American Chemical Society.*

monly used in many multicomponent distillation separation systems. The total energy consumption in the two reboilers is 5.25×10^6 kcal/hr for the feed flow rate of 300 kg-mole/hr and product purities of 99.9 mole percent. (See H. C. Cheng and W. L. Luyben, *Industrial and Engineering Chemistry Process Design and Development*, Washington, D.C.: American Chemical Society, 1985, Vol. 24, p. 707.)

Method 2: LOF (Vacuum Operation) But suppose we consider some alternatives to this conventional configuration. The simplest thing we might do is reduce the operating pressures in the columns. We can do this and still use cooling water in the condensers because the top temperatures of the two columns at atmospheric pressure are 80°C and 111°C. With 32°C cooling water, we could fairly easily drop these overhead temperatures to about 50°C by using bigger heat exchangers (more heat transfer area in the condenser). This would give pressures of 264 and 90 mm Hg in the two LOF columns. Figure 2–9 gives the process conditions for vacuum operation.

At these lower pressures and temperatures, separation of the components is easier (as we will study in chapter 7). Therefore, less energy is required: 4.8×10^6 kcal/hr, to be precise. Note that we are making exactly the same products from the same feed, but using less energy. Of course, the capital investment in equipment may be larger because of the bigger heat exchangers. But the energy reduction may easily justify this additional investment. Note

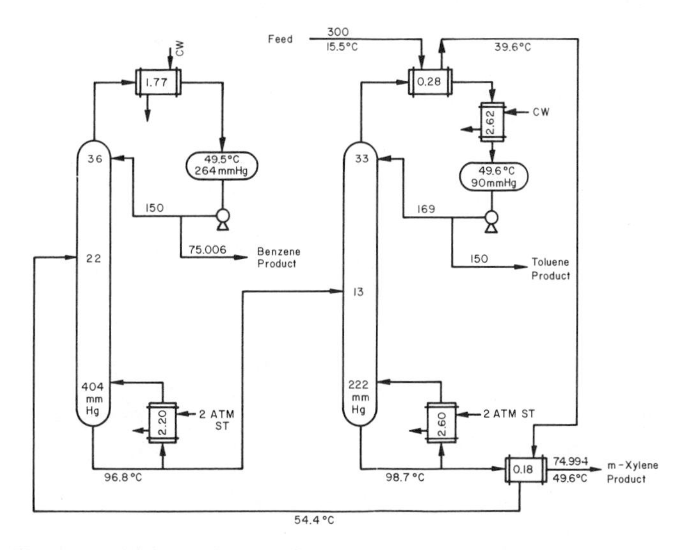

Figure 2–9 Light-product-out-first method of fractional distillation with reduced pressures. *Reprinted with permission from H.C. Cheng and W.L. Luyben,* Ind. Eng. Chem. Process Des. & Devel. *24: 707. Copyright 1985 American Chemical Society.*

also that the temperatures in the base of both columns are lower. This means that lower pressure, less expensive steam can be used. So vacuum operation saves money in two ways: less energy is needed and lower priced energy can be used.

Method 3: Prefractionation Another alternative, which is not so obvious, is shown in Figure 2–10. This two-column scheme uses a first column that is a "prefractionator" and a second column that has two feed streams and three product streams. This complex distillation configuration uses less energy (4.38 $\times$ 10^6 kcal/hr) than the conventional LOF configuration for the BTX separation.

The function of the first column in the *prefractionation* (PF) scheme is to get almost all of the lightest component (the benzene) to go out of the top and almost all of the heaviest component (the xylene) to go out of the bottom. The intermediate component (the toluene) is split more or less equally between top and bottom.

The top and bottom product streams from the first column are fed into the second column at two different feed points. Benzene product is taken out the top of the second column. Toluene product is taken as a sidestream product from a tray located between the two feed points in the second column. Xylene product is produced from the bottom of the second column. Both columns are run under vacuum for easier separation.

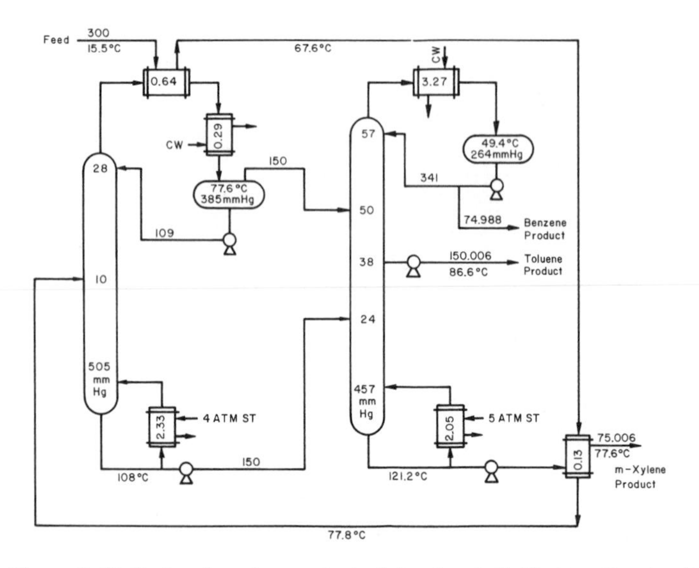

Figure 2–10 Prefractionation method of fractional distillation. *Reprinted with permission from H.C. Cheng and W.L. Luyben,* Ind. Eng. Chem. Process Des. & Devel. *24: 707. Copyright 1985 American Chemical Society.*

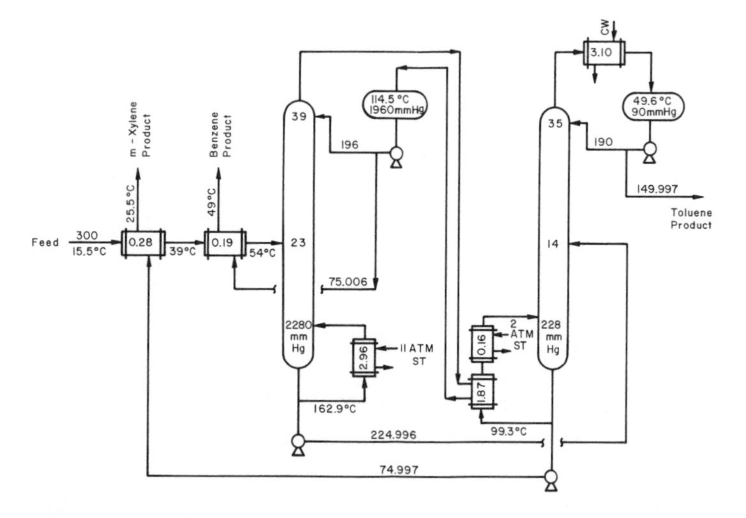

Figure 2–11 Light-product-out-first with heat integration. *Reprinted with permission from H.C. Cheng and W.L. Luyben,* Ind. Eng. Chem. Process Des. & Devel. *24: 707. Copyright 1985 American Chemical Society.*

Methods 4 and 5: LOF and PF with Heat Integration Still other alternatives are possible. For example, we can use "heat integration" and save additional energy. Figures 2–11 and 2–12 show two heat-integrated configurations, one using the LOF scheme and one using the PF scheme.

In heat-integrated columns, the overhead vapor from one high-temperature column is used to add heat to the base of a lower temperature column. This double use of energy saves much money. We will study the application of heat integration to both distillation and evaporation in detail later on in the book.

In Figure 2–11 the operating pressure in the first LOF column is raised to 1,960 mm Hg, giving a temperature of 114°C. Since this is a higher temperature than the base temperature in the second column (99°C), heat integration can be used. Most of the energy is added in the first column; only a small amount of additional energy is needed in the second column. With this technique, total energy consumption drops to 3.12×10^6 kcal/hr.

In Figure 2–12 the operating pressure in the second PF column is raised to 1,620 mm Hg, giving a temperature of 107°C. Since this temperature is higher than the temperature in the first column (91°C), heat integration can be used. Total energy consumption is only 2.36×10^6 kcal/hr in this configuration. However, steam pressures of about 18 atmospheres must be used, compared to the two-atmosphere-pressure steam that can be used in the LOF (vacuum operation) case.

We can see from this example that we have been able to reduce the energy

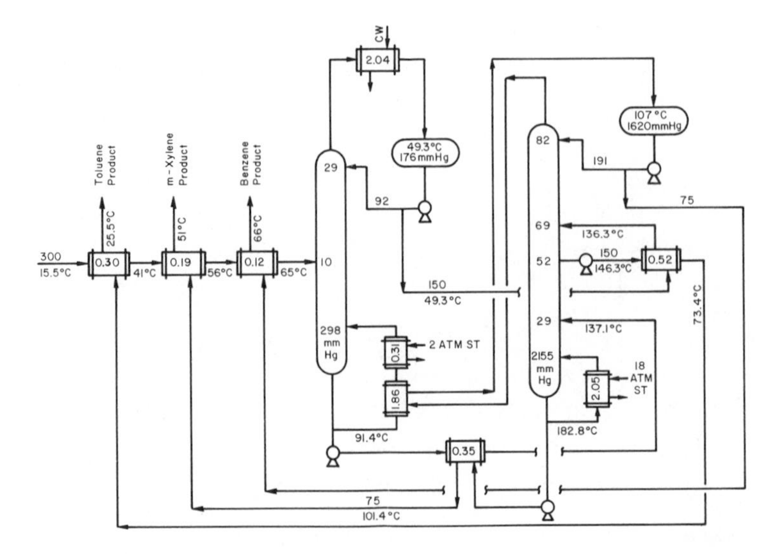

Figure 2–12 Prefractionation with heat integration. *Reprinted with permission from H.C. Cheng and W.L. Luyben,* Ind. Eng. Chem. Process Des. & Devel. *24: 707. Copyright 1985 American Chemical Society.*

consumption in the BTX separation by more than 50 percent by coming up with better and better alternative processing schemes. Thus, it is plain that the innovation, skill, artistry, intuition, creativity, and hard work of a good chemical engineer can produce great potential benefits.

2.6.2 Diesel Fuel from Coal

As a more sweeping example, let us consider the development of a process for the conversion of coal to liquid fuel that is usable as a diesel fuel. To devise a workable process, we need to know the chemical and physical properties of coal and the specifications for diesel fuel. Coal, of course, varies markedly with its source, but generally it is a complex group of large molecules mainly based on benzene ring structures. In this tangled network-like molecular structure are scattered a few atoms of oxygen, sulfur, nitrogen, vanadium, and other elements. But coal consists mostly of carbon and hydrogen atoms and might be characterized as $C_nH_{0.8n}$, where n is a number between 1,000 and 10,000. Often, the S and N atoms are so plentiful that straightforward burning of the coal would produce stack gases with illegal quantities of SO_2 and NO_x. The heavy-metal content makes it unlikely that this material could be used with catalysts developed for petroleum cracking, because these catalysts are readily poisoned by such metals.

Diesel engines will burn any liquid fuel that can be injected into the engine, provided it lubricates the pistons, burns at a controllable rate, supplies enough heat, and is grit free. Typical fuels are No. 2 and No. 5 fuel oils, but No. 5 oil must be heated to 250°F to make its viscosity low enough.

No. 2 fuel oil has a boiling range of 300 to 700°F, a kinematic viscosity of about 3 centistokes (kinematic viscosity = viscosity/specific gravity), and a density of about 33°API [°API = 141.5/(specific gravity) − 131.5]. It has a carbon-to-hydrogen ratio of 6.93 on a weight basis. Thus, its equivalent formula might be $C_nH_{1.73n}$. Its molecular weight is roughly 200. This fuel oil is petroleum based and has a large fraction of paraffins (C_nH_{2n+2}).

Our coal-derived fuel would have a larger fraction of aromatics and lower hydrogen content. Thus, our conversion process must reduce or remove sulfur and nitrogen, crack the coal molecules into smaller molecules, and incorporate more H_2 into the molecular structure. All of this can be accomplished through hydrogenation. So the heart of our new process will be a hydrogenation reactor.

The sulfur and nitrogen will react with the hydrogen to form H_2S and NH_3. The oxygen in the coal will form H_2O. Heavy metals will be unaffected. The gross hydrogenation reaction might be summarized as

$$C_nH_{0.8n} + 0.25n\ H_2 \longrightarrow \frac{n}{m}\ C_mH_{1.3m} \tag{2-1}$$

where n ranges between 1,000 and 10,000, and m is about 15.

Methods for carrying out this reaction might involve grinding the coal in an inert environment and placing it in contact with H_2 directly at an elevated temperature and pressure. An alternative would be to first dissolve the solid coal in some solvent and then contact the liquid with H_2. In fact, this would appear to be an improvement because it eliminates the need to feed a solid reactive powder into a region at a high temperature and pressure, which could be very dangerous. Explosions that occur in grain elevators are often caused by the explosively rapid burning of fine dust particles ignited by static electricity.

The hydrogenation might be done without a catalyst, or a soluble or solid catalyst might be used in a fixed or fluidized bed. The choice of the best method depends upon the cost and performance of the available catalysts. A solid catalyst might be inconvenient because of potential poisoning and the complexities of reacting two fluid phases over such a catalyst.

The products of the hydrogenation will include solid ash, liquid, and a gaseous stream containing H_2S, NH_3, H_2O, H_2, and perhaps some light hydrocarbons. These products must be separated and handled individually. The solid and liquid materials are easily separated from the gases, but separating solids from liquids will probably require filtration or centrifugation. Since temperatures are high, centrifugation is probably best. In any case, the removal and disposal of the solids from the system must be accomplished economically, safely, and without significant ecological impact. Since the ash will carry some liquid with it, the solids can probably be used as fuel.

The most obvious, and probably the cheapest, source of hydrogen is that

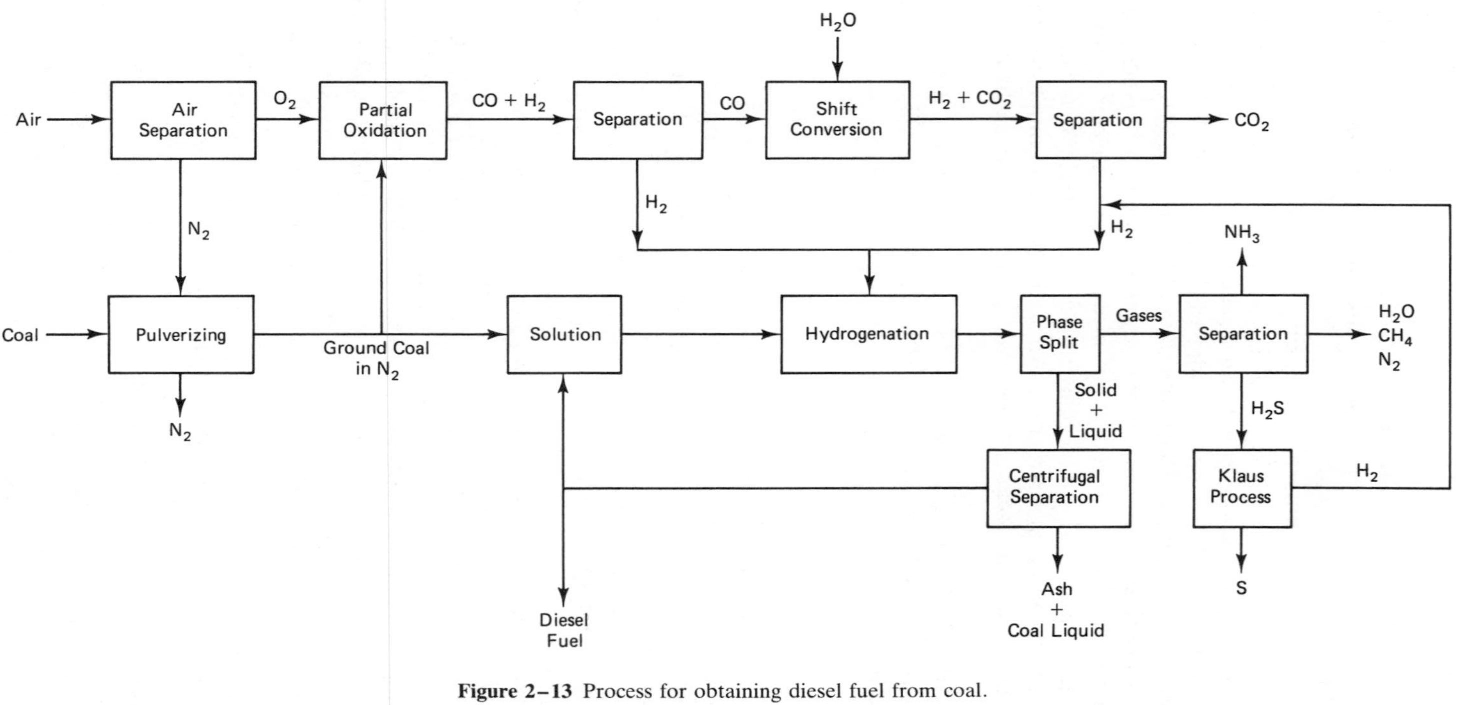

Figure 2–13 Process for obtaining diesel fuel from coal.

produced from partial oxidation of the coal followed by shift conversion. These process steps of "POX" (partial oxidation) and "shift" involve conventional, well-known technology that has been used for many years. The coal would be oxidized with pure oxygen to avoid the presence of nitrogen. The oxygen typically would come from the separation of air using cryogenic (low-temperature) distillation. An equation describing the POX reaction is

$$C_nH_{0.8n} + \frac{n}{2} O_2 \longrightarrow n\,CO + \frac{0.8n}{2} H_2 \tag{2-2}$$

The quantity of oxygen is regulated to avoid the formation of CO_2 and H_2O. The CO and H_2 could be separated in another cryogenic distillation column. The CO is made to react with H_2O in a "shift converter" to produce more H_2 as follows:

$$CO + H_2O \longrightarrow CO_2 + H_2 \tag{2-3}$$

The CO_2 could be removed from the H_2 by being absorbed in a suitable solvent.

A block diagram of the process we have envisaged to convert coal to diesel fuel is shown in Figure 2–13. Each of the blocks can represent an entire plant or a single, simple process step. Some of these plants and steps involve standard, conventional, well-known technology (air separation, partial oxidation, shift conversion, Klaus process for sulfur removal). Others (hydrogenation, centrifugal separation) are multiple-process steps that require a great deal of development effort. The separation steps may be simple, conventional distillations. Note that the feed materials are coal, air, and H_2O. The products are diesel fuel, ash, coal liquid, NH_3, sulfur, and a gas stream containing water and light hydrocarbons. The example illustrates the many complex steps and decisions that must be made to develop a complete process.

PROBLEMS

2–1. Develop a history and current profile for a chemical company, e.g., DuPont, Eastman Kodak, Rohm and Haas, Monsanto, Dow, or ICI. The annual report to the stockholders would be a good starting point. Your Placement Office or library probably has copies of these and other background reports that might be useful.

2–2. Report on one of the following chemical products: ammonia, nitric acid, urea, styrene, carbon dioxide, terephthalic acid, isopropyl alcohol, propylene. Your report should include raw material sources, common production methods (including flow sheets and typical operating conditions), major companies producing the chemical, and major uses. If possible, plant locations, plant sizes, and market price should be included.

2–3. Visit your Chemical Engineering Department undergraduate laboratory and answer the following questions:

(1) How many centrifugal pumps are there?

(2) How many positive displacement pumps are there?

(3) Are there distillation, extraction, or absorber columns in the laboratory? What kind of internals are used?

(4) Is there an evaporator, and, if so, how many stages does it have?

(5) What types of heat exchangers are used?

Typical Chemical Processes 3

This chapter describes three classical continuous large-volume processes involving organic or inorganic chemical manufacturing, and two batch processes that are typical of small-volume, high-value products.

3.1 PRODUCTION OF METHANOL

Methanol has one of the highest annual production rates of all industrial chemicals. It is used as a solvent and a chemical intermediate. In recent years there has been much discussion of the use of methanol as fuel for automobiles. Methanol can be made from any hydrocarbon source—natural gas, oil, biomass, or coal. It is possible that modern civilization's needs for portable fuel may be satisfied by methanol instead of gasoline in the twenty-first century.

There has also been some discussion of converting natural gas into methanol as an alternative to transporting liquefied natural gas (LNG) by ship from places like Indonesia and Libya to destinations like Japan and the United States. Special cryogenic (low-temperature) tankers are required to transport LNG because it has a boiling point of $-258°F$ at atmospheric pressure. But these tankers are expensive to build and also to operate if long distances are involved because the leakage of heat results in significant losses of LNG. Converting the natural gas into methanol would permit the use of conventional tankers.

The methanol production process we discuss uses natural gas, which is mostly methane (CH_4) as the hydrocarbon source. The gas is combined with steam to produce a "synthesis gas" (syngas)—a mixture of CO, CO_2, and H_2. The syngas is then fed into a methanol reactor where the CO and the CO_2 react with the H_2 to form methanol (CH_3OH) and some water:

$$CO + 2H_2 \longleftrightarrow CH_3OH \tag{3-1}$$

$$CO_2 + 3H_2 \longleftrightarrow CH_3OH + H_2O \tag{3-2}$$

The water and methanol mixture is separated into fairly pure methanol and water products in a distillation column. Figures 3–1 through 3–3 give flow diagrams of the three parts of the methanol plant.

Figure 3–1 shows the syngas step. The CH_4 is compressed by a steam-turbine-driven compressor up to about 300 psig (pounds per square inch, as read by a gauge). Steam is added, and the mixture is heated to 1,650°F in a fired furnace. In this so-called reforming furnace, the endothermic reactions (reactions requiring heat) that produce the syngas occur.

The endothermic reactions are mechanistically quite complex, but for engineering purposes we can summarize them in two simple formulas:

$$CH_4 + H_2O \rightarrow CO + 3\,H_2$$

$$CH_4 + 2\,H_2O \rightarrow CO_2 + 4\,H_2$$

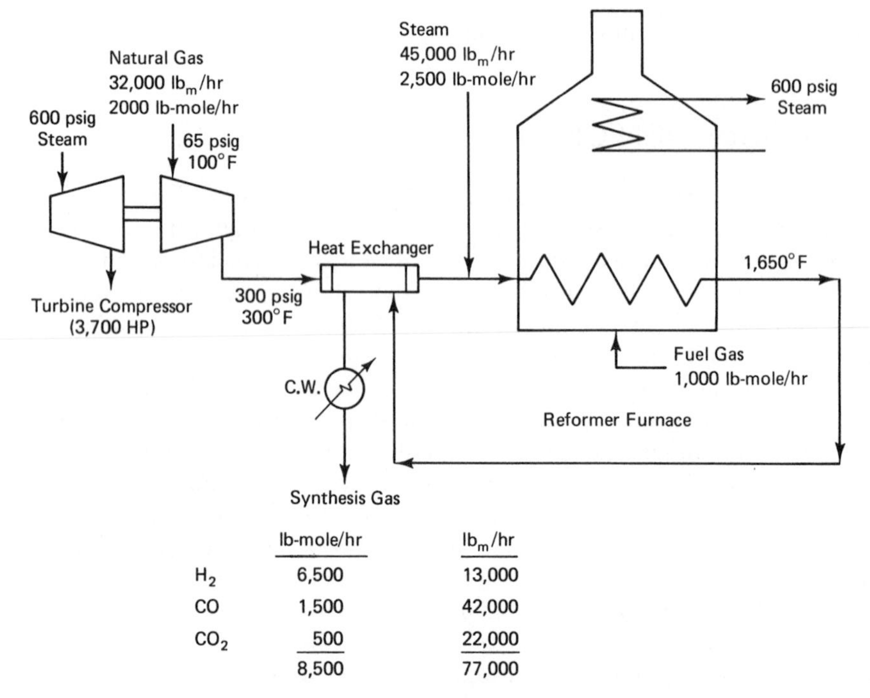

	lb-mole/hr	lb$_m$/hr
H_2	6,500	13,000
CO	1,500	42,000
CO_2	500	22,000
	8,500	77,000

Figure 3–1 Methanol process (reformer furnace).

Both CO and CO_2 are formed, but not in equal amounts due to equilibrium constraints and kinetics. Under the conditions in the syngas furnace, about three moles of CO are formed for every one mole of CO_2.

It is convenient to write the following simple overall reaction equation that describes this stoichiometry and yield constraints:

$$4\,CH_4 + 5\,H_2O \longrightarrow 3\,CO + CO_2 + 13\,H_2 \qquad (3\text{-}3)$$

The syngas leaving the furnace is a mixture of CO, CO_2, and H_2. Not all of the carbon is converted into CO_2 because there is an equilibrium between CO and CO_2 due to the reversible reactions involved. We would like to be able to make only CO, because this would make only methanol in the next step in the process, i.e., in the methanol reactor. Any CO_2 produces H_2O as well as methanol and also uses three moles of H_2 per atom of carbon instead of the two moles that CO uses. (See equations 3–1 and 3–2.)

The amount of fuel gas used in the reformer furnace is about half as much as the methane gas that is fed into the process. This combustion of natural gas provides the heat that is needed for the syngas reactions. It also produces high-pressure steam, which is used in the process in the steam turbine driving the compressors. Exhaust steam from the compressors is then further used in the distillation section of the plant to provide heat to the reboiler. This type of energy system is quite common in most chemical plants. High-pressure (1,000 psig) steam is generated in a conventional fired boiler burning fuel oil, gas, or coal. Steam is also produced at various pressures in process steam generators where heat is available from exothermic reactions and heat recovery steps. The high-pressure steam is used in turbines, on process gas compressors and pumps, and on electric generators. Exhaust steam from the turbines is taken off at several values of pressure (e.g., 300, 100, and 25 psig) and used for process heat in distillation columns, endothermic reactors, and heated tanks and buildings.

The effluent from the reformer furnace is cooled in a series of heat exchangers: first, in a process-process heat exchanger to preheat the feed gas to the furnace (this heat exchanger is sometimes called an "economizer"), and then in a water-cooled heat exchanger.

The syngas is compressed up to about 1,500 psig in a two-stage compressor with an intercooler between stages. You will learn about gas compression in your engineering thermodynamics course and will understand why multiple-stage compression with intercooling is desirable. Basically, the reason is because it reduces the energy required in the compressor. The size of compressor required is 16,500 horsepower!

The compressed syngas is preheated and fed into a fixed-bed catalytic reactor in which the methanol reactions occur. High pressure is used to drive the reversible reactions (equations 3–1 and 3–2) towards the methanol side. There are fewer moles shown on the right sides of these equations than on the left. According to LeChatelier's principle, the yield of methanol will be favored by high pressure. Achieving this high pressure is very expensive, since gas compression is an expensive operation. Pumping liquids to high pressure is not nearly as expensive as compressing gases.

The methanol reactions are exothermic: they give off heat. This heat is removed in some methanol reactors by generating steam. The reactor shown in Figure 3–2 operates adiabatically: no heat is added or removed. The heat of reaction heats up the gases as they go through the reactor. The gases enter at about 500°F and leave at about 725°F. The hot exit gases are used to preheat the feed gas and then to generate some low-pressure steam. Finally, the gases are cooled in a water-cooled heat exchanger.

At the temperature and pressure leaving the cooler, the methanol and water condense, but the excess H_2 does not. Instead, it is recompressed in the second stage of the compressor and recycled back through the reactor. This recycling is necessary because a large excess of hydrogen is used in the reactor. The high partial pressure of H_2 drives the reaction to the methanol side. The hydrogen recycle also limits the temperature rise that the gas undergoes as it passes through the adiabatic reactor. If the recycle flow were lower, the temperature rise through the reactor would be larger, and temperatures might become so high that thermal damage to the solid catalyst in the reactor would result.

There are 6,500 lb-mole/hr of H_2 in the syngas. The 1,500 lb-mole/hr of

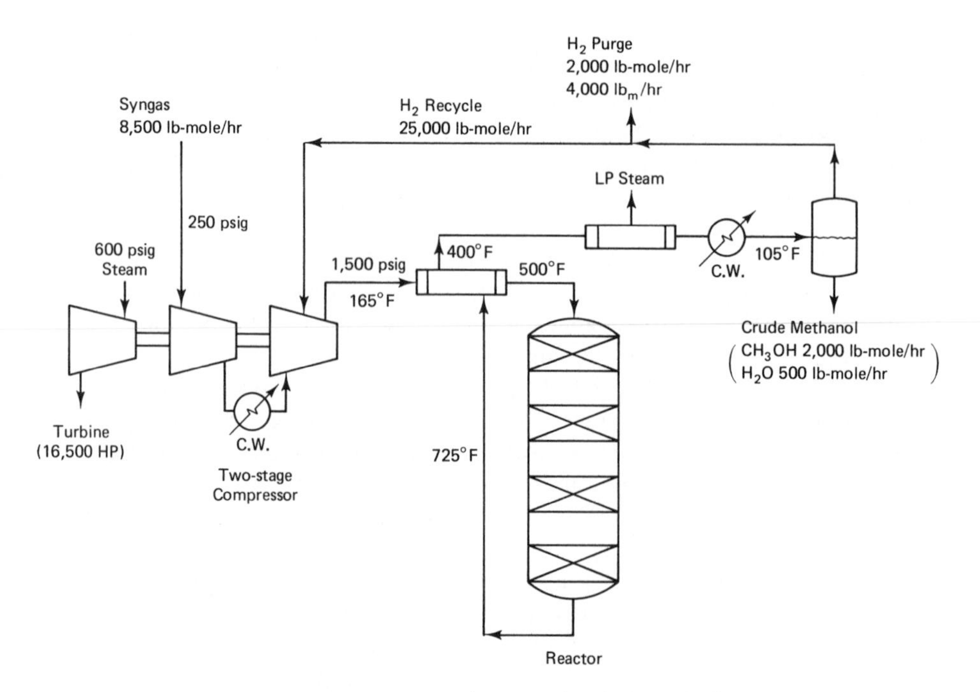

Figure 3–2 Methanol process (synthesis section).

CO consume 3,000 lb-mole/hr of H_2 to make methanol. The 500 lb-mole/hr of CO_2 consume 1,500 lb-mole/hr of H_2. The total H_2 consumed is 4,500 lb-mole/hr. Since there are 6,500 lb-mole/hr of H_2 fed in, the difference (6,500 − 4,500 = 2,000 lb-mole/hr of H_2) must be removed from the system. This is done through the purge gas stream.

The process shown in Figures 3–1 through 3–3 is a greatly simplified and idealized version of a real methanol plant. In a real reactor, the CO and CO_2 would not be completely converted in one pass through the reactor; there would be some CO and CO_2 in the recycle stream and in the purge gas. The CO and CO_2 in the purge gas constitute a loss of carbon atoms. More realistic examples will be discussed in chapter 4.

The last step in the process is the separation of the methanol/water mixture ("crude methanol") into fairly pure methanol and water products as shown in Figure 3–3. The sales specification for the methanol is 99.9 percent purity. The water is discarded, but to prevent water pollution problems, it would typically be sent to a biological waste pond where the bacteria would eat the methanol. The "bugs" cannot take too high methanol concentrations, or they

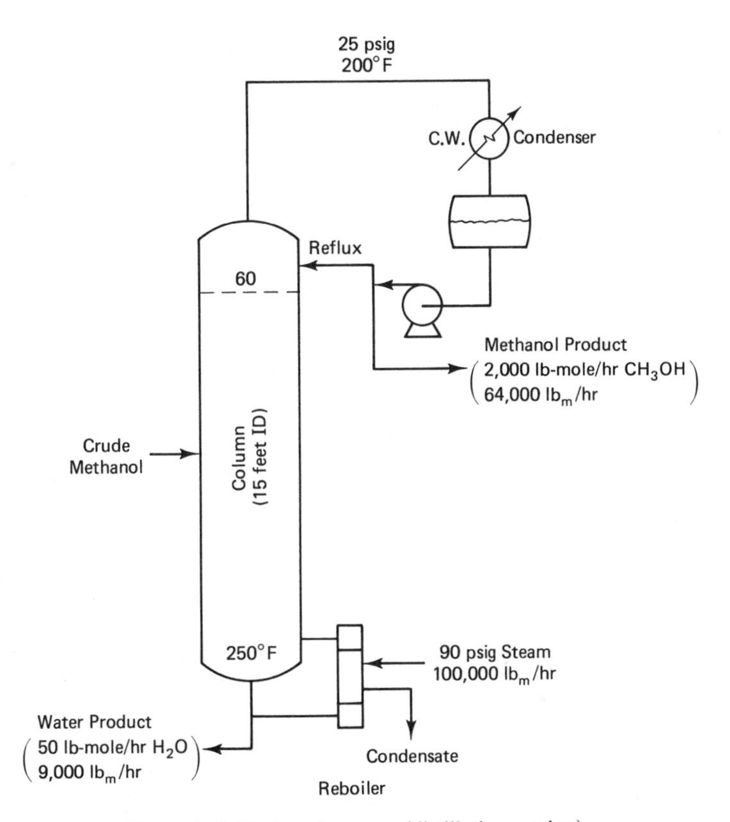

Figure 3–3 Methanol process (distillation section).

will die. In addition, any methanol that is in the water product is a yield loss for the process. So the water product from the separation step is also fairly pure (about 0.1 percent methanol).

The separation is conventionally achieved in a distillation column. Since the "normal boiling points" (boiling points at atmospheric pressure) of methanol and water are 64.5°C and 100°C, respectively, methanol tends to boil at a lower temperature. Therefore the methanol will preferentially concentrate in the vapor stream rising up through the distillation column while the water will preferentially concentrate in the liquid stream flowing down through the column. Therefore, the overhead distillate product is rich in methanol and the bottom product is rich in water. We will discuss vapor-liquid equilibrium and distillation in subsequent chapters.

The vapor that is needed in the column is produced in the reboiler, where steam is condensed on the shell side of the tube-in-shell heat exchanger and process liquid is vaporized on the tube side. The liquid that is needed in the column is produced in the condenser, which is typically a tube-in-shell heat exchanger with cooling water from a cooling tower circulating through either the shell or tube side and process vapor from the top of the column condensing on the other side. The cooling water receives the heat of condensation of the process vapor.

Other types of methanol plants will be discussed in later chapters as examples and in problems.

3.2 ALKYLATION

Alkylation is widely used in petroleum refineries to produce high-octane gasoline. The process involves the reaction of isobutane (iC_4) with unsaturated butylenes ($C_4^=$) to produce iso-octane (iC_8).

$$iC_4 \quad + \quad C_4^= \quad \longrightarrow \quad iC_8 \tag{3-4}$$

```
H3C—CH—CH3        H2C=C—CH3        H3C—CH—CH2—CH—CH2—CH3
     |                 |                 |         |
    CH3               CH3               CH3       CH3
```

The reaction is catalyzed by sulfuric acid (H_2SO_4), and the reactor operates at about 50°F.

The butylenes are produced in a catalytic cracking process. Some of the isobutane also comes from this "cat" cracking, and some comes directly from crude oil.

The reaction is heterogeneous, i.e., involves two phases. The sulfuric acid aqueous phase and the organic phase must be kept intimately mixed, so the liquids in the reactor are vigorously agitated. Since the reaction is exothermic, heat must be removed. To prevent undesirable side reactions involving the polymerization of the butylenes, the reactor temperature must be maintained at about 50°F, and a large excess of iC_4 is used.

The low temperature required for the process means that cooling water *cannot* be used to remove heat from the reactor. Clearly, the temperature of the coolant must be lower than the temperature of the reactor: *heat does not flow uphill*! The temperature of the hot stuff must be higher than the temperature of the cold stuff. Therefore, the alkylation reactor requires some type of refrigeration. We will come back to this aspect later.

The process is sketched in Figure 3–4. The reactor is a large horizontal, cylindrical vessel, about 4 meters in diameter (ID) and 20 meters long. There are several vertical baffles inside the vessel that divide it into between 20 and 30 compartments. Each compartment has a mechanical agitator to mix the two liquid phases.

The "BB" (butane-butylene) feed is a mixture of propane, isobutane, butylenes, and normal butane (C_3, iC_4, $C_4^=$ and nC_4). This stream is split into small streams that are fed into each compartment. The reason for splitting the feed is to keep the concentration of butylenes very low at all points in the reactor to prevent undesirable polymerization. A large recycle stream of iC_4, as well as a large acid recycle stream, is fed to the front of the reactor.

The last section of the reactor is not agitated, because it serves as a decanter to separate the organic and acid phases. The heavier acid phase is taken off the bottom and recycled. A small percentage of the acid is withdrawn as "spent acid," and fresh 98 weight percent H_2SO_4 is added to maintain the inventory of acid in the reactor at a concentration of about 90 weight percent. The acid becomes spent because there is a little water in the feed streams and because some undesirable side reactions occur that contaminate the acid.

The lighter organic phase from the decanter is scrubbed with NaOH and washed with water to remove any traces of acid. If the acid were not removed, all the downstream equipment would have to be very expensive acid-resistant material. With the acid removed, less expensive carbon steel can be used for the distillation column.

The organic phase from the reactor is a mixture of C_3, iC_4, nC_4, and iC_8; all the butylenes have reacted completely. This stream is fed into a large (4-meter-ID, 50-tray) distillation column. The isobutane is taken out of the top of the column, called a *deisobutanizer* (DIB). The overhead vapor from the column is condensed using cooling water, so the temperature in the reflux drum is about 120°F. This means that the DIB operates at about 115 psig, since isobutane condenses at this pressure when the temperature is 120°F. The operating pressures of most distillation columns are set by the available cooling water temperatures. We will discuss this in more detail in chapters 7 and 8.

The overhead distillate product from the DIB is recycled back to the

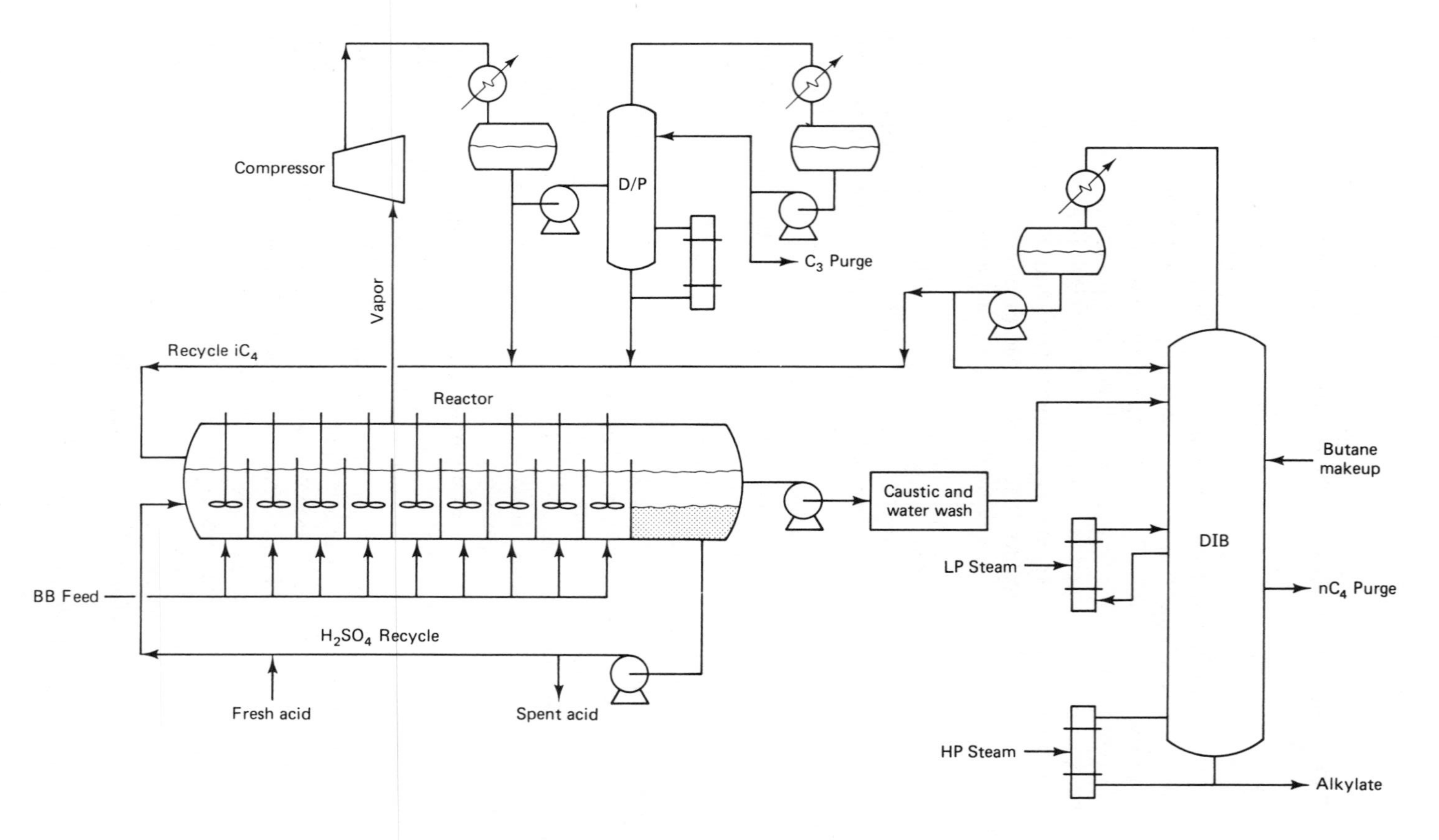

Figure 3–4 Alkylation process.

	Flow Rates (lb-mole/hr)					
	BB Feed	Butane makeup	nC_4 Purge	Alkylate	C_3 Purge	iC_4 Recycle
C_3	20	30	–	–	50	100
iC_4	100	320	19	–	1	2000
$C_4^=$	400	–	–	–	–	–
nC_4	50	300	340	10	–	200
iC_8	–	–	–	400	–	–
	570	650	359	410	51	2300

Figure 3–4 *(Con't.)*

reactor. A makeup stream that contains some of the light material in the crude oil is also fed into the DIB at a lower feed tray. This stream contains a little C_3, but is mostly iC_4 and nC_4. A lower feed tray is used because the concentration of iC_4 in the makeup stream is lower than that in the reactor effluent. Simply mixing the two feeds together would result in higher energy consumption to make the same separation.

A vapor sidestream of mostly nC_4 is removed lower down in the column, purging the nC_4 from the system. Since nC_4 is not involved in the reaction, it would build up in the system over a period of time if not removed. The notion of purging to remove an inert material is a very general one. We will see it used many times in the examples and problems we consider, especially in chapter 4.

Exactly the same problem occurs with the small amount of propane that comes into the system in the two feeds: unless purged, it would build up and fill the whole system. We will come back to this problem later.

The bottom product from the DIB is the heaviest component, the iso-octane or "alkylate." Since the high-boiling component is concentrated in the base of the column, the temperature is quite high there, thus necessitating the use of high-pressure steam in the base reboiler. A second reboiler is shown in Figure 3–4 that is located somewhat above the base. At this point in the column the nC_4 has built up enough so that the temperature is considerably lower than in the base. Now lower pressure, and thus cheaper steam can be used in this intermediate reboiler. This use of an intermediate reboiler is a good example of energy conservation. Since low-pressure steam is less expensive than high-pressure steam, it sometimes pays to invest the additional capital to build two reboilers. Note that the total energy consumption actually increases when two reboilers are used instead of one. But the cost of this energy can decrease if less expensive steam is used in the intermediate reboiler. The same arguments can be made for feed preheaters. We will talk more about the topic in chapter 8.

We now return to the problem of cooling the reactor. In the process shown in Figure 3–4, cooling is achieved by "autorefrigeration." The pressure in the reactor is kept at about 20 psig by pulling off vapor through a compressor. The latent heat of vaporization of the boiling iC_4 in the reactor removes the exothermic heat of reaction. The vapor is compressed to a pressure that is high enough (130 psig) so that it can be condensed using cooling water in the compressor condenser. The condensate is then sent back into the reactor. The liquid "flashes" (forms some vapor) as it drops in pressure going back into the reactor.

A portion of the condensate liquid is pumped into another distillation column called a *depropanizer* (D/P) that removes the propane from the system. The D/P feed is at this location because it is the point in the process where the concentration of C_3 is highest. Propane is the lightest component and therefore builds up preferentially in the vapor phase. Chapter 7 has some problems that require calculation of temperatures, compositions, and pressures throughout this system.

The D/P operates at about 210 psig because that is the pressure required to condense propane at the 120°F temperature attainable in the water-cooled overhead condenser.

Other alkylation processes use different types of refrigeration systems. Some use heat exchangers with refrigerant doing the chilling. (Freon and propane are commonly used refrigerants.) Other types of alkylation processes use other catalysts. One important catalyst is hydrogen fluoride (HF). It is more expensive and difficult to handle than sulfuric acid, but the reactor can be run at higher temperatures and therefore refrigeration is not required.

3.3 PRODUCTION OF PENICILLIN

The processes just described are both continuous-flow processes producing large-volume commodity chemicals. However, many chemical products are made in batch processes. Usually these are relatively small-volume products, often of high unit value. Batch processes are also used when the chemistry of a product is imperfectly known and moving from the laboratory batch mode to a continuous process is not straightforward. Products from batch processes include paints and pigments, pharmaceuticals, fine organic intermediates, polymers, fermentation products, foods such as ice cream and cake mixes, cosmetics, and compounded rubber.

One batch product is penicillin. Originally discovered by Fleming in a bread mold in 1929, penicillin was found to kill a wide range of bacteria in experiments carried out through the 1930s. It was used clinically just before World War II and has revolutionized the treatment of bacterial infections through widespread use since.

Actually, penicillin is a whole family of compounds that occur naturally and are also generated synthetically. They have the common formula

```
        R   H   H   H          CH3
                             S
        |   |   |   |   /        \   |
O  =  C - N - C - C              C - CH3
              |       \          /
              |        \        /
          O = C   —   N  —  C - C = O
                            |   |
                            H   OH
```

The individual penicillin compounds have different chemical groups in the R position. For example, penicillin-G has the attached R group

```
              H   H
              |   |
              C - C           H
            //      \\        |
     H - C              C - C -
         |              |   |
     H - C              C   H
            \\      //  |
              C - C     H
              |   |
              H   H
```

whereas penicillin-K has the attached R group

```
      H   H   H   H   H   H   H
      |   |   |   |   |   |   |
  H - C - C - C - C - C - C - C -
      |   |   |   |   |   |   |
      H   H   H   H   H   H   H
```

The various types of penicillin have particular antibiotic properties and are formed in various concentrations from the action of a particular nutrient under particular conditions of temperature and acidity.

We would expect that molecules as simple as these could be synthesized easily in the laboratory and produced by the methods of organic chemical production. However, the artificial formation of the crucial five-membered thiazolidine ring has been particularly difficult. The result is that although many forms of penicillin have been made synthetically, the cost of duplicating the process on a large scale has been prohibitive. So most penicillins in clinical use are made by fermentation in large-scale batch processes.

The process begins with the development of a culture of the penicillin-forming mold in the laboratory. This is done initially in test tubes with the mold growing on the surface of the nutrient mixture. The grown mold can be kept frozen until it is needed, for years if necessary. The laboratory mold is transferred to several larger vessels containing growth media. Growth continues through several enlarged steps until enough of the mold is ready to add to the production fermenter.

The production process is shown in the flow sheet of Figure 3–5. The first four units picture the laboratory development of the seed mold culture and are successive batch procedures. In the first three of these, glassware vessels are loaded with the culture and nutrients and are placed in a shaker at constant temperature for the desired time. The last stage is a tank-sized "seed" fermenter where the mycelia are grown, but in which little penicillin is formed. In the fermenter, air is blown through a violently agitated mixture containing the culture formed earlier, corn steep, lactose, and mineral salts. The products from this tank are fed to a larger fermenter after about a day of operation. This fermenter is perhaps 50,000 gallons in volume and is agitated with a stirrer driven by a 700 horsepower (HP) motor. Corn steep, lactose, and mineral salts are added. The first day of cultivation is used to build up the concentration of penicillin-forming mold. Then the penicillin precursor, the compound that is altered to form the penicillin side chain, is added. For penicillin-G this would be phenylacetic acid; to get penicillin-V, phenoxyacetic acid would be used.

The penicillin forms in this fermenter during the next three to seven days. Figure 3–6 shows the buildup of mycelia during the first day of fermentation and the increasing penicillin concentration during the next three days. As the process goes on, the sugar content of the mix drops and the pH increases modestly. Three to five main fermenters might be fed from the output of a single seed fermenter.

When the fermentation is complete, the fermentation mixture is fed to a filter where the mold mycelia are filtered from the broth in a continuous rotary vacuum filter. The liquid (called beer) is cooled to about 40°F by a refrigerated cooler and sent to a storage tank.

Up to this point the process has operated completely in a batch mode. From here on, a semicontinuous mode is used. Acid and solvent (amyl acetate) are added to the cold beer, and the solvent collects the penicillin transferred from the beer in the mixing tank. Then the solvent and beer are separated in a disc centrifuge. The beer is contacted with a second batch of solvent to complete the removal of the penicillin. This second batch of solvent, containing small quantities of penicillin, is used in the first contact stage with the fresh beer. The depleted beer is sent to a distillation unit, which is not shown in the figure, for the removal of traces of solvent.

The solvent/penicillin mixture from the first extraction is fed to a mix tank, where water and alkali are added. In this basic environment ($pH > 7$) the penicillin is more soluble in water than it is in the solvent. Figure 3–7 shows

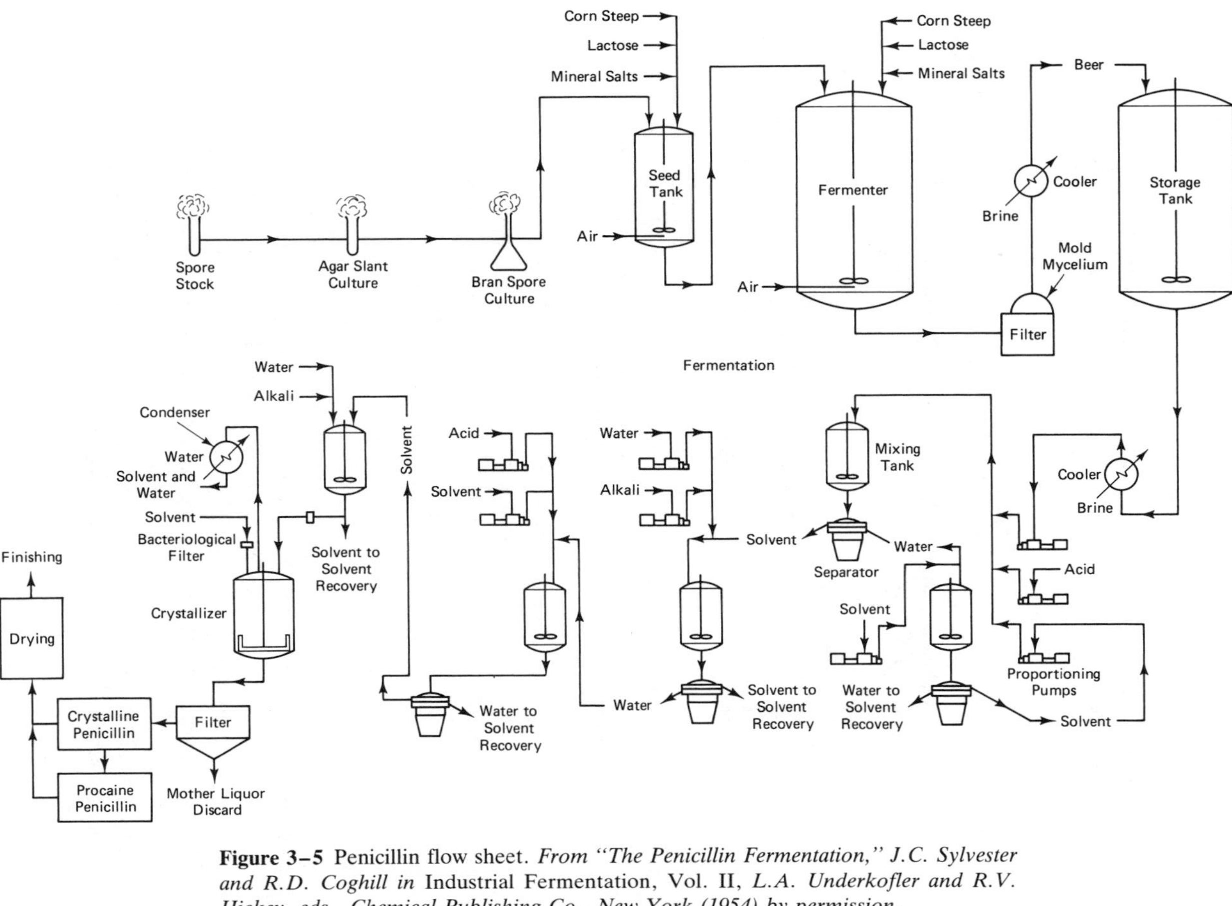

Figure 3–5 Penicillin flow sheet. *From "The Penicillin Fermentation," J.C. Sylvester and R.D. Coghill in* Industrial Fermentation, Vol. II, *L.A. Underkofler and R.V. Hickey, eds., Chemical Publishing Co., New York (1954) by permission.*

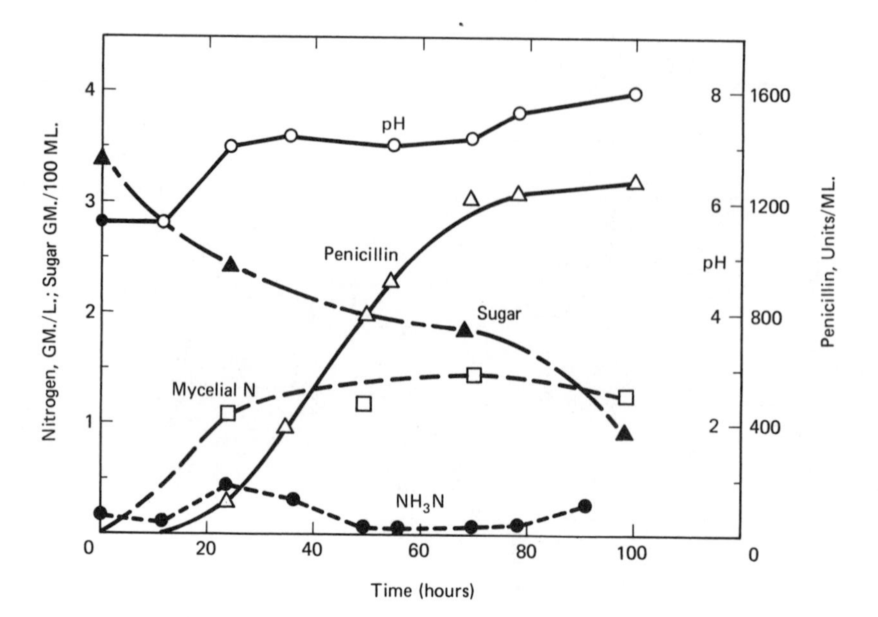

Figure 3–6 Chemical changes with a typical fermentation of penicillin. *From "The Penicillin Fermentation," J.C. Sylvester and R.D. Coghill in* Industrial Fermentation, Vol. II, *L.A. Underkofler and R.V. Hickey, eds., Chemical Publishing Co., New York (1954) by permission.*

the effect of pH on the distribution of penicillin between amyl acetate and water. Here again, the two phases are first intimately mixed and then separated by centrifugation. The solvent is sent to a solvent recovery unit, where it is purified for reuse. The water/penicillin phase is again mixed with acid and solvent, with the penicillin moving back to the solvent phase. After separation of phases, a final transfer is made from solvent to water phase. These repeated transfers between water and solvent phases purify the penicillin.

The penicillin-water phase is then sent through a charcoal clarifier where color bodies are removed. Next it goes to a crystallizer where a precipitating agent is used to force crystallization of the penicillin. The crystals are collected in a basket centrifuge. The liquid removed from the centrifuge is recycled to the crystallizer, and the solids are sent to a drier where hot air passing over pans of the crystals removes the remaining water. Fine solid particles picked up by this air stream are removed in a cyclone separator. The solids from the cyclone and those from the drier are packaged in drums and sold as crude penicillin.

Table 3–1 shows the production and price of penicillin in the U.S. from 1943 to 1951. The figures are perhaps extreme because early use of penicillin was restricted to the armed forces. But the trend of decreasing price as capacity increases is typical of that experienced when any new chemical product begins to be widely used and generally available.

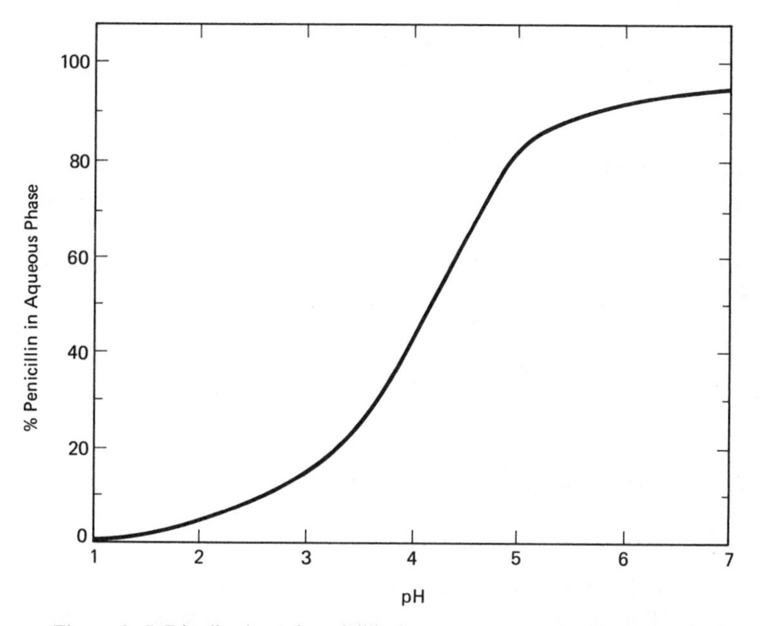

Figure 3–7 Distribution of penicillin between water and amylacetate as a function of pH. *From "The Penicillin Fermentation," J.C. Sylvester and R.D. Coghill in* Industrial Fermentation, Vol II, *L.A. Underkofler and R.V. Hickey, eds., Chemical Publishing Co., New York (1954) by permission.*

Fermentation broth concentrations vary from manufacturer to manufacturer and are generally considered proprietary. One process has been documented to produce 25–35 gram/liter (g/l) of dry cell weight and about 30 g/l of penicillin G-K from the following ingredients: 21.9 g/l of corn steep liquor (for nitrogen supply), 10.7 g/l of $(NH_4)_2SO_4$, 2.74 g/l of KH_2PO_4, 15.9 g/l of phenylacetate, and 200–282 g/l of glucose (carbon and energy supply). These

TABLE 3–1
PRODUCTION AND PRICE OF PENICILLIN, 1943–51

	Production (units × 10^8)	*Wholesale price (\$/10^5 units)*
1943	21	20.00
1944	1,633	7.00
1945	7,125	0.78
1946	25,809	0.51
1947	30,640	0.36
1948	90,501	0.31
1949	138,100	0.20
1950	219,903	0.13
1951	324,293	0.15

ingredients cost about \$454/1,000 gal of fermenter volume (in 1978 dollars). The fermentation yield was about 0.13g of penicillin G-K per g of glucose fed.

A typical modern fermentation plant uses five 50,000 gallon fermenters, each with a 700 HP agitator and with air fed from three 10,000 ft^3/minute compressors. The recovery process would use continuous centrifugal extraction, rotary vacuum filtration, continuous crystallization, and vacuum drying. The capital investment in 1978 dollars would be about \$40 million, of which \$10 million represents the cost of the process equipment itself. Direct production costs would be about \$1/gallon of broth. Fixed charges would be about 40¢/gallon of broth. Including overhead, the total manufacturing cost would be about \$1.8/gallon of broth, or \$20/kg of bulk penicillin crystals.

3.4 PRODUCTION OF TETRAMETHYL LEAD

Another example of a typical batch process is the production of tetramethyl lead (TML). This product was widely used as a gasoline additive to increase octane numbers until the ban on the use of lead for environmental reasons dried up the TML market. Its value per pound was quite high and the reaction times were quite long, so a batch reactor was used.

The reactants are a solid metallic alloy of sodium and lead and liquid methylchloride. Two parallel reactions occur, one producing the desired TML and the other producing the by-product ethane. The ethane is a gas under the conditions of pressure and temperature in the reactor. The reaction formulas are

$$Pb(Na)_4 + 4\ CH_3Cl \longrightarrow Pb(CH_3)_4 + 4\ NaCl \tag{3-5}$$

$$2\ Na + 2\ CH_3Cl \longrightarrow C_2H_6 + NaCl \tag{3-6}$$

The reaction is catalyzed by $AlCl_3$, and a small amount of toluene is added to inhibit the detonation of TML, which has an explosive capacity greater than TNT when not inhibited.

The solid flakes of NaPb alloy (which look like gravel) and the catalyst are charged to a horizontal batch reactor 1 meter in diameter and 2 meters long. The material in the reactor is mixed by a slowly rotating (10 rpm) agitator which scrapes the walls of the cylindrical reactor. The vessel is sealed, and methylchloride and toluene are pumped into the system. The amount of CH_3Cl fed is about 10 percent greater than the stoichiometric amount required to react with all the sodium.

The methyl chloride pressurizes the reactor up to about 150 psig, since the vapor pressure of CH_3Cl is about 150 psig at ambient (normal outside) temperatures. Then steam is introduced into the shell surrounding the reactor to heat up the contents. At about 100°C the reactions begin. Both reactions are exothermic, so it is now necessary to switch from heating to cooling the reactor. This transition can be quite tricky and requires an effective control system.

Heat is removed by running cooling water through a jacket surrounding the reactor and through a tube-in-shell heat exchanger (condenser) located above the reactor. Vapor from the boiling liquid mixture in the reactor goes into the condenser and is returned to the reactor as liquid. Figure 3–8 shows the reactor and condenser system. The liquid flows back into the reactor by gravity, i.e., the reactor is at a higher pressure than the condenser but the hydraulic height of liquid in the reflux line overcomes this pressure difference.

The material in the reactor is a liquid/solid slurry of solid NaPb alloy, solid salt (NaCl), and liquid methylchloride, toluene, and TML. Since methylchloride is much more volatile than TML (their normal boiling points are −24°C and 110°C, respectively), the vapor from the reactor is mostly CH_3Cl and ethane.

The ethane that is produced by the side reaction accumulates in the con-

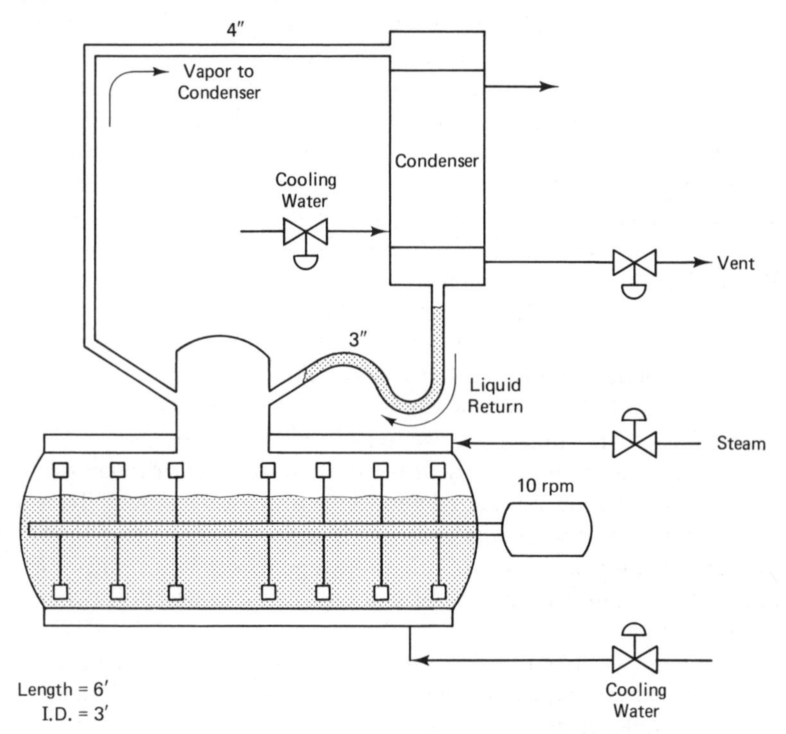

Figure 3–8 Tetra-methyl lead reactor.

denser since ethane acts like a noncondensible gas at the temperature and pressure of the system. If the ethane were not vented out of the system, the condenser would fill up with ethane (become "blanketed") and the heat removal capacity of the condenser would decrease drastically. Therefore, some gas is vented from the condenser during the batch run.

The pressure in the reactor starts out at 150 psig, but as the reactor is heated up, the pressure rises. It is controlled to follow a pressure trajectory, shown in Figure 3–9, by manipulating the cooling water flow rate to the condenser and the vent rate. Also shown in the figure are the heat removal rate and the reactor temperature as functions of time. The batch runs for about 100 minutes. After the NaPb alloy has reacted completely, the excess methylchloride is vented off to a recovery system for recycle back to the next batch.

The TML is recovered from the NaCl salt by batch steam distillation. The solids, toluene, and TML from the reactor are dropped into a vessel containing water. Steam is used to heat the mixture, driving off the toluene and TML in the vapor phase.

The control of the pressure trajectory is critical in this batch reactor. In order to minimize batch times and maximize capacity, the pressure and temperature should be as high as possible. However, if the pressure is allowed to rise too rapidly, it will raise the temperature too rapidly, increasing the rate of the exothermic reactions and the heat generation rate. If the heat cannot be removed in the jacket and the condenser as rapidly as it is being produced, a runaway reaction can occur which will blow a rupture disk on the reactor and blow the flammable and toxic reaction mass out into the atmosphere.

Batch-to-batch variations in the rates of reaction can occur due to variations in the particle size distribution of the solid NaPb alloy and to slight differences in the amount of methylchloride initially charged to the reactor. Thus, cooling rates vary from one batch to the next.

How to regulate the venting rate is another problem. If too little gas is vented, the condenser becomes blanketed with ethane, and heat removal capacity drops off. If too much gas is vented, methylchloride is lost in the vent gas. This means that more methylchloride has to be used and recycled in the process. More importantly, it means that the reactor temperatures will be higher than when more methylchloride is present because TML is a high-boiling component and methylchloride is a low-boiling component. Since higher temperatures increase the rate of heat generation, and since the region of operability is fairly narrow, you can easily find yourself between a rock and a hard place.

The dynamic control problems seen in connection with the production of tetramethyl lead are typical of many batch reactors. The TML reactor is an excellent example of the many-faceted aspects of running a batch system.

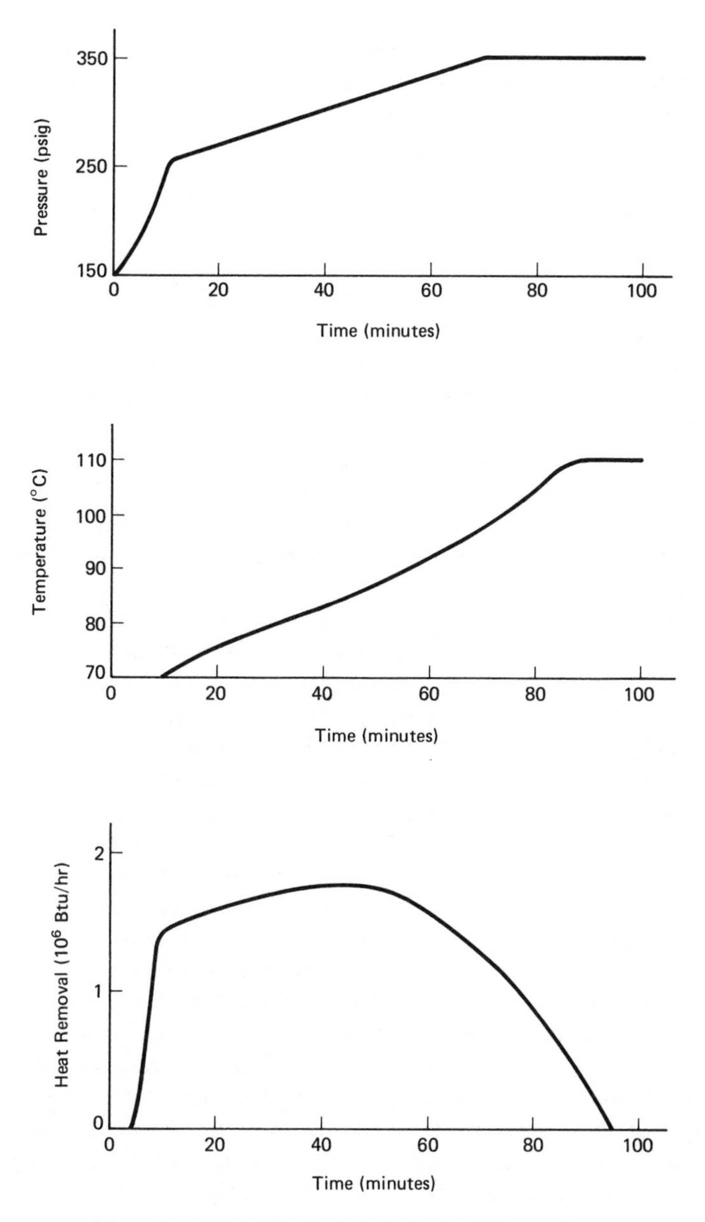

Figure 3–9 Typical batch trajectories in TML reaction.

3.5 PRODUCTION OF SULFURIC ACID

The basic equations for the formation of sulfuric acid show the oxidation of sulfur to SO_2, the further catalytic oxidation of SO_2 to SO_3, and finally the absorption of SO_3 into water to form H_2SO_4:

$$S + O_2 \longrightarrow SO_2 \tag{3-7}$$

$$SO_2 + \frac{1}{2} O_2 \longrightarrow SO_3 \tag{3-8}$$

$$SO_3 + H_2O \longrightarrow H_2SO_4 \tag{3-9}$$

Sources of sulfur include iron pyrites ore (FeS), SO_2 from stack gases, sulfur obtained from mine sources, sulfur removed from petroleum products, and other process sources. Today, by-product sulfur is so plentiful and cheap that sulfur is the dominant source of raw material.

Figure 3–10 shows the flow sheet of a modern double-absorption sulfuric acid plant. This plant would be capable of meeting present environmental limitations and yet could produce commercial quantities of sulfuric acid in a range of concentrations.

In the production process, sulfur, perhaps delivered by rail, is melted, pumped through a filter, and delivered to a sulfur burner. The burner is a horizontal, brick-lined furnace into which molten sulfur and dried air are fed at one end. The air is dried by passing it through a tower in which concentrated sulfuric acid (about 98 weight percent H_2SO_4) flows countercurrently to the rising air stream. The column is packed with ceramic rings.

The sulfur is sprayed through a pressure nozzle in the furnace. Since the degree of sulfur atomization is critical to the efficiency of burning, the nozzle pressure must be 150 psig or more. Gases from the burner will contain 9–10 mole percent SO_2 along with O_2 and N_2. Since sulfur in air ignites spontaneously, no ignition source is needed if the combustion chamber has been preheated to about 400°C. The gases leave the chamber at about 1,000°C.

The combustion gases are cooled to about 415°C in a waste heat boiler which generates steam. The gases are then fed to a catalyst bed containing the catalyst vanadium pentoxide. Here the SO_2 is oxidized to form SO_3. There are actually four catalyst beds, with the process gas being cooled as it flows from one of these beds to the next. Figure 3–11 shows the history of the temperature and conversion as the gases flow through these beds and as the conversion approaches completion. Note that the reaction is limited by the equilibrium conversion and the ultimate conversion is raised by keeping the reaction temperature low. The equilibrium temperature curve shown is for a gas mixture originally containing 10 mole percent SO_2. In some processes, additional air is fed in the second and third stages to allow for further conversion of the SO_2, but this effect is small.

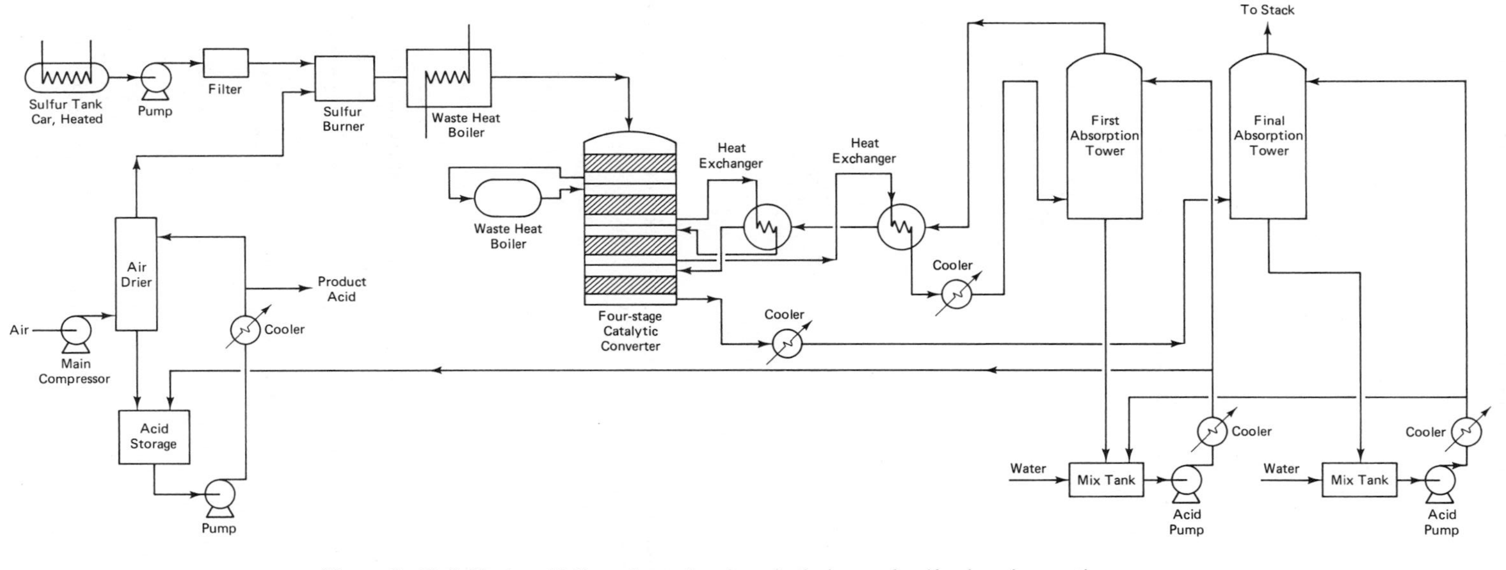

Figure 3–10 Sulfuric acid flow sheet showing air drying and sulfur burning section, sulfur trioxide converter section, and absorption section.

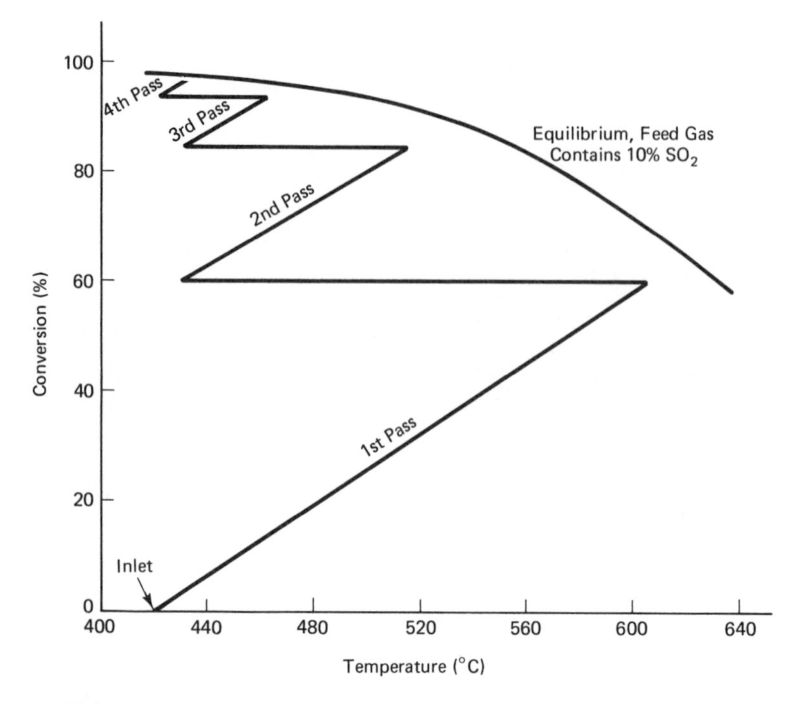

Figure 3–11 Equilibrium and reactor conversion during catalytic SO_2 oxidation.

The gas from the last bed of the reactor is cooled to about 200°C and then fed to an absorption tower where the SO_3 is absorbed into concentrated H_2SO_4. In this tower, which is packed with ceramic rings, 95 weight percent H_2SO_4 is fed to the top of the tower at about 75°C. The SO_3 is absorbed into this solution so that 98.5 weight percent H_2SO_4 leaves the column at about 120°C. This acid may be sold as product, may be used for drying the air feed, or may be diluted for sale at other concentrations. In the flow sheet shown in Figure 3–10, two absorption columns are used. Two columns are necessary in order both to remove the SO_3 almost completely and to give an exit gas stream leaving the plant that meets modern environmental limits. SO_2 stack emission limits in the U.S. require at least 99.7 percent efficiency of conversion of SO_2 to SO_3.

It is interesting to note that an inventory of H_2SO_4 is necessary to start up this contact process. Thus, some other method for making H_2SO_4 must have existed before the first contact H_2SO_4 plant was built. The old plants used to make sulfuric acid by using lead chambers to cause SO_2 and NO_x mixtures to react with water. The old process was quite costly, and no such plants exist today. The need for the presence of some amount of the product at plant startup is not unusual in today's chemical industry.

PROBLEMS

3–1. Homemade ice cream is made by preparing a custard of milk, eggs, gelatin, and sugar and cooking this while stirring until it simmers. The custard is cooled and then poured into a deep metal chamber fitted with a wooden dasher. Cream, both light and heavy, vanilla, and dissolved sugar are added, and the vessel is covered and set into a chest filled with ice and salt. A drive rotates the vessel while holding the dasher still until the ice cream thickens and swells as a result of air inclusion. The can is opened and pitted cherries, cubed peaches in juice, or some other flavoring is added.

a. Draw a flow sheet for the continuous, large-scale manufacture of ice cream using the recipe just given.

b. What additional information would be needed to design the process equipment specified in your flow sheet, and how would it be used?

3–2. Moonshine is made in the hollows of Kentucky woods, they say, by mashing corn, adding a dash of yeast to it, and letting it ferment. When fermentation is completed, the mash is cooked over a wood or charcoal fire. The fermentation and cooking are done in a large closed copper pot with an air-cooled coil connected to the top outlet. During cooking, the vapors from the mash condense on the walls of the coil and drip out into a collection vessel. Cooking is stopped when the first succession of droplets stops coming, even though further cooking could produce more condensate.

a. Draw a flow sheet for making moonshine continuously in high volume.

b. Formulate questions you might ask your favorite moonshiner which would help you fix unknowns in your design.

3–3. Draw a flow sheet for the batch iron-making process occurring in a blast furnace. Here the main raw materials are iron ore (Fe_2O_3 plus other iron oxides and impurities), coke (C), and limestone ($CaCO_3$). The air "blast" is raised to reaction temperature by heat exchange with the outlet gases as well as heat produced from exothermic reactions

$$C + \frac{1}{2} O_2 \rightarrow CO$$

$$C + O_2 \rightarrow CO_2$$

The carbon also reacts with the Fe_2O_3:

$$Fe_2O_3 + 3C \rightarrow 2Fe + 3CO$$

$$2Fe_2O_3 + 3C \rightarrow 4Fe + 3CO_2$$

The limestone acts as flux. (It liquefies at reaction temperature and controls the pH of the reaction mix.) Periodically, Fe is drawn out of the furnace as a liquid, and the three main reactants are fed in at the top of the furnace. Off-gases are largely N_2 and CO with small amounts of CO_2.

3–4. A surprisingly large-volume chemical product is printing ink. Use the library to determine how much printing ink is made and sold in the U.S. each year, who makes it, and what the flow sheet of the process by which it is made looks like.

3–5. The value of gold (and silver) has increased by a factor of 10–20 over the past five years (1975–80). As a result, abandoned mines in ghost towns of the U.S. west are being reopened and reworked. Sometimes this is being done by some of the largest U.S. corporations. For instance, the famous Comstock mine in Virginia City, Nevada, is being worked by Houston Oil & Mineral, Inc. Of course, modern processes must be used to extract the minute fraction of available gold. Draw a flow sheet of a modern gold concentrating process.

3–6. One chemical product desperately sought by our colonial forebears, and often purchased at exorbitant price, was salt. Outline a process for making salt using resources available to the colonists. Discuss raw materials, fuel supply, transportation to market, quality of product, competition, and any other relevant thing you can think of.

Total Mass and Component Balances 4

4.1 LAW OF CONSERVATION OF MASS

The engineering analysis of any chemical process starts by application of the law of conservation of mass. This law says that, barring nuclear conversions, the mass of material entering any system must either accumulate within the system or leave the system. In equation form the law is

$$\text{Input} - \text{Output} = \text{Accumulation} \tag{4-1}$$

Equation 4–1 implies that a system has been chosen. The system or process is the entity to which mass is being added, from which mass is being removed, and in which mass is accumulating with time. The rate of accumulation can be either negative (mass in the system decreasing with time) or positive (mass increasing with time). The system chosen for the mass balance can be an entire plant or any small section of a plant. It is important that the system be very precisely and explicitly defined. Usually this is best done with a flow sheet or some other schematic drawing.

To make the total mass balance, we must specify the basis on which our calculations will be made. This is usually a segment of time (e.g., inflow and outflow during one hour), but in some problems it is easier to choose a given mass of inlet or exit material (e.g., one ton of feed).

In this book, all material balances will be made on systems that are assumed to be operating under steady-state conditions. If we are at steady state, the mass inside the system cannot be changing, and equation 4–1 reduces to

$$\text{Mass In} = \text{Mass Out} \tag{4-2}$$

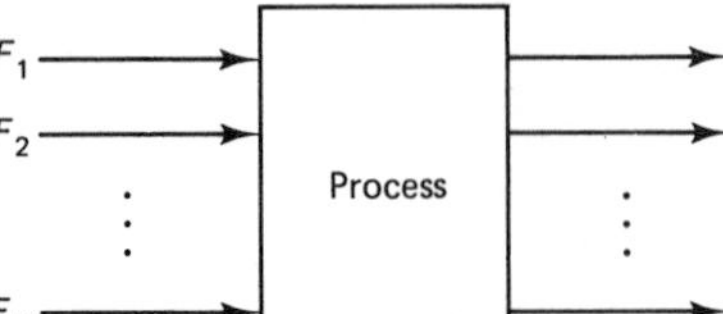

Figure 4–1 Feeds and products from a process.

In most problems we will choose a time basis, so we can equate the rate of flow of mass into the system to the rate of flow of mass out of the system, i.e.,

$$\text{Total Mass Flow Rate In} = \text{Total Mass Flow Rate Out} \qquad (4\text{–}3)$$

A process is shown in Figure 4–1 that has several feed streams entering and several product streams leaving. For this system at steady state, equation 4–4 applies:

$$\sum_{n=1}^{N} F_n = \sum_{k=1}^{M} P_k \qquad (4\text{–}4)$$

where

F_n = total mass flow rate of nth feed stream

P_k = total mass flow rate of kth product stream

Total mass flow rates can be in any mass-per-time units (lb_m/hr, kg/min, tons/day, etc.).

It should be emphasized that *mass* is conserved, not *moles*. As we will see in section 4.5, the molar flow rates of individual components into a system are *not* equal to the molar flow rates out of the system if chemical reactions are occurring inside the system.

Example 4.1.

Carry out a material (total mass) balance calculation using the total methanol plant for which the flow sheet was given in Figures 3–1 through 3–3.

Solution:

Figure 4–2 sketches the inflows and outflows for the entire methanol plant. Natural gas (CH_4) and steam (H_2O) are the two feeds. The methanol product, the water product, and the hydrogen purge gas are the three products.

Input

Natural gas (methane, CH_4)	= 32,000 lb_m/hr
Steam	= 45,000 lb_m/hr
Total Input	= 77,000 lb_m/hr

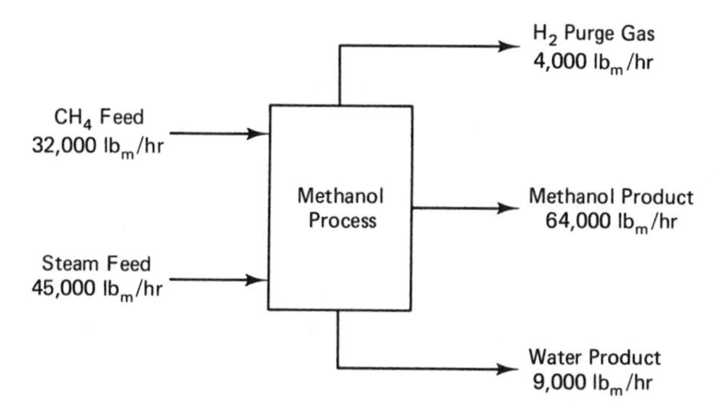

Figure 4–2 Overall mass flow rate for a methanol plant.

Output

Hydrogen purge gas (H_2)	=	4,000 lb_m/hr
Methanol product	=	64,000 lb_m/hr
Water product	=	9,000 lb_m/hr
Total Output	=	77,000 lb_m/hr

Clearly, the sum of the mass flow rates into the system does equal the sum of the mass flow rates out of the system.

Example 4.2.

Check the flow diagram for the methanol synthesis given in chapter 3 for mass flowrate consistency around the reformer furnace, the synthesis-gas compressor, and the distillation column.

Solution:

(a) Reformer furnace (Figure 3–1)

Inputs

CH_4 gas	=	32,000 lb_m/hr
H_2O	=	45,000 lb_m/hr
Total	=	77,000 lb_m/hr

Outputs

H_2 in syngas	=	13,000 lb_m/hr
CO in syngas	=	42,000 lb_m/hr
CO_2 in syngas	=	22,000 lb_m/hr
Total out	=	77,000 lb_m/hr

(b) Compressor (Figure 3–2)

Inputs

Syngas = 77,000 lb_m/hr

H_2 recycle gas = 50,000 lb_m/hr

Total in = 127,000 lb_m/hr

Output

Total reactor feed = 127,000 lb_m/hr

(c) Distillation column (Figure 3–3)

Input

Crude methanol feed = 73,000 lb_m/hr

Outputs

Methanol product = 64,000 lb_m/hr

Water product = 9,000 lb_m/hr

Total out = 73,000 lb_m/hr

4.2 UNITS

Chemical engineering calculations make use of three systems of units: the metric system, the English system, and the SI system. The metric "cgs" (centimeter–gram–second) system and the SI system are familiar to you from your courses in chemistry and physics. Most chemical data are taken and reported in metric units. The English system (pounds–feet–hours) has been traditionally used for engineering calculations and for commerce. Its use requires care because of the confusion between pounds force (lb_f) and pounds mass (lb_m), for which conversion factors are required. However, most operators, mechanics, managers, and senior engineers (including the authors) still think and talk in English units. Therefore, you have to learn the language.

The engineering community is slowly converting to the use of SI (Système International) units as a result of universally accepted agreements to do international business only in SI units. This system is based on the metric system, but uses kilograms, meters, and seconds as the basic units. Temperature is reported in Celsius or Kelvin degrees, energy in joules, force in newtons, and pressure in pascals (newtons/square meter).

Table A.1–1 in Appendix A.1 presents conversion factors between SI units and the other unit systems discussed. Table A.1–2 gives other conversion factors and fundamental constants.

The conversion of units is an unpleasant but essential part of all technical calculations. You will often have to look up the needed conversion factors in tables. Special care must always be exercised to assure the consistency of units. You should always carry along the units with the numbers in your calculations, particularly when you are learning a new design procedure or going through a calculation for the first time. After using most of the common conversion factors (e.g., 454 grams/pound, 252 calories/Btu, and 7.48 gallons/cubic foot) several times, you will memorize them.

Example 4.3.

Chlorine gas is flowing in a 4 inch diameter pipe at a velocity of 15 ft/sec under temperature and pressure conditions of 350°F and 175 psig, respectively. What is the mass flow rate of the chlorine?

Solution:

We will assume that chlorine is a perfect gas under the given conditions; that is,

$$PV = nRT$$

and

$$\text{Gas density} = \frac{nM}{V} = \frac{MP}{RT} \tag{4–5}$$

where

P = pressure

V = volume

n = number of moles

R = Perfect Gas Law constant

T = absolute temperature

M = molecular weight

Since the data are given in English units, we will do the problem with pressure in psia (lb_f/in^2 absolute), volume in cubic feet, temperature in °Rankine (absolute temperature in the English system: °R = °F + 460), and molecular weight in pounds per lb-mole.

Before we do the problem, let us discuss a few of the English units with which you may not be familiar. Pressure is reported in either psig (lb_f/in^2 gauge) or psia (lb_f/in^2 absolute). The atmosphere at sea level exerts a pressure of about 14.7 psia, or 0 psig. Thus, *gauge* pressure is the pressure above 1 atmosphere. The relationship between psig and psia is

$$\text{psia} = \text{psig} + 14.7$$

Temperature is reported in °F (Farenheit) or °R (Rankine). The relationships among the various temperature scales are:

$$°F = (°C)\,(1.8) + 32$$

$$°R = (K)\,(1.8)$$

$$°R = 460 + °F$$

$$K = 273 + °C$$

The molecular weight that you are familiar with is grams per gram-mole (the mass in grams of 0.6023×10^{24} atoms). In English units "lb-moles" are used instead. The molecular weight of a chemical component in lb_m/lb-mole is exactly the same as its molecular weight in grams/gram-mole. For example, the molecular weight of O_2 is 32 g/g-mole and 32 lb_m/lb-mole. The relationship between lb-moles and gram-moles is

$$1 \text{ lb-mole} = 454 \text{ gram-moles}$$

In English units the Perfect Gas Law constant is 1,545 when pressure is in lb_f/ft^2 (remember, if pressure is in psia, multiply by the factor 144 in^2/ft^2 to convert in^2 to ft^2), temperature is in °R, volume is in ft^3, and we are dealing with lb-moles. In SI units the Perfect Gas Law constant is 8.314 when pressure is in kilopascals (kPa), temperature is in K, volume is in cubic meters, and we are dealing with kilogram-moles. In metric units the Perfect Gas Law constant is 82.06 when pressure is in atmospheres, temperature is in K, volume is in cubic centimeters, and we are dealing with gram-moles.

Now let's solve the problem. The molecular weight of chlorine is 71 lb_m/lb-mole. The gas density is

$$\text{Density} = \frac{MP}{RT} = \frac{(71\ lb_m/\text{lb-mole})(175 + 14.7\ lb_f/in^2)(144\ in^2/ft^2)}{(1{,}545\ ft\ lb_f/\text{lb-mole}\ °R)(350 + 460\ °R)}$$

$$= 1.55\ lb_m/ft^3$$

$$\text{Cross-sectional area of the pipe} = \pi \frac{D^2}{4}$$

$$= \pi \frac{[(4 \text{ inches})(1 \text{ foot}/12 \text{ inches})]^2}{4} = 0.08727\ ft^2$$

$$\text{Volumetric flowrate} = (\text{velocity})\ (\text{area})$$

$$= (15\ ft/sec)\ (0.08727\ ft^2)\ (3{,}600\ sec/hr)$$

$$= 4{,}712\ ft^3/hr$$

$$\text{Mass flow rate} = (\text{volumetric flow rate})(\text{density})$$

$$= (4{,}712\ ft^3/hr)\ (1.55\ lb_m/ft^3) = 7{,}303\ lb_m/hr$$

Example 4.4.

How much mass of a perfect gas with molecular weight equal to 30 kg/kg-mole is contained in a vertical cylindrical tank 2 meters in diameter and 5 meters high, at 100°C and 2,000 kPa absolute pressure?

Solution:

$$\text{Volume of tank} = (\text{area})(\text{height}) = \frac{\pi D^2}{4}(H)$$

$$= \frac{\pi\,(2\text{ m})^2\,(5\text{ m})}{4} = 15.71\text{ m}^3$$

$$\text{Density of gas} = \frac{MP}{RT} = \frac{(30\text{ kg/kg-mole})(2{,}000\text{ kPa})}{(8.314\text{ kPa m}^3\text{/kg-mole K})(100 + 273\text{ K})}$$

$$= 19.35\text{ kg/m}^3$$

$$\text{Mass in tank} = (\text{volume})(\text{density})$$

$$= (15.71\text{ m}^3)(19.35\text{ kg/m}^3) = 304\text{ kg}$$

Another very common type of conversion is the transformation of concentrations from weight percents to mole percents and vice versa. You have probably done this in your chemistry courses, but to refresh your memory, an example is given.

Example 4.5.

Stack gas from a furnace has the following composition in mole percent (note that mole percentages are the same with either g-moles or lb-moles):

$$CO_2 = 12.41\text{ mole percent}$$

$$H_2O = 8.83\text{ mole percent}$$

$$N_2 = 74.20\text{ mole percent}$$

$$O_2 = 4.56\text{ mole percent}$$

Convert these concentrations to weight percentages.

Solution:

Choose a basis of 100 g-moles of gas. The weight percentages are calculated from the ratio of the grams of each component to the total number of grams (see table on page 82). For example, the weight percentage of CO_2 is

$$\frac{546.0\text{ grams }CO_2}{2{,}928.4\text{ grams total}} = 0.1864\text{ weight fraction}$$

$$= 18.64\text{ weight percent}$$

	Gram-moles	*Molecular weight*	*Grams = (g-mole)(mol. wt.)*	*Weight percent*
CO_2	12.41	44	546.0	18.64
H_2O	8.83	18	158.9	5.43
N_2	74.20	28	2,077.6	70.95
O_2	4.56	32	145.9	4.98
Total number of grams =			2,928.4	

Note that we could have chosen any number of moles as the basis and calculated exactly the same weight percents. Also, we could have used lb-moles and gotten the same results.

4.3 ORGANIZATION OF MATERIAL BALANCES

As you work through the material balance calculations on a complex flow sheet, it is helpful to number all the process points that appear on the sheet. Then a table can be made showing each stream number and its flow rate, composition, temperature, pressure, etc. Table 4–1 gives such a tabulation for the front end of the methanol plant discussed in chapter 3 and shown schematically in Figure 4–3.

Example 4.6.

A light-ends "gas plant" associated with a catalytic cracking unit in a petroleum refinery is fed 114,244 lb_m/hr of gas and 68,976 lb_m/hr of liquid hydrocarbons. The sketch in Figure 4–4 shows the absorber/stripper system used for recovering C_4 and heavier components. The lighter components are sent to fuel gas. A two-column sequence is used to produce gasoline and a C_4 "BB" (butane-butylene) stream used for alkylation feed stock to make high-octane aviation gasoline. (See chapter 3.) How much gasoline is being produced?

Solution:

Calculations will be made on a one-hour basis, taking the entire plant as the system.

Streams flowing in

Gas feed = 114,244 lb_m/hr

Liquid feed = 68,976 lb_m/hr

Lean oil makeup = 1,238 lb_m/hr

Total in = 184,458 lb_m/hr

TABLE 4–1
TABULATED PROCESS CONDITIONS FOR METHANOL PLANT

Stream	*1* *Natural gas*	*2* *Compressor exit*	*3* *Heat exch. exit*	*4* *Steam*	*5* *Reformer furnace feed*	*6* *Reformer furnace exit*
Temp (°F)	100	300	600	570	580	650
Pressure (psig)	65	300	295	290	290	260
Mass flow rate ($1,000\ lb_m/hr$)						
H_2	—	—	—	—	—	13
CO	—	—	—	—	—	42
CO_2	—	—	—	—	—	22
CH_4	32	32	32	—	32	—
H_2O	—	—	—	45	45	—
CH_3OH	—	—	—	—	—	—
Total	32	32	32	45	77	77
Component flow rate (lb-mole/hr)						
H_2	—	—	—	—	—	6,500
CO	—	—	—	—	—	1,500
CO_2	—	—	—	—	—	500
CH_4	2,000	2,000	2,000	—	2,000	—
H_2O	—	—	—	2,500	2,500	—
CH_3OH	—	—	—	—	—	—
Total	2,000	2,000	2,000	2,500	4,500	8,500

Streams flowing out

Gas to fuel $= 30{,}730\ lb_m/hr$

D/P overhead $= 10{,}077\ lb_m/hr$

B.B. $= 33{,}158\ lb_m/hr$

Gasoline = unknown

Total out must equal total in; therefore, the gasoline flow rate must be

$$\text{Gasoline out} = 184{,}458 - 30{,}730 - 10{,}077 - 33{,}158$$
$$= 110{,}493\ lb_m/hr$$

4.4 COMPONENT BALANCES WITHOUT REACTION

If there are no chemical reactions occurring in the system, the molar flow rate of each component coming into the system must be equal to the molar

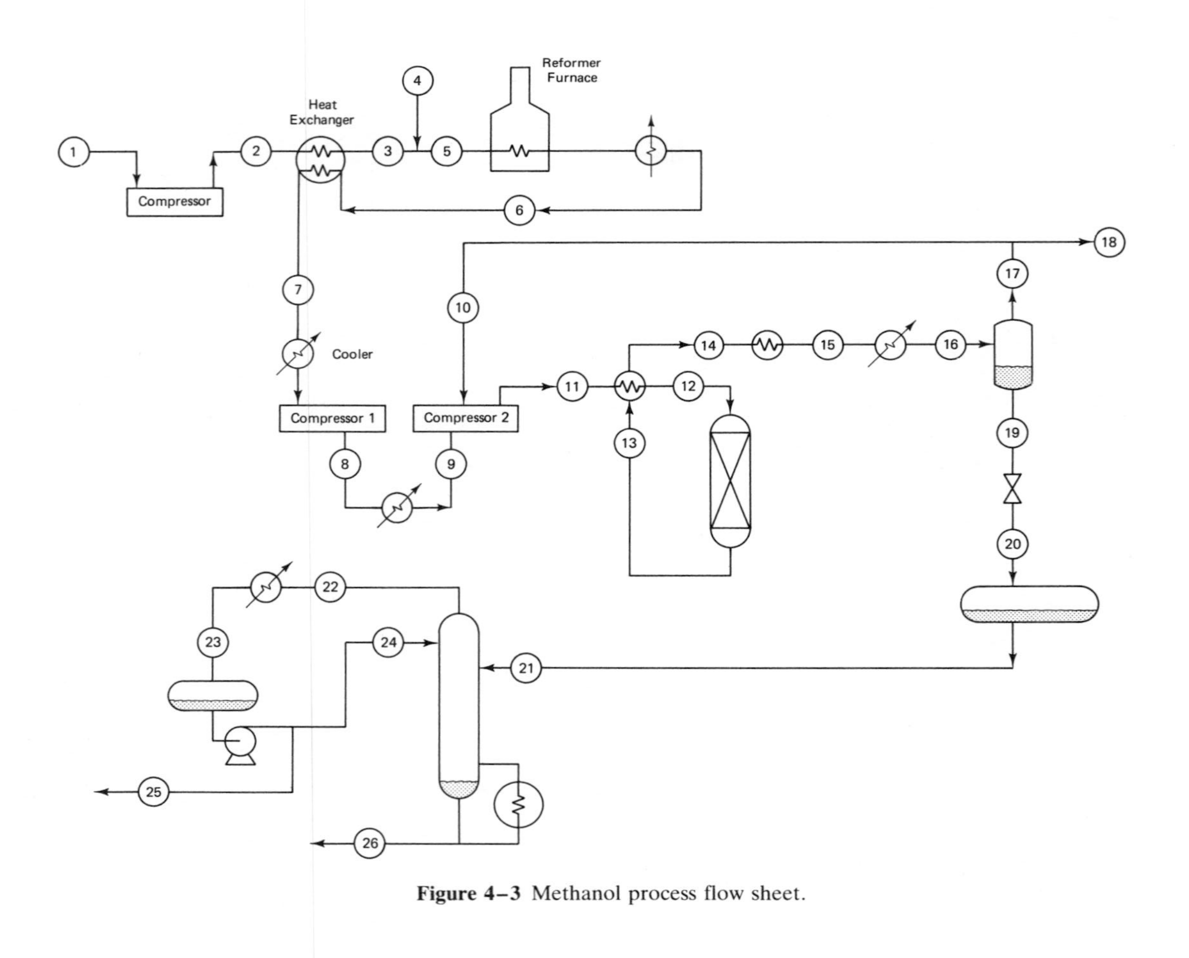

Figure 4–3 Methanol process flow sheet.

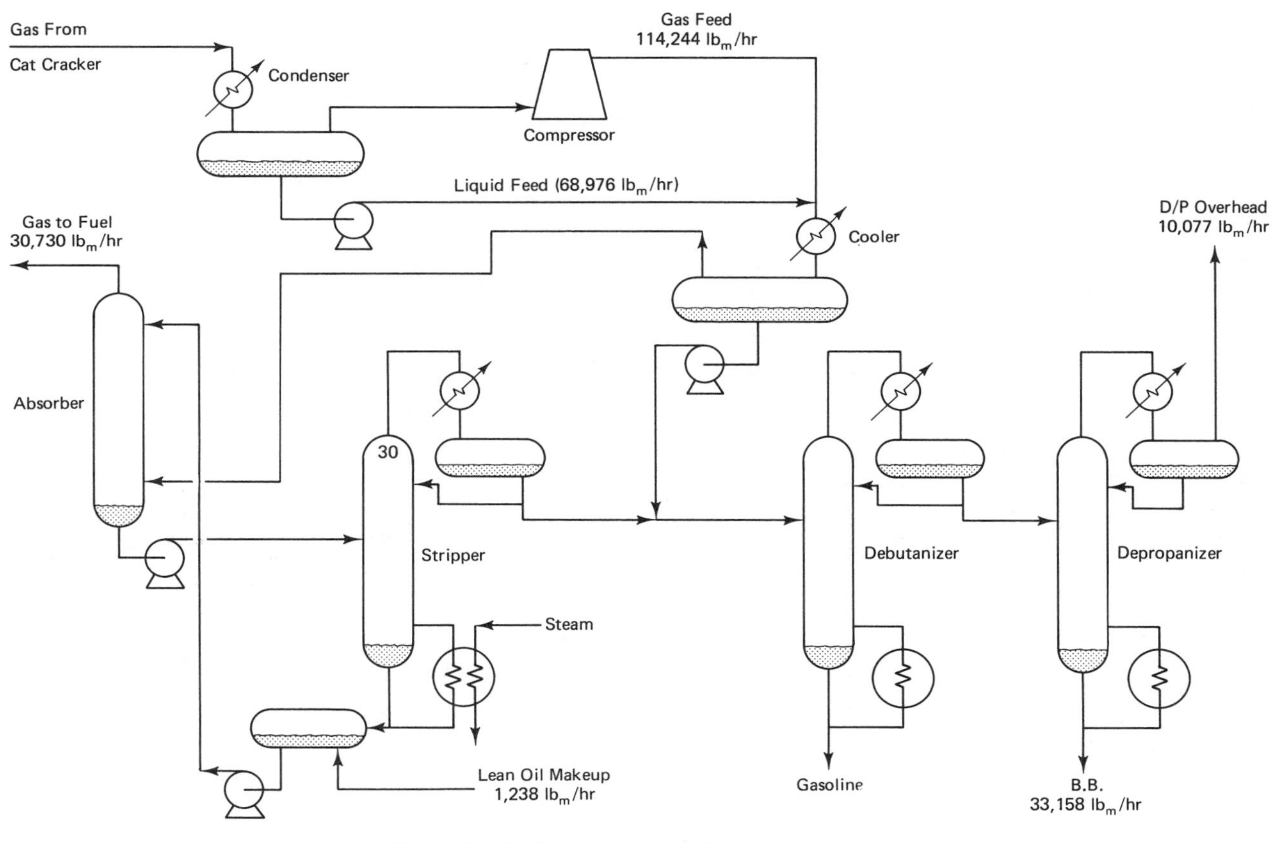

Figure 4–4 Light gas processing from a catalytic cracker.

flow rate of that same component leaving the system *if* the system is at steady state. Thus, *moles* are conserved in systems where no chemical reactions are occurring. The flow rates of each component into and out of the system must be equal if we use either molar flow rates (moles/hr) or mass flow rates (mass/hr) since the two flow rates differ only by the molecular weight of the particular component being considered.

Component balances can be written for each chemical component in a *nonreacting* system as

$$\sum_{n=1}^{N} F_n z_{nj} = \sum_{k=1}^{M} P_k x_{kj} \tag{4-6}$$

where

F_n = total molar flow rate of nth feed stream entering

z_{nj} = mole fraction of jth component in nth feed stream

P_k = total molar flow rate of kth product stream leaving

x_{kj} = mole fraction of jth component in kth product stream

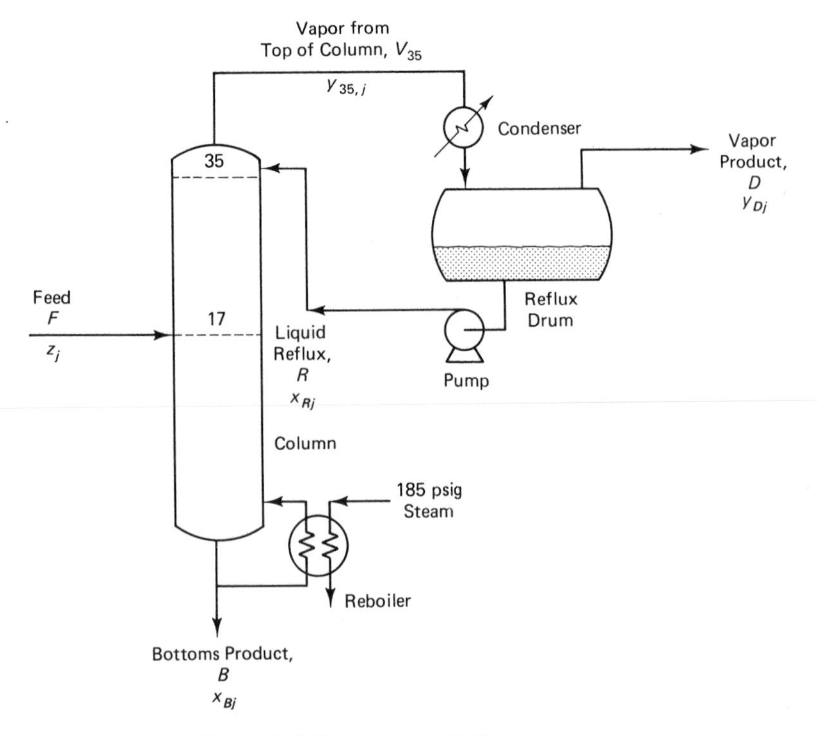

Figure 4–5 Depropanizer distillation column.

If the flow rates are in total g-mole/minute, the units of Equation 4–6 are g-moles of component j per minute. That is

$$\left|\frac{\text{total g-moles}}{\text{minute}}\right| \left|\frac{\text{moles of component } j}{\text{total g-moles}}\right| = \left|\frac{\text{moles of component } j}{\text{minute}}\right|$$

If the flow rates are in total lb-moles per hour, the units of equation 4–6 are lb-moles of component j per hour. We could of course use mass flow rates in equation 4–6, but then we *must* use weight fractions instead of mole fractions. Note that the total flow rates of feed (F_n) and products (P_k) are sometimes mass flow rates and sometimes molar flow rates. Either is perfectly acceptable as long as the two are used consistently throughout the calculations.

Example 4.7.

Plant test data from a depropanizer (D/P) distillation column (see Figure 4–5) are given below. Calculate

1. the total molar feed rate and the composition of the feed, and
2. the flow rate and the composition of the vapor leaving the top of the column.

		Molar flow rates (lb-mole/hr)		
Component	*Formula*	*Vapor distillate product*	*Bottoms product*	*Reflux*
Methane	CH_4	9.5	—	27.2
Ethylene	C_2H_2	9.5	—	14.7
Ethane	C_2H_6	18.8	—	19.4
Propylene	C_3H_6	138.4	16.8	408.0
Propane	C_3H_8	53.1	15.0	153.3
Isobutane	iC_4H_{10}	2.2	125.0	12.4
Isobutylene	iC_4H_8	2.5	145.7	14.9
N-Butylene	nC_4H_8	3.5	221.0	22.4
N-Butane	nC_4H_{10}	0.8	59.4	6.2
Isopentane	iC_5H_{12}	—	1.5	—
Pentylenes	C_5H_{10}	—	4.2	—
N-Pentane	nC_5H_{12}	—	0.4	—

Solution:

The feed composition and feed flow rate can be found from a material balance around the overall distillation column system. Molar balances can be made because there are no chemical reactions occurring in the distillation column. Accordingly, we have

$$Fz_j = Dy_{Dj} + Bx_{Bj}$$

where F, D, and B are molar flow rates and z_j, y_{Dj}, and x_{Bj} are mole fractions of the jth component.

The total feed rate F is calculated by adding up the molar flow rates of all components, where NC is the total number of components. In this example, NC = 12.

$$F = \sum_{j=1}^{NC} Fz_j = \sum_{j=1}^{NC} [Dy_{Dj} + Bx_{Bj}]$$

For example, for propane, there are 53.1 lb-mole/hr in the distillate product and 15.0 lb-mole/hr in the bottoms product. Therefore, the amount of propane coming into the depropanizer in the feed must be 53.1 + 15.0 = 68.1 lb-mole/hr. Note that the reflux stream is *within* the overall system and hence does not appear in the material balances for the system. Repeating the analysis for each component gives the following results:

Component	*Petroleum industry notation*	*Feed molar flow rate (lb-mole/hr)*	*Feed composition (mole fraction)*
CH_4	C_1	9.5	0.0115
C_2H_4	$C_2^=$	9.5	0.0115
C_2H_6	C_2	18.8	0.0227
C_3H_6	$C_3^=$	155.2	0.1876
C_3H_8	C_3	68.1	0.0823
iC_4H_{10}	iC_4	127.2	0.1538
iC_4H_8	$iC_4^=$	148.2	0.1791
nC_4H_8	$nC_4^=$	224.5	0.2714
nC_4H_{10}	nC_4	60.2	0.0728
iC_5H_{12}	iC_5	1.5	0.0018
C_5H_{10}	$C_5^=$	4.2	0.0051
nC_5H_{12}	nC_5	0.4	0.0005
Total Feed Rate		827.3lb-mole/hr	

The feed compositions were calculated by dividing the molar flow rate of each component in the feed by the total feed rate. For example, for propane, there were 68.1 lb-mole/hr in the feed. Hence, the mole fraction of propane in the feed is 68.1/827.3 = 0.0823.

The same approach is taken in order to calculate the composition of the overhead vapor from the top of the column. A component balance around the condenser and the reflux drum shows that the vapor from the column is the sum of the vapor distillate product and the reflux, i.e.,

$$V_{35}y_{35,j} = Dy_{Dj} + Rx_{Rj}$$

where V_{35} is the total molar flow rate from the top tray in the column (Tray 35), R is the total molar flow rate of the reflux, $y_{35,j}$ is the mole fraction of the jth component in the vapor from Tray 35, and x_{Rj} is the mole fraction of the jth component in the reflux. The following table results:

Component	*Vapor* V_{35} *molar flow rates (lb-mole/hr)*	*Composition of vapor from Tray 35* ($y_{35,j}$)
C_1	9.5 + 27.2 = 36.7	0.0400
$C_2^{=}$	9.5 + 14.7 = 24.2	0.0264
C_2	18.8 + 19.4 = 38.2	0.0417
$C_3^{=}$	138.4 + 408.0 = 546.4	0.5960
C_3	53.1 + 153.3 = 206.4	0.2251
iC_4	2.2 + 12.4 = 14.6	0.0159
$iC_4^{=}$	2.5 + 14.9 = 17.4	0.0190
$nC_4^{=}$	3.5 + 22.4 = 25.9	0.0283
nC_4	0.8 + 6.2 = 7.0	0.0076
C_5 + Heavier	0 0	0

Total = 916.8 lb-mole/hr

Example 4.8.

A 5 weight percent formaldehyde-in-water solution is fed to a concentrating unit that produces two product streams. One is 37 weight percent formaldehyde, the other 99.5 weight percent water. How much of the two streams is produced per 100 kg of feed?

F, z → Process → P_1 x_{P1}; P_2 x_{P2}

Solution:

Let F be the feed (100 kg) and z be the weight fraction of formaldehyde in the feed (0.05). Let P_1 and P_2 be the amounts of the two products in kg. Let x_{P1} and x_{P2} be the weight fractions of formaldehyde in the two products.

We know from the problem statement that $x_{P1} = 0.37$. We also know that the other product is 99.5 weight percent water. This means that it is 0.005 weight fraction formaldehyde. Therefore, $x_{P2} = 0.005$.

A total mass balance gives

$$100 = P_1 + P_2$$

A formaldehyde component balance gives

$$(100)(0.05) = P_1\,(0.37) + P_2\,(0.005)$$

Solving these two equations for the two unknowns P_1 and P_2 yields

$$P_1 = 12.3 \text{ kg} \qquad P_2 = 87.7 \text{ kg}$$

Example 4.9.

Wet lumber (5 weight percent H_2O) is dried to 1 weight percent H_2O in a hot-air drier. Air fed to the drier contains 0.5 weight percent water. Moist air leaving the drier contains 2 weight percent water. How much air is required to dry 1 ton/hr of wet lumber?

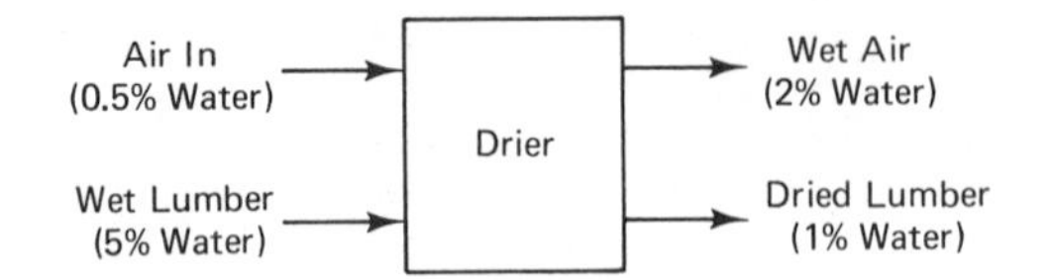

Solution:

Basis: 1 hour of operation

$$\text{1 ton/hr of wet lumber} = \text{2,000 lb}_m\text{/hr of dry wood plus water}$$

$$\text{Dry (water-free) wood fed} = (2{,}000)(0.95) = 1{,}900 \text{ lb}_m\text{/hr}$$

The dry wood is an inert material in this system. All of it coming in with the wet lumber must leave in the dried lumber stream. None of it leaves in the wet-air stream. Thus, we can do an inert balance on the dry wood.

Since the dried lumber is 1 weight percent water, the amount of dried lumber leaving is

$$[1{,}900 \text{ lb}_m\text{/hr dry wood}]\left[\frac{100 \text{ lb}_m \text{ dried lumber}}{99 \text{ lb}_m \text{ dry wood}}\right]$$

$$= 1{,}919.2 \text{ lb}_m\text{/hr dried lumber leaving}$$

Let A = lb_m/hr of dry (water-free) air fed to the process. Then

$$\text{Total air flow rate in} = (A \text{ lb}_m \text{ dry air/hr})\left|\frac{100 \text{ lb}_m \text{ total air}}{99.5 \text{ lb}_m \text{ dry air}}\right|$$

Now we perform a water balance on the process:

Water In

$$\text{Water in wet lumber fed} = (0.05)(2{,}000 \text{ lb}_m\text{/hr}) = 100 \text{ lb}_m\text{/hr}$$

$$\text{Water in air fed} = (0.005)\left[\frac{A}{0.995}\right]$$

$$\text{Total water in (lb}_m\text{/hr)} = 100 + 0.00503A$$

Water Out

$$\text{Water in dried lumber out} = (0.01)(1{,}919.2 \text{ lb}_m\text{/hr})$$

$$= 19.19 \text{ lb}_m\text{/hr}$$

$$\text{Water in air out} = (0.02)\left[\frac{A}{0.98}\right]$$

$$\text{Total water out (lb}_m\text{/hr)} = 19.19 + 0.0204A$$

Equating the water in to the water out gives

$$100 + 0.00503A = 19.19 + 0.0204A$$

$$A = 5{,}258 \text{ lb}_m\text{/hr of dry air}$$

Since the inlet air is 0.5 weight percent water, the total inlet air flow rate is

$$(5{,}258 \text{ lb}_m \text{ dry air/hr}) \left[\frac{100 \text{ lb}_m \text{ total air}}{99.5 \text{ lb}_m \text{ dry air}}\right] = 5{,}284 \text{ lb}_m\text{/hr of total inlet air}$$

Notice that in this example it was convenient to look at the flow rates of air and wood on a dry (water-free) basis. This gimmick is often useful in situations where one component remains entirely in one stream. In the example, all of the dry wood stayed in the lumber stream, whether the lumber was wet or dried. Also, all of the dry air (N_2 plus O_2 without any H_2O) stayed in the gas stream, whether it was feed air or wet air.

4.5 COMPONENT BALANCES WITH REACTIONS

When a chemical reaction occurs in a process, the number of moles of any component entering may not equal the number of moles leaving, even at steady state. Component balances must then be modified to include the rate of consumption or production of each component:

$$\left|\begin{array}{c}\text{Flow rate of}\\ j\text{th component}\\ \text{into system}\end{array}\right| = \left|\begin{array}{c}\text{Flow rate of}\\ j\text{th component}\\ \text{out of system}\end{array}\right| + \left[\begin{array}{c}\text{Rate of consumption}\\ \text{of } j\text{th component}\\ \text{inside of system}\\ \text{due to chemical}\\ \text{reactions}\end{array}\right]$$

The steady-state component balance for the jth component in a reacting system is

$$\sum_{n=1}^{N} F_n z_{nj} = \sum_{k=1}^{M} P_k x_{kj} + [\text{Consumption in Reactions}] \tag{4-7}$$

Several items are worth noting at this point:

1. One total mass balance and NC − 1 component balances can be written for any system, where NC is the number of components in the system. These NC equations are independent. We *cannot* write NC component balances and one total mass balance, because one of the equations would then be dependent on the others. For example, if we add up all of the component balances, we get the total mass balance.

2. Any inert compounds that are present are not involved in any reactions, so component balances like equation 4–6 can be used.
3. Component balances can be written in mass or molar units.
4. *Atom* balances can be written for any component. For example, the number of atoms of oxygen entering a burner in the combustion air of a furnace will equal the number of atoms of oxygen leaving the furnace in the stack gas. Of course, some of the oxygen atoms will be in compounds like CO, CO_2, and H_2O.

Let us examine several examples of component balances with reactions occurring. Some of the most interesting examples are combustion problems.

Example 4.10.

A coal containing 81 weight percent total carbon and 6 weight percent unoxidized hydrogen is burned in dry air. The rest of the coal is solid inert. The amount of air used is 30 percent more than is theoretically required to completely oxidize all of the carbon to CO_2 and all of the hydrogen to H_2O. Calculate the number of kg of air per kg of coal and the composition of the stack gas leaving the furnace, assuming this gas contains no CO.

Solution:

Reactions:

$$C + O_2 \longrightarrow CO_2$$

$$H_2 + \frac{1}{2} O_2 \longrightarrow H_2O$$

Basis: 1 kg of coal burned

Component	*kg*	*M*	*kg-mole*	*kg-mole O_2 consumed (from stoichiometry)*
C	0.81	12	0.0675	0.0675
H_2	0.06	2	0.0300	0.0150
		Total O_2 consumed		0.0825

$$\text{Air supplied at 30\% excess} = (1.3)(0.0825)\left[\frac{100\text{ moles air}}{21\text{ moles }O_2}\right] = 0.5107\text{ kg-mole air}$$

$$O_2\text{ in air} = (0.5107)(0.21) = 0.1073\text{ kg-mole of }O_2\text{ fed}$$

$$N_2\text{ in air} = (0.5107)(0.79) = 0.4035\text{ kg-mole of }N_2\text{ fed}$$

$$\text{Mass of }O_2\text{ fed} = (0.1072\text{ kg-mole }O_2)(32\text{ kg/kg-mole})$$

$$= 3.43\text{ kg }O_2$$

$$\text{Mass of } N_2 \text{ fed} = (0.4035 \text{ kg-mole } N_2)(28 \text{ kg/kg-mole})$$

$$= 11.30 \text{ kg } N_2$$

$$\text{Total weight of air fed} = 14.73 \text{ kg/kg coal fired}$$

$$O_2 \text{ unburned} = O_2 \text{ fed} - O_2 \text{ reacted}$$

$$= 0.1073 - 0.0825 = 0.0248 \text{ kg-mole } O_2$$

Material in stack gas

	kg-moles	*M*	*kg*	*Weight percent*
CO_2	0.0675	44	2.97	19.04
H_2O	0.0300	18	0.54	3.46
N_2	0.4035	28	11.30	72.44
O_2	0.0248	32	0.79	5.06
Totals	0.5258		15.60	100.00

Note that in this problem the chemical reactions were written in the manner of classical chemistry. That is, incomplete reactions, inert components, and an excess or deficiency of any reactant were not indicated in the equation. Instead, these factors were handled later. The equations indicated the stoichiometry (combining weights) alone. Even if there are several reaction paths or products, all the reaction equations must be written and dealt with, even though one may be dominant. These reaction equations must be balanced. (The number of atoms of each element must be the same on both sides of the equation.)

Example 4.10 also illustrated the concept of an excess reactant. In any reaction, it is probable that the reactants will not be in the exact stoichiometric ratio. The *critical* or *limiting* reactant is the one which will be consumed first as the reaction progresses. The other reactants will be present *in excess*. The *percent excess* is defined as the quantity of reactant beyond that required for complete consumption of the limiting reactant divided by that required for complete consumption. In example 4.10, 30 percent excess air or oxygen was used.

Example 4.11.

Fuel oil (say, $C_{18}H_{36}$) is burned in 50 percent excess dry air (excess above that required for complete combustion to CO_2 and H_2O). The products of combustion are dried to remove all the water. Analysis of the flue gas shows a ratio of CO_2 to CO of 2, on a molar basis. What are the flue gas composition and volumetric flow rate at 300°F and 18 psia after drying if 5,000 lb_m/hr of fuel oil is burned?

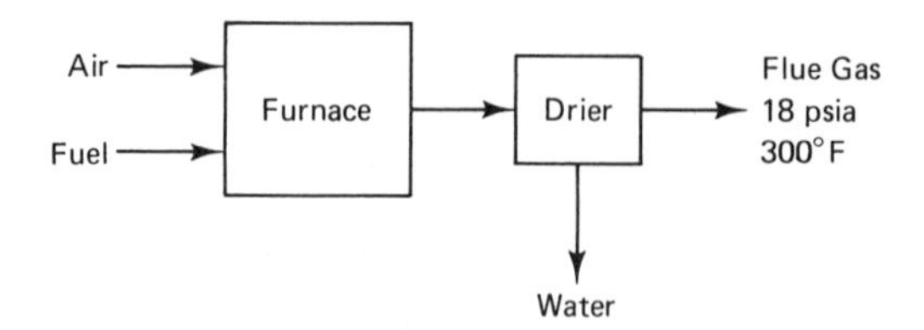

Solution:

Basis: 1 hour of operation
Molecular weight of $C_{18}H_{36}$ fuel = 252 lb_m/lb-mole
For complete combustion, the reaction is

$$C_{18}H_{36} + 27\ O_2 \longrightarrow 18\ CO_2 + 18\ H_2O$$

$$\text{Fuel} = \frac{5{,}000\ lb_m/hr}{252\ lb_m/lb\text{-mole}} = 19.84\ lb\text{-mole/hr}$$

$$O_2\ \text{required} = (27\ lb\text{-mole}\ O_2/lb\text{-mole fuel})(19.84\ lb\text{-mole/hr fuel})$$

$$= 535.7\ lb\text{-mole/hr}\ O_2$$

$$\text{Actual}\ O_2\ \text{fed @ 50\% excess} = (1.5)(535.7) = 803\ lb\text{-mole/hr}$$

$$N_2\ \text{in air} = (803\ lb\text{-mole}\ O_2)(79\ \text{mole}\ N_2/21\ \text{mole}\ O_2)$$

$$= 3{,}023\ lb\text{-mole/hr}\ N_2$$

But remember that complete combustion does not occur in this system since there is some CO in the flue gas. The molar ratio of CO_2 to CO is given as 2. There are two reactions occurring, one going to CO and the other to CO_2. The CO reaction is

$$C_{28}H_{36} + 18\ O_2 \longrightarrow 18\ CO + 18\ H_2O$$

and the CO_2 reaction is as given above.

Since the product gas contains 2 moles of CO_2 per mole of CO, the CO_2 reaction consumes two-thirds of the fuel, while the CO reaction consumes the other one-third. Thus, we have:

O_2 consumed by CO_2 reaction

$$= \frac{2}{3}(19.84\ \text{lb-mole/hr fuel})(27\ \text{mole}\ O_2/\text{mole fuel})$$

$$= 357.1\ \text{lb-mole/hr}\ O_2$$

O_2 consumed by CO reaction

$$= \frac{1}{3}(19.84\ \text{lb-mole/hr fuel})(18\ \text{mole}\ O_2/\text{mole fuel})$$

$$= 119.0\ \text{lb-mole/hr}\ O_2$$

Unreacted O_2 in flue gas = O_2 fed − total O_2 reacted

$$= 803 - (357.1 + 119.0) = 326.9 \text{ lb-mole/hr } O_2 \text{ in flue gas}$$

$$CO_2 \text{ in flue gas} = \frac{2}{3}(19.84 \text{ lb-mole/hr fuel})(18 \text{ mole } CO_2/\text{mole fuel})$$

$$= 238 \text{ lb-mole/hr } CO_2$$

CO in flue gas

$$= \frac{1}{3}(19.84 \text{ lb-mole/hr fuel})(18 \text{ mole CO/mole fuel})$$

$$= 119 \text{ lb-mole/hr CO}$$

	Total flue gas without water *lb-mole/hr*	*mole %*
N_2	3,023	81.55
O_2	327	8.82
CO	119	3.21
CO_2	238	6.42
Total	3,707	

To calculate the volumetric flow rate, we need the density. Assuming perfect gases, we have

$$\text{Molar Density} = \frac{P}{RT} = \frac{(18 \text{ psia})(144 \text{ in}^2/\text{ft}^2)}{(1{,}545 \text{ ft lb}_f/\text{lb-mole °R})(300 + 460 \text{ °R})}$$

$$= 0.00221 \text{ lb-mole/ft}^3$$

The volume of one lb-mole under these conditions is 453 ft^3/lb-mole. A useful number to remember is that one lb-mole of any perfect gas at standard conditions of 32°F and 1 atmosphere is 359 ft^3. (Be careful about standard conditions, however: some industries use standard temperatures of 60°F or 70°F, not 32°F.) Thus, we have

$$\text{Volumetric flow rate} = \frac{\text{molar flow rate}}{\text{molar density}}$$

$$= (3{,}707 \text{ lb-mole/hr})(453 \text{ ft}^3/\text{lb-mole})(\text{hr}/60 \text{ min})$$

$$= 27{,}980 \text{ ft}^3/\text{min}$$

Example 4.12.

Coal is partially oxidized with pure oxygen in the presence of steam to form a mixture of CO, CO_2, and H_2 (synthesis gas). The ratio of carbon to hydrogen atoms in the coal is 1. The ratio of moles of oxygen used per mole of carbon is 0.25. The ratio of moles of CO to CO_2 in the synthesis gas is 3.

A portion of the synthesis gas is sent to a "shift" reactor where the CO reacts completely with more steam to form additional H_2 and CO_2. All the CO_2 and water are removed from the "shifted gas" by scrubbing and drying.

The hydrogen stream from the shift process is combined with the rest of the original synthesis gas and fed into a methanol reactor. The CO and CO_2 are completely converted to methanol by the reactions

$$CO + 2H_2 \longrightarrow CH_3OH$$

$$CO_2 + 3H_2 \longrightarrow CH_3OH + H_2O$$

(a) What fraction of the synthesis gas must be sent to the shift reactor in order to provide exactly enough hydrogen to react completely with all of the remaining CO and CO_2 in the methanol reactor?
(b) How many moles of methanol are produced per mole of carbon in the coal?
(c) What is the concentration of water in the water-methanol mixture leaving the methanol reactor?

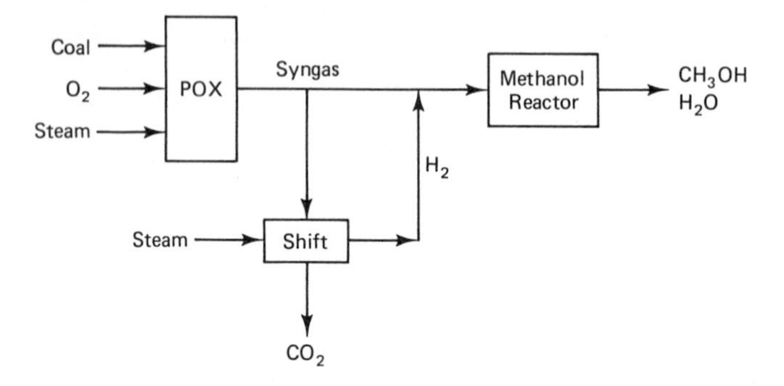

Solution:

In the POX (partial oxidation) reactor, we must have an O_2-to-C ratio of $\frac{1}{4}$ and a CO-to-CO_2 ratio of 3. The following balanced overall reaction can be written for the POX unit:

$$4CH + O_2 + 3H_2O \longrightarrow 3CO + CO_2 + 5H_2$$

This equation combines reaction stoichiometry, equilibrium constraints, and material balances representing the relative extents of the simple basic reactions occurring, viz.,

$$2\,CH + O_2 \longrightarrow 2\,CO + H_2$$

and

$$2\,CH + 2\,O_2 \longrightarrow CO_2 + H_2$$

We use the pseudocompound "CH" to describe the coal because the ratio of C to H is 1. In reality, coal is made up of hundreds of complex compounds. But on average, they have the formula C_nH_n. Notice that we have 3 moles of CO for every 1 mole of CO_2. We've also used 1 mole of O_2 for every 4 atoms of carbon. The amount of steam (H_2O) needed was back-calculated to provide enough oxygen atoms to balance the equation. This then sets the amount of hydrogen produced.

We will choose as our basis of calculation 4 atoms of carbon. Let S = the number of moles of CO fed to shift. Then the number of moles of CO_2 fed to shift

$= \frac{1}{3}S$. Now, every mole of CO fed to shift will produce one mole of H_2, since the reaction in the shift reactor is

$$H_2O + CO \longrightarrow CO_2 + H_2$$

Thus, we have:

$$\text{moles of } H_2 \text{ produced in shift} = S$$

moles of CO_2 removed in the shift process

$$= CO_2 \text{ in syngas fed to shift} + CO_2 \text{ produced in shift}$$

$$= \frac{1}{3} S + S = \frac{4}{3} S$$

The feed stream to the methanol reactor will contain CO, CO_2, and H_2 in the following amounts:

$$\text{moles of CO} = 3 - S \quad \text{since } S \text{ moles are removed in shift}$$

$$\text{moles of } CO_2 = 1 - \tfrac{1}{3}S \quad \text{since } \tfrac{1}{3}S \text{ moles are removed in shift}$$

$$\text{moles of } H_2 = 5 + S \quad \text{since } S \text{ moles are produced in shift}$$

Every mole of CO requires 2 moles of H_2 to produce methanol. Every mole of CO_2 requires 3 moles of H_2 to produce methanol and water. Hence, the number of total moles of H_2 required can be computed as follows.

$$\text{For CO:} \quad 2(3 - S)$$

$$\text{For } CO_2\text{:} \quad \underline{3(1 - S/3)}$$

$$\text{Total} = 9 - 3S$$

Since the hydrogen in the feed to methanol is $(5 + S)$ moles, and $(9 - 3S)$ moles are required, we can equate the two to solve for S:

$$5 + S = 9 - 3S \longrightarrow S = 1 \text{ mole of CO fed to shift}$$

Thus, we have

(a)

$$\text{Fraction fed to shift} = \frac{1}{3}$$

(b)

$$\text{Methanol formed from CO} = 3 - S = 2$$

$$\text{Methanol formed from } CO_2 = 1 - \frac{S}{3} = \frac{2}{3}$$

$$\text{Total methanol} = \frac{8}{3}$$

$$CH_3OH \text{ produced/atom of carbon in coal} = \frac{(8/3)}{4} = \frac{2}{3}$$

(c)

$$\text{Water from } CO_2 = 1 - \frac{S}{3} = \frac{2}{3}$$

$$\text{Mole fraction water in crude methanol product} = \frac{\text{moles water}}{\text{moles water} + \text{moles methanol}}$$

$$= \frac{2/3}{2/3 + 8/3} = 0.2$$

Notice in the example that there is a shortage of hydrogen in the syngas, i.e., there is not enough hydrogen to react with all of the CO and CO_2 to produce methanol. This is because of the small amount of hydrogen in the coal that is used as the hydrogen source, and it is the reason for running the shift reactor. The shift process generates more hydrogen and at the same time removes some of the CO_2 from the system.

Example 4.12 should be compared with the process for producing methanol described in chapter 3. In that process, the hydrocarbon source was natural gas (CH_4), and there was an excess of hydrogen available. Some H_2 was purged from the methanol reactor system, and no shift reactor was needed.

4.6 RECYCLE STREAMS

As we have already observed in chapter 3, recycle streams are often encountered in chemical processes. They are used for a variety of reasons: to increase the conversion of reactants, to increase the yield of desired products, and to provide concentration and temperature diluent in reactor feed. As always, material balances can be written around the entire plant or around any individual part of the process. It is often useful to look at the plant as a whole first, considering just the fresh feeds into the process and the final products leaving. After these are determined, the internal streams and recycles can be calculated.

In considering the internal recycle streams, it is often convenient to deal with branch points in a piping system as if they were separate pieces of process equipment. Material balances can be written around such branch points as well as around larger units or groups of units. A drier with recycle of moist air is considered in the following example.

Example 4.13.

In a wood drier the hot air must contain at least 2 weight percent water to prevent the wood from drying too rapidly and splitting or warping. The original fresh air fed contains 1 weight percent water. Wood is dried from 20 weight percent water to 5 weight percent water. The wet air leaving the drier contains 4 weight percent water. Calculate the amount of wet air that must be returned to the drier if one ton per hour of wet wood is dried.

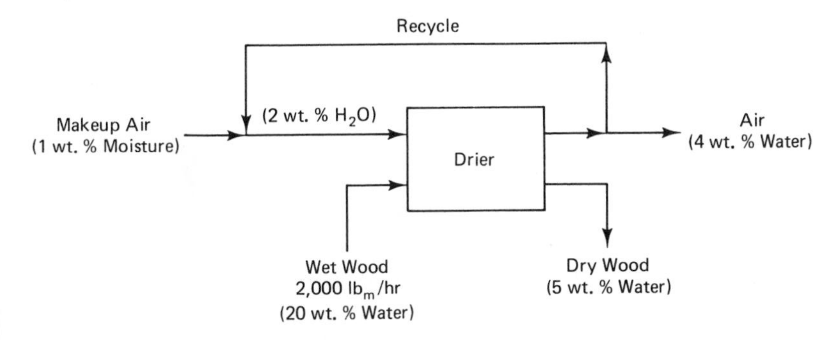

Solution:

Basis: 1 hour of operation

First we do the overall material balances on bone-dry (water-free) wood and on water to determine the flow rates of material in and out of the overall system. The amount of water that must be removed from the wood can be calculated using a bone-dry wood material balance:

Dry wood produced =

$$(2{,}000 \text{ lb}_m/\text{hr wet wood})\left(\frac{80 \text{ lb}_m \text{ bone-dry wood}}{100 \text{ lb}_m \text{ wet wood}}\right)\left(\frac{100 \text{ lb}_m \text{ dry wood}}{95 \text{ lb}_m \text{ bone-dry wood}}\right)$$

$$= 1{,}684.2 \text{ lb}_m/\text{hr dry wood}$$

$$\text{Water removed from wood} = \text{Water added to the air stream}$$
$$= 2{,}000 - 1{,}684.2 = 315.8 \text{ lb}_m/\text{hr}$$

The flow rate of makeup air into the process may be determined from a water balance. Let A equal the lb_m/hr of makeup air. Then the total flow rate of air leaving the unit must be (A + 315.8), and we have

$$\text{Water in makeup air} = 0.01\,A$$

$$\text{Water from wood} = 315.8 \text{ lb}_m/\text{hr}$$

$$\text{Total water in} = 0.01\,A + 315.8$$

$$\text{Water out in air leaving unit} = (0.04)(A + 315.8)$$

Equating water in to water out gives

$$(0.04)(A + 315.8) = 0.01\,A + 315.8$$
$$A = 10{,}106 \text{ lb}_m/\text{hr of makeup air}$$

Now we do a water component balance on the piping branch point where the fresh makeup air and the recycle air mix together. To produce an air stream entering the drier that contains 2 weight percent water, we will need to mix some of the 4 weight percent moist recycle air with the 1 weight percent makeup air. Accordingly, let R be the lb_m/hr of recycle. Then

$$\text{Water into the mix point in makeup air} = (0.01)(10{,}106\ lb_m/hr)$$

$$\text{Water into the mix point in the recycle} = (0.04)R$$

$$\text{Total in} = 0.04R + 101.06\ lb_m/hr\ \text{water}$$

$$\text{Water out of the mix point} = (0.02)(R + 10{,}106)$$

Equating water in to water out gives

$$R = 5{,}053\ lb_m/hr\ \text{of moist air recycled}$$

Some other realistic examples of systems with recycle streams are given below.

Example 4.14.

Methanol is produced from synthesis gas by the reaction

$$CO + 2H_2 \longrightarrow CH_3OH$$

Since only 15 mole percent of the CO entering the reactor is converted to methanol, a large recycle of unreacted gases is necessary. The methanol produced is condensed and separated from the unreacted gases. The ratio of H_2 to CO in the fresh-feed gas stream to the plant is exactly 2, the stoichiometric amount. Calculate the molar flow rates of fresh feed and recycle required to produce 1,000 gal/hr of liquid methanol at 70°F when the density is 49.3 lb_m/ft^3 (specific gravity = 0.791).

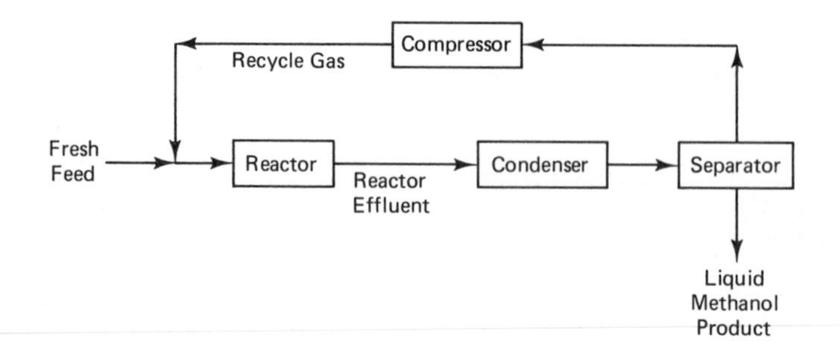

Solution:

Basis: 1 hour of operation

$(1{,}000\ gal/hr)(ft^3/7.48\ gal)(49.3\ lb_m/ft^3)(\text{lb-mole}\ CH_3OH/32\ lb_m)$

$$= 206\ \text{lb-mole/hr methanol product}$$

First we do an overall balance around the entire plant:

$$\text{Fresh feed} = \text{methanol product}$$

Each mole of CH_3OH produced comes from 1 mole of CO, which is 33.33 mole percent of the feed gas.

(206 lb-mole/hr CO)(1 mole fresh feed/0.3333 mole CO)

$$= 618 \text{ lb-mole/hr fresh feed}$$

The composition of the recycle stream will be the same as that of the fresh feed in this system (note that this is *not* generally true) because the H_2/CO in the feed is assumed to be exactly stoichiometric. Thus, every time 1 mole of CO reacts, it consumes 2 moles of H_2. The remaining gases will still have a H_2/CO ratio of 2.

Let R be the lb-mole/hr of recycle gas flow. Then

$$\text{Total reactor feed} = \text{fresh feed} + \text{recycle gas}$$

$$= 618 + R \text{ lb-mole/hr}$$

The reactor exit gas must contain 206 lb-mole/hr of CH_3OH. Since only 15 percent of the CO entering the reactor is converted into methanol, total reactor feed must be

$$(206 \text{ lb-mole/hr } CH_3OH)\left(\frac{1 \text{ mole CO reacted}}{1 \text{ mole } CH_3OH \text{ formed}}\right)\left(\frac{1 \text{ mole CO fed}}{0.15 \text{ mole CO reacted}}\right)\left(\frac{3 \text{ mole total gas}}{\text{mole CO}}\right)$$

$$= 4{,}120 \text{ lb-mole/hr total reactor gas feed}$$

The recycle flow rate R is the total reactor feed minus the fresh feed:

$$R = 4{,}120 - 618 = 3{,}502 \text{ lb-mole/hr}$$

Example 4.15.

The production of carbon tetrachloride takes place according to the reaction

$$CH_4 + 4\,Cl_2 \longrightarrow CCl_4 + 4HCl$$

Assume that the reaction is irreversible and goes to completion (i.e., at least one reactant is totally consumed). Give the molar flow rates of all components at each point numbered in Figure 4–6.

Solution:

From the flow rates given in Figure 4–6, stream 4 must contain 500 kg-mole/hr CH_4, 2,000 kg-mole/hr Cl_2, and 7,500 kg-mole/hr HCl.

Since 500 kg-mole/hr of CH_4 will react with exactly 2,000 kg-mole/hr of Cl_2 to produce 500 kg-mole/hr of CCl_4 and 2,000 kg-mole/hr of HCl, the reactor effluent will contain nothing but HCl and CCl_4. Therefore, stream 5 will contain 7,500 + 2,000 = 9,500 kg-mole/hr HCl and 500 kg-mole/hr CCl_4. Also, stream 6 will contain 500 kg-mole/hr CCl_4, stream 7 will contain 9,500 kg-mole/hr HCl, and stream 8 will contain 9,500 − 7,500 = 2,000 kg-mole/hr HCl.

Notice that the HCl product stream calculated by taking the difference between stream #7 and #3 is exactly the same as you would calculate from an overall balance around the unit, disregarding the recycle stream and any other internal flow rates. Since the reaction is producing 2,000 kg-mole/hr of HCl, it must be removed from the system somewhere. The only place HCl leaves the system is in stream #8.

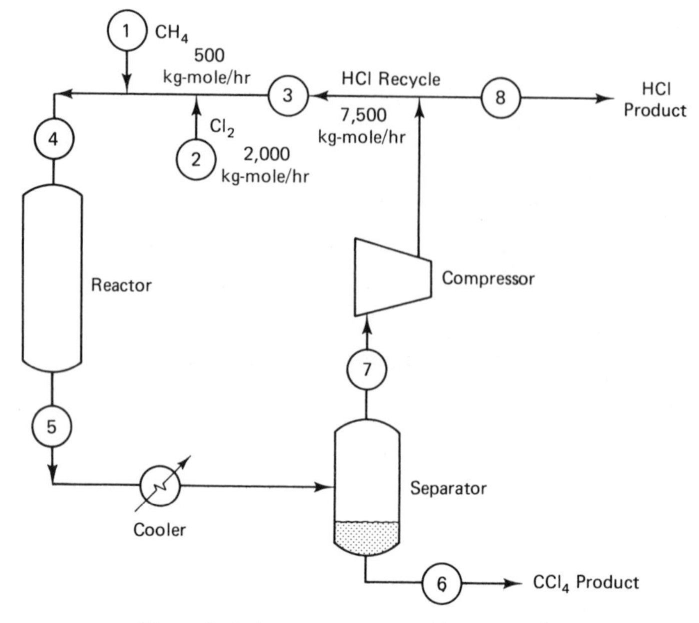

Figure 4–6 A reactor system with gas recycle.

Example 4.16.

Repeat Example 4.15 when the HCl recycle stream is no longer pure HCl, but instead contains 7,000 kg-mole/hr of HCl and 500 kg-mole/hr of Cl_2. In this case the total reactor feed has an excess amount of Cl_2. The CH_4 feed is still 500 kg-mole/hr.

Solution:

To consume 500 kg-mole/hr of CH_4, the plant must produce 2,000 kg-mole/hr of HCl which has to leave the unit. This stream, which now contains Cl_2, will carry (500) (2,000/7,000) = 143 kg-mole/hr of Cl_2 with it. Therefore, the chlorine feed will have to be increased from 2,000 to 2,143 kg-mole/hr, and we have the following table.

	Molar flow rates (kg-mole/hr)							
Stream	*1*	*2*	*3*	*4*	*5*	*6*	*7*	*8*
CH_4	500	—	—	—	—	—	—	—
Cl_2	—	2,143	500	2,643	643	—	643	143
HCl	—	—	7,000	7,000	9,000	—	9,000	2,000
CCl_4	—	—	—	—	500	500	—	—

Example 4.17.

The commercial FMC process for making hydrogen peroxide (H_2O_2) uses a complex organic compound *A* which reacts with hydrogen in a hydrogenator according to the equation

$$A + H_2 \longrightarrow B$$

The product B is then oxidized by air in the oxidizer to produce H_2O_2 and regenerate A:

$$B + O_2 \longrightarrow A + H_2O_2$$

Eighty-five percent of the oxygen fed to the oxidizer reacts with the B fed to the oxidizer. The stream from the oxidizer is then flashed to remove all the unreacted oxygen and nitrogen from the air. The H_2O_2 is extracted from the organic phase by water in a liquid/liquid extractor such that the concentration of H_2O_2 in the aqueous solution leaving the extractor is 25 weight percent. No organic material is lost in the flash drum or from the liquid/liquid extractor.

The concentration of the organic stream fed to the hydrogenator is 90 mole percent A and 10 mole percent B. The concentration of the organic stream leaving the hydrogenator is 25 mole percent A and 75 mole percent B.

(a) Calculate the flow rates of H_2, air, and water in lb_m/hr in a plant that produces 1,000 tons per day of 25 weight percent H_2O_2 in a water solution.
(b) Calculate the flow rates in lb-mole/hr of the organic streams flowing into and out of the hydrogenator.

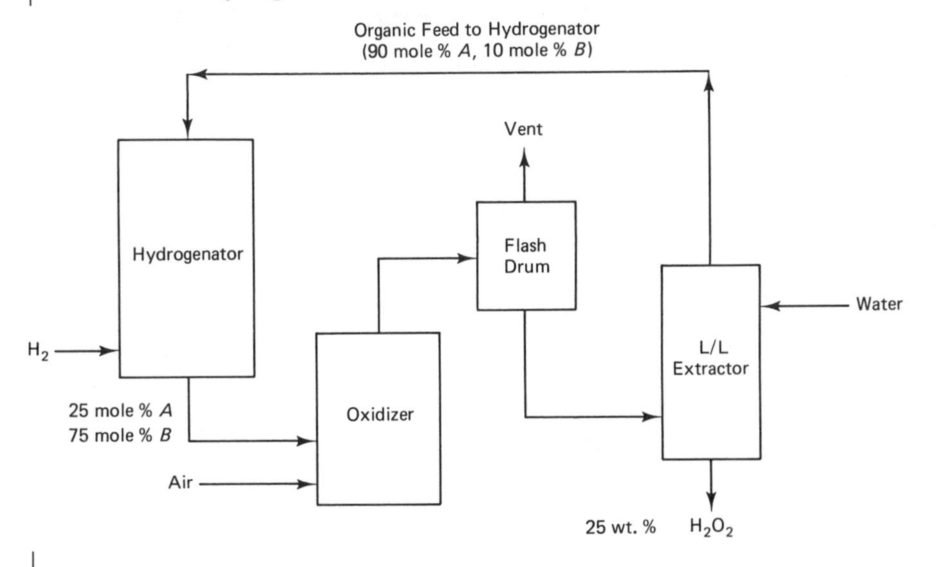

Solution:

To produce 1,000 ton/day of solution, 250 ton/day of H_2O_2 must be produced. Thus, we have

$$(250 \text{ T/D})(\text{D}/24 \text{ hr})(2{,}000 \text{ lb}_m/\text{T})(\text{lb-mole H}_2\text{O}_2/34 \text{ lb}_m \text{ H}_2\text{O}_2)$$

$$= 612.8 \text{ lb-mole/hr H}_2\text{O}_2 \text{ produced}$$

Each mole of H_2O_2 requires one mole of O_2 and one mole of H_2. Hence,

$$(612.8 \text{ lb-mole/hr H}_2)(2 \text{ lb}_m/\text{lb-mole}) = 1{,}226 \text{ lb}_m/\text{hr H}_2 \text{ fed to hydrogenator}$$

Since only 85 percent of the oxygen fed to the oxidizer reacts, the total amount of oxygen that must be fed in is

$$(612.8 \text{ lb-mole/hr } O_2 \text{ reacted})(1 \text{ mole } O_2 \text{ fed}/0.85 \text{ mole } O_2 \text{ reacted})$$

$$= 720.9 \text{ lb-mole/hr } O_2 \text{ fed in air stream to oxidizer}$$

$$N_2 \text{ in air feed} = (720.9 \text{ lb-mole/hr } O_2)\left[\frac{79 \text{ moles } N_2}{21 \text{ moles } O_2}\right]$$

$$= 2{,}712.0 \text{ lb-mole/hr } N_2$$

$$\text{Air flow rate} = (720.9 \text{ lb-mole/hr } O_2)(32 \text{ lb}_m/\text{lb-mole})$$

$$+ (2{,}712 \text{ lb-mole/hr } N_2)(28 \text{ lb}_m/\text{lb-mole}) = 99{,}000 \text{ lb}_m/\text{hr air}$$

Water fed to the liquid/liquid extractor to make a 75 weight percent water solution will be

$$(0.75)(1{,}000 \text{ T/D})(2{,}000 \text{ lb}_m/\text{T})(\text{D}/24 \text{ hr}) = 62{,}500 \text{ lb}_m/\text{hr water}$$

Thus, the answers for part (a) are:

$$H_2 = 1{,}226 \text{ lb}_m/\text{hr}$$

$$\text{Air} = 99{,}000 \text{ lb}_m/\text{hr}$$

$$\text{Water} = 62{,}500 \text{ lb}_m/\text{hr}$$

For part (b), let R be the lb-mole/hr of organic recycle flow from the oxidizer into the hydrogenator. Then

$$\text{Flow rate of } A \text{ into hydrogenator} = 0.90\, R \text{ lb-mole/hr}$$

In the hydrogenator, 612.8 lb-mole/hr of A react with the H_2. So we have

$$\text{Flow rate of } A \text{ leaving hydrogenator} = 0.90\, R - 612.8 \text{ lb-mole/hr}$$

The reaction in the hydrogenator of A with H_2 produces one mole of B for every one mole of A reacted. Therefore, there is no change in the total number of moles of organic mixture of A and B in the hydrogenator. Thus, the number of lb-mole/hr of organic material leaving the hydrogenator is also R lb-mole/hr.

The concentration of the stream leaving the hydrogenator is 25 mole percent A. Therefore,

$$\text{Flow rate of } A \text{ in hydrogenator effluent} = 0.25\, R \text{ lb-mole/hr}$$

But from our previous calculation, this must be equal to $0.90\, R - 612.8$. Equating the two gives

$$0.25\, R = 0.90\, R - 612.8$$

$$R = 942.8 \text{ lb-mole/hr of organic recycle}$$

Thus the flow rates of the organic streams flowing into and out of the hydrogenator are 942.8 lb-mole/hr.

PROBLEMS

4–1. A pressure vessel in a petroleum refinery operates at 325 psig and 465°F.
  (a) Convert the pressure into psia, mm Hg, atmospheres, inches of water, and inches of mercury.
  (b) Convert the temperature into °C, K, and °R.

4–2. Oil is flowing through a 6-inch-diameter pipe at a rate of 50,000 barrels per day (B/D) (42 gallons per barrel) and has a density of 35 pounds per cubic foot (lb_m/ft^3). Calculate the flow rate of oil in
  (a) Pounds per minute.
  (b) Liters per second.
  (c) gpm (gallons per minute).
  (d) CFM (cubic feet per minute).
  (e) Calculate the velocity of oil flow in the pipe in ft/sec.

4–3. A "perfect" gas ($PV = nRT$) is contained in a vertical cylindrical vessel 10 meters in diameter and 20 meters high. Process conditions are 325 psig and 465°F. The molecular weight of the gas is 50.
  (a) What are the mass and molar densities of the gas in lb_m/ft^3 and lb-mole/ft^3 and in gram/liter and g-mole/liter?
  (b) How many pounds of gas are in the vessel?

4–4. The steam turbine driving a natural gas compressor is rated at 3,700 horsepower.
  (a) How many Btu/hr, calories/minute, and joules/second is this energy demand?
  (b) At 3¢ per kilowatt-hour, what is the annual energy cost of driving the compressor?

4–5. Butylene (C_4H_8) is chlorinated in a reactor to produce dichlorobutane ($C_4H_8Cl_2$). The reaction proceeds as follows:

$$CH_2{=}CH{-}CH_2{-}CH_3 + Cl_2 \rightarrow CH_2Cl{-}CHCl{-}CH_2{-}CH_3$$

An excess of butylene is used to suppress undesirable side reactions. Chlorine is completely reacted. Excess butylene is separated from the reaction product in a distillation column and recycled back through the reactor. The recycle butylene stream is pure butylene. See figure at top of page 106.
  (a) If 9,000 lb_m/hr of chlorine are fed in, how much butylene is needed and how much dichlorobutane is produced?
  (b) The molar ratio of total butylene to chlorine fed into the reactor is 5 to 1. Determine the total reactor feed, reactor effluent, and butylene recycle flow rates and compositions.

4–6. A 500 million lb_m/yr methanol plant uses a 40-tray, $9\frac{1}{2}$ ft I.D. distillation column to separate methanol from water. Parameters of the column are as follows:

Feed Rate	2,300 lb mole/hr
Feed Composition	0.80 mole fraction methanol
Distillate Composition	0.999 mole fraction methanol
Bottoms Composition	0.001 mole fraction methanol

Calculate the distillate and bottoms flow rates in lb mole per hr.

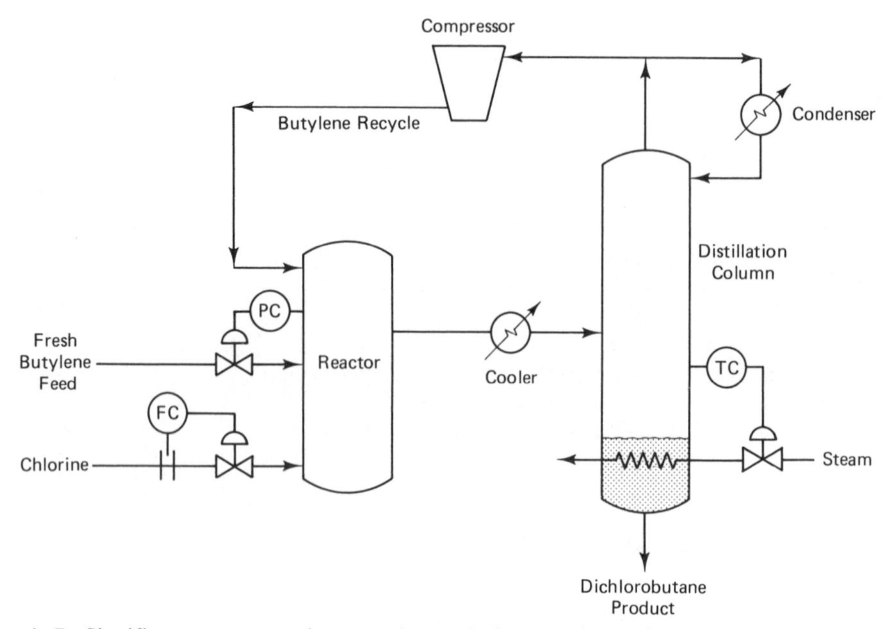

4–7. Significant energy savings can be made by using two "heat-integrated" distillation columns instead of one column in the methanol process considered in problem 4–6. The two-column sequence takes half the methanol product (at 99.9 mole percent methanol) off the top of the first column and the rest off the top of the second column.

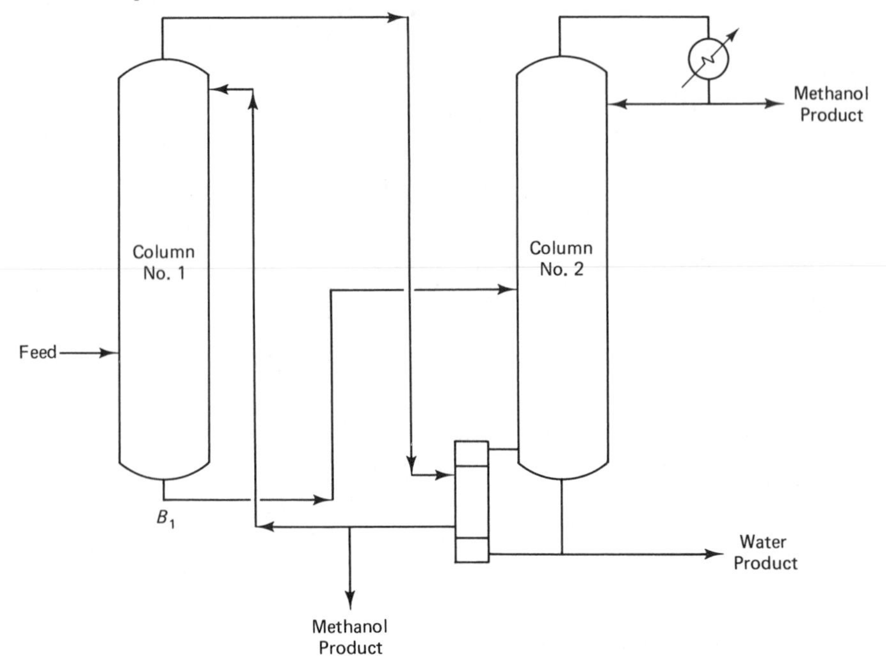

Calculate the flow rate and composition of the bottoms from the first column (the feed to the second column) using the same feed and product specifications as in problem 4–6.

4–8. In the manufacture of sulfuric acid from iron pyrites, the following two reactions occur simultaneously when the pyrites are burned in air:

$$FeS_2 + \frac{5}{2} O_2 \rightarrow FeO + 2SO_2 \quad (1)$$

$$2FeS_2 + \frac{11}{2} O_2 \rightarrow Fe_2O_3 + 4SO_2 \quad (2)$$

From such a burner, the flue gases are analyzed as 10.2 mole percent SO_2, 7.8 mole percent O_2, and 82.0 mole percent N_2.

(a) In what ratio do reactions 1 and 2 take place?

(b) How much excess air is fed to the burner if all the pyrites are burned?

(c) If the flue gases are at 600°C and 780 mm Hg total pressure, and if the pyrites fed contained 10 weight percent inerts, how many cubic meters of flue gas are generated per ton of solids fed to the burner?

4–9. The air used to burn S or FeS_2 in making sulfuric acid must first be dried. This is done by passing it counter to a stream of concentrated H_2SO_4 flowing through a packed tower. The acid enters the tower at 93 weight percent and is diluted as it absorbs H_2O. Some 92.5 weight percent H_2SO_4 is withdrawn from the system as the acid recycles, and the acid concentration is returned to 93 weight percent by adding 99 weight percent H_2SO_4.

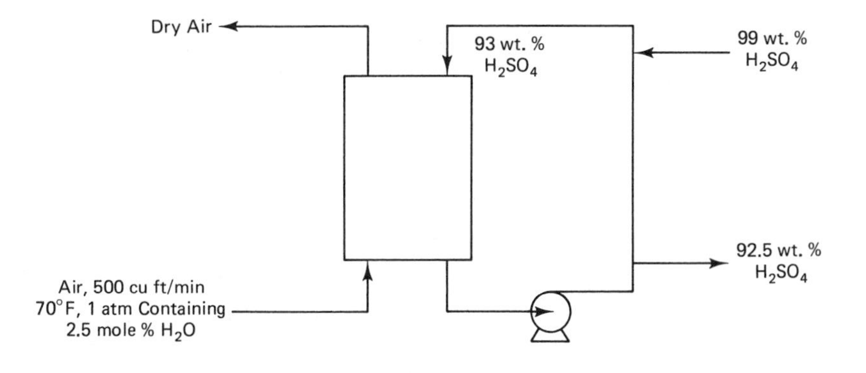

From these data, calculate:

(a) The mass of 99 weight percent H_2SO_4 added per hour.

(b) The rate of flow of 93 weight percent H_2SO_4 entering the top of the column.

4–10. The SOHIO process for producing acrylonitrile ($CH_2{=}CH{-}C{\equiv}N$) involves feeding air, ammonia, and propylene ($CH_3{-}CH{=}CH_2$) into a fluidized-bed catalytic reactor. If 10 percent excess air is used with stoichiometric amounts of ammonia and propylene, calculate the flow rates of reactor feed streams and the flow rate and composition of the reactor exit gas for a 50 million lb-per-yr acrylonitrile plant. Water is the other product of the reaction. Everything leaves the reactor in the gas phase, and the nitrogen in the air is inert.

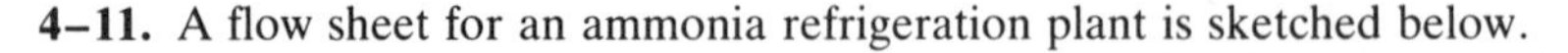

4–11. A flow sheet for an ammonia refrigeration plant is sketched below.

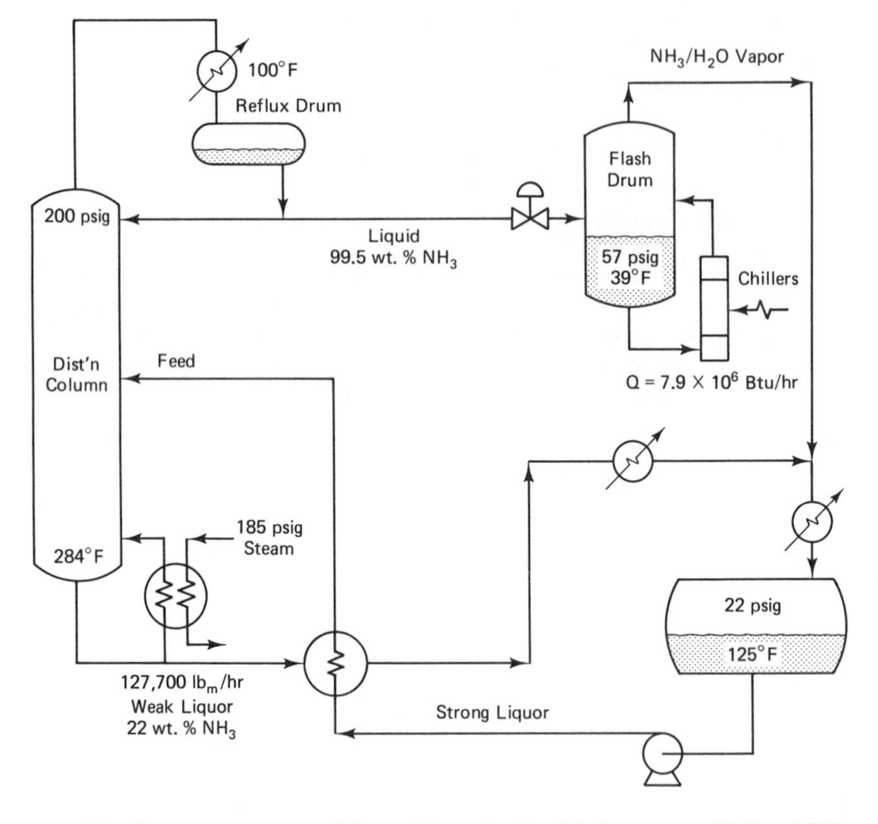

Heat from a process requiring refrigeration is added at a rate of 7.9 × 10^6 Btu/hr to the system in a number of "chillers." The heat is removed by boiling a liquid in the flash tank.

The vapor generated is dissolved in 127,700 lb_m/hr of "weak liquor" (22 weight percent NH_3, 78 weight percent H_2O), producing "strong liquor," which is fed into a distillation column. Ammonia is concentrated at the top of this column. Distillate composition is 99.5 weight percent NH_3.

As liquid enters the flash drum, it drops in pressure from 200 psig to 57 psig, and 12 weight percent of it flashes immediately to form vapor. This cools the remaining liquid down to 39°F.

The heat of vaporization of the ammonia-rich liquid in the flash tank is 460 Btu/lb_m.

(a) How much vapor leaves the top of the flash drum, and what is its composition?

(b) How much "strong liquor" is fed into the distillation column, and what is its composition?

4–12. One form of coal gasification that was used all across America before the advent of natural gas pipelines was called a "producer gas" plant. In this process, a bed of coal was heated by passing air through it to burn some of the coal. When the bed became hot enough, air flow was stopped and steam was fed into the bed. The reactions that occurred were:

$$C_nH_m + n\,H_2O \rightarrow n\,CO + \left(n + \frac{m}{2}\right) H_2 \qquad (A)$$

$$CO + H_2O \rightarrow H_2 + CO_2 \qquad (B)$$

Since these reactions are endothermic, the bed would cool, and eventually air flow would have to be resumed to reheat the bed.

Let us assume that the steam-flow step can be analyzed as a steady-state process over a short period of time. Analysis of the coal is 80 weight percent C, 7 weight percent H, and 13 weight percent Si and other solid inerts.

If the steam that is fed in is 70 percent of that required for complete conversion of the C to CO_2 (Reaction A + Reaction B), and if all of the carbon reacts with steam, forming either CO or CO_2, determine

(a) The composition of the producer gas.

(b) The volume (cubic feet at 1 atm and 60°F) of the producer gas delivered per ton of coal reacted with H_2O.

4–13. A gaseous mixture of chemical components (90 mole percent *A* and 10 mole percent *B*) is fed to a chemical reactor. Two reactions occur:

$$A \rightarrow C$$

$$A \rightleftarrows B$$

60 percent of the *A* fed in reacts to form product *C*.

The reversible reaction involving *A* and *B* has an equilibrium constant of 0.10, i.e., the ratio of the mole fraction of *B* to the mole fraction of *A* in the reactor effluent is 0.1.

Calculate the composition of the reactor effluent gas.

4–14. A simplified dimethylamine (DMA) process consists of a reactor and a separation unit. Recycle flows of NH_3, MMA, and TMA combine with fresh NH_3 and CH_3OH feeds before entering the reactor. The recycle streams, pure DMA, and pure H_2O are produced in the separation unit. No water is recycled.

The reactor effluent contains all three methylamines (MMA, DMA, and TMA) in the molar ratio of 1:2:1, as well as water and ammonia. It is 50 mole percent NH_3. Methanol (CH_3OH) is completely consumed in the reactor by the reactions

$$NH_3 + CH_3OH \rightleftarrows H_2O + CH_3NH_2 \qquad (MMA)$$

$$NH_3 + 2\,CH_3OH \rightleftarrows 2\,H_2O + (CH_3)_2NH \qquad (DMA)$$

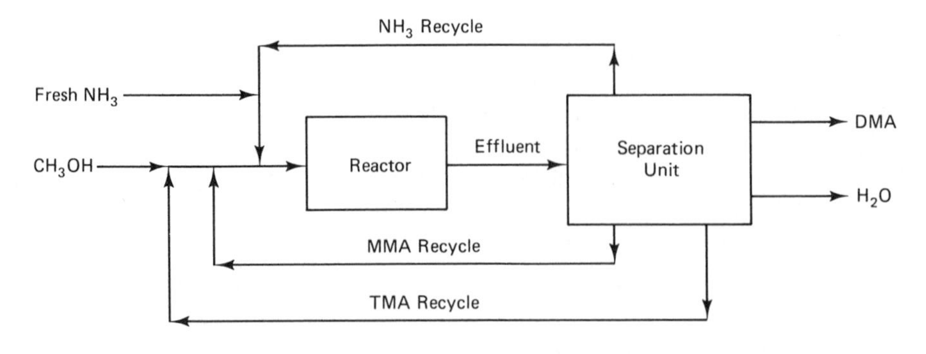

$$NH_3 + 3\ CH_3OH \rightleftarrows 3\ H_2O + (CH_3)_3N \qquad \text{(TMA)}$$

(a) What is the composition (mole percent) of the reactor effluent?
(b) What are the fresh feed rates of ammonia and methanol per mole of DMA produced?
(c) Calculate the ammonia recycle, the MMA recycle, and the TMA recycle flow rates in a plant producing 20,000 lb_m/day of DMA.

4–15. The methylamines-production process considered in problem 4–14 is to be modified so that not only is DMA produced, but also TMA and MMA. The reactions proceed as follows:

$$NH_3 + CH_3OH \rightleftarrows H_2O + MMA$$

$$NH_3 + 2\ CH_3OH \rightleftarrows 2\ H_2O + DMA$$

$$NH_3 + 3\ CH_3OH \rightleftarrows 3\ H_2O + TMA$$

The molar ratio of MMA to DMA to TMA in the reactor effluent is 1:2:1, and the reactor effluent is 50 mole percent NH_3.

The plant is to produce 1,000 lb-mole/day of MMA, 5,000 lb-mole/day of DMA, and 2,000 lb-mole/day of TMA.

DMA and water are not recycled, but NH_3, MMA, and TMA are recycled back to the reactor section. Methanol is completely consumed in the reaction section.

Calculate the methanol and ammonia fresh-feed rates, the NH_3, MMA, and TMA recycle rates, and the composition of the reactor effluent.

4–16. Air separation is typically done in two distillation columns connected as shown in the following diagram.
Typically, R_2 will equal 3 moles for every mole of R_1. Assume air is a binary mixture of oxygen and nitrogen, and determine
(a) The flow rates R_1 and R_2 for a plant producing 2,500 cubic feet/hr of oxygen gaseous product at 350 psia and $-100°F$. Assume perfect gases.
(b) The flow rate of "crude oxygen" on the same basis.
(c) The fraction of the oxygen in the feed gas that is collected in the oxygen product.

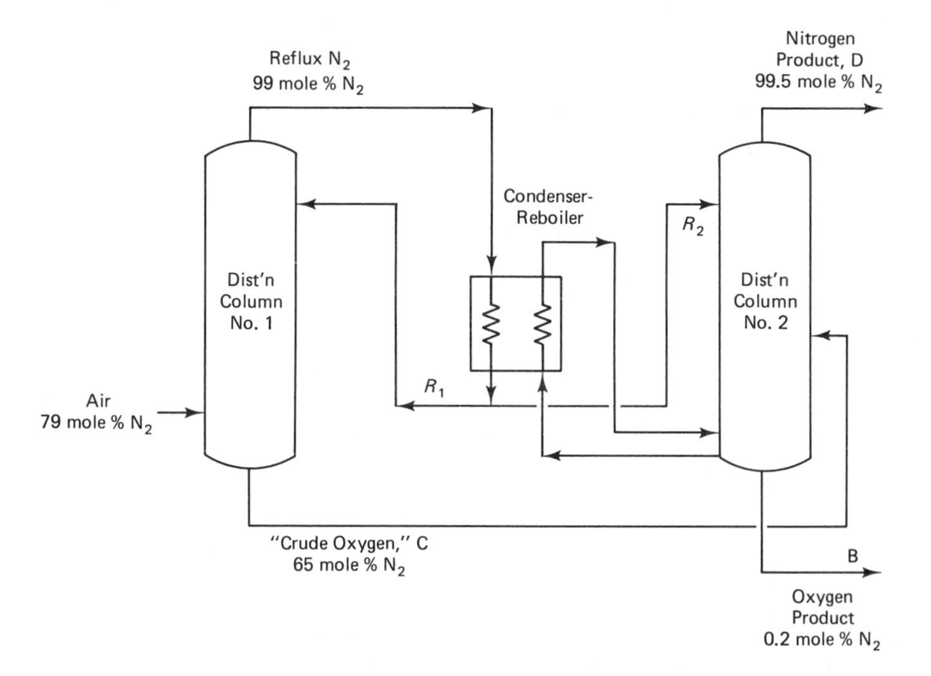

4–17. 8,000 lb_m/hr of ethylene (C_2H_4) is fed into a chemical reactor. Hydrogen chloride gas (HCl) is added as required by a pressure controller to maintain reactor pressure at 350 psig.

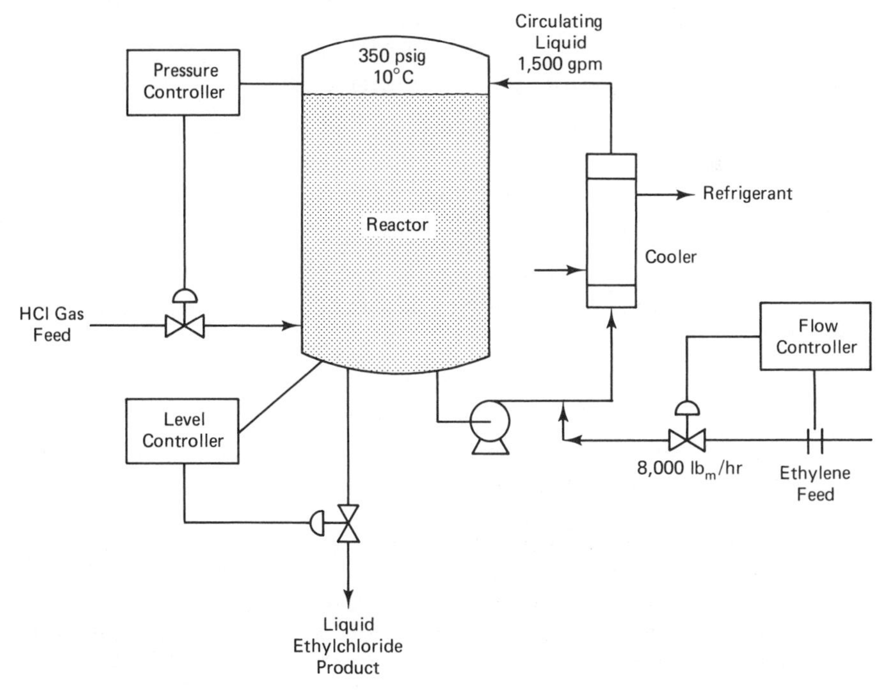

Ethylene reacts with HCl to form liquid ethyl chloride, which is removed from the reactor by the level controller to maintain a constant liquid level in the reactor. The reaction equation is

$$C_2H_4 + HCl \rightarrow C_2H_5Cl$$

Under steady-state conditions, what are the molar and mass flow rates of HCl gas and ethyl chloride liquid?

4–18. Nitric acid is made by oxidation of ammonia and absorption of the product NO in water. The reaction cycle is as shown in the following diagram:

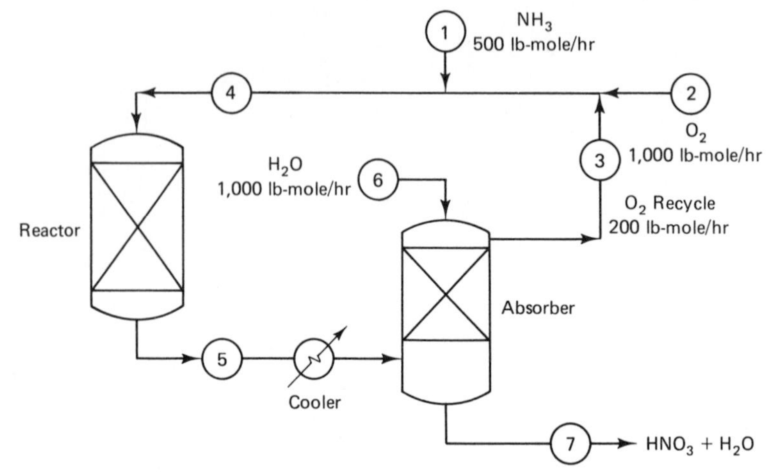

In the reactor, the oxidation of ammonia proceeds by the reaction

$$4NH_3 + 5O_2 \rightarrow 4NO + 6H_2O$$

The reactions in the absorber are complex, but may be summed up as

$$2NO + H_2O + \frac{3}{2}O_2 \rightarrow 2HNO_3$$

If these reactions go to completion and are irreversible, and if no side reactions occur, determine the number of lb-moles of each component at each numbered point around the cycle. What is the concentration of HNO_3 solution produced in weight percent HNO_3?

4–19. In the production of synthesis gas for ammonia, air separation is used to obtain both O_2 and N_2. An air separation system is shown in the diagram on page 113.

After filtration, drying, and CO_2 removal, the air contains 78 mole percent N_2, 1 mole percent Ar, and 21 mole percent O_2. Stream C from the high-pressure column contains 65 mole percent N_2. Stream D contains 99.5 percent N_2 with the rest Ar. The N_2 product is 99.9 mole percent N_2 + Ar, the rest being O_2. The O_2 product is 99.9 percent O_2, with the remainder being N_2. Determine the moles of air feed required per 1,000 tons/day of O_2 product, and the flow rates of streams C and D. Determine, also, the volumetric flow rate of N_2 product at 20 psia and 70°F and of O_2 product at 15 psia at 60°F.

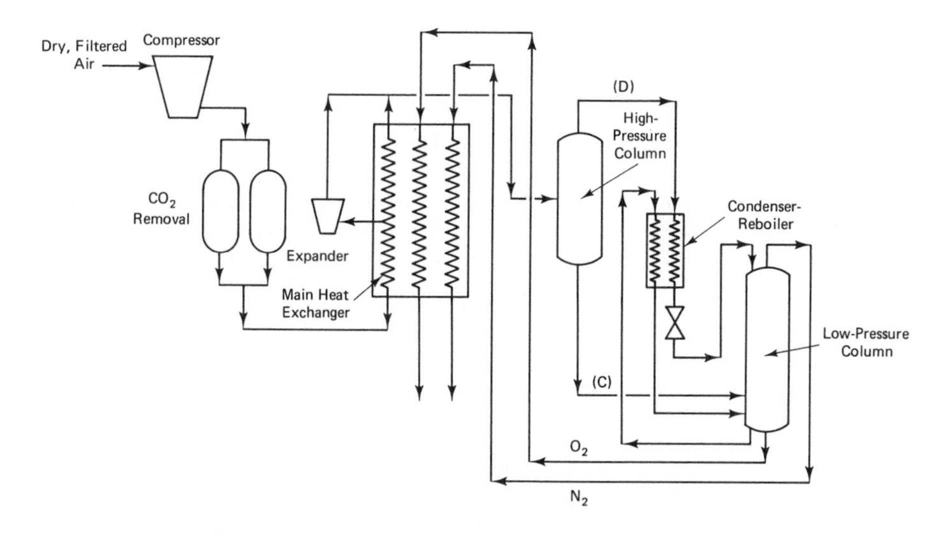

4–20. In the manufacture of sulfuric acid by the contact process, iron pyrites, FeS_2, is burned in dry air, the iron being oxidized to Fe_2O_3. The sulfur dioxide thus formed is further oxidized to the trioxide form by conducting the gases mixed with air over a catalytic mass of platinum-black at a suitable temperature. It will be assumed that in the operation sufficient air is supplied to the pyrites burner that the oxygen will be 40 percent in excess of that required if all the sulfur actually burned were oxidized to the trioxide. Of the pyrites charged, 15 percent is lost by falling through the grate with the "cinder" and is not burned.

(a) Calculate the weight of air to be used per 100 kg of pyrites charged.

(b) In the burner and a "contact shaft" connected with it, 40 percent of the sulfur burned is converted to the trioxide. Calculate the composition, by weight, of the gases leaving the contact shaft.

(c) By means of the platinum catalytic mass, 96 percent of the *sulfur dioxide remaining* in the gases leaving the contact shaft is converted to the trioxide. Calculate the total weight of SO_3 formed per 100 kg of pyrites charged.

(d) Assuming that all gases from the contact shaft were passed through the catalyzer, calculate the composition by weight of the resulting gaseous products.

(e) Calculate the overall degree of completion of the conversion of the sulfur in the pyrites charged to SO_3 in the final products.

4–21. A gaseous mixture of benzene and air passes through a 500 lb_m batch of mixed acid (51.8 weight percent H_2SO_4, 40 weight percent HNO_3, and 8.2 weight percent H_2O). Benzene is completely consumed by the reaction, and nitrobenzene is produced.

$$C_6H_6 + HNO_3 \rightarrow C_6H_5NO_2 + H_2O$$

Two liquid layers are formed: a nitrobenzene layer, and a mixed acid layer in which the water of reaction is absorbed. No water is lost in the exit air stream.

The sulfuric acid content of the mixed acid layer at the end of the batch cycle is 58.1 weight percent. Calculate the fraction of the nitric acid which has been consumed.

4–22. In a sulfuric acid plant, the gas stream leaving the last converter stage goes to an adsorption column in which 96 percent of the SO_3 is absorbed into 75 weight percent H_2SO_4 to generate 98 weight percent H_2SO_4 solution. Part of the concentrated H_2SO_4 is diluted with H_2O to generate 75 weight percent H_2SO_4, which is recycled to the column. The composition of the absorber feed gas is

$$SO_2 = 0.65 \text{ mole }\%$$

$$SO_3 = 7.52 \text{ mole }\%$$

$$O_2 = 2.02 \text{ mole }\%$$

$$N_2 = 89.81 \text{ mole }\%$$

If the plant is to produce 300 tons/day of sulfuric acid (100 percent equivalent), calculate

(a) The flow rate of 75 weight percent H_2SO_4 to the absorber in lb_m/hr.
(b) The flow rate of 98 weight percent H_2SO_4 from the absorber in lb_m/hr.
(c) The amount of water to be added to part of the 98 weight percent H_2SO_4 to redilute it to 75 weight percent H_2SO_4.
(d) The composition of the exit gases in mole percent. (Assume these gases are dry.)

A flow diagram of the process is as follows:

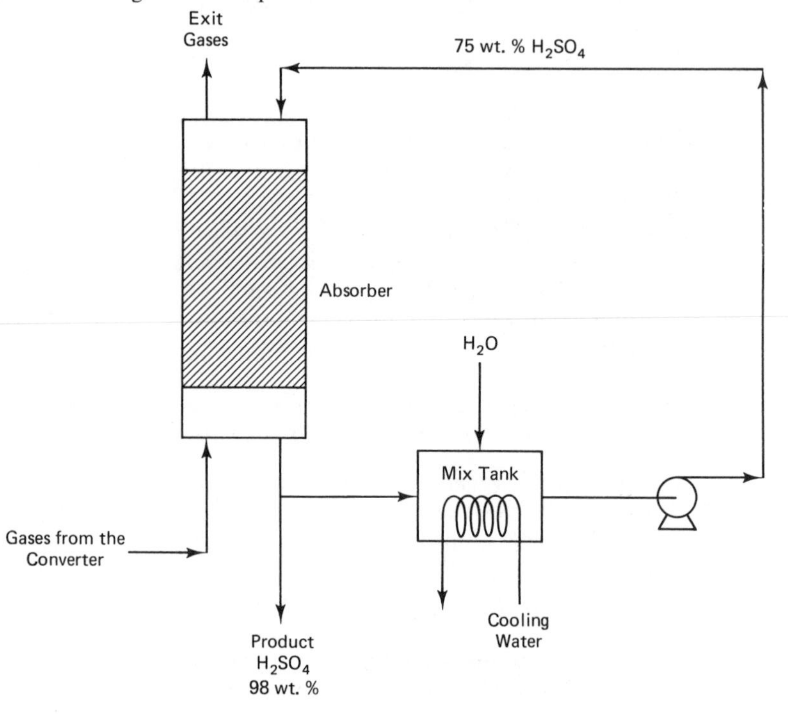

4–23. The Chevron refinery at Perth Amboy feeds 50,000 B/D (42 gal/B) of high-sulfur "sour" gas oil (2.5 weight percent sulfur) to its Isomax unit to reduce the sulfur content of the gas oil to 0.3 weight percent. The specific gravity of the gas oil is 0.6.

Hydrogen gas is fed into the Isomax unit, where it reacts with sulfur in the gas oil to form hydrogen sulfide (H_2S). The H_2S is sent to a Claus unit to recover elemental sulfur by oxidizing the H_2S to water and sulfur.

(a) How many tons per day of sulfur are produced?

(b) How many moles per day of hydrogen must be fed to the Isomax unit?

The hydrogen is produced in a naphtha reformer. The naphtha contains 10 mole percent cyclohexane (C_6H_{12}), which is converted to benzene (C_6H_6) in the reformer as follows:

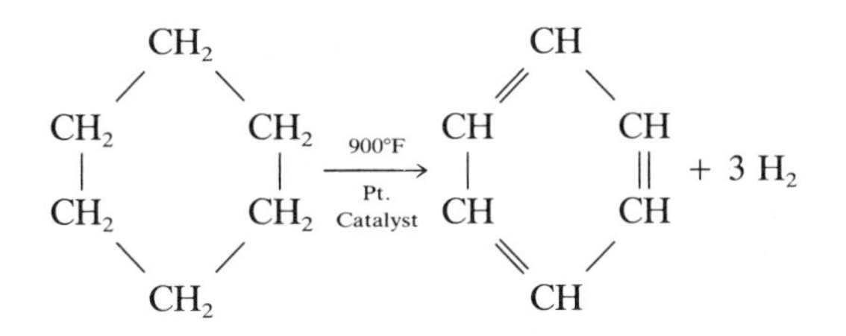

The specific gravity of naphtha is 0.55, and its molecular weight is 100.

(c) How many barrels per day of naphtha must be fed into the reformer to generate the hydrogen required in the Isomax unit?

4–24. Tertiary butyl alcohol (TBA) is produced by causing isobutane (iC_4) to react with oxygen at 350°F and 600 psia as follows:

$$CH_3-\underset{}{\overset{\displaystyle CH_3}{\overset{|}{C}H}}-CH_3 + \frac{1}{2}O_2 \longrightarrow CH_3-\underset{\displaystyle CH_3}{\underset{|}{\overset{\displaystyle CH_3}{\overset{|}{C}}}}-OH$$

The molecular weights of iC_4 and TBA are 58.1 and 74.1, respectively.

A large recycle stream provides an excess of iC_4 to prevent undesirable side reactions. The iC_4 is separated from the TBA in a distillation column (debutanizer) and recycled back to the reactor. A portion of the debutanizer overhead is fed to another distillation column (C_4 splitter) from which normal butane impurity is purged from the system.

The normal butane (nC_4) comes into the reactor with the butane feed, which is 90 mole percent iC_4 and 10 mole percent nC_4. The nC_4 (molecular weight 58.1) is not involved in the reaction. The purge stream from the C_4 splitter is 95 mole percent nC_4 and 5 mole percent iC_4. The TBA product removed from the debutanizer is pure TBA. The composition of the recycle stream is 85 mole percent iC_4 and 15 mole percent nC_4. A recycle-to-fresh-butane feed ratio of 5 is used. Oxygen is completely consumed in the reactor.

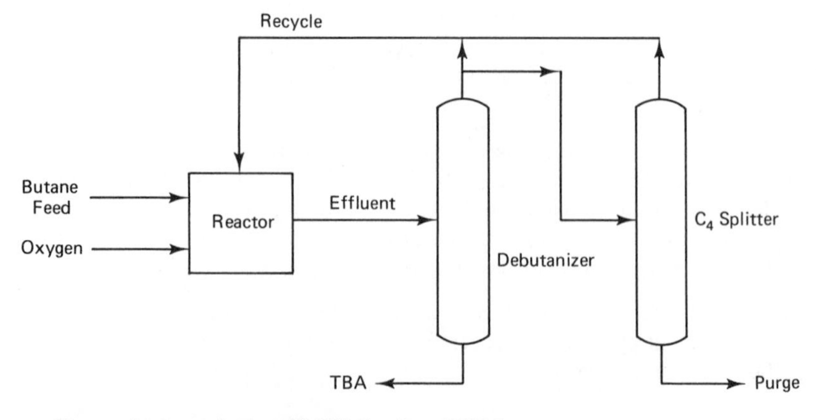

For a plant producing 50,000 lb_m/hr of TBA:

(a) Calculate the purge flow rate from the C_4 splitter.
(b) Calculate the total butane feed rate to the reactor.
(c) Calculate the oxygen feed rate.
(d) Calculate the reactor effluent composition.
(e) If half of the debutanizer overhead is fed to the C_4 splitter, what is the composition of the C_4 splitter overhead?

4–25. A coffee-drying process takes roasted beans (40 percent solubles and 60 percent inerts by mass), dissolves the solubles in hot water, and removes the water by spray drying. Two pounds of hot water are used per pound of coffee beans. The solids are removed from the solution by centrifugation before the solution is sent to the spray drier. If (a) 80 weight percent of the solubles is dissolved in the hot water, (b) the inerts leave the extraction step with 0.5 lb_m of solution per pound of undissolved solids, and (c) the spray-dried product contains all the coffee solubles, no inerts, and 10 weight percent water, determine the number of pounds of original roasted beans required per pound of dried coffee.

4–26. Cyclohexane (C_6H_{12}) is reformed to produce benzene (C_6H_6) and hydrogen in the reforming process sketched on page 117.

30,000 lb_m/hr of pure cyclohexane fresh feed is mixed with recycle gas. There are seven lb-moles of recycle gas per lb-mole of fresh feed. The composition of the recycle gas is 90 mole percent H_2, 5 mole percent cyclohexane, and 5 mole percent benzene. The overall conversion of fresh cyclohexane to benzene in the process is 80 percent. The liquid product from the drum is a mixture of benzene and cyclohexane. Some of the gas leaving the compressor is purged from the system.

(a) Calculate the purge gas flow rate.
(b) Calculate the flow rate and composition of the liquid product.
(c) Calculate the molar flow rates of all components in the streams entering and leaving the reactor.

4–27. Terephthalic acid ($C_8H_6O_4$) is used to produce polyester fibers. It is made by oxidizing para-xylene (C_8H_{10}) with air. Water is the other product formed.

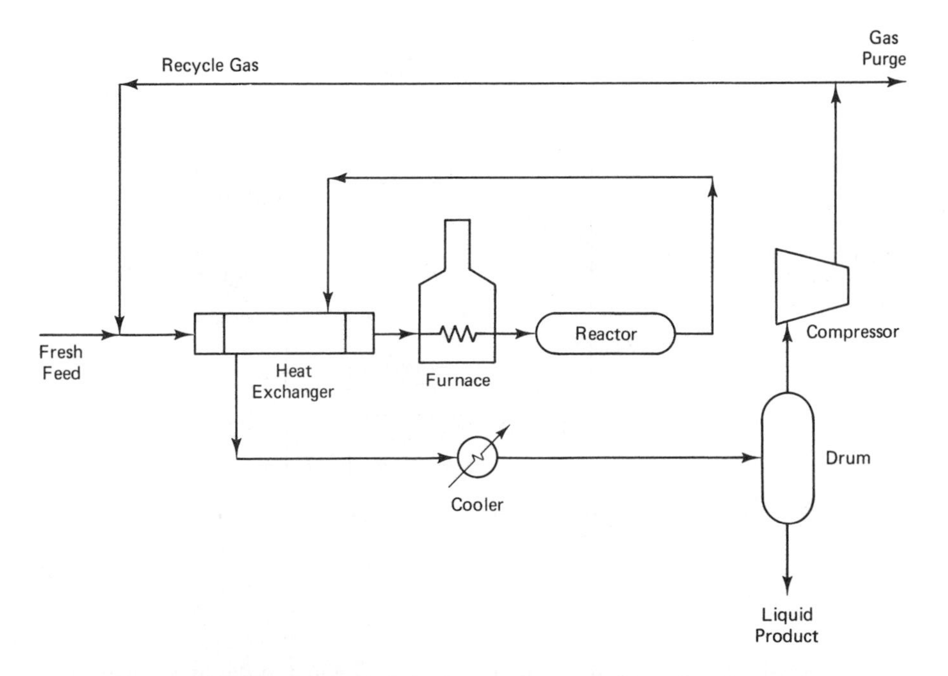

Xylene and air are fed into a reactor. Gas containing 5 mole percent oxygen and 95 mole percent nitrogen leaves the top of the reactor. Water and terephthalic acid leave the bottom of the reactor. All of the xylene reacts. The feed rate of xylene is 50,000 lb_m/hr.

What is the air feed rate?

4–28. Nitric acid is made by oxidizing ammonia in air, converting NO to NO_2 catalytically, and absorbing the NO_2 into H_2O. The resulting dilute HNO_3 cannot be concentrated to commercial strength by direct evaporation because it forms a constant boiling concentration (an azeotrope) at about 68 percent HNO_3. This concentrating can be done by adding a component that suppresses the vapor pressure of H_2O. A proposed process using Mg $(NO_3)_2$ is shown on page 118.

Determine, per ton of 99 weight percent HNO_3 per hr, the flow rate (lb_m/hr) of

(a) HNO_3 feed solution
(b) water evaporated
(c) 55 weight percent Mg $(NO_3)_2$ solution leaving the distillation unit
(d) 75 weight percent Mg $(NO_3)_2$ solution returning to the acid feed

4–29. The Rohm and Haas process for producing ethyl acrylate ($C_5H_8O_2$) uses two reactors in series. In the first reactor, ethylene (C_2H_4) reacts with component B to form component D:

$$C_2H_4 + B \longrightarrow D$$

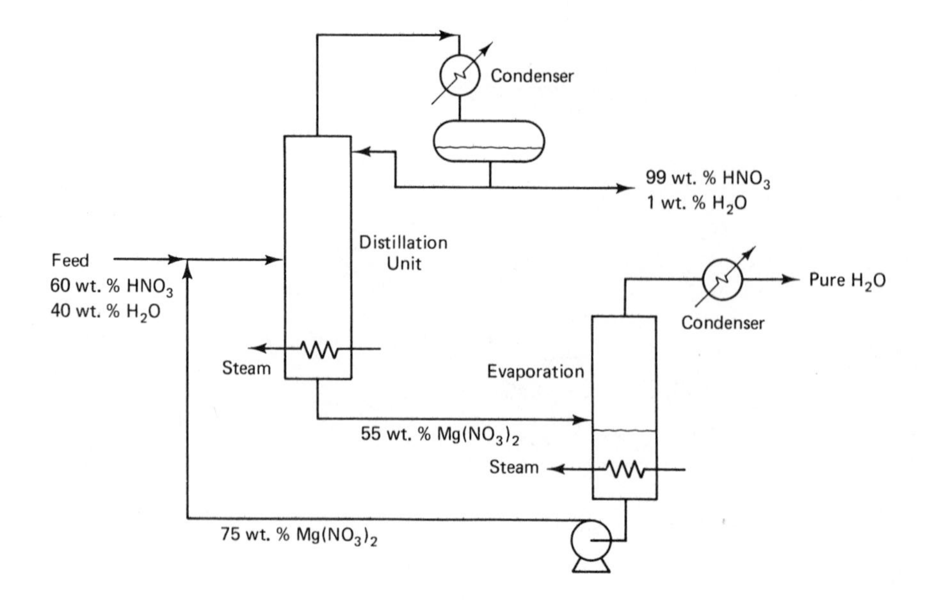

The ethylene reacts completely. Component B is fed to the first reactor in a recycle stream containing 95 mole percent B and 5 mole percent D. The effluent stream from Reactor No. 1 contains 10 mole percent B and 90 mole percent D.

In the second reactor, the effluent from the first reactor is mixed with a feed stream of fresh acrylic acid ($C_3H_4O_2$). The acrylic acid reacts with component D to form ethyl acrylate ($C_5H_8O_2$) and component B:

$$C_3H_4O_2 + D \longrightarrow C_5H_8O_2 + B$$

The effluent from Reactor No. 2 is fed into a distillation column. All of the ethyl acrylate is recovered in the distillation product. All the components B and D go out the bottom and are recycled to Reactor No. 1.

For a plant making 10,000 lb_m/hr of ethyl acrylate, calculate

(a) The ethylene and acrylic acid feed rates.

(b) The recycle flow rate from the distillation column to the reactor, in lb-mole/hr.

4–30. A binary mixture of 30 mole percent acetic acid and 70 mole percent water is to be separated in a sequence of three heat-integrated distillation columns. The total feed rate is 100 lb-mole/hr. The overhead distillate products from all three columns are 97 mole percent water. The distillate product flow rates from all three columns are identical.

The bottoms product from the last column is 5 mole percent water. Calculate the flow rates and compositions of the bottom streams from the first and second columns.

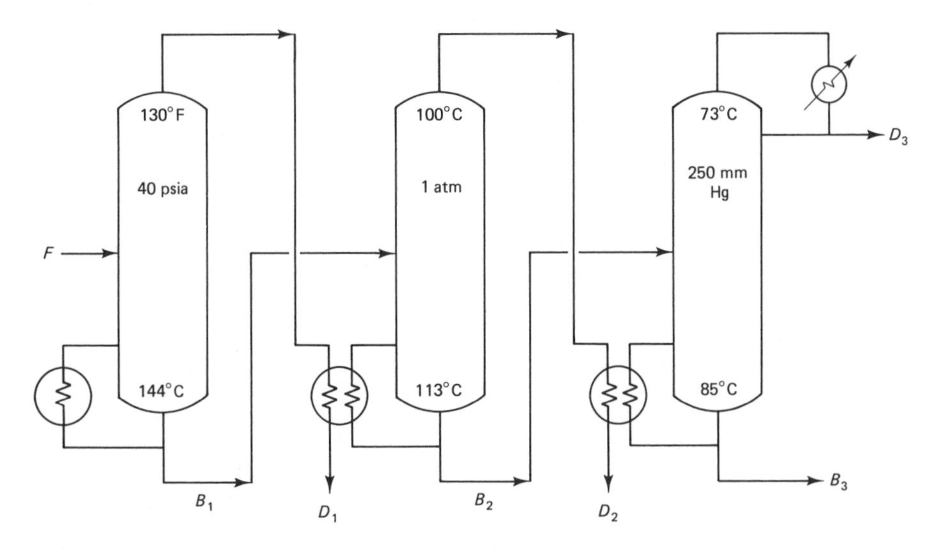

4–31. The product *B* is obtained using the gas phase reaction

$$2A \rightarrow B$$

in a reactor that will convert 25 percent of the reactant *A* in the total feed to the reactor. The product *B* is then removed by condensation, and the unconverted *A* is recycled to the reactor. Unfortunately, the feed of almost pure *A* contains a nonreactive gas *C* (0.4 mole *C*/100 moles *A*) that will accumulate in the system if not purged. If we can accept a maximum of 10 moles *C*/100 moles *A* at the reactor entrance, what is the yield of product *B* (moles of *A* converted to *B*)/(moles of *A* fed in fresh feed)?

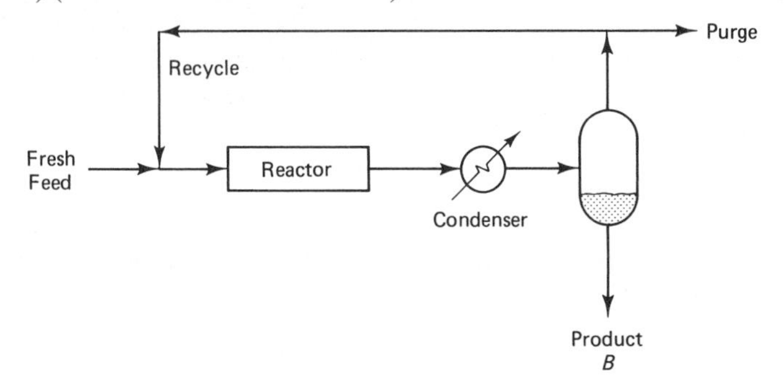

4–32. The synthesis gas from a coal partial oxidation process (POX) contains CO, CO_2, and H_2 in molar ratios of 3, 1, and 5, respectively. The reaction is

$$4\ CH + O_2 + 3\ H_2O \longrightarrow 3\ CO + CO_2 + 5\ H_2$$

There is not enough H_2 to react with all the CO and CO_2 to produce methanol:

$$CO + 2\ H_2 \longrightarrow CH_3OH$$

$$CO_2 + 3\ H_2 \longrightarrow CH_3OH + H_2O$$

Therefore, a portion of the "syn gas" is sent to a "cold box" unit. First, all of the CO_2 is removed from the gas stream. Then the remaining CO/H_2 mixture is partially separated in a cryogenic distillation column. The composition of the overhead distillate product is 80 mole percent H_2. The composition of the bottoms product from the column is essentially 100 percent CO. This bottoms stream is used in another process.

The overhead distillate product is added to the portion of the syn gas that was not fed to the cold box. The total is then fed into a methanol reactor.

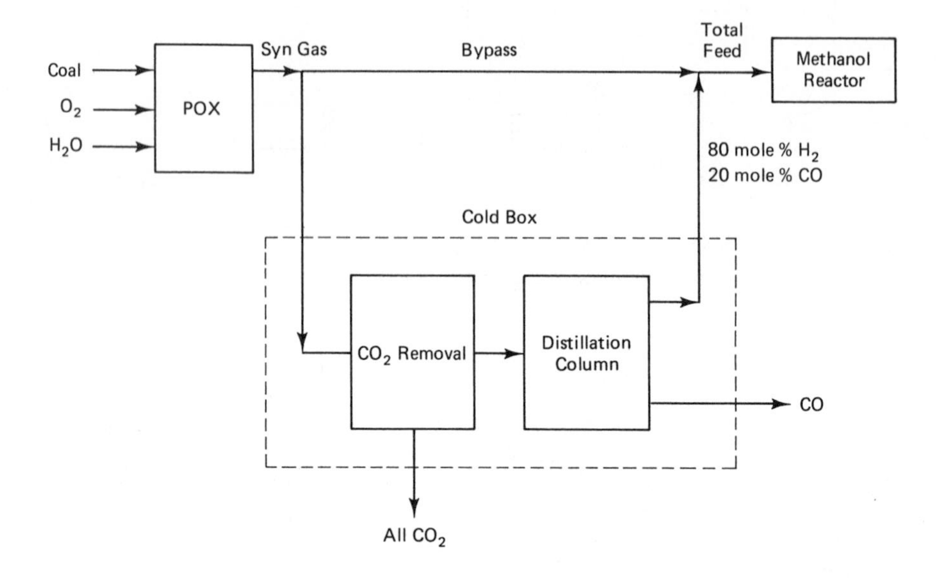

Assuming steady-state conditions, calculate what fraction of the syn gas must be fed to the cold box in order to have just enough H_2 in the total feed to the methanol reactor to react with the CO and CO_2 in the total methanol reactor feed.

4–33. Formaldehyde is made by the partial oxidation of methanol with air according to the reaction equation

$$CH_3OH + \frac{1}{2} O_2 \longrightarrow HCHO + H_2O$$

The gas mixture fed to the reactor contains 8 mole percent methanol and 10 mole percent oxygen (stream 4 in the figure below). The methanol is completely converted to formaldehyde in the reactor, which contains a bed of $Fe_2O_3 \cdot MoO_3$ catalyst particles. The gas stream from the top of the separator contains only O_2 and N_2.

Calculate the flow rates (lb-mole/min) of streams 1, 2, 3, 4, and 5 needed to make 3,000 lb_m/min of 37 weight percent formaldehyde/water solution.

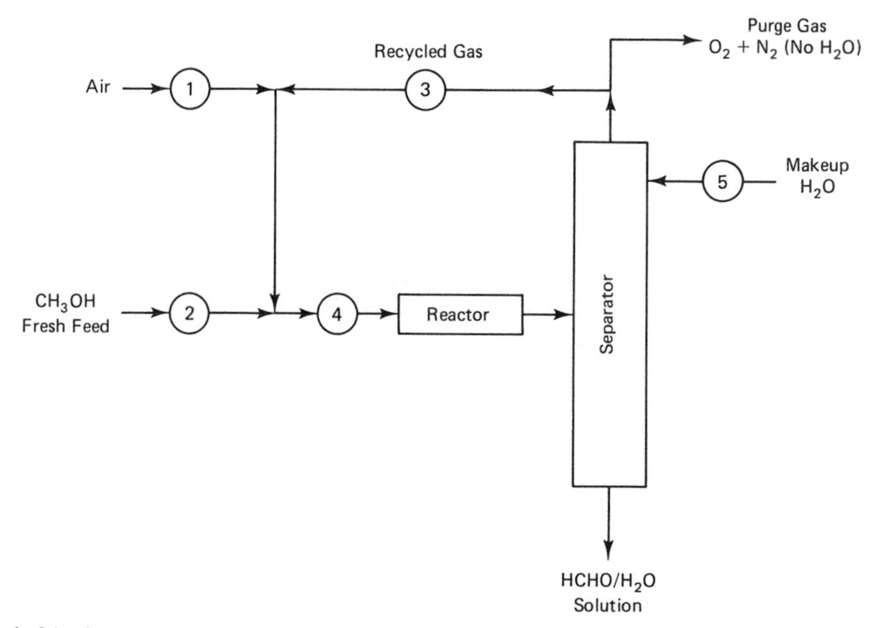

4–34. A gas stream has a composition of 65 mole percent CO_2, 30 mole percent N_2, and 5 mole percent He. One million cubic feet per day of this gas at 70°F and 150 psig is cooled so that the bulk of the CO_2 and some N_2 are condensed. Ninety percent of the CO_2 originally present is condensed in a liquid stream that has a composition of 95 mole percent CO_2 and 5 mole percent N_2. What is the composition of the remaining gas stream? What is the volumetric production rate (cubic feet/day) of this gas at −100°C and 10 atmospheres absolute pressure?

4–35. Synthesis gas for a methanol process is produced in a partial oxidation (POX) unit that is fed steam, oxygen, and both natural gas and coal. The ratio of CO to CO_2 in the syn gas is 3. The ratio of O_2 fed to the total number of atoms of carbon in the coal (CH) and natural gas (CH_4) is $\frac{1}{4}$.

Taking 3 moles of CO as a basis in the syn gas, how many moles of CH and CH_4 must be fed to POX to produce exactly the amount of H_2 in the syn gas that is required to react with the CO and CO_2 to make the methanol product?

4–36. Alfalfa hay contains 30 weight percent water when it is baled. It is then dried to 10 weight percent water in an LPG-fired air drier. Each pound of water evaporated requires 1,000 Btu of energy. Each pound of LPG produces 23,300 Btu when it is burned and costs 31 cents/pound. The overall efficiency of the drier, given by

$$\text{Efficiency} = \frac{\text{Heat transferred into the hay}}{\text{Heat produced by burning LPG}}$$

is 50 percent.

How much does it cost ($/ton) to dry one ton of baled hay?

5 Simple Countercurrent Processes

In chapter 4 we applied total and component balances to a variety of processes. In this chapter we will combine these balance equations with phase equilibrium equations to analyze simple countercurrent processes. This simultaneous use of material balance and phase equilibrium equations is characteristic of many of the calculations made by chemical engineers to design and analyze the operation of chemical processes. We will use the same approach in chapter 8 to look at binary (two-component) distillation columns, and in chapter 9 to look at ternary (three-component) liquid-liquid extraction.

5.1 EQUILIBRIUM STAGE

Many chemical processes involve operations where two phases (liquid-vapor, liquid-liquid or gas-solid) are made to contact each other in order to transfer chemical components from one phase to the other. *Phase equilibrium* is said to exist when two or more phases are kept in contact with each other for a long enough period of time so that all compositions of all phases stop changing. At this steady-state equilibrium condition, the chemical components will be distributed between the phases in different concentrations. If the differences in composition are big enough, it may be possible to achieve a separation of the chemical components by using a series of contact stages.

Consider for example, a mixture of methanol and water held in a closed vessel at atmospheric pressure. If samples are taken of the vapor and liquid phases, the vapor phase analysis will show a higher concentration of methanol than the liquid phase analysis. This occurs because methanol is a lighter, i.e., more volatile, component, boiling at 64.5°C at 1 atmosphere whereas water boils at 100°C under atmospheric pressure. Normally, the larger the difference in boiling points, the more separation is achieved in a single equilibrium stage (with a single contact between the two phases at phase equilibrium).

Chemical engineering thermodynamics deals extensively with the subject of equilibrium. One of its principles is that thermal, chemical, and pressure potentials are in balance at equilibrium. To see how this operates, suppose we have three phases in equilibrium with each other: a vapor phase V, liquid phase L, and solid phase S. Then for thermal equilibrium, the temperatures of all three phases must be the same, i.e,

$$T^V = T^L = T^S \tag{5-1}$$

Similarly, for pressure equilibrium, the pressures of all phases must be the same, i.e,

$$P^V = P^L = P^S \tag{5-2}$$

Finally, for chemical equilibrium, the chemical potentials of all components must be the same in all three phases, i.e.,

$$\mu_j^V = \mu_j^L = \mu_j^S \qquad j = 1, 2, 3, \ldots, \text{NC} \tag{5-3}$$

where

μ_j = the chemical potential of the jth component, and

NC = the total number of chemical components

Equation 5–3 must hold true for all components. Note carefully that it is the *chemical potentials* that are the same in all phases, *not* the concentrations of the chemical; in general, the concentrations will not be the same in all phases.

At this point you probably have learned only a little about chemical potentials. In later thermodynamics courses you will study the formulation of μ_j and its interpretation in terms of measurable system variables like temperature and concentration.

We will cover some of the simplest and most useful concepts of phase equilibrium in more detail in chapter 6. For now, we assume that we are given the necessary phase equilibrium relationships and that we normally know the functional relationship between the composition of one phase and the composition of the other phase. Thus, if we know the composition of the liquid phase and its temperature, we can calculate (or look up on a graph) the composition of the vapor phase that would be in equilibrium with liquid of the given composition. This is called a *bubblepoint* calculation and includes calculation of the operating pressure of the system.

The composition of the liquid phase will be called x_j (mole fraction of the jth component). If we have a three-component system and know two compositions (e.g., x_1 and x_2), the mole fraction of the third component can be calculated from the relationship

$$x_1 + x_2 + x_3 = 1 \tag{5-4}$$

Similarly, the composition of the vapor phase will be called y_j (mole fraction of the jth component), and the sum of the y_j's must be unity:

$$y_1 + y_2 + y_3 = 1 \tag{5-5}$$

We use the symbols T for temperature and P for total system pressure.

An equilibrium, or ideal, stage involving vapor and liquid phases is a unit in which (1) two or more process liquid and vapor streams are mixed together, (2) chemical components are transferred between phases until equilibrium is reached, and (3) the two new liquid and vapor phases are separated. We can represent this diagrammatically as

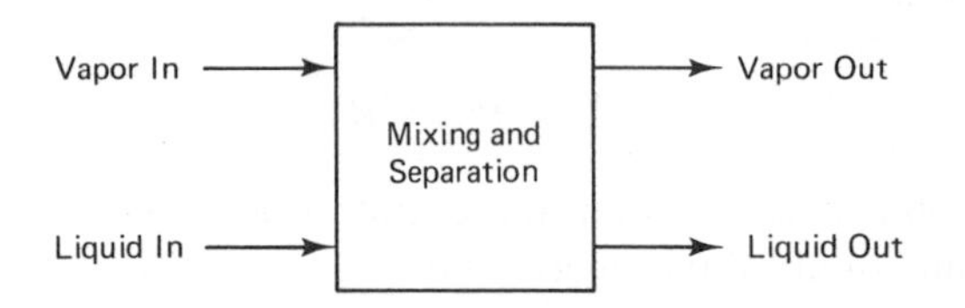

For example, on the nth tray of a distillation column (see diagram below), liquid enters from the tray above at a total molar flow rate L_{n+1} and with a composition $x_{n+1,j}$. Vapor enters from the tray below at a total molar flow rate V_{n-1} and with a composition $y_{n-1,j}$. The two phases leaving the tray are liquid (L_n and x_{nj}) and vapor (V_n and y_{nj}). Under ideal conditions, the vapor and liquid are in phase equilibrium with each other. That is to say, the compositions x_{nj} and y_{nj} are related to each other through vapor-liquid equilibrium (VLE) relationships.

Total molar flow rates and mole fractions are normally used because VLE relationships usually use mole fractions. However, mass flow rates and weight fractions can also be used. Note that we number trays from the bottom of the column up, starting with tray number 1 at the bottom, tray number 2 second from the bottom, and so on up to the top.

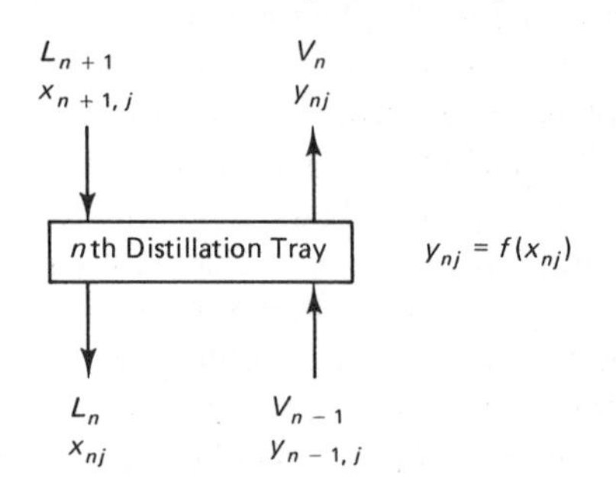

Example 5.1.

A vapor stream entering a sieve plate in a distillation column contains 40 mole percent benzene and 60 mole percent toluene. Its flow rate is 550 kg-mole/hr. Liquid flows into the tray at a rate of 700 kg-mole/hr and with a composition of 45 mole percent benzene and 55 mole percent toluene. Under the conditions on the plate, the composition of the vapor leaving the tray (y mole fraction benzene) is related to the composition of the liquid leaving the tray (x mole fraction benzene) by the simple equation

$$y = 1.3\,x$$

The liquid and vapor streams leaving the plate have the same total molar flow rates as those entering the plate. Calculate the compositions of the streams leaving the plate.

Solution:

Since no chemical reactions are occurring, a benzene component balance on the plate gives the following results:

In

$$\text{Benzene in liquid} = (700 \text{ kg-mole/hr})(0.45 \text{ m.f. benzene})$$

$$= 315 \text{ kg-mole/hr benzene}$$

$$\text{Benzene in vapor} = (550 \text{ kg-mole/hr})(0.40 \text{ m.f. benzene})$$

$$= 220 \text{ kg-mole/hr benzene}$$

$$\text{Total benzene in} = 535 \text{ kg-mole/hr benzene}$$

Out

$$\text{Benzene in liquid} = 700\,x$$

$$\text{Benzene in vapor} = 550\,y$$

$$\text{Total benzene out} = 700\,x + 550\,y$$

Now we use the VLE relationship, $y = 1.3\,x$ to obtain

$$535 = 700\,x + 550\,(1.3\,x)$$

$$x = 0.378$$

$$y = 0.492$$

Example 5.2.

At one stage in a distillation column separating methane (CH_4) and propane (C_3H_8), operating conditions are

$$T = 50°\text{F}$$

$$P = 300 \text{ psia}$$

With this temperature and pressure, the "K values" (ratio of y to x) for the two components are

$$K_j = \frac{y_j}{x_j}$$

$$K_{CH_4} = 7.2$$

$$K_{C_3H_8} = 0.38$$

The vapor and liquid streams entering the plate total 200 kg-mole/hr with an overall composition of 50 mole percent methane and 50 mole percent propane. What are the compositions and flow rates of the liquid and vapor streams leaving the tray?

Solution:

The VLE relationships are

$$y_{\text{methane}} = 7.2\, x_{\text{methane}}$$

$$y_{\text{propane}} = 0.38\, x_{\text{propane}}$$

Also,

$$x_{\text{methane}} + x_{\text{propane}} = 1$$

$$y_{\text{methane}} + y_{\text{propane}} = 1$$

Combining these with the VLE equations gives

$$(1 - y_{\text{propane}}) = 7.2\,(1 - x_{\text{propane}})$$

$$1 - 0.38\, x_{\text{propane}} = 7.2\,(1 - x_{\text{propane}})$$

$$x_{\text{propane}} = 0.9091$$

$$y_{\text{propane}} = 0.3455$$

$$x_{\text{methane}} = 0.0909$$

$$y_{\text{methane}} = 0.6545$$

Notice that the lighter methane component has a higher concentration in the vapor phase than in the liquid phase. The reverse is true for the heavier propane component, since more of it is in the liquid phase.

Now a total molar balance and a methane component balance can be used to find the vapor flow rate V and the liquid flow rate L leaving the plate. From the former, we get

$$L + V = 200 \text{ kg-mole/hr}$$

And from the latter (since no reactions occur), we obtain

$$0.0909\, L + 0.6545\, V = (200 \text{ kg-mole/hr})(0.5 \text{ m.f. methane})$$

$$V = 145.2 \text{ kg-mole/hr}$$

$$L = 54.8 \text{ kg-mole/hr}$$

Examples 5.1 and 5.2 deal with phase equilibrium when vapor and liquid phases are present. Similar treatment can be applied to liquid-liquid systems (see chapters 6 and 9), to solid-liquid systems, to gas-solid systems, and to systems with any number of phases present.

5.2 BASIC CONFIGURATION

Separation processes usually require several stages to achieve the desired changes in concentrations of process streams. Each stage enhances the separation. Countercurrent flow of the two phases is usually employed, because this gives the highest efficiency of operation. One stream flows in one direction through the cascade, while the other flows in the opposite direction.

To achieve countercurrent flow, vertical columns are usually used. The heavier (higher density) phase flows down through the column; the lighter (lower density) phase rises up through the column. Figure 5–1 shows the basic setup for a vapor-liquid system. Vapor flows up the column because the pressure in the bottom is higher than in the top. Liquid flows down the

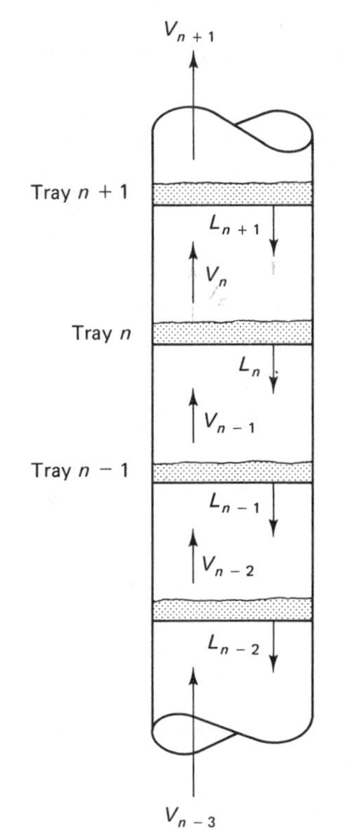

Figure 5–1 Simple countercurrent liquid-vapor contacting column.

column, despite the pressure gradient that opposes its downward flow, because the liquid builds up enough hydraulic pressure head to counterbalance the pressure difference.

Total vapor flow rates are denoted by V and are usually given in molar units. Thus, V_n is the total molar flow rate of vapor leaving the nth stage in the cascade. Similarly, L_n is the total molar flow rate of liquid leaving the nth stage. Compositions x_{nj} and y_{nj} are mole fractions of component j in the liquid and vapor streams, respectively, from tray n. The symbol y is used for vapor compositions, and x is used for liquid compositions.

Assuming that there are no chemical reactions occurring, we can write total molar and component balances around the nth tray.

$$\text{Total molar balance: } L_{n+1} + V_{n-1} = L_n + V_n \tag{5-6}$$

jth component balance:

$$L_{n+1}\, x_{n+1,j} + V_{n-1}\, y_{n-1,j} = L_n\, x_{nj} + V_n\, y_{nj} \tag{5-7}$$

For ideal stages, the vapor leaving each tray is in phase equilibrium with the liquid leaving that same tray. That is, in VLE,

$$y_{nj} = f(x_{nj}, T_n, P_n)$$

where T_n and P_n are the temperature and pressure on the nth tray.

A gas absorber provides a simple example. Figure 5–2 shows a seven-

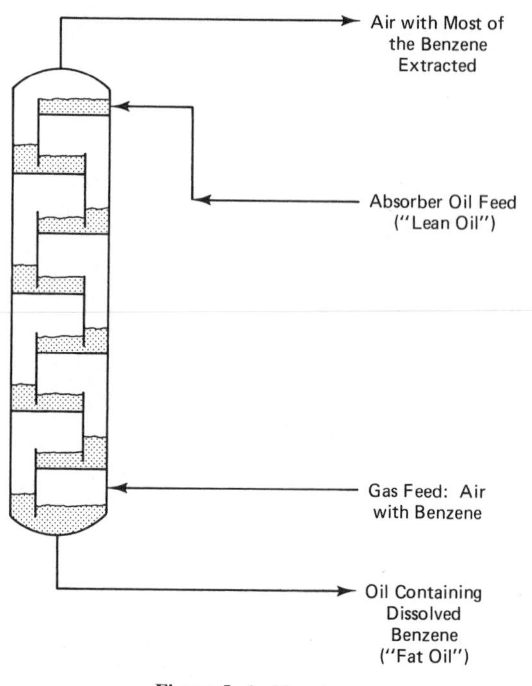

Figure 5–2 Absorber.

tray absorber in which a heavy "lean oil" is fed to the top tray. Gas containing air and benzene is fed into the base of the column. The heavy oil "soaks up" the benzene as it flows down through the column. The gas loses benzene as it flows up through the column.

5.3 EQUIPMENT

A brief discussion of some of the typical countercurrent cascade equipment might be useful for you at this point before we get into the mathematical analysis and design of these systems.

5.3.1 Columns

The industrial columns used for countercurrent operations are tall, vertical, cylindrical vessels, ranging in diameter from a few inches to 30 feet and in height from 6 to 150 feet. Some columns require so many trays and would be so tall, that they are separated into two columns to prevent interference with airplane traffic at the local airport.

Materials of construction range from carbon steel and stainless steel to copper, inconel, and tantalum. The material required for corrosion resistance drastically increases the cost of the column, which runs over a million dollars for some columns.

Figures 5–3 and 5–4 show typical industrial-sized columns with their associated piping, heat exchangers, pumps, and tanks. Figure 5–3(a) is an aerial photograph of a typical chemical plant. Note the layout of the product and raw material tanks, process units, office and laboratory buildings, and docks for ships and barges.

5.3.2 Column Internals

The purpose of the column is to effect contact between the liquid and vapor streams. Some columns are "packed" columns and some are "tray" columns. Packed columns are filled with a variety of devices to promote liquid-vapor contact and minimize pressure drop. Some of the most popular packings are shown in Figure 5–5. The entire cross-sectional area of the column is randomly filled with these little (1 inch by 1 inch) devices. Some packed columns use structured wire-mesh packing instead of the randomly dumped packing. These packings are quite expensive, but perform well in vacuum columns where pressure drop is important. Plugging of the packing and corrosion of the fine metal structure can cause problems.

The height of a bed of packing is typically 15 to 20 feet. Tall columns have several beds with liquid redistribution plates between the beds. Packing is not widely used for large-diameter columns because of problems with liquid maldistribution.

Figure 5–3 Views of the Showa Denko plant at Oita, Japan. (a) Aerial view (b) Pipe rack in foreground, distillation columns in background. *Courtesy Showa Denko K. K.*

Figure 5–4 Distillation columns, reactors, and heat exchangers in an alkylation unit. *Courtesy Stratco, Inc.*

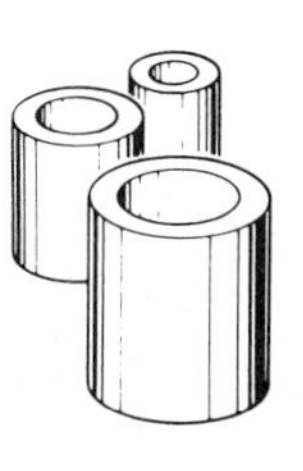

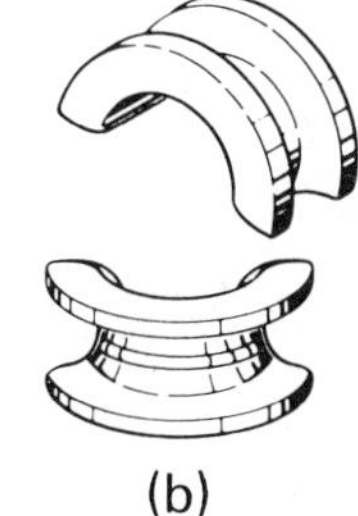

(a) (b)

(c)

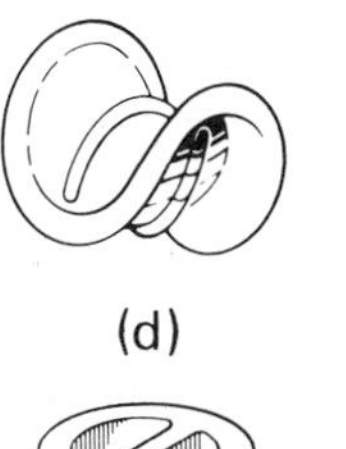

(d) (e)

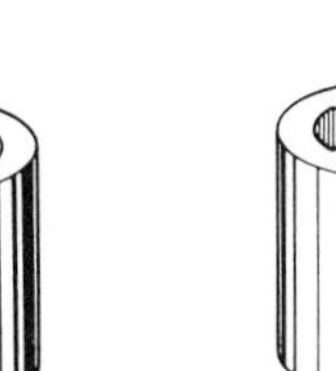

(f) (g)

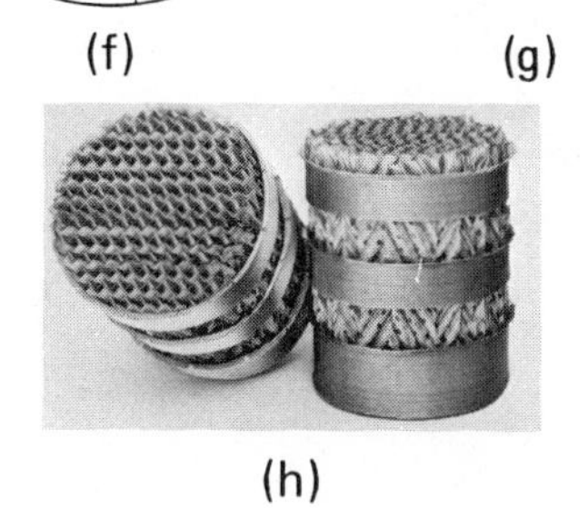

(h)

Figure 5–5 Various column packing materials. (a) Raschig rings (b) Intalox saddles (c) Pall rings (d) Cyclohelic spiral ring (e) Berl saddles (f) Lessing ring (g) Cross-partition ring (h) Woven metal packing. *Courtesy Glitsch, Inc.*

The majority of industrial columns have flat, horizontal internal trays to provide the liquid-vapor contact. Small openings in the tray permit the vapor to bubble up through the liquid that is held on the tray by an overflow weir. The liquid flows over the weir, down the "downcomer," and onto the lower tray.

Figure 5–6 shows types of tray arrangements. Many types of trays have

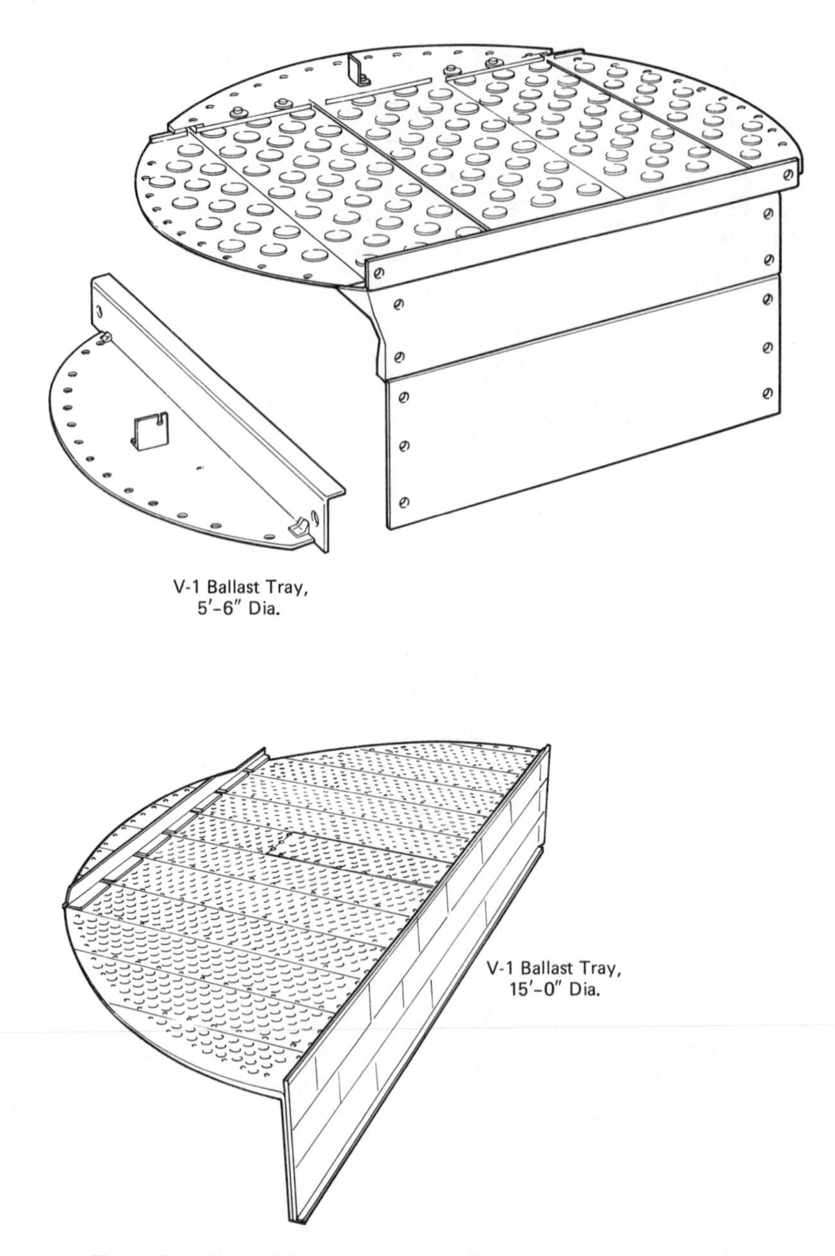

Figure 5–6 Types of tray arrangements. (a) Single-pass trays (b) Double-pass trays. *Courtesy Glitch, Inc.*

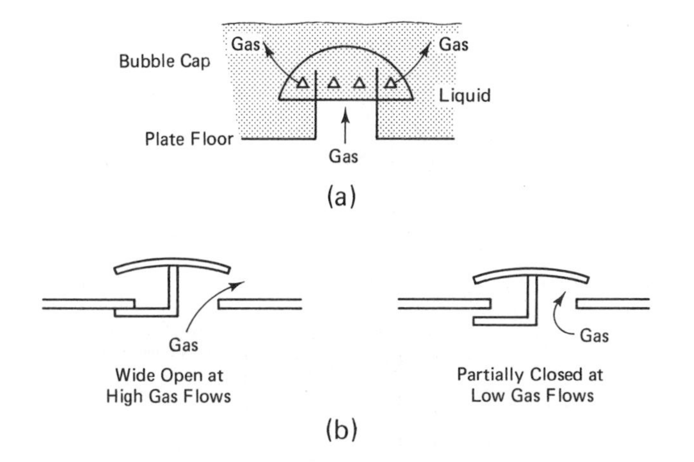

Figure 5–7 Bubblecap and valve trays.

been developed. Most older columns used bubblecap trays. (See Figure 5–7(a).) Liquid is held on the tray by a chimney under each cap. Vapor flows up through the chimney, out through the slots in the bubblecap, and up through the liquid. Bubblecap trays are expensive, but operate well under "turndown" conditions, i.e., when the vapor and liquid rates are reduced to low values. This is because the chimney prevents the liquid from dumping down the holes through which the vapor is rising.

A sieve tray is simply a flat metal plate with a large number of small holes ($\frac{1}{4}$ to 1 inch ID) drilled through it. Sieve trays are less expensive than bubblecap trays. However, the turndown of a sieve plate is limited to about 60 percent of the design vapor rate. At lower vapor rates there is not enough vapor pressure drop to keep the liquid from dumping, or "weeping," down through the holes. This bypassing of the liquid results in poor vapor-liquid contact and low tray efficiencies.

Valve trays are probably the most widely used trays. They have some type of movable flapper that varies the effective hole area as the vapor rate changes. (See Figure 5–7(b).) Valve trays are somewhat more expensive than sieve plates but have much better turndown.

Tray spacing is the distance between trays and varies from 6 inches to 4 feet, depending on the chemical components being separated and the diameter of the column. A typical tray spacing is 2 feet, which is big enough to allow a person to crawl between the trays for repairs, cleaning, inspection, or modification. In low-temperature cryogenic systems where the whole column must be encased in a "cold box" (well-insulated container), it is more economical to have smaller tray spacing to minimize the height of the column. The trays may be designed so that they can be installed and removed from the top of the column as a group or cartridge of trays.

Figure 5–8 shows typical liquid flow patterns down through the trays.

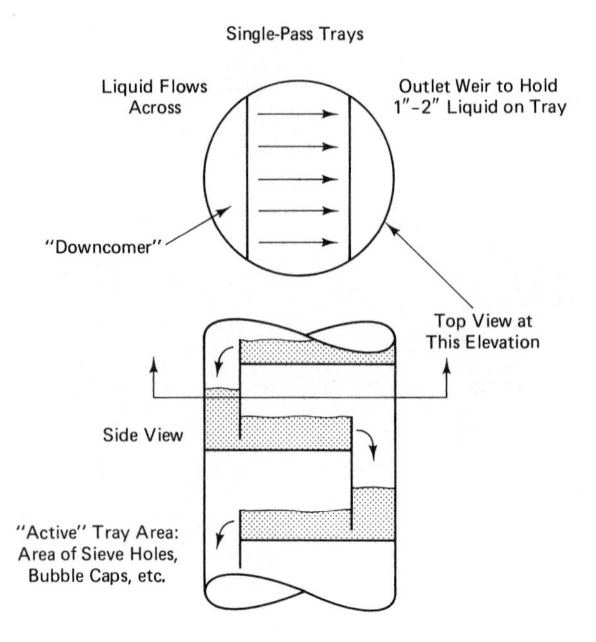

Double-Pass Trays (for High Liquid Flow Rates)

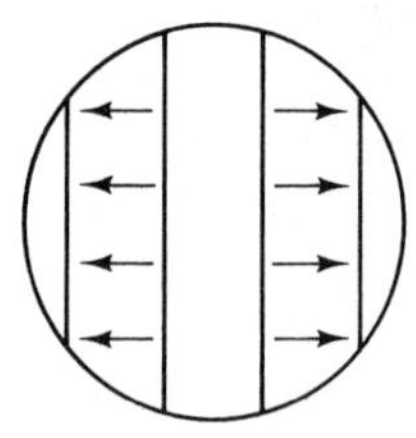

Alternating Inflow and Outflow Trays

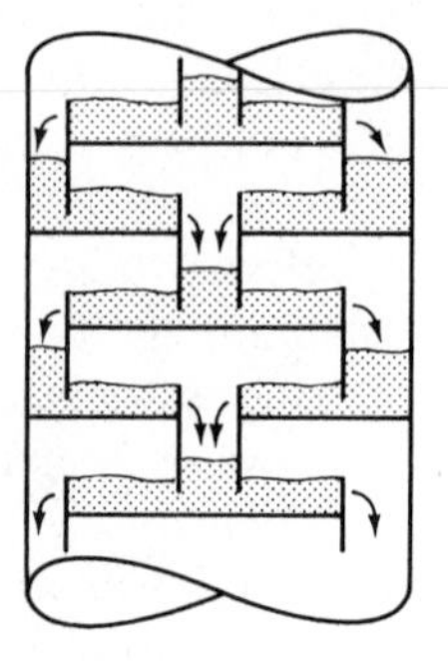

Figure 5–8 Liquid flow patterns.

Liquid flows across the tray, over an overflow weir, and down into the downcome. The liquid height in the downcomer builds up to a point high enough so that the liquid can flow onto the lower tray despite the fact that this lower tray has a higher pressure. If the height of liquid in the downcomer becomes greater than the tray spacing, the liquid cannot flow down the column correctly, and flooding occurs. Large-diameter columns often have double-pass trays as shown in the figure. The path that the liquid has to flow across the tray is reduced because there are two outlet weirs and two downcomers. Alternate trays have inflow and outflow of liquid.

5.4 SIMPLE PHASE EQUILIBRIUM

To analyze equilibrium-stage processes, we need phase equilibrium data. In some special cases, these data can be represented quite simply by algebraic equations or on graphs.

When the concentrations of a component are low, the VLE relationship can be as simple as

$$y_j = K_j x_j \tag{5-8}$$

where K_j is the distribution coefficient or "K value" of the jth component and is approximately constant at low concentrations of j. Figure 5–9(a) illustrates this in an x-y coordinate system. The ordinate is y_j, the vapor composition in mole fraction of component j. The abscissa is x_j, the liquid composition in mole fraction of component j.

If K_j is not constant, the equilibrium or VLE graph will not be a straight line. Rather, it will have some curvature to it, as illustrated in Figure 5–9(b). Whatever the shape of this VLE line, we will assume in this chapter that the

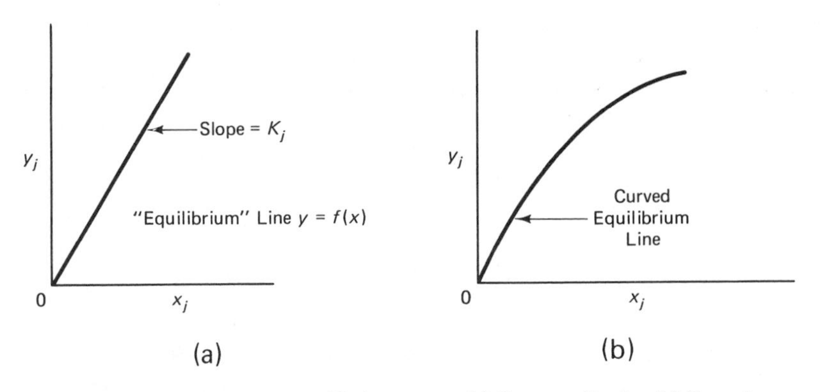

Figure 5–9 Liquid-vapor equilibrium curves. (a) Constant K-value (b) Curved equilibrium line.

VLE relationship is given and use it to relate the compositions of the liquid and vapor streams leaving each tray. These streams are assumed to be in vapor-liquid equilibrium with each other. Therefore, the VLE line allows us to determine the composition of the vapor leaving a tray if we know the liquid composition on that tray. Or it lets us determine the composition of the liquid on the tray if we know the composition of the vapor leaving the tray. These calculations are called *bubblepoint* and *dewpoint* calculations, respectively. You will learn how to do them both analytically and numerically for complex multicomponent systems in chapter 7. Here, we use only simple equations or graphs to relate y and x.

One of the most frequently used VLE relationship in binary systems is an equation that relates y to x through a *relative volatility* α.

$$y = \frac{\alpha x}{1 + (\alpha - 1)x} \tag{5-9}$$

The vapor composition y is a nonlinear function of the liquid composition x. This relationship is used in several of the problems at the end of the chapter. We will discuss relative volatilities in more detail in chapter 6.

5.5 EQUIMOLAL OVERFLOW

For distillation columns, the liquid and vapor flow rates up and down the column on all stages are often approximately constant if we use molar flow rates. This is because molar heats of vaporization (Btu/lb-mole or Joules/kg-mole) vary only slightly from component to component, particularly if the components are similar in nature. Thus, while the latent heats of vaporization of methanol and water are 474 and 970 Btu/lb_m, respectively, a very large difference, because their molecular weights are 32 and 18, the latent heats of vaporization on a molar basis are 15,170 and 17,500 Btu/lb-mole, respectively, not a big difference at all.

If components have fairly similar heats of vaporization, then every time a mole of vapor condenses a mole of liquid vaporizes. This "equimolal overflow" situation permits considerable simplification of the equations describing a cascade of equilibrium stages. We have

$$\begin{aligned} L_n &= L_{n+1} = L_{n+2} = \ldots = L \\ V_n &= V_{n+1} = V_{n+2} = \ldots = V \end{aligned} \tag{5-10}$$

The molar flow rates of liquid and vapor are denoted by L and V, respectively. We will assume in this chapter that they are the same on all trays. In chapter 13 we relax this assumption and use energy balances on each tray to calculate the different liquid and vapor rates on the trays.

5.6 STRIPPING COLUMN

One of the most common examples of a simple cascade of equilibrium stages is a stripping column, shown in Figure 5–10. Light material in the liquid feed is stripped out of the liquid as it moves down the column. The vapor that does the stripping is produced by boiling some of the liquid in the base of the column, using some hotter material like steam. The figure shows a four-tray stripper with a "partial reboiler" in the base of the column. The objective is to have less of the lighter component in the bottoms liquid product than was in the original feed.

The feed flow rate is F (for example, kg-mole/hr). The feed composition is z (mole fraction light component). Throughout this book we will deal only with binary distillation (two components). Therefore, we need to know only one mole fraction to define the compositions completely.

The vapor rate through the column and out the top is V (kg-mole/hr). The composition of the vapor leaving the top tray is y_4 since it is in equilibrium with the liquid on tray 4, whose composition is x_4. The flow rate of the bottoms product is B (kg-mole/hr), and its composition is x_B (mole fraction light component). If the stripper is doing its job, x_B must be less than z, i.e., there will be less light component in the bottoms product than in the feed. The liquid flow rate on each of the trays in the stripper is equal to the liquid feed rate ($L = F$).

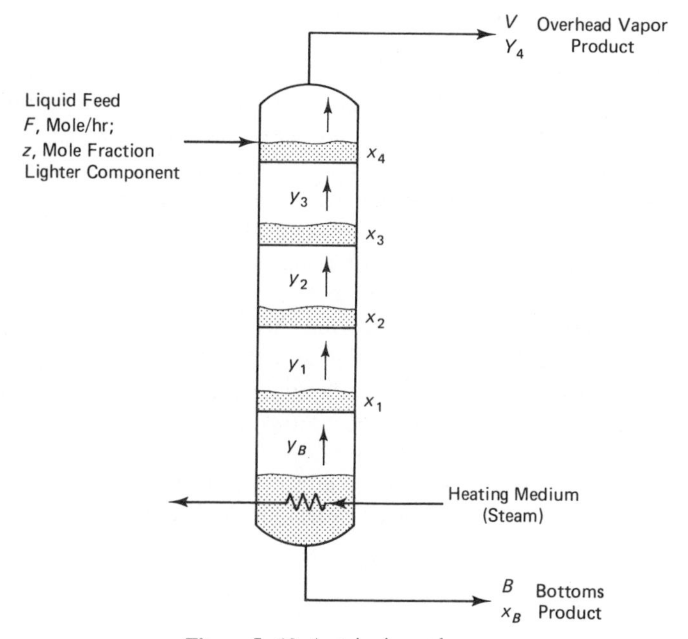

Figure 5–10 A stripping column.

In this chapter, we will use simple VLE relationships. More realistic VLE relationships will be developed in chapters 6 and 7. The simplest VLE equation is $y = Kx$, where the K value for the light component is constant.

5.6.1 Equations

The following equations hold for the stripping column:

Base of column

$$\text{Light component balance: } Lx_1 = Vy_B + Bx_B \tag{5–11}$$

$$\text{VLE: } y_B = Kx_B \tag{5–12}$$

Tray 1

$$\text{Light component balance: } Lx_2 = Vy_1 + \mathrm{B}x_B \tag{5–13}$$

$$\text{VLE: } y_1 = Kx_1 \tag{5–14}$$

Tray n

$$\text{Light component balance: } Lx_{n+1} = Vy_n + Bx_B \tag{5–15}$$

$$\text{VLE: } y_n = Kx_n \tag{5–16}$$

Notice that equation 5–15 can be rearranged to give

$$y_n = \left|\frac{L}{V}\right| x_{n+1} + \left|\frac{-Bx_B}{V}\right| \tag{5–17}$$

This is the equation of a straight line ($y = mx + b$) which, in chemical engineering jargon, is called the *operating line*. Operating line equations are simply component balances. Drawing an operating line on a graph in an x-y coordinate system will give a straight line with slope L/V, the liquid-to-vapor flow rate ratio in the column. When we discuss more complex columns in chapter 8, we will find that we will have several operating lines, one for each section of the column. The slopes of these operating lines are *always* the liquid-to-vapor ratios in the different sections.

Figure 5–11 gives an x-y plot showing three straight lines: the operating line (OL), the equilibrium line (VLE), and the 45° line ($y = x$). This last line turns out to be useful in graphically analyzing the stripping column. To see why, let us solve for the point of intersection of the operating line and the 45° line.

Let x_{int} equal the values of x and y at the point of intersection of the OL line and the 45° line. Substituting x_{int} into equation 5–17 gives

$$x_{int} = \frac{L}{V} x_{int} - \frac{Bx_B}{V}$$
$$x_{int} = \frac{Bx_B}{L - V} \tag{5–18}$$

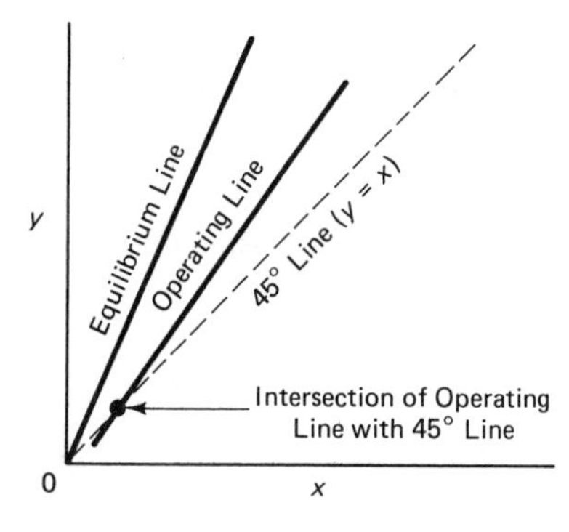

Figure 5–11 Operating line, equilibrium line, and 45° reference line on xy coordinates.

Now an overall total mass (or molar) balance around the stripper gives

$$F = V + B = L \tag{5–19}$$

Substituting for $(L - V)$ in Equation 5–18 yields

$$x_{\text{int}} = \frac{Bx_B}{B} = x_B \tag{5–20}$$

Thus, the operating line and the 45° line intersect at the composition of the bottoms product x_B! This is a very important and useful finding, because it permits us to draw the operating line quickly. We simply start at x_B on the 45° line and draw a straight line with a slope equal to the liquid/vapor ratio in the column.

Now we can graphically step our way up from the bottom of the column, alternately using the OL line and the VLE line. In effect, we are doing "tray-to-tray" calculations graphically.

Certainly, we can solve these OL and VLE equations directly—sometimes analytically, but usually numerically. Indeed, we will do a numerical solution in section 5.9 using the computer. However, at this stage it is vital to first learn to do these tray-to-tray calculations graphically. The insight it will give you into the effects of design and operating variables is invaluable.

So get out your graph paper, a long straightedge, and a French curve and roll up your sleeves. The only way you will learn this material is to do it. The problems at the end of this chapter should help you. If you understand the material in this chapter, you will find the material in chapter 8 very easy since the same idea of applying component balances (OL) and vapor-liquid equilibrium (VLE) is used. The columns, however, are a little more realistic and more complex than the simple single-section cascades considered here.

5.6.2 Design Procedure

The verb "to design" in cascade processes means to determine the number of stages required to make a given separation. Typically, the feed flow rate and composition are given. We then establish the operating pressure (usually as set by cooling-water temperatures), and this sets the VLE relationships. The desired product compositions are also specified. The design problem then is to determine N_T, the total number of trays in the column.

The design procedure for the stripping column is as follows:

1. Draw the VLE line.
2. Draw the OL line. This is a straight line starting at x_B on the 45° line and having a slope equal to L/V.
3. Draw a vertical line up from x_B until it intersects the VLE line. The value of y at this point is y_B, since x_B and y_B are in phase equilibrium, i.e., they must lie on the VLE line.
4. Draw a horizontal line toward the right from y_B until it intersects the OL line. The intersection is a solution of equation 5–11 graphically for x_1. Both x_1 and y_B lie on the OL line as given in equation 5–17, since equation 5–11 is just a specific case of that equation.
5. Draw another vertical line at x_1 up to the VLE line to get y_1.
6. Draw another horizontal line at y_1 over to the OL line to get x_2.

This procedure of stepping up the column is continued, alternately using the VLE line and the OL line, until the feed composition z is reached on the OL line. The number of steps required to go from x_B to z is equal to the number of theoretical stages that must be built to achieve the specified separation at the specified L/V ratio. The total number of stages includes N_T theoretical trays and one step for the partial reboiler.

If we change the L/V ratio, the OL line will shift, and a different number of trays will be required to make the same separation. If we change the operating pressure of the column, the VLE line may shift and require a different number of trays. Thus, we can see clearly the effects of many different parameters on the pictures that we get from the graphs. To misquote Confucius, a picture is worth a thousand numbers!

Example 5.3.

Hydrogen is stripped from a liquid binary mixture of 2.5 mole percent H_2 and 97.5 mole percent carbon monoxide by feeding 55 kg-mole/hr of liquid to the top plate of a stripping column in a cryogenic (low-temperature) "cold box." Vapor is generated in the base of the column by an internal heating coil. Liquid and vapor rates can be assumed to be the same on all trays. The K-value for H_2 out of CO under the low-temperature and high-pressure conditions in the stripper is 4, i.e.,

$$y_{\text{hydrogen}} = 4x_{\text{hydrogen}}$$

How many theoretical plates are required to achieve a purity of 0.25 mole percent H_2 in the liquid bottoms product when the vapor boilup rate is 20 kg-mole/hr? What is the composition of the overhead vapor product from the stripper?

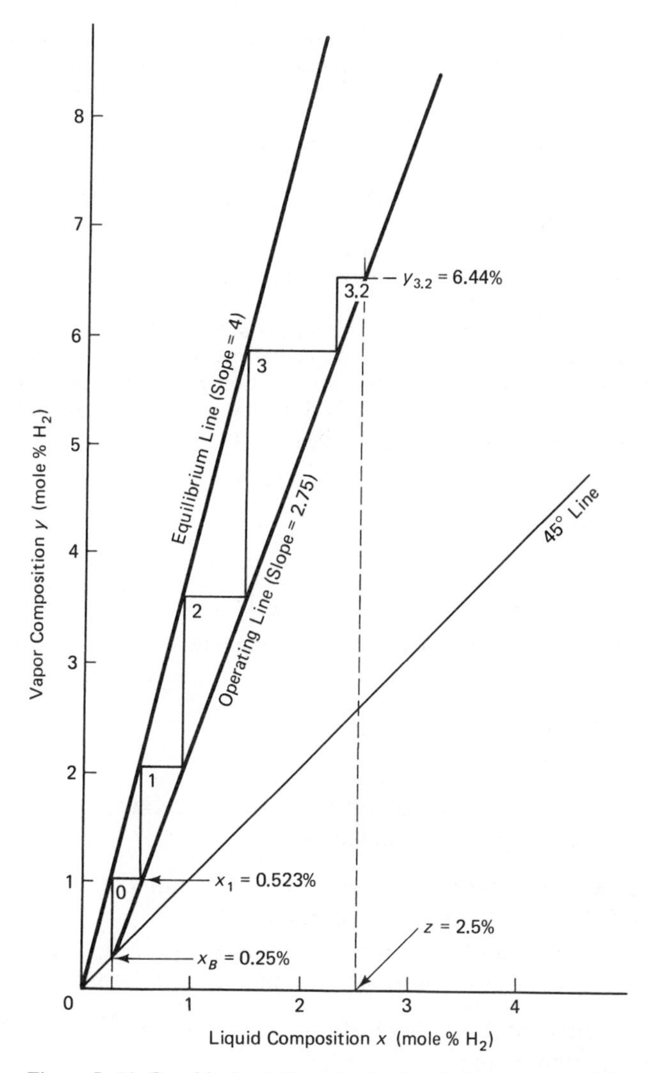

Figure 5–12 Graphical solution of stripping design (example 5.3).

Solution:

First we draw the VLE line, starting at the (0,0) point on an x-y diagram (see Figure 5–12), with a slope of 4.

Next we draw the operating line. Its slope is equal to the liquid-to-vapor ratio in the stripper:

$$\frac{L}{V} = \frac{55}{20} = 2.75$$

The OL line starts on the 45° line at $x_B = 0.25$ mole percent.

Then we step up the column. The first step (labeled "0" in the figure) represents the partial reboiler in the base of the column. If x_B equals 0.25 mole percent, y_B equals 1.0 mole percent (on the VLE line). If y_B is 1.0 mole percent, x_1 is 0.523 mole percent (on the OL line). Then y_1 is 2.09 mole percent (on the VLE line).

A total of 4.2 steps (3.2 trays plus the reboiler) is required to reach the feed composition (z = 2.5 mole percent H_2). In general, the number of trays required will not be an integer. Of course, in a real column we can only build an integer number of trays. But this need not be a problem, because we always put in more trays than the theoretical number. Since real trays seldom achieve phase equilibrium, we have to use an efficiency number to relate them to theoretical trays. This will be discussed further in chapter 8.

The composition of the overhead vapor product can be read off the graph as 6.44 mole percent H_2. A horizontal line at the overhead composition always intersects the operating line at the feed composition z in a stripping column. You can prove this by substituting z into the OL line equation (equation 5–17) and comparing the result with a component balance around the entire column.

$$Bx_B + Vy_{NT} = Fz = Lz \tag{5–21}$$

Rearranging gives

$$y_{NT} = \frac{L}{V}z + \frac{(-Bx_B)}{V} \tag{5–22}$$

Thus, the point (z, y_{NT}) must lie on the operating line.

5.6.3 Effects of Design and Operating Parameters

Anything that makes the equilibrium and operating lines move *closer* together results in *more* stages. Included are

1. Smaller K, representing more difficult separation
2. Bigger L/V ratio, representing less vapor boilup
3. Smaller x_B, representing higher bottoms purity (less light component)
4. Larger z, representing more light component in the feed

We will discuss more of these effects in chapter 8 when we look at a complete distillation column.

5.6.4 Limiting Conditions

There are limits to the concentrations that can be achieved in any stripping column. These limits occur when the number of stages required to make the separation becomes extremely large. The absolute limit is when the number of stages goes to infinity.

As the vapor flow rate V in the stripper is reduced, the OL line slope gets bigger, and the OL line moves toward the VLE line. This means that the number of stages increases. A limit occurs when the operating line intersects the equilibrium line at the feed composition z as shown in Figure 5–13. At

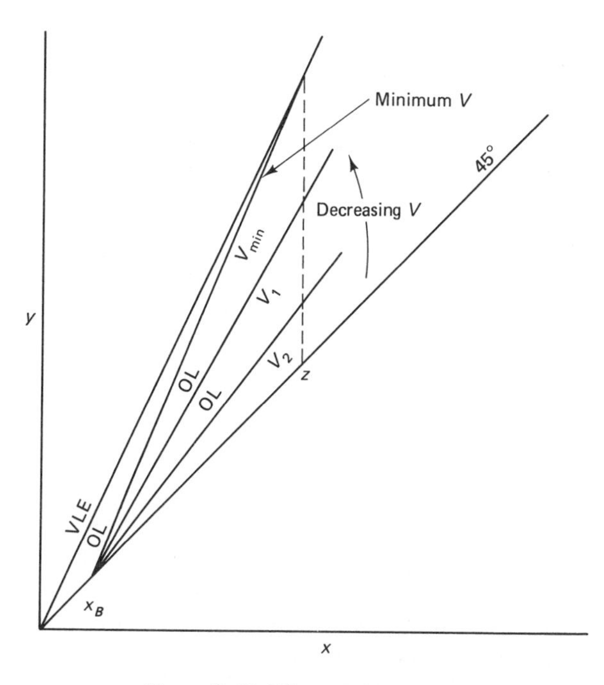

Figure 5–13 Effect of changing V.

this minimum vapor rate, it would take an infinite number of stages to get to z because the operating and VLE lines converge at that point.

In a complete distillation column, we will find a similar limiting condition called the *minimum reflux ratio*. The concept is exactly the same as in the stripper. (See chapter 8.) An infinite number of trays are required at this reflux ratio.

5.7 ABSORBER

Another commonly encountered simple cascade process is a gas absorber. Some of the components in the gas that is fed into the bottom of the column are absorbed by the liquid stream that is fed into the top of the column. Figure 5–14 shows an absorber with the nomenclature that we will use.

The feed gas flow rate is V. We assume equimolal overflow, so the vapor flow rate throughout the entire column and overhead is V. The equimolal overflow assumption is not very good in most practical absorbers, but we want to keep things as simple as possible for the moment.

The composition of the feed gas is called y_0 (mole fraction of the component that is being absorbed). This is usually the heaviest component in the gas. The lightest components go through the column without being absorbed very much. For example, we could absorb benzene out of an air/benzene gas

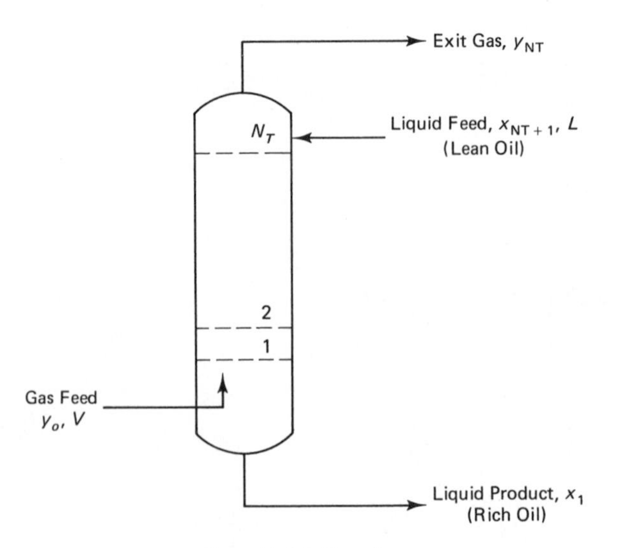

Figure 5–14 Absorber.

mixture using a heavy-oil liquid stream. The benzene will be absorbed in the oil, and very little air will be absorbed.

Since the heavy benzene component in the air is being absorbed, its K-value in the absorber is usually less than unity, and the VLE line lies below the operating line. Just the reverse is true in the stripping column, where the lightest component is being removed. The K-value here is greater than unity, and the operating line lies below the VLE line.

A component balance can be written around any section of the column, yielding an operating line equation for the absorber. Assuming constant L and V flow rates and cutting the column above tray n, as shown in Figure 5–15, we obtain

$$Vy_0 + Lx_{n+1} = Vy_n + Lx_1 \tag{5–23}$$

Rearranging to put this equation in the slope-intercept form of a straight line, we have

$$y_n = (L/V)\, x_{n+1} + (y_0 - Lx_1/V) \tag{5–24}$$

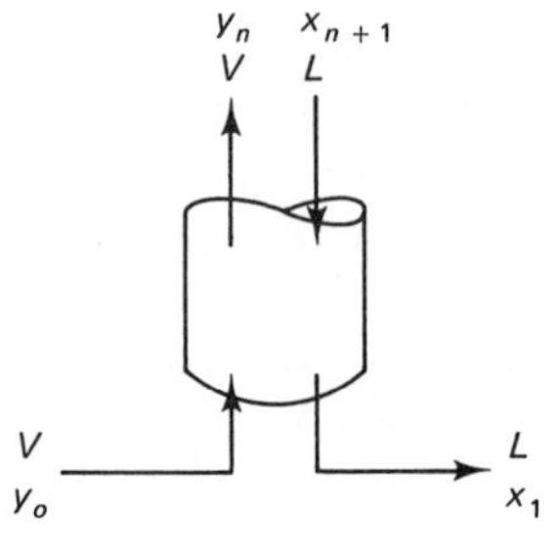

Figure 5–15 Bottom section of an absorber.

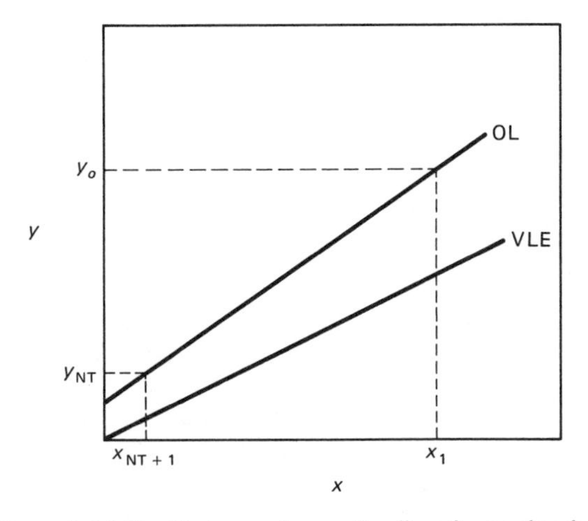

Figure 5–16 Equilibrium and operating lines for an absorber.

Note that the slope of the operating line is the ratio of the liquid-to-vapor flow rates in the column, just as it is in the stripper example.

This operating line can be plotted on an x-y diagram together with the VLE line, and stages can be stepped off by moving back and forth between the VLE and operating lines. The operating line must go through the points that correspond to the concentrations of streams that *pass* each other in the column. The liquid stream dropping down from tray $(n + 1)$ with composition x_{n+1} passes the vapor stream from tray n, which has composition y_n. The operating line equation (equation 5–24) relates the compositions of these passing streams.

Remember that the passing vapor and liquid streams are *not* in phase equilibrium. However, the vapor and liquid streams leaving the *same* tray (x_n and y_n) are in phase equilibrium. The streams considered on the operating line are the liquid from tray $n + 1$ and the vapor from tray n.

The very top and the very bottom of the column have streams that also pass each other. At the top, the streams are the exit gas y_{NT} and the liquid feed x_{NT+1}. At the bottom, the passing streams are the feed gas y_0 and the liquid product x_1. Therefore, the easy way to draw the operating line is to locate the points (x_{NT+1}, y_{NT}) and (x_1, y_0). A straight line connecting these two points is the operating line and has a slope of L/V. (See Figure 5–16.)

Example 5.4.

The absorber shown in Figure 5–14 is used to recover benzene from air. Gas feed flow rate to the absorber is 100 kg-mole/hr. The feed gas contains 10 mole percent benzene. The liquid heavy oil fed to the top tray of the absorber contains no benzene. Its flow rate is 80 kg-mole/hr. The VLE relationship for benzene in the absorber is $y_{\text{benzene}} = 0.5\, x_{\text{benzene}}$. Equimolal overflow can be assumed. It is desired

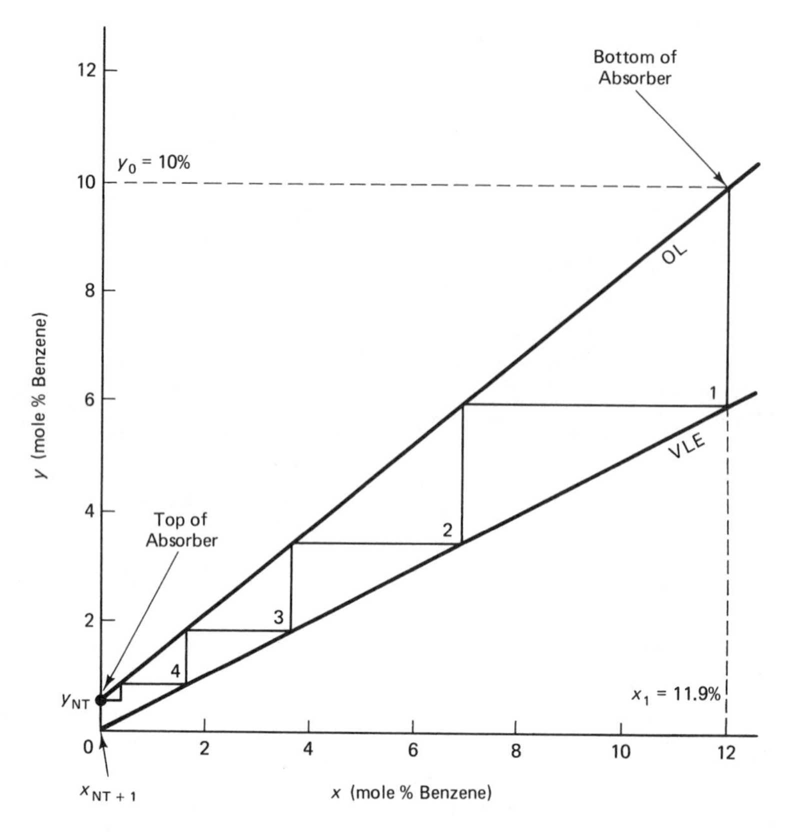

Figure 5–17 Graphical solution of example 5.4.

to attain an exit gas composition of 0.5 mole percent benzene at the top of the absorber. What is the composition of the liquid leaving the bottom of the absorber, and how many equilibrium stages are required?

Solution:

Figure 5–17 gives the graphical solution. First, the VLE line is drawn from the origin with a slope of 0.5. Next, the operating line is drawn from the *x*-*y* values that we know exist at the top of the column, viz.,

$$x_{NT+1} = 0 \qquad y_{NT} = 0.5 \text{ mole } \%$$

Then the OL line is drawn from this point with a slope equal to the L/V ratio in the column, i.e.,

$$\frac{L}{V} = \frac{80}{100} = 0.8$$

Notice that the OL line intersects a horizontal line through the feed composition

(y_0 = 10 mole percent) at a value of x equal to 11.9 mole percent. This x must be x_1, the composition of the liquid from the bottom of the column.

Let us check our graphical solution by doing an overall benzene component balance around the system. We have

Benzene in

$$(100 \text{ kg-mole/hr feed})(0.10 \text{ m.f. benzene}) + (80 \text{ kg-mole/hr oil in})(0 \text{ m.f. benzene}) = 10 \text{ kg-mole/hr}$$

Benzene out

$$(100 \text{ kg-mole/hr exit gas})(0.005 \text{ m.f. benzene}) + (80 \text{ kg-mole/hr oil out})(x_1) = 0.5 + 80\, x_1$$

Equating benzene in to benzene out gives x_1 = 11.88 mole percent, which is the same value we obtained graphically.

Stepping up the column from the point (x_1, y_0) on the OL line to the point (x_{NT+1}, y_{NT}) requires about four and one-half steps.

5.8 KREMSER EQUATION

If the ratio VK/L is constant in a series cascade of N equilibrium stages, the compositions of the feed and product streams can be analytically related by the Kremser Equation,

$$\frac{x_{N+1} - x_0}{x_1 - x_0} = \frac{1 - E^{N+1}}{1 - E} \tag{5-25}$$

where

$$E = \frac{VK}{L} \tag{5-26}$$

$$x_0 = \frac{y_0}{K} \tag{5-27}$$

Note that E cannot be equal to 1. (Problem 5–8 requests a derivation of the Kremser Equation for the case where E is 1; see also the discussion below.)

The Kremser Equation permits you to solve the tray-to-tray calculation problem analytically if the operating line and the equilibrium line are both straight. It does not give you the same insight as the graphical solution, but it is useful when very small concentrations and a large number of trays are involved. The graphical method can get tedious when you have to make a large number of steps. Perhaps the most common use of the Kremser Equation is in absorption and liquid-liquid extraction calculations.

The derivation of the Kremser Equation is as follows.

Component balance on stage 1

$$Lx_2 + Vy_0 = Lx_1 + Vy_1$$

$$L(x_2 - x_1) = VKx_1 - VKx_0$$

$$x_2 - x_1 = \frac{VK}{L}(x_1 - x_0) = E(x_1 - x_0) \tag{5-28}$$

Component balance on stage 2

$$Lx_3 + Vy_1 = Lx_2 + Vy_2$$

$$x_3 - x_2 = E(x_2 - x_1) = E^2(x_1 - x_0) \tag{5-29}$$

For subsequent stages,

$$x_4 - x_3 = E^3(x_1 - x_0)$$

$$\vdots$$

$$x_{N+1} - x_N = E^N(x_1 - x_0) \tag{5-30}$$

Adding all these equations together gives

$$(x_{N+1} - x_N) + (x_N - x_{N-1}) + \ldots + (x_3 - x_2) + (x_2 - x_1)$$

$$= x_{N+1} - x_1 = (E^N + E^{N-1} + \ldots + E^2 + E)(x_1 - x_0)$$

Adding $(x_1 - x_0)$ to both sides yields

$$x_{N+1} - x_0 = (1 + E + E^2 + E^3 + \ldots + E^N)(x_1 - x_0) \tag{5-31}$$

But the Nth-order power series in E can be expressed in closed form,

$$\frac{(1 - E^{N+1})}{(1 - E)}$$

Dividing both sides by $(x_1 - x_0)$ gives the Kremser Equation.

Example 5.5.

Solve example 5.4 using the Kremser Equation.

Solution:

$$E = \frac{VK}{L} = \frac{(100)(0.5)}{80} = 0.625$$

$$x_0 = \frac{y_0}{K} = \frac{0.10}{0.5} = 0.22$$

Plugging these values into the Kremser Equation gives

$$\frac{x_{N+1} - x_0}{x_1 - x_0} = \frac{1 - E^{N+1}}{1 - E}$$

$$\frac{0 - 0.20}{0.1188 - 0.20} = \frac{1 - (0.625)^{N+1}}{1 - 0.625}$$

Solving for N gives $N = 4.47$, which is the same answer as obtained graphically.

5.9 COMPUTER SOLUTIONS

The equations describing the sequence of equilibrium stages are very conveniently solved on a digital computer. A sample FORTRAN program for the design of a stripping column is given in Table 5–1, with numerical results

TABLE 5–1
FORTRAN STRIPPING COLUMN DESIGN PROGRAM

```
C**************************************************
C    STRIPPING COLUMN DESIGN PROGRAM
C***************************************************
C
C GIVEN:
C      LIQUID FEED FLOWRATE = F (KG-MOLE/HR)
C      FEED COMPOSITION = Z (MOLE FRACTION LIGHT COMPONENT)
C      BOTTOMS COMPOSITION = XB (MOLE FRACTION LIGHT COMPONENT)
C      VAPOR BOILUP FLOWRATE = V (KG-MOLE/HR)
C
C CALCULATE:
C      OVERHEAD VAPOR COMPOSITION = Y(NT)
C      TOTAL NUMBER OF TRAYS = NT
C
C NOTES:
C      (1) A PARTIAL REBOILER IS ASSUMED, SO THE TOTAL NUMBER OF
C          EQUILIBRIUM STAGES IS NT + 1.
C      (2) A SIMPLE VLE RELATIONSHIP Y=KX IS USED IN THIS PROGRAM.
C      (3) OTHER ASSUMPTIONS INCLUDE EQUIMOLAL OVERFLOW, EQUILIBRIUM
C      STAGES AND BINARY SYSTEM.
C
C*********************************
C MAIN PROGRAM
C*********************************

      DIMENSION X(100),Y(100)
      REAL L,K
      DATA F,Z,XB/55.,0.025,0.0025/
      WRITE(6,1)F,Z,XB
    1 FORMAT( ' F = ',F6.1,'   Z =',F6.4,'     XB =',F6.4)
      WRITE(6,6)
      V=20.
C LOOP TO CHANGE VAPOR FLOWRATE
      WRITE(6,3)
    3 FORMAT(7X,'  V     K       NT')
      DO 100 LV=1,6
      WRITE(6,6)
    6 FORMAT(/)
      L=F
      B=F-V
      K=4.
C LOOP TO CHANGE K-VALUE
      DO 50 LK=1,4
```

TABLE 5-1 CON'T.

```
C CALCULATE YB FROM XB
      CALL VLE(XB,YB,K)
C COMPONENT BALANCE ON BASE
      X(1)=(V*YB+B*XB)/L
      LOOP=0
      N=1
C STEP UP TRAYS
   10 LOOP=LOOP+1
C TEST TO SEE IF N IS GREATER THAN 100
      IF(LOOP.GT.99)THEN
      WRITE(6,2)
    2 FORMAT(1X,'TRAYS EXCEED 100')
      STOP
      ENDIF
      CALL VLE(X(N),Y(N),K)
      X(N+1)=(V*Y(N)+B*XB)/L
C TEST TO SEE IF X(N+1) IS GREATER THAN Z
      IF(X(N+1).GT.Z)GO TO 20
      N=N+1
      GO TO 10
   20 NT=N
      WRITE(6,5)V,K,NT
    5 FORMAT(3X,2F7.3,3X,I3)
   50 K=K-0.3
  100 V=V-1.5
      STOP
      END
C******************
C VLE SUBROUTINE
C******************
      SUBROUTINE VLE(X,Y,K)
      REAL K
      Y=K*X
      RETURN
      END
```

TABLE 5-2
RESULTS OF STRIPPING COLUMN DESIGN PROGRAM

```
F =   55.0    Z =  .0250     XB =  .0025

         V       K        NT

    20.000  4.000      4
    20.000  3.700      4
    20.000  3.400      5
    20.000  3.100      7

    18.500  4.000      4
    18.500  3.700      5
    18.500  3.400      7
    18.500  3.100     10

    17.000  4.000      5
    17.000  3.700      6
    17.000  3.400      9
    17.000  3.100     20

    15.500  4.000      7
    15.500  3.700      9
    15.500  3.400     19
TRAYS EXCEED 100
```

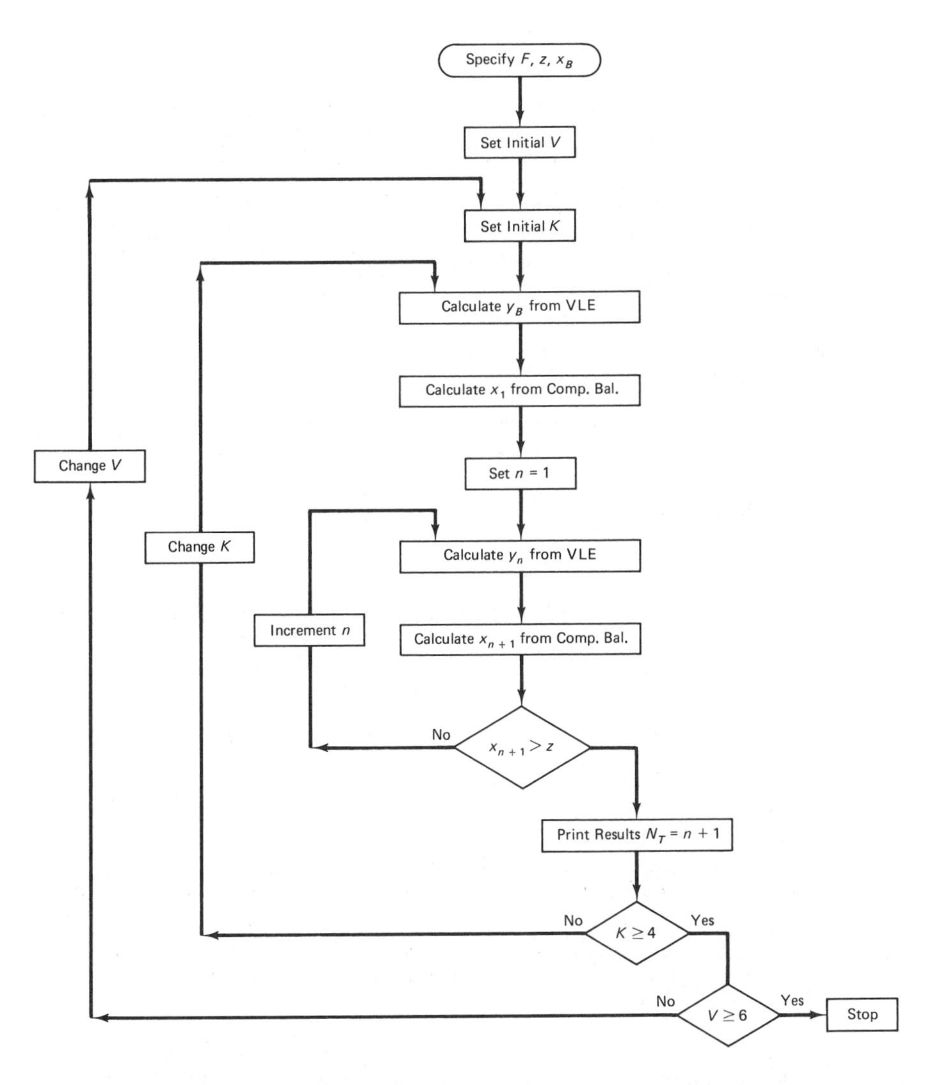

Figure 5–18 Computer flowchart for stripping column design program.

shown in Table 5–2. The results include data for the stripper designed graphically in example 5.3. Other designs are also given using different K-values and different vapor flow rates. Note the effect of these parameters on the number of trays required.

A subroutine is used to handle the VLE bubblepoint calculations. A known value of liquid composition x is fed into the subroutine along with the K-value. The subroutine returns the calculated value of vapor composition y that is in equilibrium with the liquid.

Figure 5–18 gives a flowchart for the program.

PROBLEMS

5–1. A sieve plate in a distillation column has a vapor stream of 500 lb-mole/hr entering it with composition of 45 mole percent methanol and 55 mole percent water. The liquid stream flowing down onto this plate flows at 700 lb-mole/hr and has a composition of 50 mole percent methanol and 50 mole percent water. The distribution coefficient for methanol (K-value) is 2.2. The liquid and vapor streams leaving the plate have the same molar flow rates as those approaching the plate. Calculate the composition of the streams leaving the plate, assuming that equilibrium occurs on the plate (i.e., the plate is an ideal stage).

5–2. Defining the relative volatility of A to B, α_{AB}, as the distribution coefficient of A divided by that of B, draw the x-y equilibrium diagram for a binary system in which $\alpha_{AB} = 2$. *Note*: The diagram will show x_A plotted against y_A (y_A on the ordinate).

5–3. Off-gases from the wedge roaster of a zinc plant come from the oxidation of ZnS with 100 percent excess air. The reaction

$$\text{ZnS} + \text{O}_2 \rightarrow \text{Zn} + \text{SO}_2$$

goes to completion. The gases are then passed over a V_2O_5 catalyst where the reaction

$$\text{SO}_2 + \frac{1}{2}\,\text{O}_2 \rightarrow \text{SO}_3$$

takes place. Several stages of catalyst are used with intercooling, so that 98 percent of the SO_2 reacts to form SO_3.

(a) What are the compositions (mole percent) of the gaseous components after reaction to form SO_3?

This gas is then passed through an absorption column countercurrent to a stream of sulfuric acid. 99.9 percent of the SO_3 is absorbed into the H_2SO_4 solution. The K_{SO_3} is 0.3, defined as the mole fraction of SO_3 in the gas phase divided by the mole fraction of SO_3 in the acid phase assuming that no chemical reactions occur. Assume equimolal overflow throughout the column. (Acid evaporates into the gas phase.)

(b) What is the minimum flow rate of acid entering the top of the column?

(c) If the acid flow rate is set at 1.5 times the minimum, determine the number of ideal stages required for the separation specified and the composition of the unabsorbed gas.

5–4. It is proposed to separate acetic acid from the water in which it is produced by countercurrent extraction with pure di-n-butyl ether. The distribution coefficient D for the acid between the two phases is 5.0 on a mass ratio basis. (That is, at equilibrium, the mass ratio of acetic acid to di-n-butyl ether in the ether layer is five times the mass ratio of acetic acid to water in the water layer.) The original feed contains acid at a mole ratio of 0.2, and it is desired to remove 95 percent of the acid into the ether solvent. A three-stage mixer-settler countercurrent extraction system is available for this process.

(a) Is it possible to remove 95 percent of the acid from the water stream?
(b) If so, what is the minimum flow ratio of ether to dilute acid feed that must be used?
(c) Using this minimum ether flow, what is the final concentration of ether-acetic acid solution?

List the assumptions you made in carrying out this calculation.

5–5. Ethane (C_2H_6) in a refinery gas stream is recovered by absorbing it in a liquid, heavy-oil stream. A five-tray gas/liquid countercurrent absorber column is used. The exit gas stream from the absorber contains 3 mole percent C_2H_6. 50 moles of liquid, containing no C_2H_6, are fed into the top of the absorber for every 100 moles of gas fed to the bottom of the absorber.

The K-value for ethane under the conditions of temperature and pressure in the absorber is 0.5 mole percent C_2H_6 in gas/mole percent C_2H_6 in the liquid. Liquid and vapor rates are constant throughout the column.

Calculate the concentration of ethane in (a) the gas stream fed to the absorber and (b) the liquid stream leaving the first tray at the bottom of the column.

5–6. A stripping column is used to remove carbon dioxide (CO_2) from methanol (CH_3OH). 100 kg-mole/minute of liquid feed are introduced on the top tray of the column. Vapor is generated in a partial reboiler at the base of the column. Liquid and vapor rates are constant throughout the column. The K value for CO_2 out of CH_3OH under the operating condition in the column is 2. There are three theoretical trays plus the reboiler. Use graphical solution techniques to answer the following questions.

(a) If the liquid bottoms product from the column is 1 mole percent CO_2 and the vapor boilup is 59.8 kg-mole/minute, what are the feed and overhead vapor compositions?
(b) If the vapor boilup is increased to 100 kg-mole/minute and the feed composition is the same as in part (a), what is the overhead vapor composition and what is the composition of liquid in the column base?

5–7. A stripping column is used to concentrate isobutane (iC_4) from a 45 mole percent mixture of iC_4 and normal butane (nC_4). 2,200 kg-mole/hr of liquid feed is introduced onto the top tray of the column. Vapor is generated in the partial reboiler and goes up the column and out the top as overhead product. Liquid and vapor rates are constant throughout the column. The relative volatility between iC_4 and nC_4 is 2 at the temperatures and pressure in the column.

A liquid bottoms product containing 5 mole percent iC_4 is desired. An overhead vapor product containing 60 mole percent iC_4 is desired.

How many trays are required, and how much vapor boilup is required?

5–8. Derive the Kremser Equation for the case when $E = 1$.

5–9. You have been given the job of designing a stripping column to separate a binary mixture of 10 mole percent propane (C_3H_8) and 90 mole percent hexane (C_6H_{14}). The feed rate is 100 kg-mole/hr of liquid. Bottom liquid product from the stripper is 1.0 mole percent C_3H_8. A partial reboiler at the base of the stripper generates 37.5 kg-mole/hr of vapor.

Several trays up in the stripper, another "intermediate" reboiler vaporizes an additional 37.5 kg-mole/hr of vapor, using less expensive low-pressure steam. This reboiler is located on a tray where the liquid composition is approximately 5 mole percent propane.

Assume equimolal overflow. The K value of propane is constant at 2 over the composition range of interest.

How many theoretical trays are required in the stripper, and on what tray should the intermediate reboiler be located? What is the composition of the vapor leaving the top of the stripper?

5–10. A stripping column is used to separate a binary liquid mixture of 11.7 mole percent propylene oxide (PO) and 88.3 mole percent acetone (Ac) into a bottoms product with 2 mole percent PO and a vapor overhead product with 15.3 mole percent PO.

The stripper has a partial reboiler at the base. It also has a cooler located several trays up in the column. This heat exchanger removes heat and condenses some of the vapor coming up the column. The ratio of the vapor rate below the cooler to the vapor rate above the cooler is 3. The cooler is located at a tray where the liquid composition is approximately 8 mole percent PO.

Assume equimolal overflow and a K value of PO out of acetone of 1.5.

(a) What are the product flow rates per 100 moles of feed?
(b) On the same basis, what are the vapor and liquid rates in the two sections of the column?
(c) How many trays are required in the stripper?
(d) On what tray is the cooler located?

5–11. A stripping column is used to remove some light component from a binary mixture. Feed to the stripper is 10 mole percent light component. Assume equimolal overflow, theoretical trays, a partial reboiler, and a constant K value.

What is the maximum possible recovery of heavy component (moles of heavy in the bottoms product from the stripper per mole of heavy in feed) if the bottoms concentration is 1 mole percent light and

(a) $K = 2$.
(b) $K = 4$.

5–12. The hydrogen used in ammonia synthesis may be prepared by separating a CO-H_2 stream, removing CO by countercurrent contact with liquid N_2. The nearly pure H_2 is then mixed with gaseous N_2 to make the N_2:$3H_2$ mix fed to the reactor.

In one application, a 90 mole percent H_2, 10 mole percent CO gas stream enters the absorption column countercurrent to pure liquid N_2. The exit gas from the column is 1 mole percent CO. Assuming constant molar flows of vapor and liquid streams and a K-value of CO of 1.25, determine

(a) The minimum flow rate of liquid N_2 in moles per mole of feed gas required to allow the specified separation.

(b) The number of stages required for this separation if 1.3 times the minimum flow of liquid N_2 is used.

5–13. A process uses an absorber column and a stripper to recover tetrahydrofuran (THF) from a vapor stream. 100 kg-mole/hr of vapor, containing 15 mole percent THF, is fed into the bottom of an absorber. The liquid fed to the top of the absorber contains 1 mole percent THF, the rest being water. The gas leaving the absorber has a composition of 1 mole percent THF. The liquid leaving the bottom of the absorber contains 40 mole percent THF. Assume equimolal overflow in the absorber and a constant K-value for THF equal to 0.3 since the temperature and pressure in the absorber are low.

The 40 mole percent THF liquid from the base of the absorber is fed into a stripper which operates at higher pressure and temperature, giving a K-value for THF of 2. Vapor from the top of the stripper is 70 mole percent THF. Bottom product from the stripper is 1 mole percent THF. Assume equimolal overflow in the stripper.

The bottom liquid product from the stripper is mixed with a makeup liquid stream that is also 1 mole percent THF. This total liquid stream is fed to the top of the absorber.

Using graphical solution methods, find the following:

(a) Vapor flow rate leaving the top of the absorber.

(b) Vapor flow rate leaving the top of the stripper.

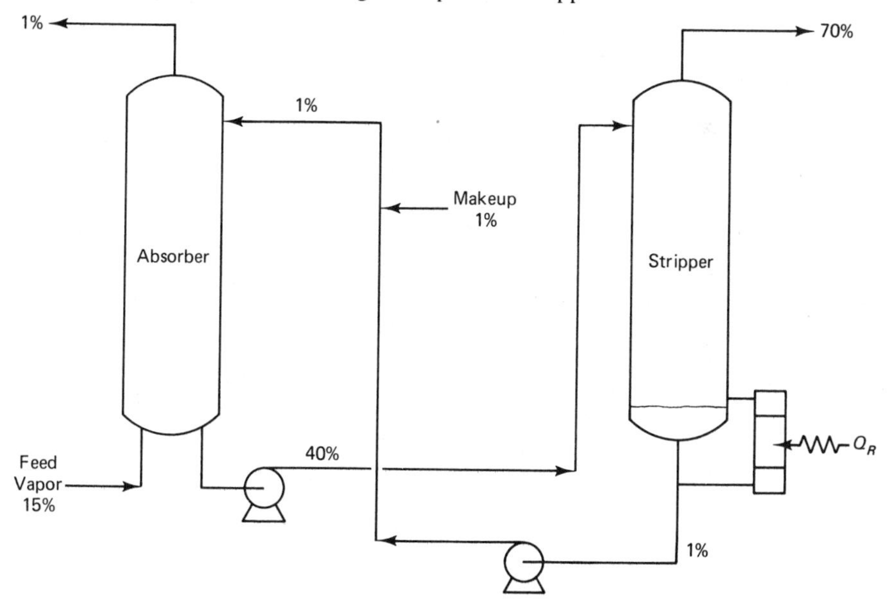

(c) Makeup liquid flow rate.
(d) Liquid flow rate in the absorber.
(e) Liquid flow rate in the stripper.
(f) Number of theoretical trays in the absorber.
(g) Number of theoretical trays in the stripper.

5–14. A three-tray column is used to concentrate a binary mixture of ethanol and water. Liquid feed, at a rate of 1,500 kg-mole/hr and composition of 13 mole percent ethanol, is fed to the top tray.

Instead of using a reboiler in the bottom of the column, "live steam" (pure water vapor) is fed in at a rate of 1,000 kg-mole/hr below the first tray.

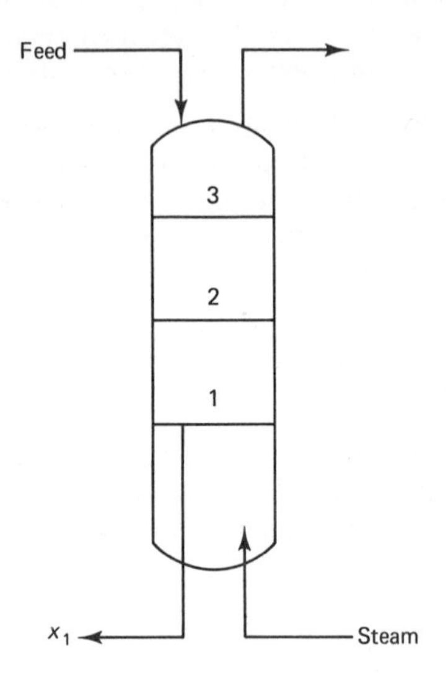

The liquid leaving the column from tray 1 has a composition of 2 mole percent ethanol. Assume equimolal overflow and theoretical trays.
(a) What is the composition of the vapor leaving the top of the column?
(b) Assuming a simple VLE relationship ($y = Kx$), what is the K-value in this system?

5–15. A countercurrent cascade receives a feed of gaseous NH_3 containing 0.1 mole of H_2O/100 moles of NH_3. It is desired to "purify" the NH_3 by removing a concentrated H_2O-NH_3 liquid from the bottom of a column taking the purified NH_3 gas off the top.

In this operation, it is expected that the K-value for water will be 0.05.
(a) If a product of 0.01 mole of H_2O/100 moles of NH_3 is desired, what is the minimum flow of waste NH_3 required in moles/mole of feed?
(b) If a liquid flow of twice the minimum is used, what is the concentration of waste NH_3?

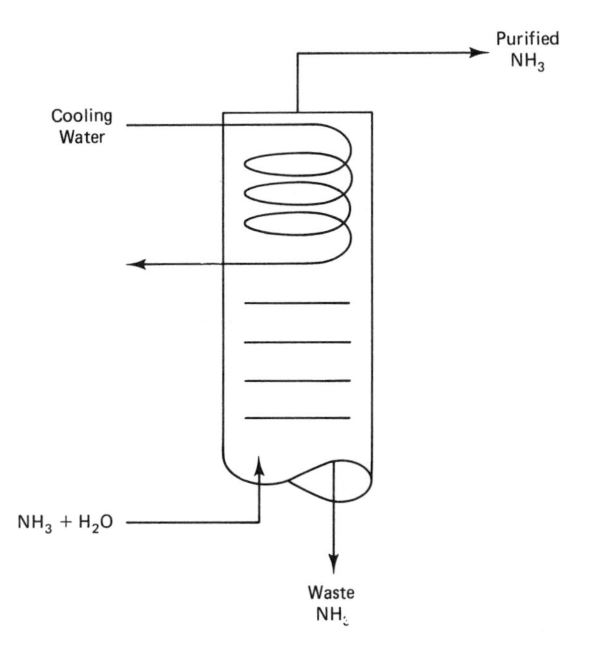

5–16. 300 kg-mole/hr of gas at 300 psig and 70°F is fed into a cryogenic separation system. The feed gas contains 80 mole percent methane (CH_4) and 20 mole percent propane. As shown in the figure on page 158, the feed gas is compressed and cooled with cooling water. Then it is used to reboil a stripping column and is further cooled in a product/feed heat exchanger. Finally, the cooled gas is dropped to 100 psig in an expander that drives the feed-gas compressor.

The stream leaving the expander is 50 percent liquid and 50 percent vapor. The two phases, which are in phase equilibrium, are separated in a flash drum. The VLE in this system can be described by a constant relative volatility of 10 for methane out of propane. The *x-y* curve for a relative volatility of 10 is given on page 159.

The liquid phase from the flash drum is fed into the top of a stripping column. The vapor from the top of the stripper is 92 mole percent methane. The liquid bottoms product from the partial reboiler in the base of the column is 5 mole percent methane. Equimolal overflow can be assumed.

(a) What are the compositions of the two phases leaving the flash drum?
(b) What is the vapor flow rate in the stripper?
(c) How many trays are required in the stripper?
(d) What are the flow rate and composition of the total gas stream leaving the process?

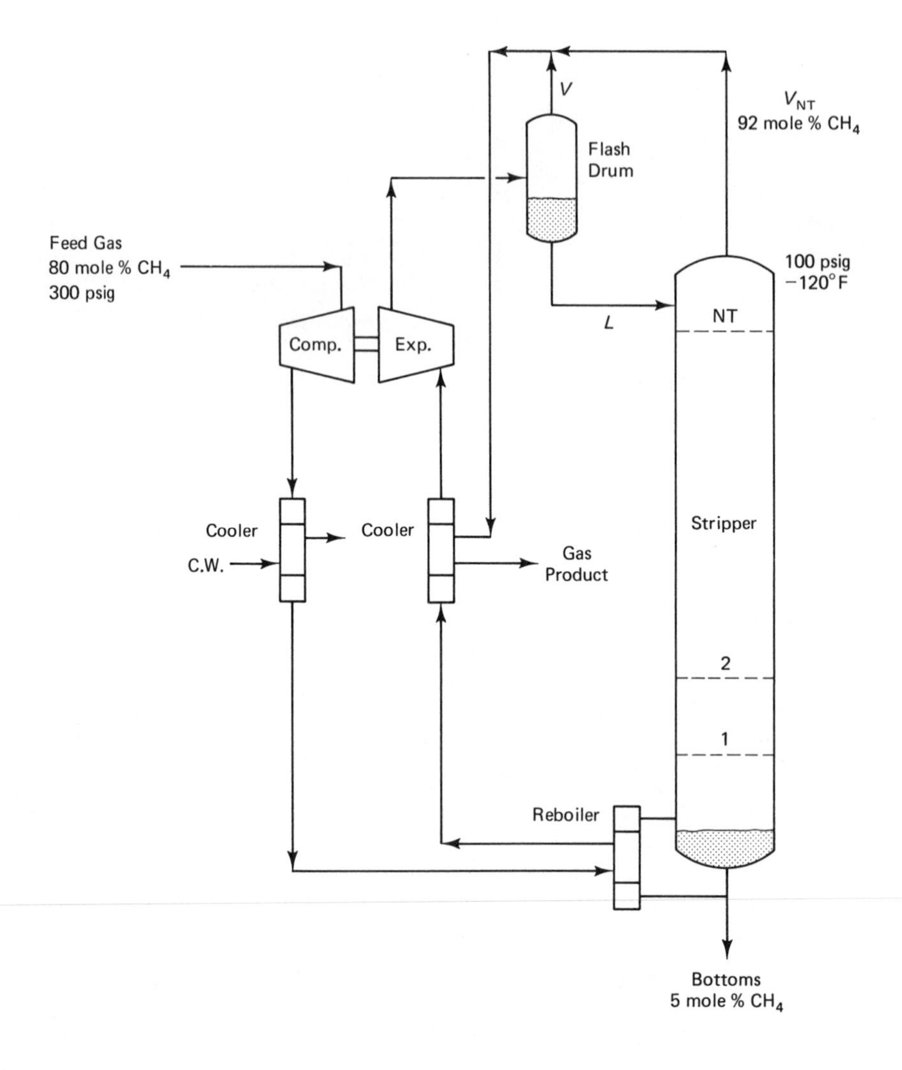

V
V_NT
92 mole % CH4
Flash
Drum
Feed Gas
80 mole % CH4
300 psig
100 psig
−120°F
L
NT
Comp.
Exp.
Cooler
Cooler
Stripper
C.W.
Gas
Product
2
1
Reboiler
Bottoms
5 mole % CH4

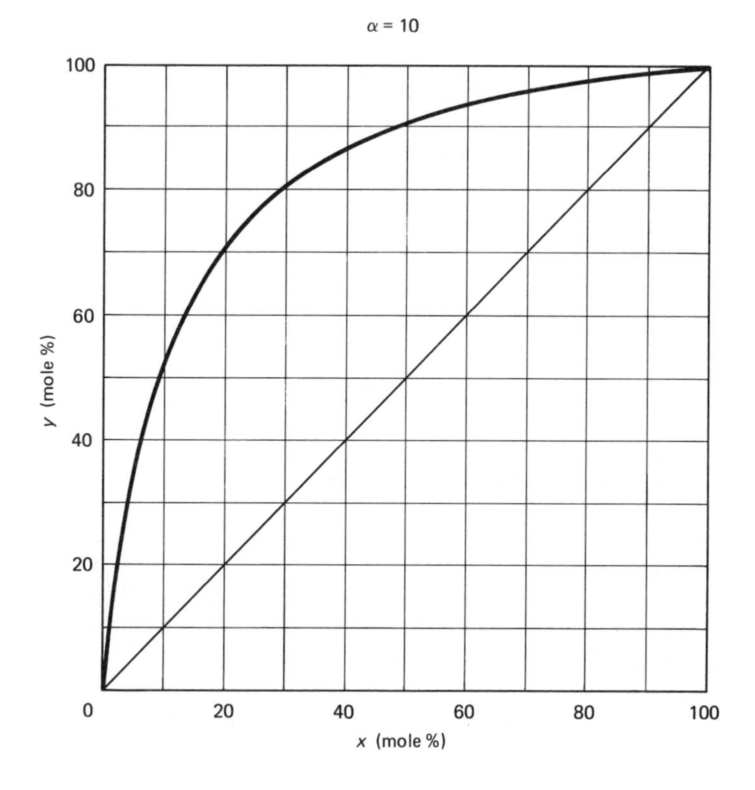
$\alpha = 10$
100
80
60
40
20
0
20
40
60
80
100
y (mole %)
x (mole %)

6 Phase Equilibrium

6.1 THERMODYNAMIC BASIS FOR PHASE EQUILIBRIUM CALCULATIONS

In chapter 5 the idea of an equilibrium stage was presented and stage calculations were made. The equilibrium relationship was expressed either as the formula $y = Kx$ or by means of a graph relating x and y. In this chapter we will deal with the thermodynamic relations that are used to calculate equilibrium constants. Our treatment will necessarily be brief; more thorough treatment is presented in courses in physical chemistry and chemical engineering thermodynamics.

The criterion of equilibrium is that all potentials for change are in balance: there are no changes in the equilibrium state and no tendency to leave that state. For three phases—vapor (V), liquid (L) and solid (S)—the temperatures and pressures must be equal at equilibrium:

$$T^V = T^L = T^S \tag{6–1}$$

$$P^V = P^L = P^S \tag{6–2}$$

As mentioned in chapter 5, the chemical potentials of each component must be the same in all phases. This criterion was developed by Willard Gibbs in the late nineteenth century. Formally,

$$\mu_j^V = \mu_j^L = \mu_j^S \qquad j = 1,2,3, \ldots ,\text{NC} \tag{6–3}$$

where

μ_j^k = chemical potential of the jth component in the kth phase

NC = number of chemical components

By summing up all of the equations describing a system and comparing them to the number of variables in the system, Gibbs derived the phase rule

$$\pi + \phi = C + 2 \tag{6–4}$$

where

π = number of phases present

ϕ = number of "degrees of freedom," i.e., the number of variables that must be specified in order to completely describe the system

C = number of chemical components

Because of calculational complexities, it is convenient to replace the chemical potential μ by the *fugacity* f, defined by the equations

$$\mu_j = RT \ln f_j + g_{(T)} \tag{6–5}$$

$$\lim f_j = p_j \text{ as pressure approaches zero} \tag{6–6}$$

where

f_j = fugacity of the jth component

$g_{(T)}$ = a function of temperature alone

p_j = partial pressure of the jth component = Py_j

where

P = total pressure of the system

y_j = mole fraction of the jth component in the vapor

Fugacity can be thought of as an effective pressure that corrects for nonideal behavior.

An alternative equation for the phase equilibrium criterion is

$$f_j^V = f_j^L = f_j^S \tag{6–7}$$

Fugacity is discussed in detail in thermodynamics courses. For this introductory course, we will use some simplifying assumptions. For liquid-phase fugacity, we will use

$$f_j^L = P_j^s x_j \gamma_j \tag{6–8}$$

where

P_j^s = vapor pressure of pure component j at the temperature of the system

x_j = mole fraction of component j in the liquid phase

γ_j = liquid-phase activity coefficient of component j

γ_j is a function of composition and temperature. We will discuss this relationship more fully later.

For vapor-phase fugacity, we will use

$$f_j^V = Py_j \tag{6–9}$$

where

P = total system pressure

y_j = mole fraction of component j in the vapor phase

Equation 6–9 is valid only if the gas phase is ideal. A correction factor must be included for nonideal gases, which typically occur when system pressures are high.

Combining equations 6–7, 6–8, and 6–9 yields the VLE relationship that we will use in this book:

$$Py_j = P_j^s x_j \gamma_j \tag{6–10}$$

This equation is applicable to many systems, but must be modified at high pressures.

A special case of equation 6–10 can be used in systems that consist of similar compounds. If the molecules are similar, they do not attract or repel each other. Therefore, the activity coefficients are unity. This is called an "ideal" VLE system and is given by

$$Py_j = P_j^s x_j \tag{6–11}$$

an equation known as Raoult's Law. It was developed empirically before the fugacity relationships had been formulated and is still extensively used. Only vapor pressure data are required to do vapor-liquid equilibrium calculations using Raoult's Law.

6.2 VAPOR PRESSURES OF PURE COMPONENTS

The phase behavior of a pure component is sketched in Figure 6–1. Each line represents a phase boundary. At conditions of pressure and temperature along any one of the lines, two phases—solid/liquid, liquid/vapor or solid/vapor—exist in equilibrium. At the "triple point" temperature and pressure,

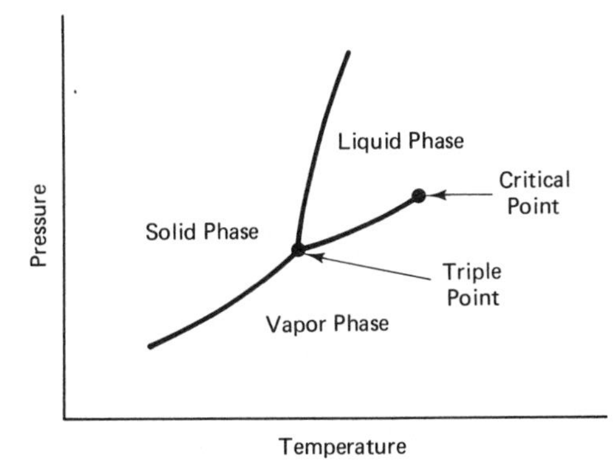

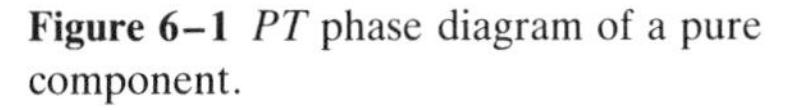
Figure 6–1 *PT* phase diagram of a pure component.

all three phases coexist. It is also possible to have several solid phases, which add additional phase boundaries to the diagram.

The solid-liquid phase boundary exists to the highest measured pressures. The solid/gas phase boundary continues to the lowest measured temperatures for all materials except helium. The liquid-gas phase boundary, however, stops at the *critical point*—the point at which one can no longer distinguish between liquid and vapor properties. Beyond this point we refer to the *fluid phase*. The critical point is determined by heating a liquid-vapor mixture and visually observing when the meniscus between the phases disappears.

The liquid and vapor phases are characterized by a lack of crystal structure and by the mobility of the molecules of the system. In a liquid, the molecules are close enough to each other to be trapped within the intermolecular force fields of neighboring molecules. In a vapor, the individual molecules are not inhibited. Molecular motion becomes more violent as the temperature increases, and the molecules in the liquid phase are spaced further apart. Eventually, the distinctions between liquid and vapor phases no longer exist.

We will restrict our attention to the liquid-gas phase boundary, though parallel developments are used for each of the other two-phase boundaries. The liquid-gas phase boundary represents the vapor pressure curve for the pure component.

6.2.1 Experimental Vapor Pressures

Vapor pressures are measured experimentally by placing the pure component in a closed vessel. The liquid and vapor phases are held at constant temperature until equilibrium is reached (a slow process). The pressure in the system is the vapor pressure of the pure component at the fixed temperature. Then the temperature is changed, and a new pressure is determined. Figure 6–2 sketches the typical exponential dependence of vapor pressure on temperature. The temperature at which the vapor pressure is equal to 1 atmosphere is called the *normal boiling point*.

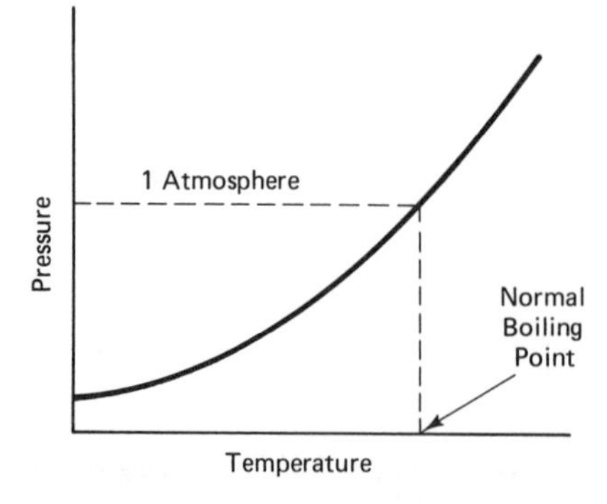

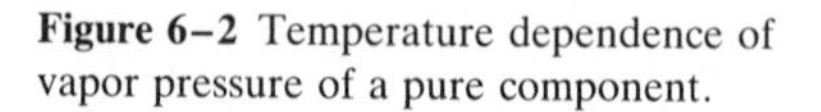
Figure 6–2 Temperature dependence of vapor pressure of a pure component.

Vapor pressure data are usually fairly easy to obtain. They are available for most chemical species in standard handbooks.

6.2.2 Correlating Vapor Pressures

Vapor pressure curves have been correlated by equations of the form

$$\ln P^s = C + \frac{D}{T} + (\text{added terms in } T^2, \ln T, \text{etc.})$$

Over a limited temperature range, the abbreviated Antoine equation gives an adequate representation of vapor pressures. This equation is given by

$$\ln P^s = C + \frac{D}{T} \qquad (6\text{–}12)$$

where

P^s = vapor pressure of a pure component (mm Hg, psia, kPa, or atmospheres)

T = absolute temperature (°R or K)

C, D = constants for the specific pure component considered

For one pure component, the constants C and D can be calculated from two data points, P_1^s at T_1 and P_2^s at T_2, by solving the simultaneous equations

$$\ln P_2^s = C + \frac{D}{T_2} \qquad (6\text{–}13)$$

$$\ln P_1^s = C + \frac{D}{T_1} \qquad (6\text{–}14)$$

The solution is

$$D = (T_1 T_2)\frac{[\ln(P_2^s/P_1^s)]}{(T_1 - T_2)} \qquad (6\text{–}15)$$

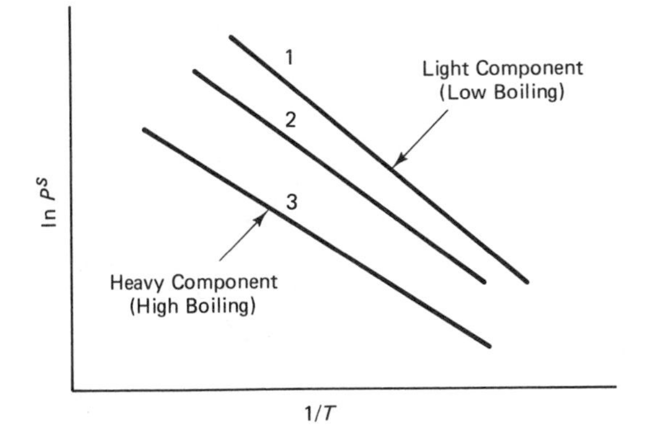

Figure 6–3 Vapor pressure vs. temperature for three components.

$$C = \ln P_2^s - \frac{D}{T_2} \qquad (6\text{–}16)$$

Remember that two vapor pressure data points must be used for each pure component, giving C and D constants for each pure component: C_j and D_j.

Vapor pressure data are typically plotted on semilog graph paper. Figure 6–3 sketches vapor pressure lines for three components. Component 1 has the highest vapor pressure at a given temperature, so it is the lightest, or lowest-boiling, component. Component 3 has the lowest vapor pressure, so it is the heaviest, or highest-boiling, component. The slopes of the vapor pressure curves are in general not the same. However, if components are chemically similar, their vapor pressure curves do have about the same temperature dependences.

It has become common practice to plot vapor pressure data on graph paper having a logarithmic vertical scale and a horizontal scale lined in intervals of reciprocal absolute temperature, but marked directly as temperature. Two examples are given in Figures 6–4 and 6–5. In Figure 6–4, vapor pressures for pure methanol and pure water are plotted. The pressure units are mm Hg, and the temperature units are °C. Figure 6–5 gives vapor pressures for a number of common compounds, with pressures in psia and temperature in °F. Note the unusual nonlinear scale used for the abscissa.

Example 6.1.

Vapor pressure data for methanol and water are given below. Calculate the Antoine constants.

methanol

760 mm Hg at 64.5°C
2600 mm Hg at 100°C

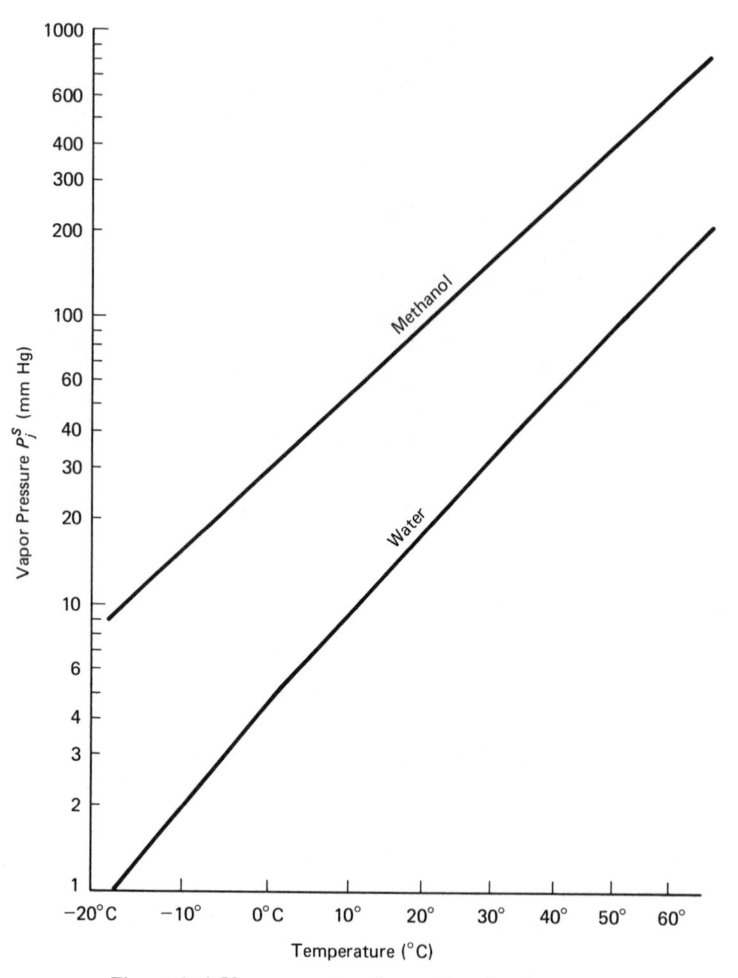

Figure 6–4 Vapor pressures for methanol and water.

water

760 mm Hg at 100°C
145.4 mm Hg at 59.4°C

Solution:

Set up two equations for each component.

methanol

$$\ln 760 = C_m + \frac{D_m}{(64.5 + 273)}$$

$$\ln 2600 = C_m + \frac{D_m}{(100 + 273)}$$

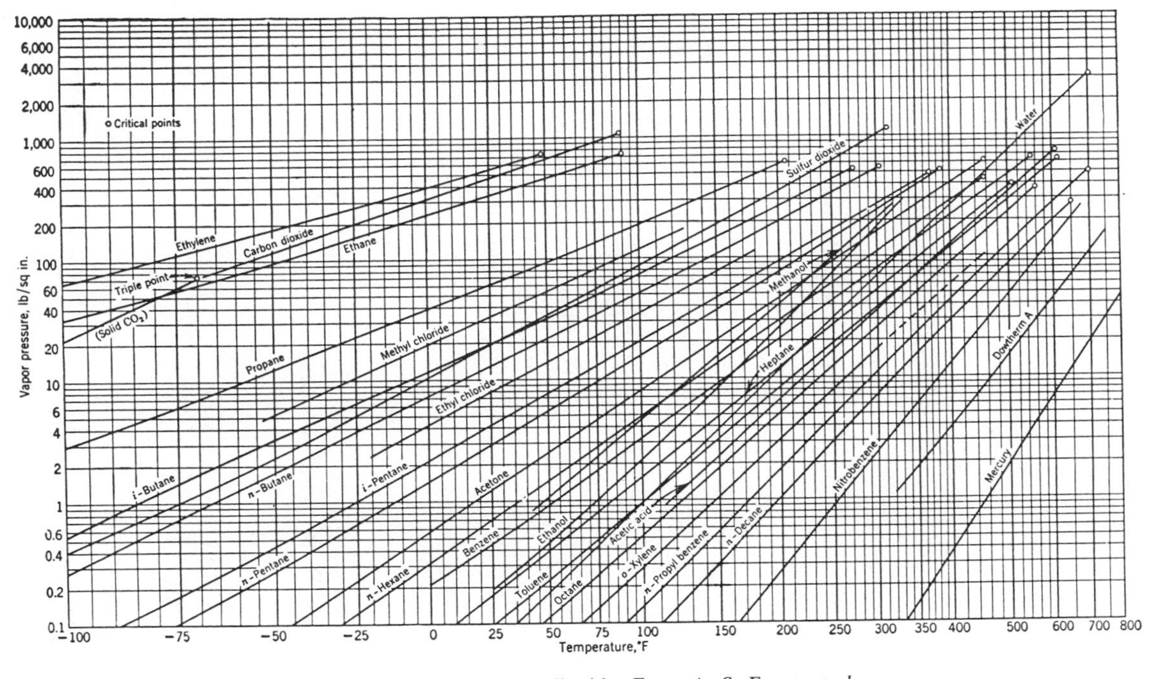

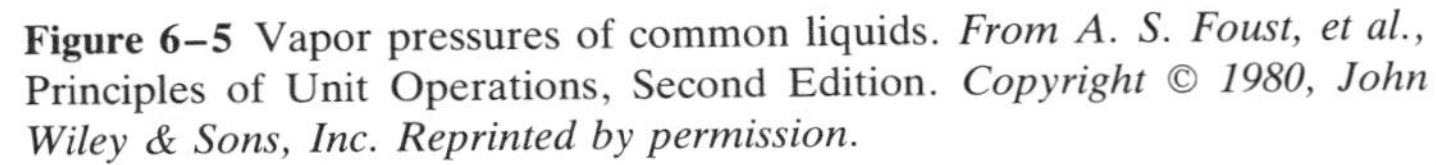

Figure 6–5 Vapor pressures of common liquids. *From A. S. Foust, et al.,* Principles of Unit Operations, Second Edition. *Copyright © 1980, John Wiley & Sons, Inc. Reprinted by permission.*

water

$$\ln 760 = C_w + \frac{D_w}{(100 + 273)}$$

$$\ln 145.5 = C_w + \frac{D_w}{(59.4 + 273)}$$

Solving each pair of equations simultaneously for C_m, D_m, C_w, and D_w gives

methanol

$$\ln\left(\frac{2600}{760}\right) = D_m\left(\frac{1}{373} - \frac{1}{337.5}\right)$$

$$D_m = -4361.5$$

$$C_m = \ln 760 - \frac{D_m}{337.5} = 19.5564$$

$$\ln P_m^s = 19.5564 - \frac{4361.55}{T}$$

water

$$\ln\left(\frac{760}{145.4}\right) = D_w\left(\frac{1}{373} - \frac{1}{332.4}\right)$$

$$D_w = -5050.50$$

$$C_w = \ln 760 - \frac{D_w}{373} = 20.1735$$

$$\ln P_w^s = 20.1735 - \frac{5050.50}{T}$$

Remember that the constants C and D depend on the units used for pressures and temperatures. Remember also that the temperatures must be absolute.

6.3 BINARY VAPOR-LIQUID EQUILIBRIUM

Vapor-liquid equilibrium data can be measured by placing a mixture of the two components in a vessel held at either constant temperature or constant pressure. Equilibrium between the liquid and vapor phases will be approached slowly, but may be promoted by agitating the vessel or by circulating the vapor phase so that it bubbles through the liquid. Even with promotion, it may take several hours for equilibrium to be established. At that point, the pressure and temperature are measured and samples are taken to determine the vapor and liquid compositions. Sampling must be done to prevent temperature and pressure changes. One method is to reduce the chamber volume as the samples are taken. After the data are measured, the composition of the vessel is changed and the procedure is repeated over the full range of compositions.

The results of these measurements may be shown on several different types of plots for binary systems:

1. *Pxy* diagram—x and y as functions of pressure at constant temperature
2. *Txy* diagram—x and y as functions of temperature at constant pressure
3. *xy* diagram—x versus y at constant pressure (temperature is a parameter along the curve)

Since most applications require data at constant pressure, *Txy* and *xy* diagrams are the most commonly used. However, many VLE data are collected at constant temperature, so *Pxy* diagrams are also quite common.

6.3.1 *Pxy* Diagrams

Figure 6–6(a) is a *Pxy* diagram for a binary mixture at constant temperature. In this diagram, compositions are given as mole fraction of the light component. The upper curve gives the liquid composition x versus pressure P. The lower curve gives the vapor composition y versus pressure. The region above the x curve corresponds to subcooled liquid, while the region below the y curve corresponds to superheated vapor. The region in between the x and y curves corresponds to the two-phase region where both saturated liquid and saturated vapor exist.

Both of the curves go through the point $x = y = 0$ and the point $x = y = 1$, since these points correspond to pure heavy and pure light components, respectively. If the liquid phase contains just one component, the vapor must also contain only that component.

The pressure P_H of the system at $x = 0$ is the vapor pressure of the pure heavy component at the temperature at which the data were collected. The pressure P_L of the system at $x = 1$ is the vapor pressure of the pure light component at the temperature at which the data were collected.

In Figure 6–6(b), a line of constant pressure P_1 is drawn. The liquid and vapor compositions x_1 and y_1 are obtained from the intersection of the phase envelope with the P_1 constant-pressure line.

The effects of changing pressure at constant temperature on a mixture of composition z can be seen by following a vertical line as shown in Figure 6–6(c) at the composition z. At low pressure below the phase envelope, there is only a superheated vapor phase with composition z. If the pressure is increased to P_{DP}, a liquid phase will begin to form. P_{DP} is called the *dewpoint* pressure, because the first drops of liquid form as "dew." The liquid droplets that form will be much leaner in the light component. The composition of

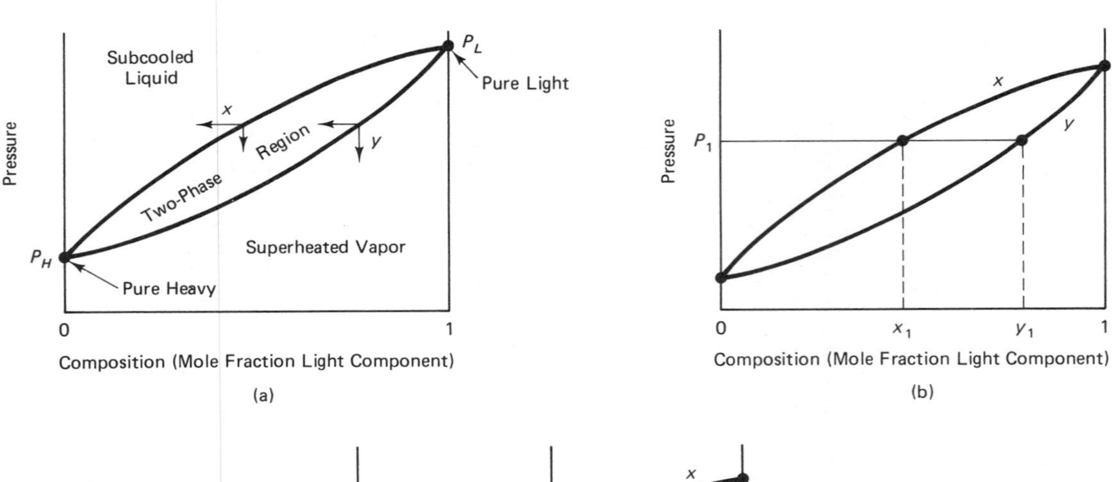

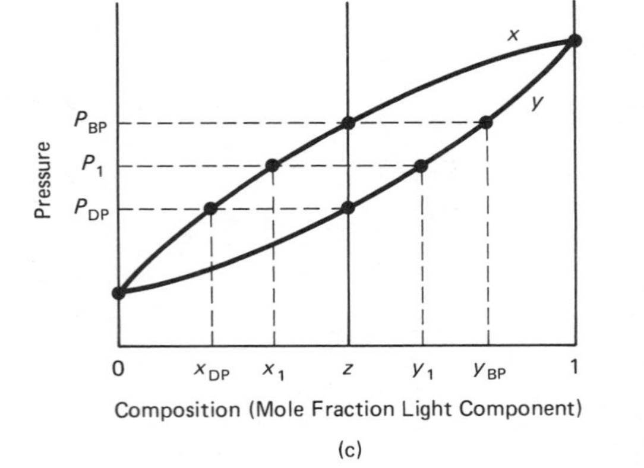

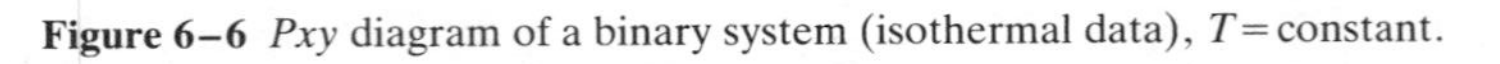

Figure 6–6 *Pxy* diagram of a binary system (isothermal data), T = constant.

the first drop of liquid formed at P_{DP} is read off the *Pxy* diagrams as x_{DP}. The composition of the vapor is still z.

As the pressure is increased further, more liquid forms. At a pressure P_1 there is almost as much liquid of composition x_1 as vapor of composition y_1. As the pressure is further increased, P_{BP} is reached. Here the last bit of the vapor phase exists as bubbles, so this pressure is known as the *bubblepoint* pressure of the mixture with composition z at the specified temperature. The final bubble of vapor has a concentration y_{BP} and is in equilibrium with the liquid of composition z. At still higher pressures, only a subcooled liquid phase exists with composition z.

The upper, x versus P, curve is known as the *bubblepoint* or *saturated liquid* curve. The lower, y versus P, curve is known as the *dewpoint* or the *saturated vapor* curve. The term "saturated" implies that both the liquid and the vapor phases are present. The region of the diagram between the bubblepoint curve and the dewpoint curve is an area of two-phase equilibrium. The compositions of the phases themselves are found along the phase envelopes.

It is particularly important to note that the light component does not vaporize first: both components are found in the vapor phase if they are both present in the liquid phase. The light component is present in larger composition in the vapor phase than it is in the liquid phase and the reverse is true for the heavy component. But some heavy component is always present in the vapor, and some light component is always present in the liquid.

6.3.2 *Txy* Diagrams

Figure 6–7 shows a *Txy* diagram for a binary system at constant pressure. The boiling point temperature of the pure light component T_L at the pressure specified in the diagram is the temperature at the point $x = y = 1$. T_H is the boiling temperature of the heavy component at the system pressure.

At some arbitrary temperature T_1, the compositions of the liquid and vapor phases are x_1 and y_1. The bubblepoint curve is the upper curve, and the dewpoint curve is the lower curve. The region below the x versus T curve is subcooled liquid. The region above the y versus T curve is superheated vapor.

Heating a mixture of composition z from a low temperature to a high temperature corresponds to following a vertical line at z in the figure. The mixture is all liquid until it reaches the bubblepoint temperature T_{BP}, where the first bubble of vapor forms and has a composition y_{BP}. Increasing the temperature further to the dewpoint temperature T_{DP} yields a mixture that is all vapor. The last drops of liquid have a composition x_{DP}.

The vapor-liquid equilibrium data are functions of both temperature and pressure. If *Txy* diagrams for the same system at different total pressures

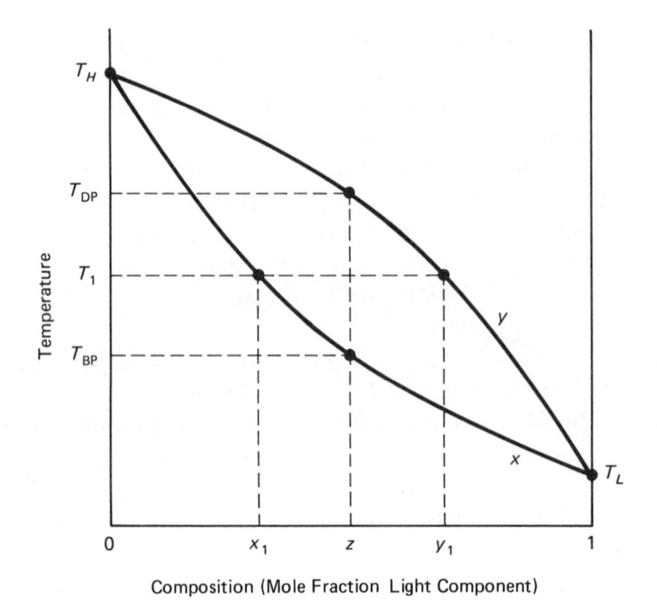

Figure 6–7 *Txy* diagram of a binary system (isobaric data), P = constant.

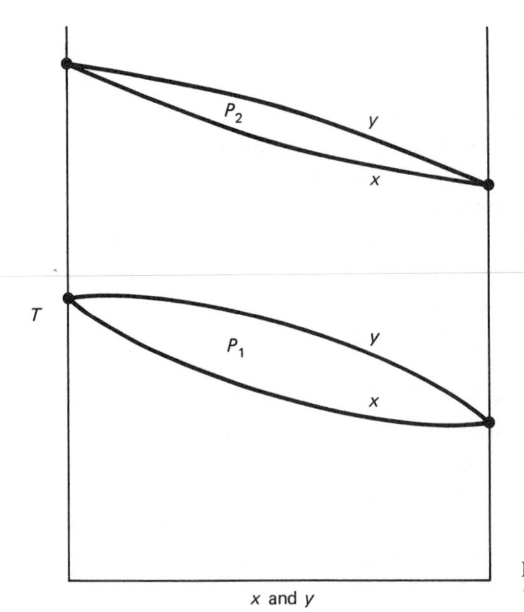

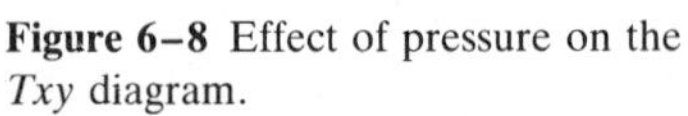

Figure 6–8 Effect of pressure on the *Txy* diagram.

were drawn on a single plot, they would look something like Figure 6–8. In this diagram, (1) pressure P_2 is greater than pressure P_1, (2) the vapor-liquid equilibrium temperatures are higher at P_2 than at P_1, and (3) the phase equilibrium lines at the higher pressure are usually closer together. Thus, the equilibrium compositions of liquid and vapor will be closer together at P_2 than at P_1. Separation of components is usually more difficult at higher temperatures and pressures.

6.3.3 *xy* Diagrams

Figure 6–9 is a typical *xy* diagram for a binary system. The *y* versus *x* curve relates the compositions of the liquid and vapor phases in equilibrium with each other. These diagrams can be generated from either constant-pressure or constant-temperature data. However, they are most commonly drawn for constant pressure, since most applications are essentially isobaric. If the diagram is plotted for a constant pressure, temperature is a parameter along the curve. Temperatures are rarely shown explicitly on *xy* diagrams, but can be obtained from a *Txy* diagram for the same pressure.

The equilibrium curve gives values of *x* and *y* that are in equilibrium. Of course, the equilibrium curve must pass through the points (0,0) and (1,1) of the diagram. The equilibrium temperature decreases as the composition of the light component increases. The 45° line (on which $x = y$) is usually put

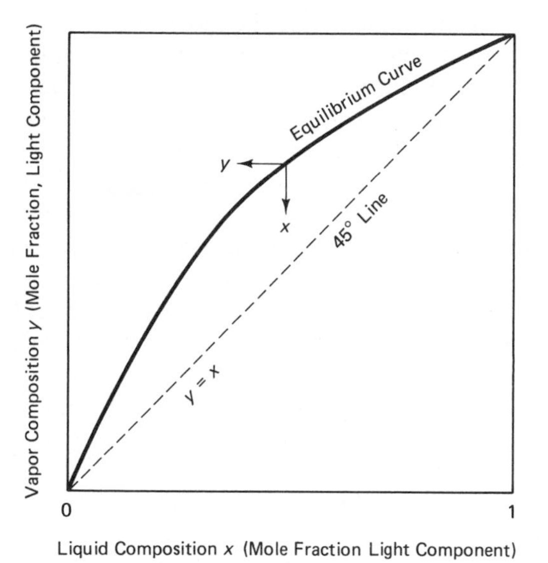

Figure 6–9 *xy* diagram for a binary system.

on the diagram for reference purposes. The farther the equilibrium curve lies from the 45° line, the easier the separation of components becomes, i.e., the bigger the differences between x and y.

6.4 NONIDEALITY IN VAPOR-LIQUID EQUILIBRIUM

At low and moderate pressures, nonideal vapor-liquid equilibrium can be described by equation 6–10, viz.,

$$Py_j = P_j^s x_j \gamma_j$$

This equation assumes that most of the nonideal behavior is caused by liquid-phase effects. In the liquid phase, molecules are much more closely spaced than in the vapor phase. Therefore, attraction and repulsion among molecules have larger effects. The topic is dealt with extensively in physical and organic chemistry.

6.4.1 Activity Coefficients

Repulsion Molecules that are dissimilar enough from each other will exert repulsive forces. For example, polar water molecules are strongly repulsed by organic hydrocarbon molecules. The repulsive forces result in activity coefficients γ_j greater than unity, since the molecules tend to leave the liquid phase. When dissimilar molecules are mixed together, a greater partial pressure is exerted, resulting in a *positive deviation from ideality.*

The activity coefficient of component j normally becomes bigger as the composition of j decreases, and must approach unity as the mole fraction of j approaches unity, since j becomes pure at that point. Figure 6–10(a) gives typical plots of activity coefficients γ_1 and γ_2 of two components in a binary mixture versus the mole fraction x_1 of the light component. The two components show positive deviations from ideality. When x_1 is zero, the pure heavy component (component 2) has an activity coefficient of unity. The activity coefficient of the light component (component 1) will be large and is called the *infinite dilution* activity coefficient γ_1^∞. At the other end, when $x_1 = 1$, the activity coefficient of component 1 must be unity and the activity coefficient of component 2 is the infinite dilution activity coefficient.

Attraction If the molecules of the components attract each other, the activity coefficients will be less than unity, since the molecules will exert lower partial pressures than if they were pure. An example of such a system is the nitric acid/water mixture. Activity coefficients less than unity are called *negative deviations from ideality.*

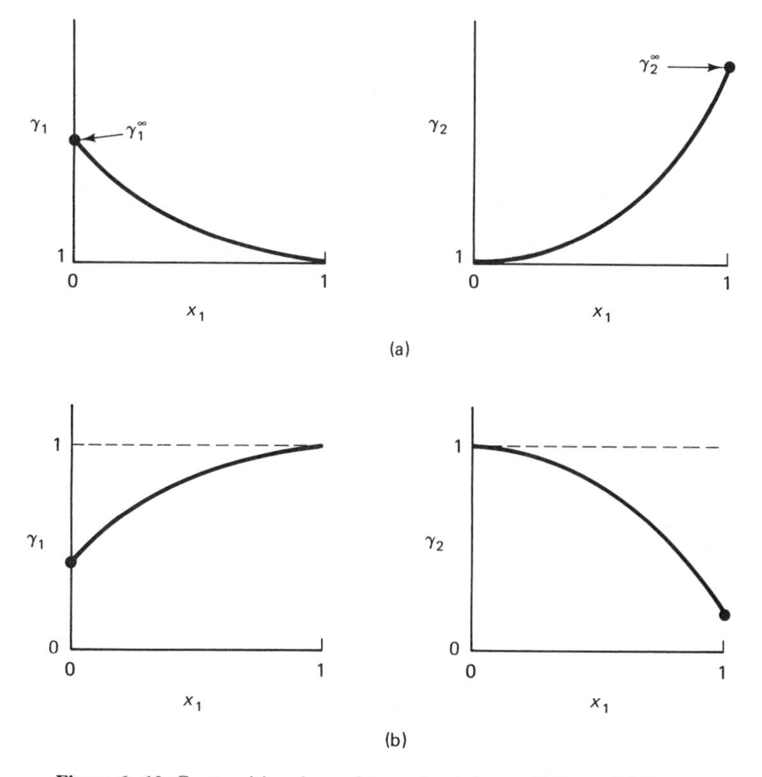

Figure 6–10 Composition dependence of activity coefficients. (a) Repulsion (b) Attraction.

Figure 6–10(b) sketches typical activity coefficient dependence on composition in the case of attraction. Note that the activity coefficients have values less than unity, but greater than zero.

Example 6.2.

Calculate the activity coefficients of methanol and water from the following data at 760 mm Hg pressure:

$$x = 0.30 \text{ m.f. methanol} = x_1$$

$$y = 0.665 \text{ m.f. methanol} = y_1$$

$$T = 78^\circ\text{C}$$

Vapor Pressure Data at 78°C (172.4°F)

$$\text{Methanol: } P_1^s = 1.64 \text{ atm}$$

$$\text{Water: } P_2^s = 0.43 \text{ atm}$$

Solution:

Equation 6–10 can be written and solved for γ for both methanol (component 1) and water (component 2) as follows:

Methanol

$$Py_1 = P_1^s x_1 \gamma_1$$

$$(1 \text{ atm})(0.665) = (1.64 \text{ atm})(0.30)(\gamma_1)$$

$$\gamma_1 = 1.35$$

Water

$$Py_2 = P_2^s x_2 \gamma_2$$

$$(1 \text{ atm})(1 - 0.665) = (0.43 \text{ atm})(1 - 0.30)(\gamma_2)$$

$$\gamma_2 = 1.11$$

The water and methanol molecules are not chemically similar since water is ionized and strongly polar while methanol is more organic in character. Therefore, the activity coefficients are greater than unity.

6.4.2 Azeotropes

In the mixture of water and methanol in example 6–2, repulsive forces are at work because the two molecules are dissimilar. Water is even less similar to the longer chain alcohols (ethanol, propanol, etc.), because these components are increasingly organic in nature. The increased repulsion between molecules can result in the formation of an *azeotrope*, which is a liquid mixture whose equilibrium vapor has the same composition as the liquid, i.e., $x_j = y_j$ for an azeotrope.

An azeotrope represents a drastic restriction on the ability to separate components using distillation. Specifically, the azeotrope prevents phase separations from producing liquid or vapor compositions on the other side of the azeotrope from the original mixture. Instead, separation based on vaporization can only extend from the pure component to the azeotrope.

When nonideality becomes great enough, mixtures that boil outside the range of boiling points of the pure components can be formed. Mixtures in which there is strong repulsion between the different species boil at temperatures that are less than the boiling point of the pure light component; mixtures in which there is strong attraction boil at temperatures that are higher than the boiling point of the pure heavy component.

Minimum-Boiling Homogeneous Azeotropes Figure 6–11 illustrates how repulsive forces can produce minimum-boiling azeotropes. Suppose we have two components in a mixture that tend to repel each other—for example, "cowboy" molecules and "Indian" molecules (the Dallas Cowboys and the

Washington Redskins if you are a football fan). Suppose further that the Indians are the light (more volatile, lower boiling) component. Figure 6–11(a) is a *Pxy* diagram showing isothermal data. The pressure at the left-hand side where $x_{\text{Indians}} = 0$ (where there are only cowboys present) is the vapor pressure of pure cowboys, P^s_{cowboys}. The pressure at the right-hand side of the figure at $x_{\text{Indians}} = 1$ is the vapor pressure of pure Indians, P^s_{Indians}. Since the light component is the Indians, P^s_{Indians} is greater than P^s_{cowboys}.

Let us start at the right side of the figure at $x_{\text{Indians}} = 1$, where there are initially nothing but Indian molecules. Now throw a few cowboy molecules into the system. Since there is strong repulsion between cowboys and Indians, the small number of cowboy molecules will be surrounded by hostile Indians and will try strenuously to escape. If the repulsion is great enough, the system

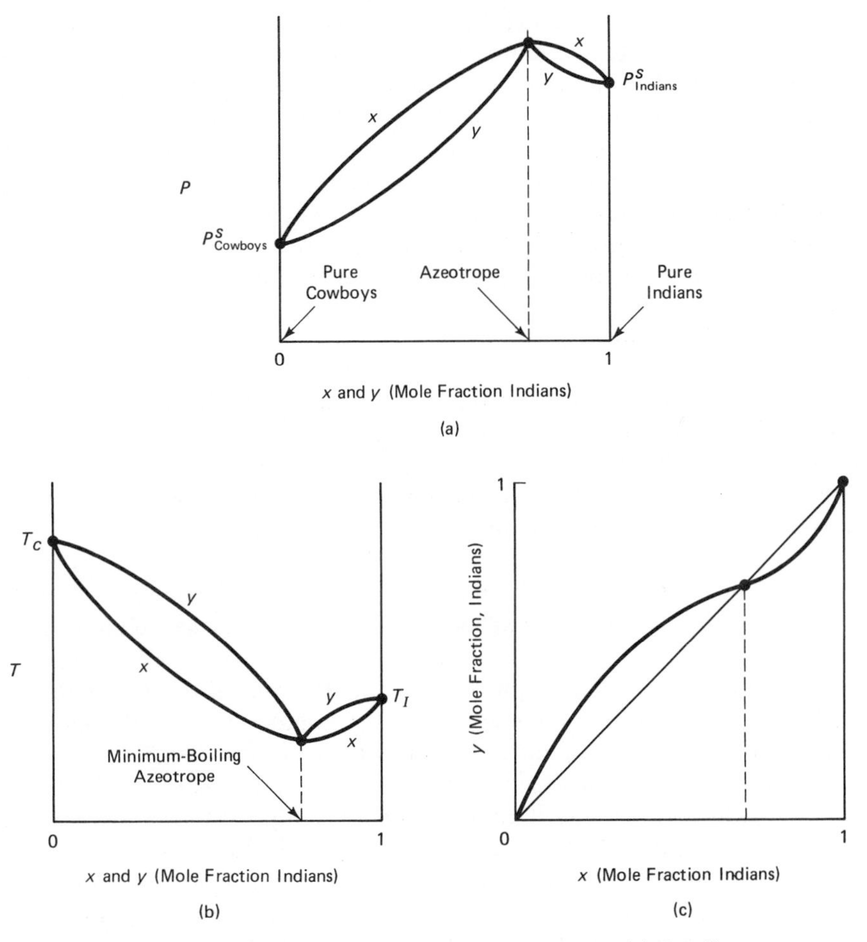

Figure 6–11 Homogeneous minimum-boiling azeotrope. (a) *Pxy* diagram (const. *T*) (b) *Txy* diagram (const. *P*) (c) *xy* diagram (either *T* or *P* const.).

pressure may actually *increase* for a while as we add more of the heavy cowboy molecules. Despite the fact that the molecules being added are heavier, the repulsive forces produce an increase in pressure in the constant-temperature system. Even though the cowboy vapor pressure is lower, the concentration of cowboys is greater in the vapor phase than in the liquid phase because the cowboys have very large activity coefficients. Therefore, x_{Indians} is greater than y_{Indians} in this region.

As more and more cowboy molecules are added, we eventually reach a maximum pressure. This occurs at the azeotropic composition, where the compositions of the vapor and the liquid phases are the same. Further increases in the number of heavy cowboy molecules then produce the expected decreases in pressure and yield vapor phases that are richer in Indians than in cowboys, i.e., y_{Indians} is greater than x_{Indians}.

Figure 6–11(b) shows a *Txy* diagram for the same system. Now pressure is held constant, and temperature changes. Throwing some cowboy molecules into the pure Indian mixture produces such repulsion that the temperature must be reduced to maintain the same system pressure. The result is a higher concentration of cowboy molecules in the vapor phase than in the liquid phase because of their high activity coefficients. Eventually, a minimum temperature is reached at the azeotropic composition. This type of system is called a *minimum-boiling homogeneous* azeotrope. Figure 6–11(c) gives the corresponding *xy* diagram. Note that the VLE curve crosses the 45° line at the azeotrope and that the *y*-values are smaller than the *x*-values above this azeotropic composition.

Maximum-Boiling Azeotropes If the molecules attract, rather than repel, each other, adding some light component to a vessel containing pure heavy component can actually decrease the pressure instead of increasing it in a constant-temperature system. This phenomenon is illustrated in Figure 6–12 for a mixture of light component F and a heavy component M.

Let us start at the left side of the *Pxy* diagram (Figure 6–12(a)) at the pure heavy component M. The pressure is P_M^s, the vapor pressure of pure M. Now we add some of the light component F. In the absence of interaction, we would expect that the pressure will increase in this constant-temperature system since we are adding light component. However, if the F and M molecules attract each other, adding some of the F molecules may decrease the pressure of the system. These cuddly little F molecules "snuggle up" to the M molecules, reducing their tendency to escape into the vapor phase. Mathematically, the activity coefficient of F is significantly less than unity, and the concentration of F in the vapor is less than in the liquid (y_F is less than x_F).

On the constant-pressure *Txy* diagram (Figure 6–12(b)), the effect of attraction is to require an increase in temperature as F molecules are added to maintain the same pressure. Eventually a maximum temperature is reached,

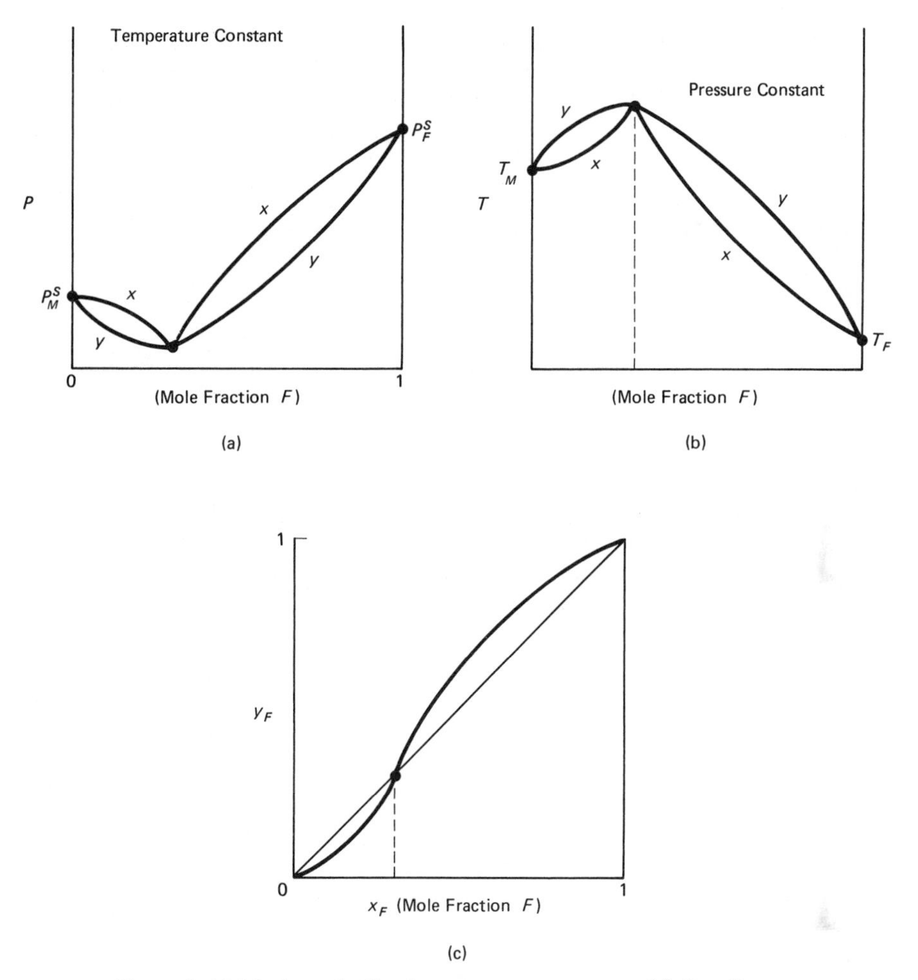

Figure 6–12 Maximum-boiling homogeneous azeotrope. (a) *Pxy* diagram (b) *Txy* diagram (c) *xy* diagram.

so this type of system is called a *maximum-boiling homogeneous* azeotrope. On an *xy* diagram, the VLE curve is below the 45° line until the azeotrope is reached.

Heterogeneous Minimum-Boiling Azeotropes If repulsive forces are extremely large, the molecules repel each other so much that they tend not even to remain together in the same liquid phase. When the activity coefficients become very large, the system may break into two liquid phases. Such a system is called a *heterogeneous* azeotropic system.

Figure 6–13 sketches typical *Txy* and *xy* diagrams for a heterogeneous system. The pure light and pure heavy components boil at temperatures T_L and T_H, respectively. At a lower temperature T_{AZ}, and at the pressure specified, there will be two liquid phases that coexist with each other. These two

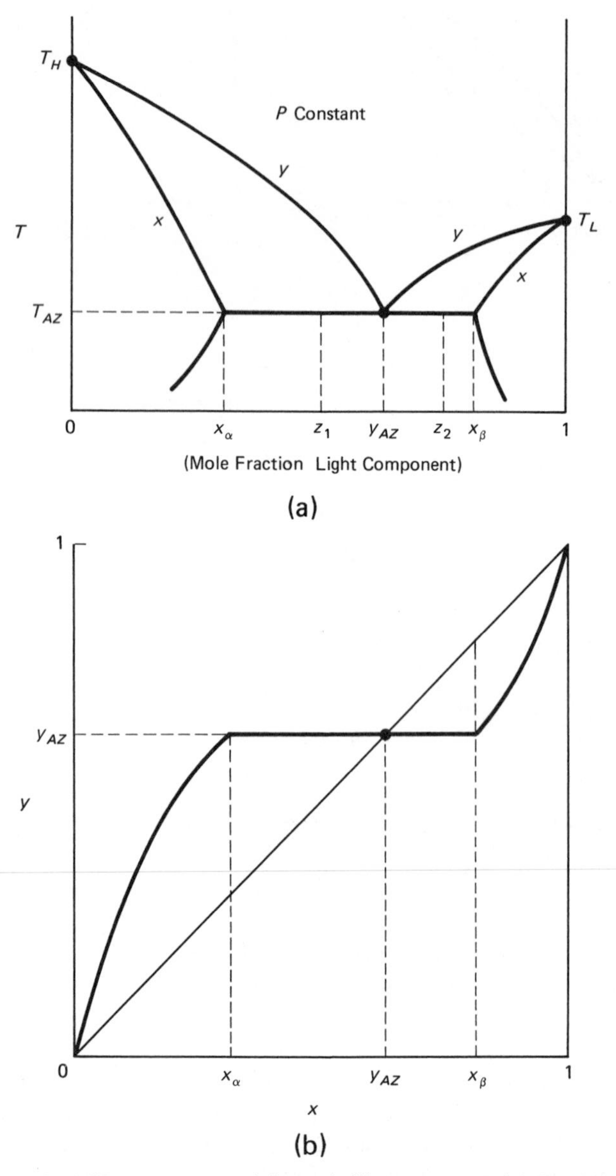

Figure 6–13 Heterogeneous minimum-boiling azeotrope. (a) *Txy* diagram (b) *xy* diagram.

liquid phases have compositions x_α and x_β that differ in their respective amounts of light and heavy component.

The two liquids are also in equilibrium with a vapor of composition y_{AZ} at temperature T_{AZ} and the specified pressure. Note that there are now three phases ($\pi = 3$) and two components ($C = 2$). Therefore, the phase rule (equation 6–2) tells us that there is only one degree of freedom, that is,

$$\pi + \phi = C + 2$$

$$\phi = 2 + 2 - 3 = 1$$

Thus, if one variable is fixed, all the others are fixed. Accordingly, since the pressure on the diagram is set, everything else is specified: the temperature must be T_{AZ}, the compositions of the liquid phases must be x_α and x_β, and the composition of the vapor phase must be y_{AZ}.

If we have a mixture that has an overall composition of z_1 as shown in Figure 6–13(a), it will separate into two liquid phases with compositions x_α and x_β if the temperature is T_{AZ}. The amounts of the two liquid phases can be calculated from total mass and component balances. Since the overall composition z_1 is closer to the x_α point, there will be more liquid formed with composition x_α. If the overall composition were z_2, there would be more of the x_β liquid phase formed.

Below the temperature T_{AZ}, there is no vapor phase present, only two liquid phases in equilibrium. The curves represent the solubilities of the two components in each other. Above T_{AZ}, there is only one liquid phase in equilibrium with a vapor phase as given on the x and y curves.

Example 6.3.

Ethanol and n-hexane form a minimum-boiling azeotrope at 33.2 mole percent ethanol at 58.68°C and 760 mm Hg pressure. The vapor pressures of ethanol and n-hexane are 6 psia and 12 psia, respectively, at 58.68°C (137.6°F). Determine the activity coefficients of ethanol and n-hexane at the azeotropic condition.

Solution:

At the azeotrope, the vapor and liquid compositions are identical. Therefore, $y_j = x_j$. Equation 6–10 can now be solved for the activity coefficients:

$$\gamma_{\text{ethanol}} = \frac{P y_{\text{ethanol}}}{P^s_{\text{ethanol}} x_{\text{ethanol}}}$$

$$= \frac{P}{P^s_{\text{ethanol}}}$$

$$= \frac{14.7 \text{ psia}}{6 \text{ psia}} = 2.45$$

$$\gamma_{\text{hexane}} = \frac{Py_{\text{hexane}}}{P^s_{\text{hexane}}x_{\text{hexane}}}$$

$$= \frac{P}{P^s_{\text{hexane}}}$$

$$= \frac{14.7 \text{ psia}}{12 \text{ psia}} = 1.23$$

Note that the activity coefficients are both greater than unity, indicating repulsion. The activity coefficient of ethanol is larger because it is at a lower concentration (33.2 mole percent).

6.4.3 DePriester Charts For Light Hydrocarbons

Figure 6–14 gives *K*-value charts for some light hydrocarbons. These charts do not assume ideal vapor-phase behavior. Some corrections for pressure effects are included.

K-values (also called equilibrium constants or distribution coefficients) are widely used in the petroleum industry. As mentioned in chapter 5, the *K*-value of component j is defined as the ratio of its composition in the vapor y_j to its composition in the liquid phase x_j, i.e.,

$$K_j = \frac{y_j}{x_j} \tag{6–17}$$

We will demonstrate the use of *K*-values in VLE calculations in chapter 7. Note that Figure 6–14(a) is used for low temperatures and Figure 6–14(b) for high temperatures. To find the appropriate *K*-values, a straight line is drawn on the diagram connecting the temperature and pressure of the system. The intersection of this line with the *K*-value curve for each hydrocarbon gives its *K*-value at this temperature and pressure.

6.5 RELATIVE VOLATILITY

A widely used measure of the difficulty of separating mixtures is *relative volatility*. The relative volatility α_{jk} of component j compared to component k is defined as

$$\alpha_{jk} = \frac{y_j/x_j}{y_k/x_k} \tag{6–18}$$

From equation 6–17, the relative volatility is the ratio of the *K*-values:

$$\alpha_{jk} = \frac{K_j}{K_k} \tag{6–19}$$

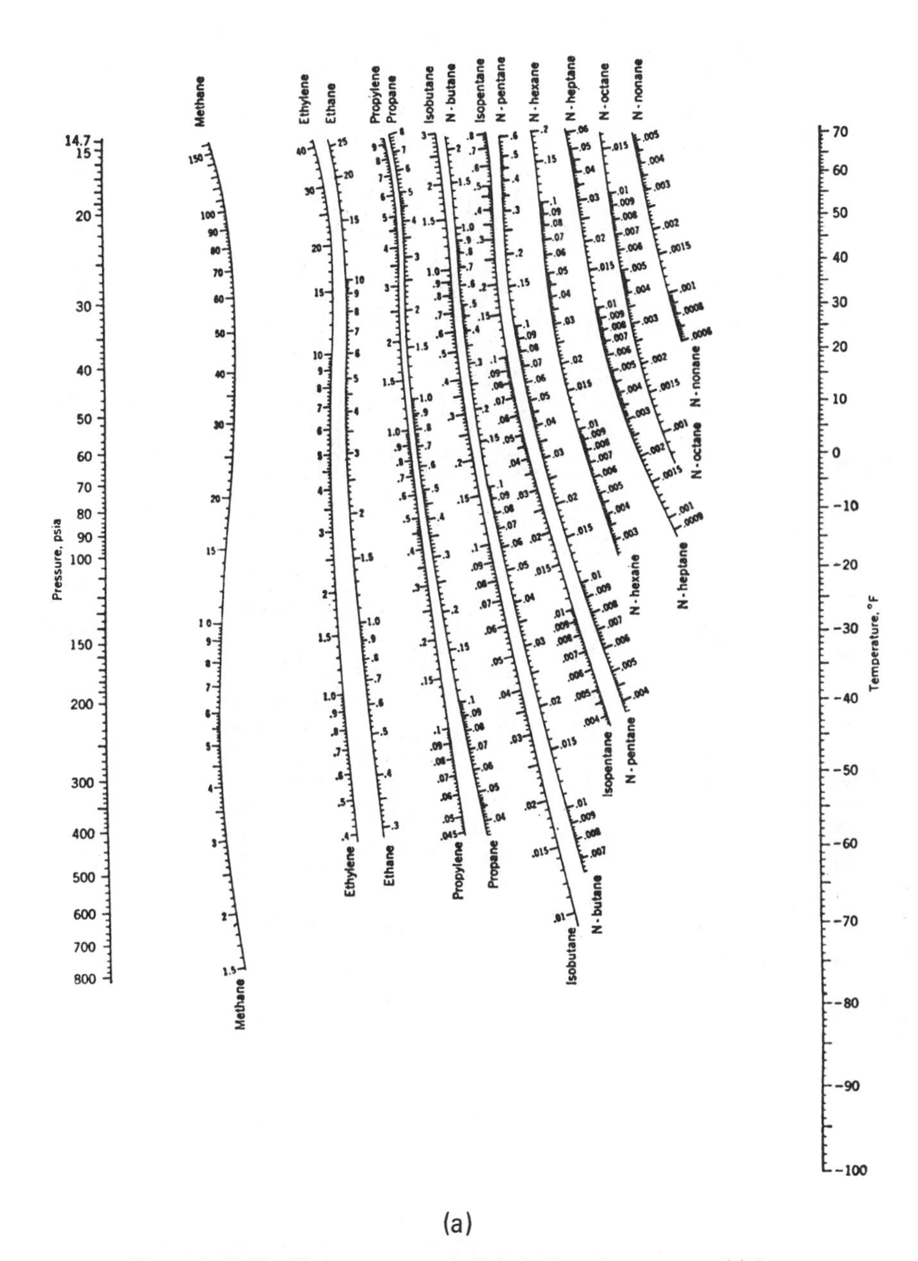

(a)

Figure 6–14 Equilibrium constants in light-hydrocarbon systems. (a) Low-temperature range (b) High-temperature range. *From C.L. DePriester,* Chem. Eng. Progr., Symposium Ser. *7:49 (1953), by permission of the American Institute of Chemical Engineers.*

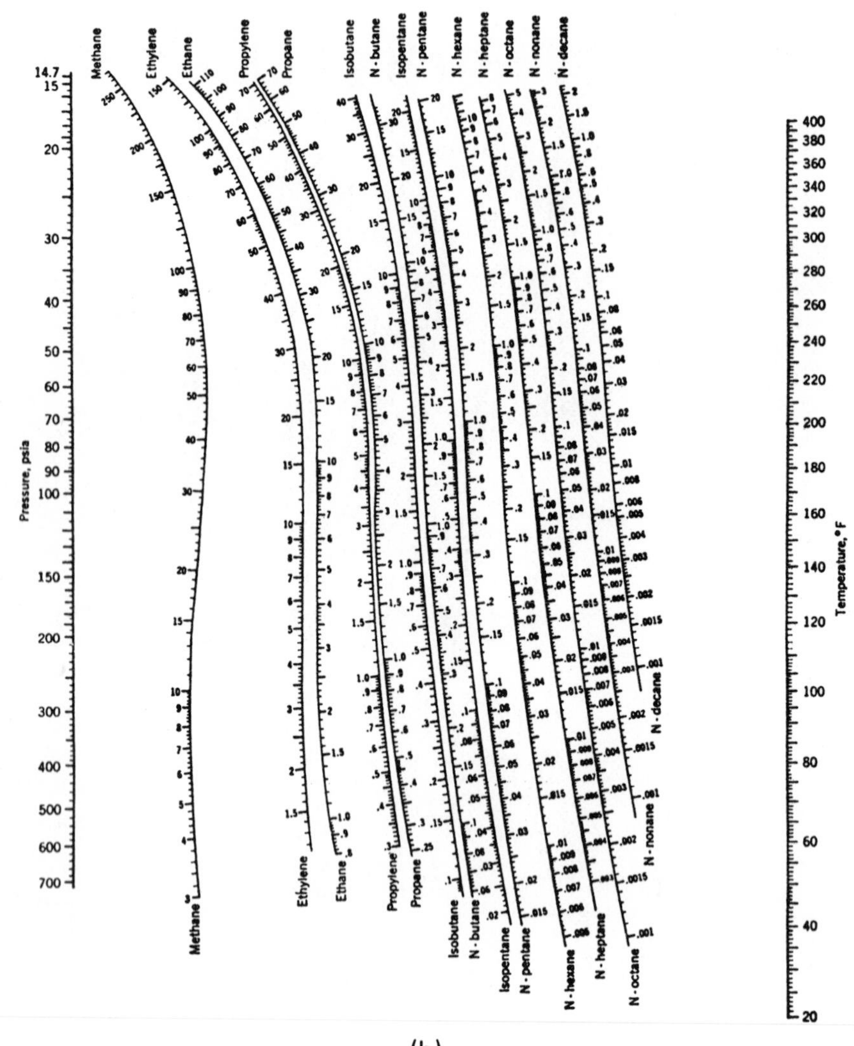

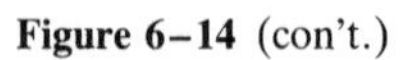
(b)

Figure 6–14 (con't.)

If the system is ideal (i.e., it obeys Raoult's Law), the relative volatility is simply the ratio of the vapor pressures:

$$\alpha_{jk} = \frac{P_j^s}{P_k^s} \tag{6–20}$$

6.5.1 Binary Systems

For a binary (two-component) system, subscripts are not necessary, and just plain α is used to mean α_{LH}, the relative volatility of the light component compared to that of the heavy component. If x and y are the mole fractions of the light component, then $(1 - x)$ and $(1 - y)$ are the mole fractions of the heavy component. Substituting into equation 6–18 gives

$$\alpha = \frac{y/x}{(1 - y)/(1 - x)}$$

Solving for y, we get

$$y = \frac{\alpha x}{1 + (\alpha - 1)x} \tag{6–21}$$

This is one of the most frequently used equations in distillation. If α is a constant, it gives a simple, explicit relationship between x and y. If you know the liquid composition x, you merely plug it into equation 6.21 and solve for y. No trial-and-error calculations are involved.

In a fair number of systems, relative volatilities are essentially constant. In general, however, they are functions of temperature and composition. In most systems, α decreases as temperature increases, which means that separation of components becomes more difficult. Therefore, it is often desirable to keep temperatures as low as possible (use low pressure) to reduce energy consumption.

Figure 6–15 shows some VLE curves on an xy diagram for various values of α. The bigger the relative volatility, the fatter the VLE curve and the easier the separation. As α approaches 1, the VLE curve approaches the 45° line $x = y$. It is impossible to separate components by distillation if the value of α is too close to unity. In fact, distillation is seldom used if α is less than about 1.05.

6.5.2 Multicomponent Systems

For a multicomponent system, the relative volatilities are defined with respect to some component, typically the heaviest one. The components are usually listed in descending order of volatility. The relative volatility of the lightest component (component 1) is written as α_1. The relative volatility of the next lightest component is written as α_2, and so on. The relative volatility of the heaviest component, α_H, equals unity if all the relative volatilities are defined with respect to the heaviest component.

The relative volatility of component j compared to component H is

$$\alpha_{jH} = \alpha_j = \frac{y_j/x_j}{y_H/x_H} \tag{6–22}$$

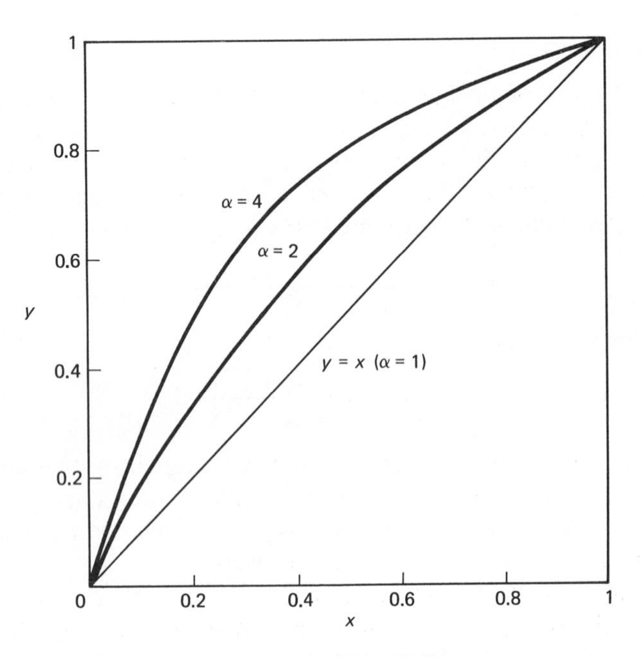

Figure 6–15 Constant relative volatility curves.

Solving for y_j yields

$$y_j = (\alpha_j x_j) \frac{y_H}{x_H} \tag{6–23}$$

Adding up all the y_j's, we must get unity:

$$\sum_{j=1}^{NC} y_j = \sum_{j=1}^{NC} (\alpha_j x_j) \frac{y_H}{x_H} = 1 \tag{6–24}$$

Solving for y_H/x_H gives

$$\frac{y_H}{x_H} = \frac{1}{\sum_{j=1}^{NC} \alpha_j x_j} \tag{6–25}$$

Substituting the right side into equation 6–23 gives

$$y_j = \frac{\alpha_j x_j}{\sum_{k=1}^{NC} \alpha_k x_k} \tag{6–26}$$

Equation 6–26 is an explicit equation for calculating all the vapor mole fractions from known liquid compositions and relative volatilities.

Example 6.4.

A multicomponent liquid mixture has the compositions and relative volatilities given in the table below. Calculate the composition of the vapor phase.

Solution:

x_j	α_j	$\alpha_j x_j$	$y_j = \alpha_j x_j \Big/ \left[\sum_{k=1}^{3} \alpha_k x_k\right]$
0.35	3.5	1.225	0.517
0.45	2.1	0.945	0.399
0.20	1	0.200	0.084
1.00		2.370	1.000

6.6 LIQUID-LIQUID PHASE EQUILIBRIUM

Up to this point, we have primarily considered equilibrium between vapor and liquid phases. Now we want to consider phase equilibrium between two liquid phases. Many chemical engineering processes involve systems with two liquid phases. These liquids may be almost immiscible (like oil and water), or there may be appreciable solubility of components in both liquid phases. The liquid-liquid extraction process that we will study in chapter 9 is the most common application of liquid-liquid equilibrium (LLE).

6.6.1 Binary Systems

Liquid-liquid systems with two components can be represented in *Tx* diagrams that are the same as those we used in VLE systems, except that now we are looking at the low-temperature part of the diagram where no vapor is present.

A variety of solubility curves are possible. A typical one is shown in Figure 6–16(a). At a given temperature, the compositions of the two liquid phases are x_α and x_β. The solubility increases with temperature (at the pressure specified) until the two liquid phases disappear at the upper critical solubility temperature. Above that temperature mutual solubility exists at all compositions, and there is only one liquid phase.

At lower pressures, the vapor-liquid equilibrium, usually including a minimum-boiling azeotrope, intercepts the liquid-liquid equilibrium. The result is shown in Figure 6–13.

In some chemical systems there may be both upper and lower critical solubility temperatures, as shown in Figure 6–16(b). For temperatures between the upper and lower critical solubilities, two liquid phases exist.

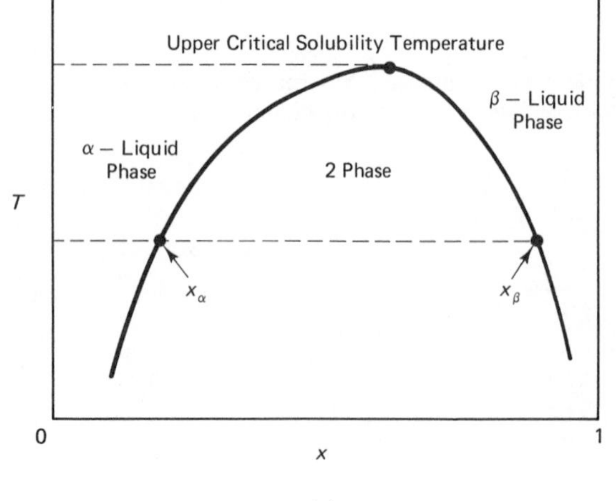

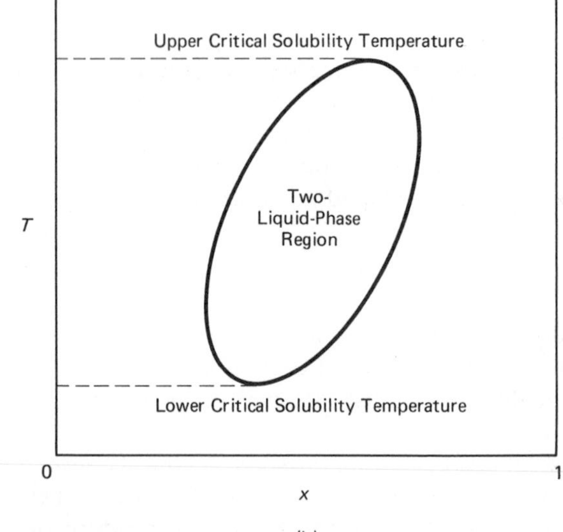

Figure 6–16 Liquid-liquid solubility for a binary system. (a) System with reduced mutual solubility at low temperature (b) System with increased mutual solubility at low temperature.

6.6.2 Ternary Systems

Most practical situations involving liquid-liquid equilibrium involve three or more components. We will deal with three component systems in discussing liquid-liquid extraction. In this process, a solute is removed from a feed stream by contacting it with a solvent. The solute is quite soluble in the solvent, while the other component in the feed is less soluble. The three-component systems that occur are represented on ternary diagrams. (See below.)

In ternary systems we have three components and two phases. From the phase rule, the number of degrees of freedom $\phi = C + 2 - \pi = 3 + 2 - 2 = 3$. Since, normally, the temperature and the pressure are specified, there is only one degree of freedom. Thus, specifying the concentration of one component in either phase completely determines all the other concentrations. We will see this later in the ternary diagrams.

Terminology The nomenclature in liquid-liquid equilibrium is not nearly as standardized as it is in vapor-liquid equilibrium. We will use the convention that the solute will be called component 1, the raffinate component 2, and the solvent component 3. The two liquid phases will be called the *solvent phase* and the *raffinate phase*. The solvent phase is rich in solvent and preferentially soaks up component 1 (the solute), which we are trying to separate from the other component in the feed (component 2, raffinate). The raffinate phase is the liquid phase which is rich in the component 2 (raffinate) and from which the solute (component 1) is being removed. The original feed is usually a mixture of solute (component 1) and raffinate (component 2).

The solvent-rich phase contains mostly solvent (component 3) and solute (component 1) and only a small amount of raffinate (component 2). The compositions of this phase, usually expressed as mass fractions, are x_1^S, x_2^S, and x_3^S. Of course, we need to specify only two of these mass fractions to specify the entire composition of the phase, since the mass fractions must add up to unity. We will use the solute and the solvent mass fractions, x_1^S and x_3^S.

The raffinate-rich phase contains mostly solute (component 1) and raffinate (component 2), but also possibly some small amount of solvent. The compositions of this phase are x_1^R, x_2^R, and x_3^R in mass fractions. Only two of these are needed, so we will use x_1^R and x_3^R.

Triangular Diagrams Ternary systems are represented on two types of triangular diagrams: equilateral triangles and right triangles. Both are sketched in Figure 6–17. The corners of each triangle correspond to 100 percent of one component; the sides of each triangle correspond to 0 percent of one component.

For equilateral triangles, lines of constant percent of component 1 are horizontal, running parallel to the base (the line between vertices 2 and 3). Lines of constant percent of component 2 run parallel to the side of the triangle with vertices 1 and 3, and lines of constant percent of component 3 run parallel to the side with vertices 1 and 2.

For right triangles, the abscissa corresponds to zero percent of component 1, the ordinate corresponds to zero percent of component 3, and the 135° line corresponds to zero percent of component 2. The vertical axis gives the

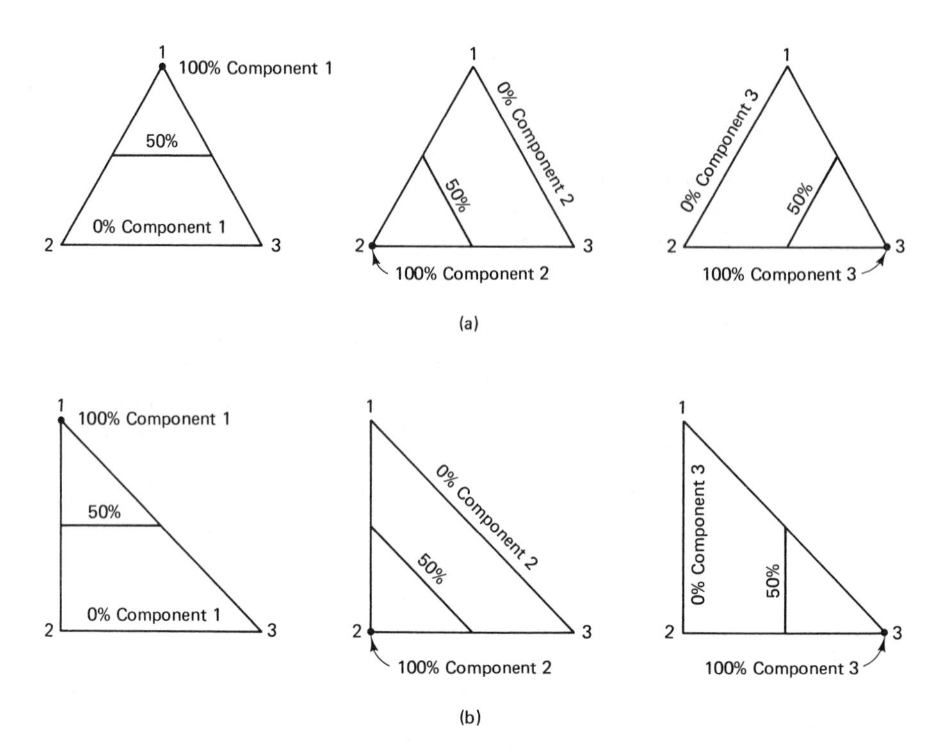

Figure 6–17 Liquid-liquid equilibrium for ternary systems. (a) Equilateral triangles (b) Right triangles.

percent solute (component 1) as x_1, and the horizontal axis gives the percent solvent (component 3) as x_3.

Figure 6–18 shows typical LLE data given on triangular diagrams. Any point within the diagram represents a ternary composition, a point along one of the sides of the diagram represents a binary composition, and the vertices represent pure components.

Since right triangles are easier to plot, we will use them in this book. However, you should be familiar with the equilateral triangular representation, too, because much of the data in the literature are plotted in this form.

LLE Tie-Lines Different chemical systems give different types of triangular diagrams. The type shown in Figure 6–18 has one region in which the system breaks into two liquid phases at equilibrium. The phase boundary, called the *solubility line*, is the solid line. Within the two-phase region, liquid-liquid equilibrium *tie-lines* (the dashed lines) connect compositions of the two phases that are in equilibrium with each other. The left side of the phase boundary gives the compositions of the raffinate-rich liquid phase (x_j^R). The right side of the phase boundary gives the compositions of the solvent-rich liquid phase

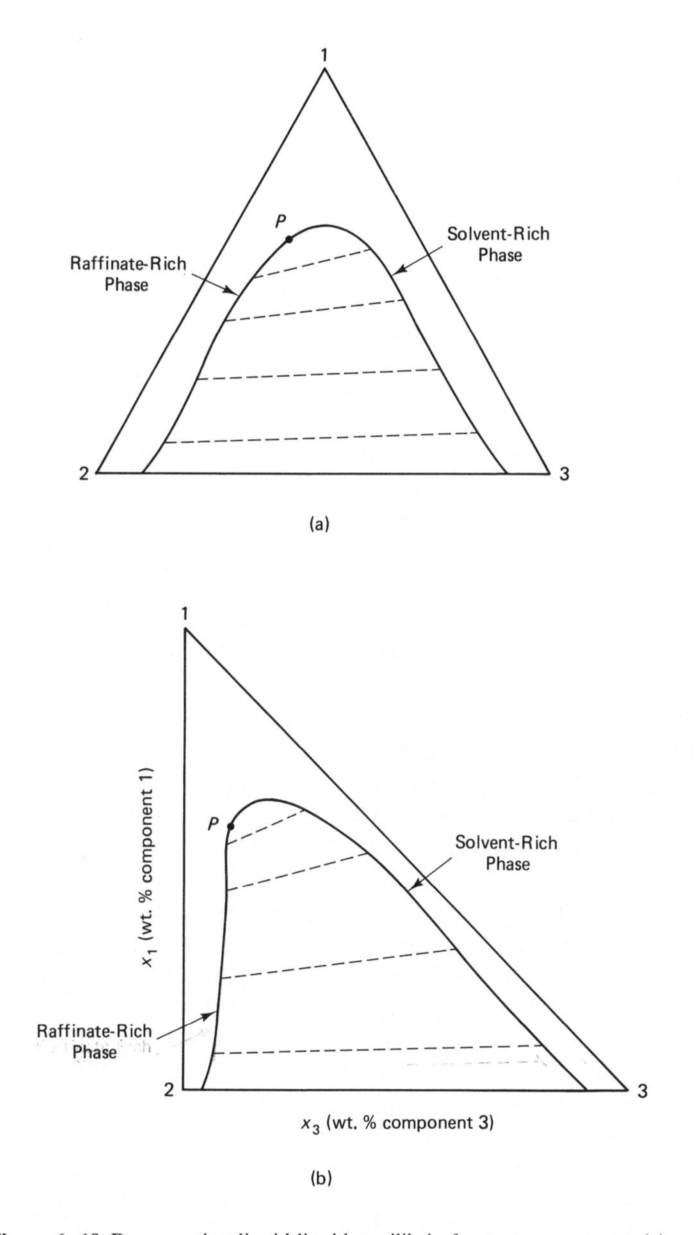

Figure 6–18 Representing liquid-liquid equilibria for ternary systems. (a) Using an equilateral triangle (b) Using a right triangle.

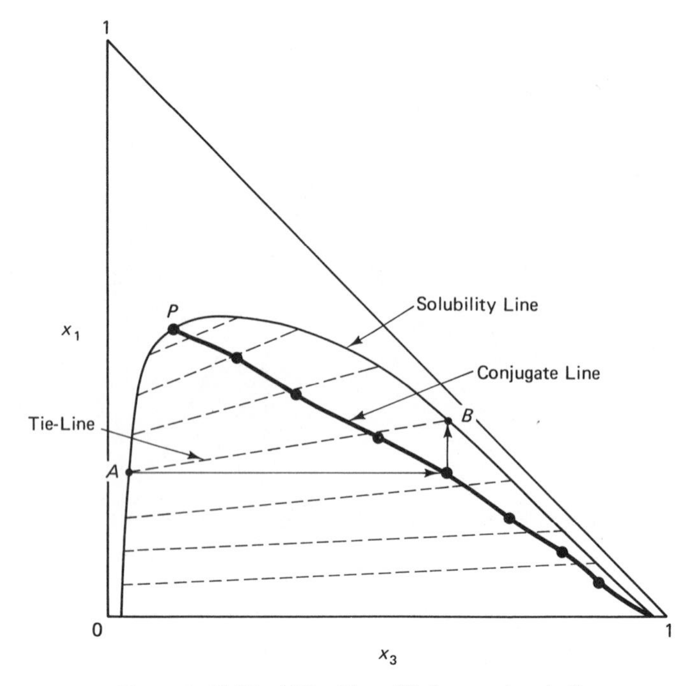

Figure 6–19 Liquid-liquid equilibrium conjugate line.

(x_j^S). The LLE tie-lines and the equilibrium phase boundary are normally found by laboratory experimentation.

A mixture that has an overall composition inside the two-phase region will split into two liquid phases with compositions given at the two ends of the LLE tie-line. There are, of course, an infinite number of tie-lines which lie between the few shown. Figure 6–19 shows how a *conjugate* line can be used to locate the tie-lines. From point A on the left phase boundary, the other end of the tie-line is found by drawing a horizontal line to the conjugate line. A vertical line is then drawn from the point of intersection to the right phase boundary. The point of intersection of this line and the right phase boundary (point B in the figure) is the other end of the tie-line.

As the system becomes richer in solute, the tie-lines get shorter and ultimately become just a point at the *plait point* P. Outside the two-phase region, a single, homogeneous liquid phase exists.

As temperature changes, so does solubility. Usually, the solubility increases as the temperature increases, as shown in Figure 6–20. For this reason, most liquid-liquid extraction systems operate at low temperatures and sometimes even require refrigeration. Pressure, on the other hand, has little effect on solubility.

Notice that if we specify only one concentration of one liquid phase, all

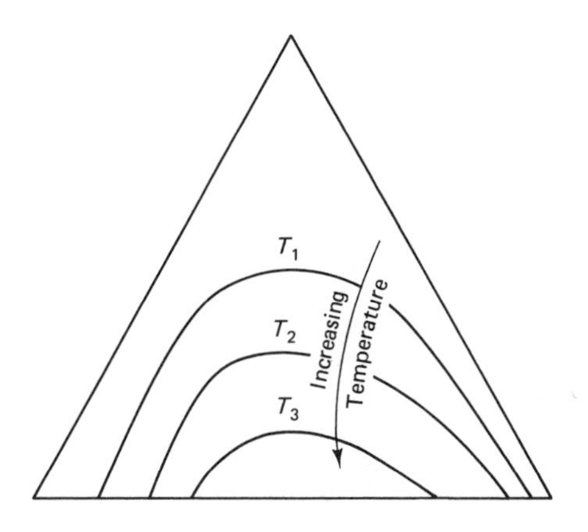

Figure 6–20 Temperature effect on ternary phase diagram.

the other concentrations can be immediately determined from the phase diagram. For example, if we fix the concentration of component 1 in the raffinate-rich phase (x_1^R), we can read from the diagram:

1. The concentration of component 3 in the raffinate-rich phase (x_3^R), by using the left side of the solubility curve.
2. The concentrations of components 1 and 3 in the solvent-rich phase that is in equilibrium with the raffinate-rich phase, by going to the other end of the LLE tie-line. The concentrations x_1^S and x_3^S are read from the right side of the solubility curve. It is clear that the system shown has only one degree of freedom.

Example 6.5.

Thirty thousand kg/hr of a ternary mixture of 19 weight percent isopropyl alcohol (IPA), 41 weight percent toluene, and 40 weight percent water are fed into a decanter operating at 25°C. Figure 6–21 gives the LLE data for the system. Determine the compositions and flow rates of the two liquid streams leaving the decanter.

Solution:

Let component 1 be the solute (IPA) and component 3 the solvent (water). Then the abscissa of the figure is in percent water, and the ordinate is in percent IPA.

The overall compositions of the feed (z_1 = 19 percent and z_3 = 40 percent) are located on the diagram. The compositions of the two liquid phases are read off the diagram at the two ends of the LLE tie-line. The raffinate-rich phase is 14 percent IPA and 2 percent water (the rest being toluene). The solvent-rich phase is 23 percent IPA and 74 percent water.

The flow rates of the two liquid phases can be calculated from a total mass balance and a component balance. Let the total mass flow rates of the solvent-rich

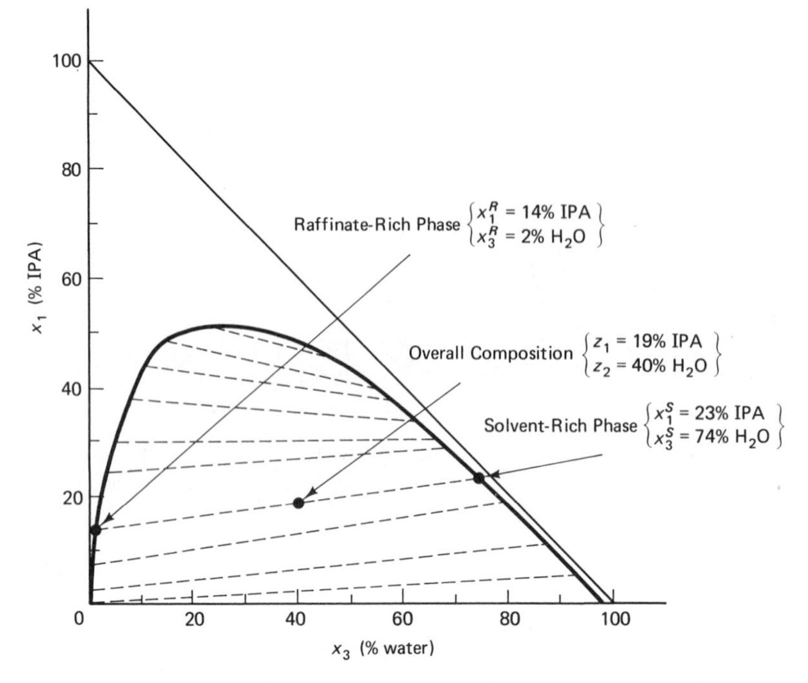

Figure 6–21 Liquid-liquid equilibrium diagram for example 6.5

phase be S (kg/hr) and of the raffinate-rich phase be R (kg/hr). First, we use a water balance to calculate S and R, and then we check the IPA balance.

$$\text{Total mass: } 30{,}000 = S + R$$

$$\text{Water: } (30{,}000)(0.40) = (S)(0.74) + (R)(0.02)$$

Solving simultaneously gives

$$S = 15{,}833 \text{ kg/hr}$$

$$R = 14{,}167 \text{ kg/hr}$$

Theoretically, the IPA should balance perfectly. However, we should recognize that we cannot read the LLE data from the diagram too accurately, so there will be some error. We have

$$\text{IPA in} = (30{,}000)(0.19) = 5{,}700 \text{ kg/hr IPA}$$

$$\text{IPA out} = (S)(0.23) + (R)(0.14)$$

$$= (15{,}833)(0.23) + (14{,}167)(0.14) = 5{,}625 \text{ kg/hr IPA}$$

This is a pretty good check considering the accuracy of reading compositions from the diagram.

PROBLEMS

6–1. A phase diagram for an isopropanol-water binary system at 380 mm Hg pressure is shown below.

(a) What is the boiling-point temperature of pure isopropanol at 380 mm Hg?
(b) What is the boiling-point temperature of pure water at 380 mm Hg?
(c) What is the bubblepoint temperature of a 20 mole percent mixture of isopropanol and water at 380 mm Hg?
(d) What is the composition of the vapor that is in equilibrium with the liquid at this temperature?
(e) What is the dewpoint temperature of a 20 mole percent mixture?
(f) What is the composition of the liquid in equilibrium with this vapor?

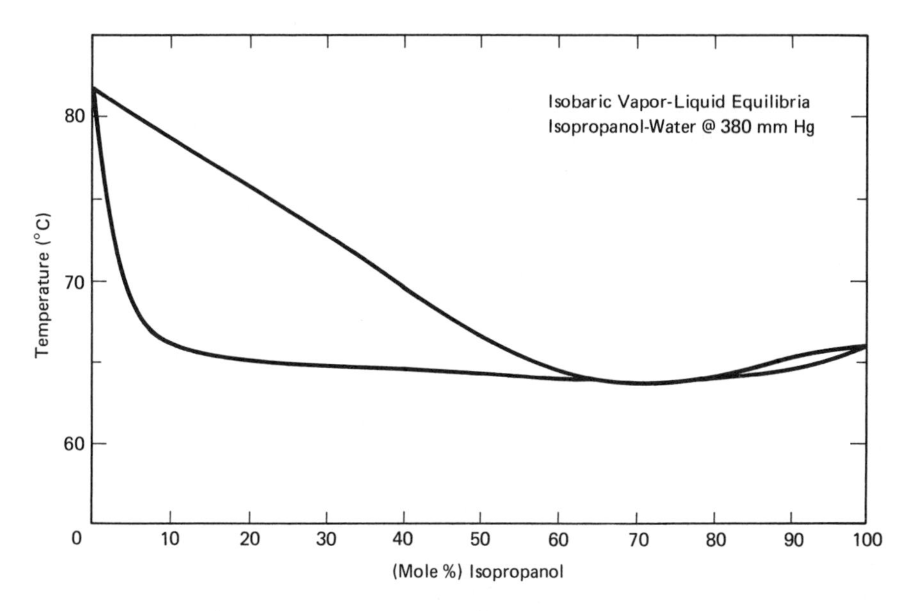

Txy diagram for the isopropanol-water system, $P = 380$ mm Hg.

6–2. A phase diagram for a methanol-water binary system at 39.9°C is shown on page 196.

(a) What are the vapor pressures of pure methanol and pure water at 39.9°C?
(b) What is the bubblepoint pressure of a 30 mole percent mixture of methanol and water at 39.9°C?
(c) What is the composition of the vapor that is in equilibrium with the liquid at this pressure?
(d) What is the dewpoint pressure of a 30 mole percent mixture?
(e) What is the composition of the liquid in equilibrium with this vapor?

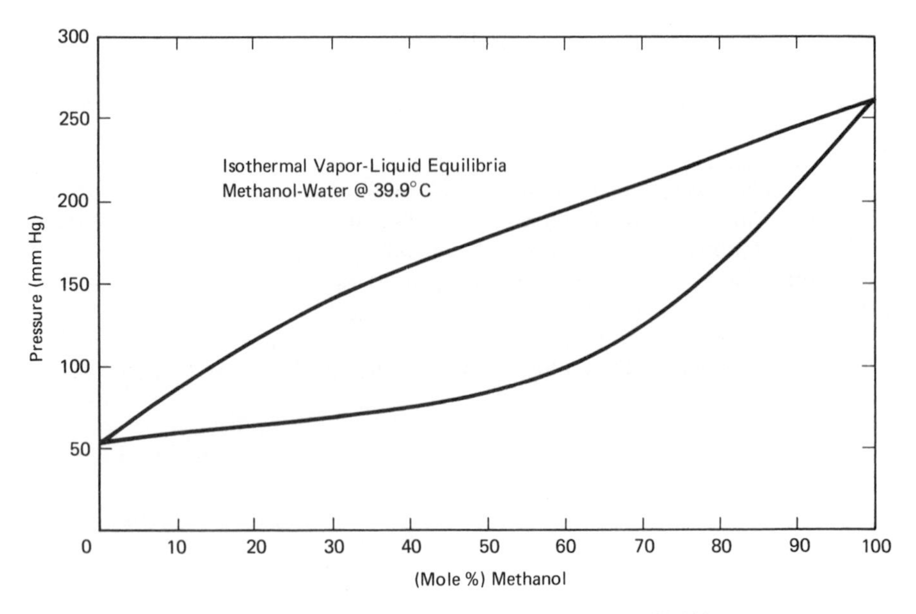

Pxy diagram for the methanol-water system, $T = 39.9$°C.

6–3. Data for a *Txy* diagram are as follows:

T, °F	*x*, *mole* %	*y*, *mole* %
120	0.00	0.00
118	0.02	0.30
116	0.04	0.52
114	0.06	0.66
112	0.09	0.76
110	0.14	0.83
108	0.22	0.87
106	0.32	0.92
104	0.46	0.95
102	0.67	0.98
100	1.00	1.00

(a) If the overhead distillate product from a distillation column separating the components listed is 90 mole percent light component, what is the temperature on the top tray in the column, and what is the temperature in the reflux drum, assuming overhead vapor from the column is totally condensed?

(b) If the temperature on the feed tray is 111°F, what are the compositions of the liquid and vapor streams leaving that tray?

(c) If the bottoms product is a liquid of 1 mole percent light component taken from a partial reboiler, what are the temperature of this liquid and the composition of the vapor stream that returns to the column from this reboiler?

6–4. Thirty thousand lb_m/hr of a ternary mixture of 30 weight percent furfural, 30 weight percent ethylene glycol, and 40 weight percent water are fed into a decanter at 25°C and 101 kPa. Liquid-liquid equilibrium data at this temperature and pressure are given below.

(a) Make equilateral and right triangular plots.

(b) Calculate the compositions and flow rates of the streams leaving the decanter.

MISCIBILITY DATA (WT. %)

Furfural	*Ethylene glycol*	*Water*
94.8	0	5.2
84.4	11.4	4.1
63.1	29.7	7.2
49.4	41.6	9.0
40.6	47.5	11.9
33.8	50.1	16.1
23.2	52.9	23.9
20.1	50.6	29.4
10.2	32.2	57.6
9.2	28.1	62.2
7.9	0	92.1

EQUILIBRIUM DATA (WT. %)

Glycol in water layer	*Glycol in furfural layer*	
49.1	49.1	(Plait Point)
32.1	48.8	
11.5	41.8	
7.7	28.9	
6.1	21.9	
4.8	14.3	
2.3	7.3	

6–5. Ethanol and water form a binary azeotrope at 654 mm Hg and 74.8°C which is 89.65 mole percent ethanol. The vapor pressures of ethanol and water at 74.8°C are 660 mm Hg and 300 mm Hg, respectively. Calculate the activity coefficients of ethanol and water at the azeotropic composition.

6–6. The Chemical Engineers Handbook reports the following data for vapor-liquid equilibrium for a chloroform-benzene system at one atmosphere pressure. (A = chloroform.)

x_A	y_A	T, °C
0	0	80.6
.15	.20	79.0
.29	.40	77.3
.44	.60	75.3
.66	.80	71.9
1.00	1.00	61.4

Make a quantitative evaluation of the ideality or nonideality of the system. Chloroform has a vapor pressure of 5 atm at 120°C; benzene has a vapor pressure of 200 mm Hg at 42.2°C.

6–7. The binary system consisting of normal propyl alcohol and water has an azeotrope at 190°F and 43.2 mole percent n-propyl alcohol at atmospheric pressure. For normal propyl alcohol, the boiling point at atmospheric pressure is 208°F and at 200 mm Hg is 152.2°F. The boiling point of water at 200 mm Hg is 151.7°F.

(a) Sketch the temperature-composition diagram for the system at atmospheric pressure, showing vapor, liquid, and two-phase vapor and liquid regions.

(b) Determine the activity coefficients of n-propyl alcohol and of water at the atmospheric azeotrope.

(c) Sketch an xy diagram for the system at atmospheric pressure.

(d) If the K-value for n-propyl alcohol in water is 15.0 at very low alcohol concentrations, determine the activity coefficient for n-propyl alcohol in water both at very low concentrations ($x \rightarrow 0$) and at very high concentrations ($x = 1.0$).

6–8. Calculate the activity coefficients of ethanol and water at 39.76°C in a liquid with composition 6.89 mole percent ethanol. Vapor-liquid equilibrium data are:

	Mole % Ethanol		
T (°C)	*x (liquid)*	*y (vapor)*	*P (mm Hg)*
39.76	0	0	54.3
39.76	6.89	45.6	81.4
39.76	100	100	129.8

6–9. A five-component mixture has the constant relative volatilities and liquid compositions given below:

Component no.	x_j	α_j
1	.05	10
2	.20	5
3	.35	2
4	.25	1
5	.15	0.2

Calculate the composition of the vapor in equilibrium with this liquid.

6–10. Calculate the activity coefficients of isopropanol and water in a liquid of composition

(a) 10 mole percent isopropanol at 760 mm Hg

(b) 13.05 mole percent isopropanol at 380 mm Hg

Vapor-liquid equilibrium data are:

	Mole % Isopropanol		
P (mm Hg)	*x (liquid)*	*y (vapor)*	*T* (°C)
760	0	0	100
760	10	50.15	82.70
760	100	100	82.25
380	0	0	81.68
380	13.05	52.55	65.59
380	100	100	66.02

6–11. The acetone-chloroform system forms an azeotrope ($P = 1$ atm) at 64.5°C and 65.5 mole percent chloroform. Pure component data are as follows:

For chloroform at $T = 61.3$°C, vapor pressure $P^s = 760$ mm Hg
at $T = 10.4$°C $P^s = 100$ mm Hg
For acetone at $T = 56.5$°C $P^s = 760$ mm Hg
at $T = 7.7$°C $P^s = 100$ mm Hg

(a) Are these compounds mutually attractive or mutually repulsive?

(b) Draw a *Txy* diagram for this system at $P = 1$ atm.

(c) What are the activity coefficients of chloroform and acetone at the 1 atm azeotrope composition?

(d) Draw a flow sheet for a continuous process to separate a 50 mole percent acetone-chloroform solution into pure acetone and pure chloroform. Explain how your process works.

6–12. Ten thousand lb_m/hr of a ternary mixture of 15.8 weight percent isopropyl alcohol (IPA), 67.8 weight percent toluene, and 16.4 weight percent water are fed into a decanter. A liquid-liquid equilibrium diagram is shown below.

(a) What are the flow rates and compositions of the streams leaving the decanter?

(b) If the feed composition is changed to 21.2 weight percent IPA, 19.2 weight percent toluene, and 59.6 weight percent water, what are the new product compositions and flow rates?

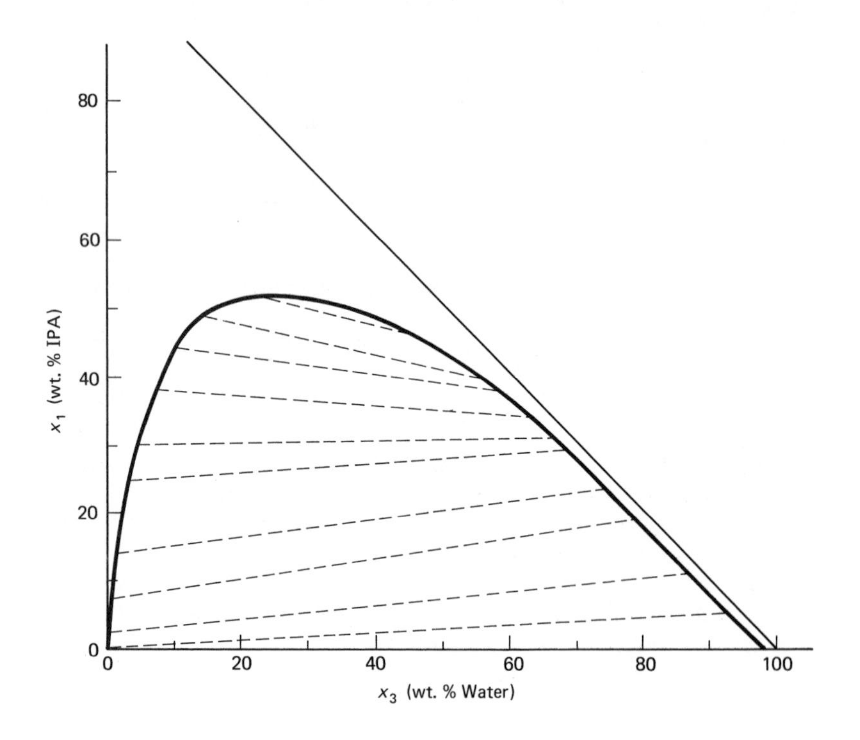

7 *Vapor-Liquid Equilibrium Calculations*

In chapter 6 we considered some of the fundamental principles of phase equilibrium. In this chapter we will illustrate the application of these principles to several kinds of very important vapor-liquid equilibrium calculations. Both ideal (Raoult's Law) and nonideal chemical systems will be considered, as applied to binary and multicomponent mixtures. The basic equation that we will use for vapor-liquid equilibrium is

$$y_j P = x_j P_j^s \gamma_j \tag{7-1}$$

7.1 BUBBLEPOINT AND DEWPOINT CALCULATIONS FOR IDEAL SYSTEMS

Since ideal systems have activity coefficients that equal unity, equation 7-1 simplifies to

$$y_j P = x_j P_j^s \tag{7-2}$$

This relationship is used in both bubblepoint and dewpoint calculations. For Raoult's Law to hold, the vapor must be an ideal gas, and both liquid and vapor phases must form ideal solutions. (That is, molecules in the mixture act as if they are all identical.) This is a stringent limitation for the liquid phase.

7.1.1 Bubblepoint Calculations

Bubblepoint calculations are performed when the composition of the liquid phase (all of the x_j's) is known and we wish to calculate the composition of the vapor phase (the y_j's). There are two cases: when the pressure of the system is known, and when the temperature of the system is known.

Bubblepoint Temperature Calculations Given the pressure P and liquid composition x_j of the system, we wish to calculate the temperature T of the system and the composition y_j of the vapor that is in phase equilibrium with the liquid. Actually, we have already done this calculation graphically for binary systems on the Txy diagrams in chapter 6. There, the Txy diagram at the given pressure was obtained, and then the temperature of the system with a specified x was read off the T-versus-x curve. Finally, the value of y was read off the T-versus-y curve at the same temperature. Here, we wish to solve this problem analytically, using equation 7–2.

Equation 7–2 can be rearranged to give

$$y_j = \frac{x_j P_j^s}{P} \tag{7–3}$$

The sum of all the y_j's must add up to unity; that is,

$$\sum_{j=1}^{NC} \left[\frac{x_j P_j^s}{P}\right] = 1 \tag{7–4}$$

where

$$\text{NC} = \text{number of components}$$

Since P is a constant, it can be pulled outside the summation, yielding

$$P - \sum_{j=1}^{NC} [x_j P_j^s] = 0 \tag{7–5}$$

Equation 7–5 is one equation in one unknown—temperature—since the pure component vapor pressures are functions only of temperature. Everything else is known.

In theory, equation 7–5 can be solved analytically for T, as long as we have Antoine-type equations for the vapor pressures. However, it is seldom possible to do so because of the exponential functions. Therefore, bubblepoint calculations almost always require iterative solutions. The procedure is:

1. Guess the temperature T.
2. Calculate (or look up) the vapor pressures $P_j^s = f_{(T)}$ of each component at T.

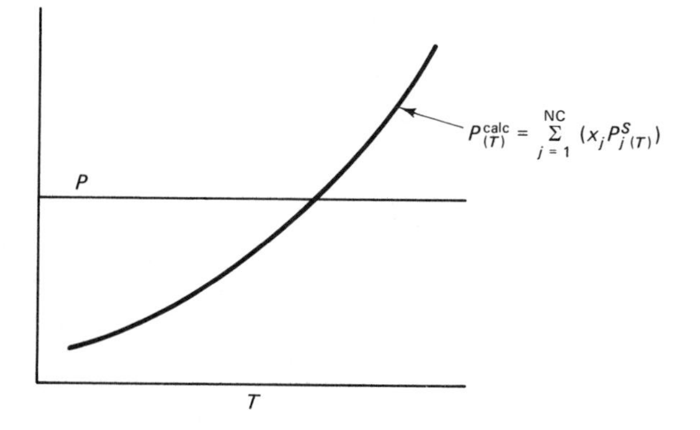

Figure 7–1 Bubblepoint calculation.

3. Calculate

$$P^{calc} = \sum_{j=1}^{NC} [x_j P_j^s] \qquad (7\text{–}6)$$

which depends only on temperature.

4. Check to see whether P^{calc} is sufficiently close to the actual pressure P. Typically, a convergence criterion of 10^{-5} or 10^{-6} relative error is used, where

$$\text{Relative error} = \left| \frac{[P^{calc} - P]}{P} \right| \qquad (7\text{–}7)$$

5. If the absolute value of the relative error is small enough, the correct temperature has been guessed, and the vapor compositions can be calculated from

$$y_j = \frac{x_j P_j^s}{P} \qquad (7\text{–}8)$$

6. If the error is too big, a new temperature guess is made and the procedure is repeated.

Of course, a good method for reguessing the temperature is required. Figure 7–1 sketches the shape of the function $P^{calc}_{(T)}$. If P^{calc} is greater than P, the temperature guessed was too high. (The large T gave large vapor pressures, and these resulted in a value for P^{calc} that was too high.) If P^{calc} was smaller than P, the temperature guessed was too low.

Example 7.1.

The liquid in a process vessel contains 40 mole percent benzene, 35 mole percent toluene, and 25 mole percent o-xylene. The total pressure in the vessel is 2 atmospheres. What is the temperature in the vessel, and what is the composition of the vapor?

Solution:

Let x_1 be the mole fraction benzene in the liquid, x_2 be the mole fraction toluene, and x_3 be the mole fraction o-xylene. The pressure is 2 atmospheres, so $P = 1{,}520$ mm Hg. A temperature of 120°C is guessed first. The vapor pressures of the pure components are looked up at 120°C, and P^{calc} is calculated.

Guess $T = 120$°C:

	Component	x_j	P_j^s *at 120°C*	$x_j P_j^s$
1	Benzene	0.40	2,300 mm Hg	920
2	Toluene	0.35	1,000	350
3	o-Xylene	0.25	380	95

$P^{calc} = 1{,}365$ mm Hg

Since P^{calc} is less than P (1,520 mm Hg), a higher temperature must be guessed.

Guess $T = 125$°C:

	Component	x_j	P_j^s *at 125°C*	$x_j P_j^s$	y_j
1	Benzene	0.40	2,600 mm Hg	1,040	0.671
2	Toluene	0.35	1,140	399	0.258
3	o-Xylene	0.25	440	110	0.071

$P^{calc} = 1{,}549$ mm Hg

For a hand calculation, this P^{calc} is close enough to the desired P. When we do these calculations on a computer later in this chapter, we will tighten up the convergence criterion to 10^{-5} or 10^{-6} relative error.

The vapor compositions are calculated from equation 7–8. Notice that the concentration of the lightest component (benzene) is higher in the vapor phase (67.1 mole percent) than in the liquid phase (40 mole percent). For the heaviest component (xylene), the reverse is true: there is 7.1 mole percent in the vapor and 25 mole percent in the liquid.

Bubblepoint Pressure Calculations If the composition of the liquid phase and the temperature are given, the job is to calculate the total pressure of the system and the composition of the vapor. This calculation is easier than a bubblepoint temperature calculation because it is not iterative. Since the temperature is known, the vapor pressures can be read off of a Coxe Chart or calculated from Antoine Equations for each component. Then equation 7–5 is used to calculate the pressure directly. The procedure is as follows:

1. Calculate (or look up) the vapor pressures of all the pure components at the given temperature.
2. Calculate the total pressure of the system from

$$P = \sum_{j=1}^{NC} x_j P_j^s \qquad (7\text{–}9)$$

3. Calculate the vapor compositions from equation 7–8.

Example 7.2.

A liquid mixture of 40 mole percent benzene, 35 mole percent toluene, and 25 mole percent o-xylene is held at 100°C in a closed vessel. What is the pressure in the vessel, and what is the composition of the vapor that is in equilibrium with the liquid?

Solution:

The vapor pressures of the pure components are looked up at 100°C. At this temperature, they are 1,360 mm Hg for benzene, 550 mm Hg for toluene, and 200 mm Hg for o-xylene. Thus, we have:

	Component	x_j	P_j^s *at 100°C*	$x_j P_j^s$	y_j
1	Benzene	0.40	1,360 mm Hg	544	0.692
2	Toluene	0.35	550	192	0.244
3	o-Xylene	0.25	200	50	0.064

$P = 786$ mm Hg

7.1.2 Dewpoint Calculations

In bubblepoint calculations, the liquid compositions are known but the vapor compositions are unknown. In dewpoint calculations, the reverse is true: the vapor compositions are known but the liquid compositions are unknown.

There are two types of dewpoint calculations: determining the dewpoint pressure when the temperature is given, and determining the dewpoint temperature when the pressure is given. The first requires no iteration; the second involves a trial-and-error solution converging on the correct temperature.

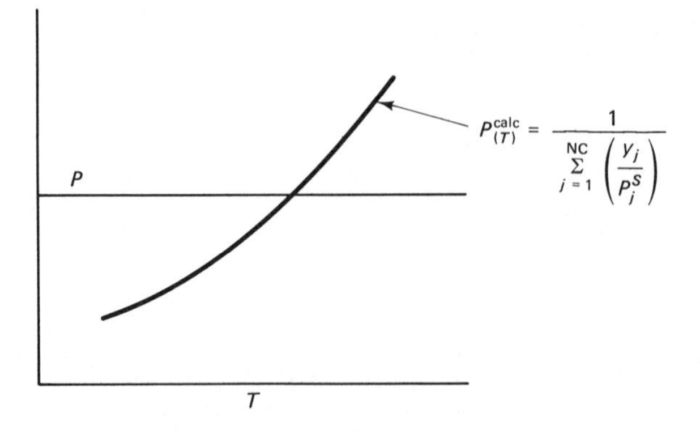

Figure 7–2 Dewpoint calculation.

Both calculations are based on equation 7–2. We rearrange the equation to solve for x_j and add up all the x_j's to get unity:

$$x_j = \frac{y_j P}{P_j^s} \tag{7-10}$$

$$\sum_{j=1}^{NC} x_j = 1 = \sum_{j=1}^{NC} \left[\frac{y_j P}{P_j^s}\right] = P \sum_{j=1}^{NC} \left[\frac{y_j}{P_j^s}\right] \tag{7-11}$$

$$P = \frac{1}{\left[\sum_{j=1}^{NC} (y_j/P_j^s)\right]} \tag{7-12}$$

Equation 7–12 is one equation with one unknown if we are given the y_j's and either pressure or temperature. If the temperature is known, the pure component vapor pressures are determined, and P is calculated directly. If the pressure is given, a temperature guess is made, and equation 7–12 is used to check whether the guess is correct.

Since a dewpoint temperature calculation is the more difficult of the two, we provide an illustration of it. The problem is, given P and y_j, we wish to calculate T and x_j. The method is as follows:

1. Guess a value of the temperature T.
2. Determine the pure component vapor pressures at this temperature.
3. Calculate P^{calc} from equation 7–12.
4. Test to see whether P^{calc} is sufficiently close to P.

5. If P^{calc} is too far from P, reguess T. If the temperature guess was too high, vapor pressures will be big, and the denominator of equation 7–12 will be small, resulting in a value of P^{calc} that is too big. If this is the case, a lower temperature must be guessed. Figure 7–2 sketches the functional dependence of P^{calc} on temperature in a dewpoint temperature calculation.

Example 7.3.

Vapor and liquid phases are present in a vessel at 760 mm Hg pressure. The composition of the vapor phase is 40 mole percent benzene, 35 mole percent toluene, and 25 mole percent o-xylene. Calculate the temperature in the vessel and the composition of the liquid phase in equilibrium with the vapor.

Solution:

Since the composition of the vapor phase is known, we make a dewpoint calculation. The pressure is specified, but the temperature must be found iteratively. First, we guess a temperature of 125°C.

Guess T = 125°C:

Component	y_j	P_j^s *at 125°C*	y_j/P_j^s
Benzene	0.40	2,600 mm Hg	0.0001538
Toluene	0.35	1,140	0.000307
o-Xylene	0.25	440	0.000568
			0.001029

$$P^{calc} = 1 \Bigg/ \left[\sum_{j=1}^{NC} (y_j/P_j^s) \right] = 1/0.001029 = 972 \text{ mm Hg}$$

P^{calc} is bigger than P (760 mm Hg), so a lower temperature must be guessed.

Guess T = 120°C:

Component	y_j	P_j^s *at 120°C*	y_j/P_j^s
Benzene	0.40	2,300	0.000174
Toluene	0.35	1,000	0.000350
o-Xylene	0.25	380	0.000658
			0.001182

$$P^{calc} = 1/0.001182 = 846 \text{ mm Hg}$$

P^{calc} is still bigger than P, so an even lower temperature must be used.

Guess $T = 115°C$:

Component	y_j	P_j^s *at 115°C*	y_j/P_j^s	x_j
Benzene	0.40	2,040	0.000196	0.142
Toluene	0.35	860	0.000407	0.294
o-Xylene	0.25	320	0.000781	0.564
			0.001384	1.000

$$P^{calc} = 1/0.001384 = 723 \text{ mm Hg}$$

P^{calc} is now fairly close to the correct P of 760 mm Hg. Accordingly, we can calculate $x_j = y_j P^{calc}/P_j^s$. This is shown in the last column of the table.

A common mistake made by students is to look at equation 7–12 and think that instead of using $1/\Sigma(y_j/P_j^s)$ to calculate P, they can use $\Sigma(P_j^s/y_j)$. This is, of course, wrong.

7.2 NONIDEAL SYSTEMS

In section 7.1, we assumed that the systems were ideal (i.e., obeyed Raoult's Law). In many systems, however, Raoult's Law cannot be used. Bubblepoints and dewpoints are still needed and can be calculated; however, K-values or activity coefficients must be available. Equation 7–1 is used if activity coefficients are available. The De Priester charts (see chapter 6) can be used to determine K-values in some light hydrocarbon systems.

7.2.1 Bubblepoint Calculations

The procedures for nonideal bubblepoint calculations are similar to those used in the ideal case. Activity coefficients are strong functions of liquid compositions and weak functions of temperature. Since the liquid compositions and the temperature are known in bubblepoint pressure calculations, the activity coefficients can be calculated directly. In bubblepoint temperature calculations, the slight temperature dependence can be included by recalculating the activity coefficients at each iteration.

If K-values are used, the bubblepoint calculations use the equations

$$y_j = K_j x_j \tag{7–13}$$

$$\sum_{j=1}^{NC} y_j = 1 = \sum_{j=1}^{NC} K_j x_j$$

where the K-values are functions of temperature, pressure, and composition. If a bubblepoint temperature is to be calculated, a guess of temperature is made, and the K-values of all components are determined (for example, from the De Priester charts—see Figure 6–14). If the sum of the $K_j x_j$ is unity, the correct temperature has been guessed.

The De Priester charts give K-values for light hydrocarbon systems. (Note that only the simplest systems are presented.) Since the charts are nomographs, a straightedge connecting the temperature and pressure for which K-values are needed will intersect curves for each compound at its K-value. These K-values have been obtained by calculating fugacities from an equation of state. They account for temperature and pressure dependence, but do not account for composition dependence. Therefore, they are only good first approximations. Of course, they are more nearly correct than simply using vapor pressures and Raoult's Law.

7.2.2 Dewpoint Calculations

Dewpoint calculations for nonideal systems are a little more difficult than bubblepoint calculations. Since the liquid compositions are unknown, the activity coefficients cannot be directly calculated. Rather, an iterative procedure is required.

First, an initial guess of the activity coefficients must be made. Perhaps we assume that they are unity (ideal). Then a temperature guess is made, and vapor pressures and x_j's are calculated. The x_j values are then used in the next round of the iteration to calculate new activity coefficients. Note that if the activity coefficients are very strong functions of composition, it is sometimes difficult to achieve convergence of the iterative calculations.

The relative ease of doing bubblepoint calculations as compared with doing dewpoint calculations is precisely the reason why most distillation calculations are started at the bottom of the column and work their way up from tray to tray. Liquid compositions are determined from component balances, and then vapor compositions are determined from bubblepoint calculations. If the calculations were started at the top of the column, vapor compositions would be determined from component balances, and then liquid compositions would have to be determined from dewpoint calculations.

7.3 COMPUTER CALCULATIONS

When using a computer for iterative calculations such as bubblepoint and dewpoint determinations, numerical methods must be used to make successive guesses and to converge to the correct solution. A host of methods have been developed to do this. We will present two of the simplest and most useful of these techniques: interval-halving and Newton-Raphson. They will be used in many of the programs presented throughout this book.

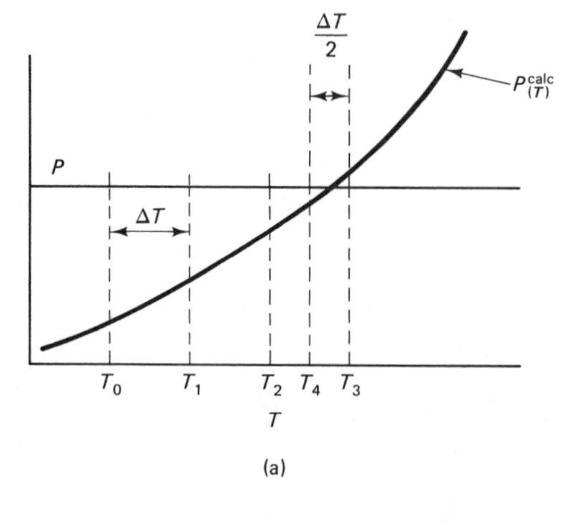

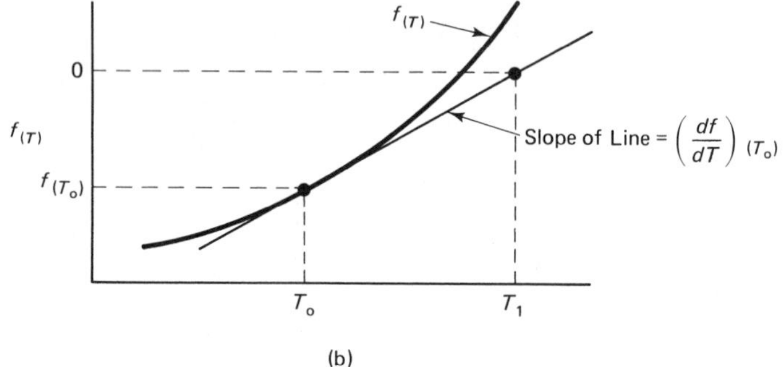

Figure 7–3 Numerical convergence methods. (a) Interval-halving (b) Newton-Raphson iteration.

7.3.1 Interval-Halving Convergence Technique

Interval-halving is a brute-force approach to the problem, but it is remarkably reliable and very useful. It requires an initial guess of the solution T_0, as shown in Figure 7–3(a). If the guess is not correct, a step in the guessed variable is made in the appropriate direction to T_1. Iterations are then continued at this constant step size (T_2 and T_3) until a test of the solution shows that the desired end point has been passed. Once this occurs, the step size is halved at each subsequent iteration, while moves are made in the appropriate directions until the solution has been obtained to the desired precision.

The method is slow in converging, but it is relatively easy to program and does converge reliably. Table 7–1 gives a FORTRAN program for doing

TABLE 7–1
FORTRAN ROUTINE TO PERFORM BUBBLEPOINT CALCULATIONS BY INTERVAL-HALVING

```
SUBROUTINE EQUIL(X,T,Y,PSIA)
C   BUBBLE POINT CALCULATIONS  FOR METHANOL-WATER USING
INTERVAL-HALVING
      D1=-337.5*373.*ALOG(2600./760.)/35.5
      C1=ALOG(760.)-D1/337.5
      D2=-332.4*373.*ALOG(760./145.4)/40.6
      C2=ALOG(760.)-D2/373.
      A=0.85
      B=0.48
      X1=X
      X2=1.-X
      GAM1=EXP(A*X2*X2/((A*X1/B+X2)**2))
      GAM2=EXP(B*X1*X1/((X1+B*X2/A)**2))
      P=PSIA*760./14.7
    3 DT=2.
      LOOP=0
      FLAGM=1.
      FLAGP=1.
   10 LOOP=LOOP+1
      IF(LOOP.GT.50) STOP
      PS1=EXP(C1+D1/(T+273.))
      PS2=EXP(C2+D2/(T+273.))
      PCALC=X1*PS1*GAM1+X2*PS2*GAM2
      IF(ABS(P-PCALC) .LT. P/10000.) GO TO 50
      IF(P-PCALC)20,20,30
   20 IF(FLAGM)21,21,22
   21 DT=DT*.5
   22 T=T-DT
      FLAGP=-1.
      GO TO 10
   30 IF(FLAGP)31,31,32
   31 DT=DT*.5
   32 T=T+DT
      FLAGM=-1.
      GO TO 10
   50 Y=X1*PS1*GAM1/P
      RETURN
      END
```

bubblepoint temperature calculations using interval-halving. The chemical system presented is the nonideal methanol-water binary mixture. Component 1 is methanol and component 2 is water. Vapor pressure data for methanol and water are read from the Coxe chart and used in the equations

$$\ln P_1^s = C_1 + \frac{D_1}{T}$$

$$\ln P_2^s = C_2 + \frac{D_2}{T}$$

The variables A and B used in the program are constants for the Van Laar equations for methanol-water that give the composition dependence of the activity coefficients. The Van Laar equations are commonly used to fit activity coefficient versus composition data. We have, for methanol,

$$\gamma_1 = \exp \left| \frac{Ax_2^2}{[x_2 + (A/B)x_1]^2} \right| \tag{7–14}$$

and for water,

$$\gamma_2 = \exp \left| \frac{Bx_1^2}{[x_1 + (B/A)x_2]^2} \right| \tag{7–15}$$

where

x_1 = mole fraction of methanol

x_2 = mole fraction of water

γ_1 = activity coefficient of methanol

γ_2 = activity coefficient of water

The iteration procedure using interval-halving starts with statement 3, where initial values are set for the step size DT and the number of iterations LOOP. FLAGM and FLAGP control, by their signs, whether or not the interval is halved.

7.3.2 Newton-Raphson Iteration

The Newton-Raphson iteration technique is the most widely used iterative convergence method. It requires the determination of the derivative of the function, which can be done either analytically or numerically. In bubblepoint temperature calculations where the temperature T is the unknown, the derivative can usually be determined analytically since the vapor pressures are known functions of temperature.

The method is easily understood if one looks at it graphically, as illustrated in Figure 7–3(b). Suppose we have an equation that can be expressed in the form

$$f_{(T)} = 0 \tag{7–16}$$

In a bubblepoint temperature calculation, this would be

$$P - \sum_{j=1}^{NC} [x_j P_j^s \gamma_j] = 0 \tag{7–17}$$

We wish to find the value of T that makes the function zero, i.e., the root of the equation $f_{(T)} = 0$. A guess of temperature T_0 is made, and the function is evaluated to see if it is zero at T_0. If it is not, the slope df/dT of the function is evaluated at T_0, and a new guess T_1 is determined from the equation

$$T_1 = T_0 - \frac{f_{(T_0)}}{[df/dT]_{(T_0)}}$$

where

$f_{(T_0)}$ = value of the function at $T = T_0$

$[df/dT]_{(T_0)}$ = value of the derivative of the function with respect to T when $T = T_0$

The function is then evaluated at T_1 to see whether it is close enough to zero. If it is not, the derivative is evaluated at T_1, and a new guess

$$T_2 = T_1 - \frac{f_{(T_1)}}{[df/dT]_{(T_1)}}$$

is made. The procedure is repeated until the function is close enough to zero. The iteration algorithm at the kth step is

$$T_{k+1} = T_k - \frac{f_{(T_k)}}{[df/dT]_{(T_k)}} \tag{7-18}$$

The Newton-Raphson method often gives rapid convergence if the function $f_{(T)}$ does not change slope markedly or if the initial guess is close to the solution. If there is a large change in slope, the solution may actually diverge.

Table 7–2 gives a FORTRAN subroutine in which the bubblepoint temperature calculation is done for an ideal multicomponent mixture using the Newton-Raphson technique. The x_j values, pressure P, and a temperature guess are fed into the subroutine, which then returns the correct temperature and the y_j values to the main program.

If the vapor pressures are given by the Antoine Equation, the value of the derivative is

$$\frac{df}{dT} = \sum_{j=1}^{NC} \left[x_j \exp\left(C_j + \frac{D_j}{T} \right) \left(\frac{-D_j}{T^2} \right) \right] \tag{7-19}$$

The subroutine is dimensioned so that up to ten components can be handled. Only ideal mixtures are used, since all the activity coefficients are unity.

7.4 ISOTHERMAL FLASH CALCULATIONS

So far, we have illustrated calculations in which we know either the liquid composition (bubblepoint) or the vapor composition (dewpoint). There is another class of cases, however, where neither the vapor nor the liquid com-

TABLE 7–2
FORTRAN ROUTINE TO PERFORM BUBBLEPOINT CALCULATIONS BY NEWTON-RAPHSON TECHNIQUE

```
C    MULTI-COMPONENT (IDEAL)BUBBLE-POINT CALCULATIONS VIA NEWTON-RAPHSON
C   D AND C PARAMETERS FOR EACH COMPONENT ARE SUPPLIED
C       THROUGH COMMON
      SUBROUTINE EQUIL(X,T,Y,PSIA,NC)
      DIMENSION D(10),C(10),X(10),Y(10),PS(10)
      COMMON D,C
C    VAPOR PRESSURES ARE GIVEN IN "MM HG"
      P=760.*PSIA/14.7
      LOOP=0
   10 LOOP=LOOP+1
      IF(LOOP.GT.50) STOP
      PCALC=0.
      DO 20 J=1,NC
      PS(J)=EXP(C(J)+D(J)/(T+273.))
   20 PCALC=PCALC+X(J)*PS(J)
      F=PCALC-P
      IF(ABS(F).LT. P/10000.) GO TO 50
      DF=0.
      DO 30 J=1,NC
   30 DF=DF-X(J)*PS(J)*D(J)/((T+273.)**2)
      T=T-F/DF
      GO TO 10
   50 DO 60 J=1,NC
   60 Y(J)=X(J)*PS(J)/P
      RETURN
      END
```

position is known. A commonly encountered example is the *flash drum*. In this process an entering stream (usually a liquid at high pressure) flashes as it comes into a vessel that is at a lower pressure. Significant quantities of both liquid and vapor phases are formed. Figure 7–4 illustrates the system. The variables used are

V = Vapor flow rate (moles/time)

L = Liquid flow rate (moles/time)

F = Feed flow rate (moles/time)

z_j, x_j, y_j = mole fractions of component j in the feed, liquid, and vapor streams, respectively

In a typical *isothermal flash calculation*, the feed flow rate, feed composition, pressure in the vessel, and temperature in the vessel are known. The flow rates and compositions of the liquid and vapor streams are unknown.

- Given: F, z_j, T, and P
- Unknown: V, L, y_j, and x_j

There are 2(NC) unknowns, where NC is the number of components. Accordingly, we need 2(NC) equations to describe the system. Since we have NC component balances, NC more equations are required. These are obtained

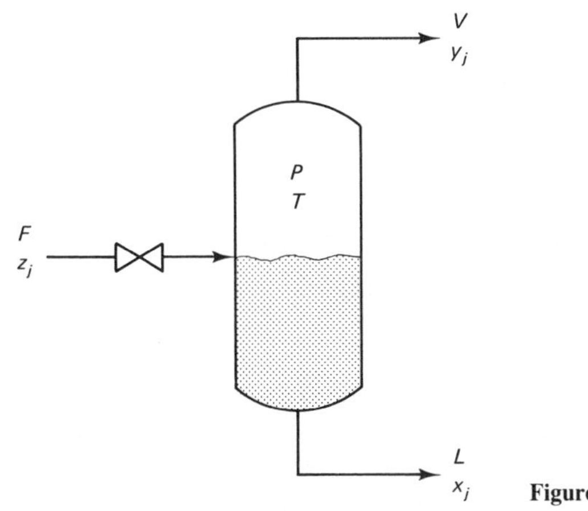

Figure 7–4 Flash drum.

from the VLE relationships, since the liquid and vapor phases are assumed to be in phase equilibrium with each other. We then have our required 2(NC) equations:

Total mass balance (1 equation):

$$F = L + V \tag{7–20}$$

Component mass balances (NC − 1 equations):

$$Fz_j = Lx_j + Vy_j \tag{7–21}$$

VLE equations (NC equations):

$$\frac{y_j}{x_j} = K_j = \frac{P_j^s \gamma_j}{P} \tag{7–22}$$

A convenient method for solving these algebraic equations simultaneously is as follows. Both equations 7–22 and 7–20 are substituted into equation 7–21:

$$z_j F = x_j L + y_j V$$

$$z_j F = (F - V)x_j + VK_j x_j \tag{7–23}$$

Solving for x_j gives

$$x_j = \frac{z_j F}{F - V + VK_j} \tag{7–24}$$

Dividing numerator and denominator by F yields

$$x_j = \frac{z_j}{1 + (V/F)[K_j - 1]} \tag{7–25}$$

If the system is ideal, $K_j = P_j^s/P$, and equation 7–25 becomes

$$x_j = \frac{z_j}{1 + \left(\frac{V}{F}\right)\left[\frac{P_j^s}{P} - 1\right]} \tag{7-26}$$

Equation 7–25 or 7–26 can be written for each of the NC components in the system. Adding up all the x_j's then gives one equation:

$$\sum_{j=1}^{NC} \left\{\frac{z_j}{1 + (V/F)\,[K_j - 1]}\right\} = 1 \tag{7-27}$$

If the K_j's are functions of temperature and pressure only (both of which are known), equation 7–27 represents one equation in one unknown (V/F, i.e., the fraction of the feed that is vaporized). We subsequently describe an iterative method for solving for this quantity. If, on the other hand, the K_j's depend on the liquid composition (i.e., the activity coefficients are not unity), a more complicated iteration is required.

A *flash calculation* is an iterative trial-and-error procedure where the problem is to guess the V/F ratio that makes the sum of the x_j's add up to unity.

7.4.1 Flash Calculation Iteration for Ideal Systems

The first thing to do in any flash calculation is to make sure that something will flash, i.e., that there will be both liquid and vapor phases. This is because there is no guarantee that the feed will produce both liquid and vapor under the conditions in the flash tank.

If the temperature specified in the tank is above the dewpoint temperature of the feed at the given pressure, all of the feed will leave as superheated vapor. If the temperature specified in the tank is below the bubblepoint of the feed at the given pressure, all of the feed will leave as subcooled liquid. So first of all, we must see whether the specified temperature and pressure put us in the two-phase region.

Bubblepoint Either bubblepoint temperature or bubblepoint pressure calculations can be used. Since the pressure calculations do not require an iterative solution, we will use them.

If we are in a region where nothing flashes, the pressure of the system must be higher than the bubblepoint pressure of the feed at the specified temperature. The vapor pressures of the pure components are determined at the temperature of the system, and the bubblepoint pressure of the feed is calculated as

$$P_{\text{BUB}} = \sum_{j=1}^{\text{NC}} [z_j P_j^s] \tag{7-28}$$

If P is greater than P_{BUB}, nothing will flash, so $V/F = 0$ and $x_j = z_j$.

Dewpoint If we are in a region where everything flashes, the pressure of the system must be lower than the dewpoint pressure of the feed at the specified temperature. The latter is given by

$$P_{\text{DEW}} = \frac{1}{\left[\sum_{j=1}^{\text{NC}} (z_j / P_j^s)\right]} \tag{7-29}$$

If P is less than P_{DEW}, everything flashes, so $V/F = 1$ and $y_j = z_j$.

Flash Calculation Once we know that two phases will be present, equation 7–26 must be solved for V/F. Figure 7–5(a) gives the typical shape of the function XSUM when plotted against V/F. The only region of interest is between $V/F = 0$ and $V/F = 1$; any mathematical solution outside this region is not real. We have

$$\text{XSUM} = \sum_{j=1}^{\text{NC}} \left\{ \frac{z_j}{1 + \left(\frac{V}{F}\right)\left[\frac{P_j^s}{P} - 1\right]} \right\} \tag{7-30}$$

This function is not monotonic. Consequently, although an iterative algorithm like interval-halving can be used, use of a derivative-based algorithm like the Newton-Raphson technique may result in convergence problems.

Note that the function is always equal to unity for any values of temperature and pressure when V/F is set equal to zero. Thus, the equation usually has two roots. One of them is the real root, the value of V/F between 0 and 1. The other is a fictitious root which satisfies the equation but does not correspond to reality.

Figure 7–5(b) shows several curves for different temperatures, indicating how the function XSUM varies with V/F at a fixed system pressure. All the curves go through the point (0,1). If the temperature is above the dewpoint, there is no other point of intersection of XSUM with the horizontal line y = 1 inside the region of interest, i.e., between $V/F = 0$ and $V/F = 1$. If the temperature is below the bubblepoint, there is also no other intersection.

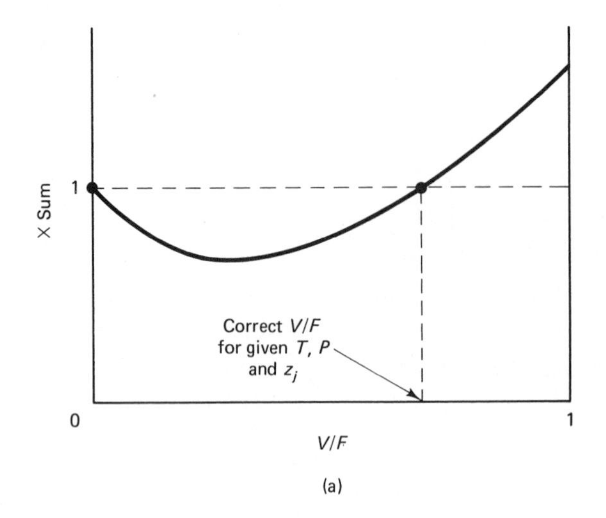

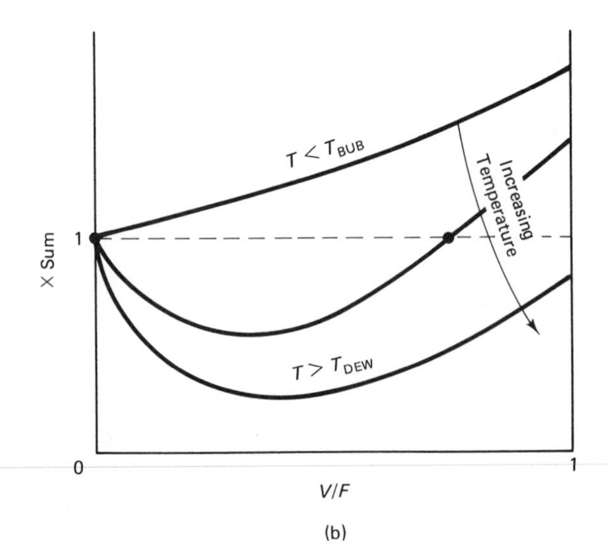

Figure 7–5 Dependence of $\sum_{j=1}^{NC} x_j$ on the ratio V/F. (a) Both V and L present (b) Dependence when either V or L is not present.

Example 7.6.

A liquid feed stream containing 40 mole percent benzene, 35 mole percent toluene, and 25 mole percent ortho-xylene is flashed to 110°C and 760 mm Hg. Determine the flow rates and compositions of the vapor and liquid streams after the flash for 100 kg-mole/hr of feed. The VLE relationships can be assumed to be ideal, i.e., the system obeys Raoult's Law.

Solution:

The first thing to do in any flash calculation is to check whether the actual pressure is between the bubblepoint and dewpoint pressures of the feed at the operating temperature. The vapor pressures at 110°C (230°F) are read from the Coxe chart given in chapter 6.

Vapor pressures at 110°C

$$\text{Benzene} = P_1^s = 33 \text{ psia}$$

$$\text{Toluene} = P_2^s = 14 \text{ psia}$$

$$\text{o-Xylene} = P_3^s = 5.6 \text{ psia}$$

$$P_{\text{BUB}} = (0.4)(33) + (0.35)(14) + (0.25)(5.6) = 19.5 \text{ psia}$$

$$P_{\text{DEW}} = \frac{1}{[(0.40/33) + (0.35/14) + (0.25/5.6)]} = 12.23 \text{ psia}$$

Since the actual pressure is 14.7 psia, we know that we are in the two-phase region. Now we can proceed with the flash calculation.

We must guess a value of V/F and see whether the sum of the x_j's is unity. Equation 7–26 is used with $P = 14.7$ psia to calculate each of the individual x_j's.

Guess $V/F = 0.5$:

Component	z_j	P_j^s *at 110°C*	x_j
Benzene	0.40	33 psia	0.247
Toluene	0.35	14	0.359
o-Xylene	0.25	5.6	0.362
			0.968

Since Σx_j is less than unity, we must guess a higher value of V/F.

Guess $V/F = 0.7$:

Component	x_j
Benzene	0.214
Toluene	0.362
o-Xylene	0.441
	1.017

Now our guess is too high, so we guess somewhat lower.

Guess $V/F = 0.65$:

Component	x_j
Benzene	0.221
Toluene	0.361
o-Xylene	0.418
	1.000

Now we can calculate the vapor composition from

$$y_j = \frac{x_j P_j^s}{P}$$

We get the following results:

Component	x_j	P_j^s *at 110°C*	y_j
Benzene	0.221	33	0.496
Toluene	0.361	14	0.344
o-Xylene	0.418	5.6	0.159
			0.999

The sum of the y_j's does not add up exactly to unity, indicating that there is a small error in the calculations. The problem is that the correct V/F is not exactly 0.65; if we were to carry more decimal places on the x_j's, we would see this clearly as follows.

Component	x_j *with* $V/F = 0.65$
Benzene	0.2211
Toluene	0.3612
o-Xylene	0.4183
	1.0006

If we were doing this calculation on a computer, we would continue to reguess V/F until the sum of the x_j's was accurate to perhaps one part in one million (0.000001).

The vapor and liquid flow rates are

$$V = (V/F)(F) = (0.65)(100) = 65 \text{ kg-mole/hr}$$

$$L = F - V = 100 - 65 = 35 \text{ kg-mole/hr}$$

7.4.2 Flash Calculations Using *K*-Values

If the *K*-values used in the VLE relationship are dependent on temperature and pressure only, and not on composition, the isothermal flash calculation is essentially the same as if the system were ideal. The equations describing an isothermal flash are then

$$F = L + V$$

$$Fz_j = Lx_j + Vy_j$$

$$y_j = K_j x_j$$

Combining gives equation 7–25:

$$x_j = \frac{z_j}{1 + (V/F)(K_j - 1)}$$

To check whether we are between the bubblepoint and the dewpoint, we calculate the quantities

$$f_1 = \sum [z_j K_j] \tag{7–31}$$

and

$$f_2 = \sum \left[\frac{z_j}{K_j}\right] \tag{7–32}$$

If the feed stream (with composition z_j) is right at its bubblepoint at the temperature and pressure given, the function f_1 (which is the sum of the y_j's) will add up exactly to unity. If the feed is above its bubblepoint temperature, the function f_1 will be greater than unity. This means that some of the feed will flash.

If the feed stream is right at its dewpoint at the temperature and pressure in the drum, the function f_2 (which is the sum of the x_j's) will add up exactly to unity. If the flash temperature is below the dewpoint temperature, some liquid will be formed. Since the K-values will be smaller than at the dewpoint temperature, the ratios z_j/K_j will be bigger. This gives a value of f_2 greater than unity, indicating that some liquid is present.

Instead of using $[\Sigma x_j] - 1$ as the function of V/F to converge to zero, some engineers prefer to use the function

$$f_{(V/F)} = \sum_{j=1}^{NC} [y_j - x_j] = 0 = \sum_{j=1}^{NC} \left\{ \frac{z_j(K_j - 1)}{1 + \dfrac{V}{F}(K_j - 1)} \right\} \tag{7–33}$$

Example 7.7.

A process stream contains 20 mole percent CH_4, 30 mole percent C_2H_6, 30 mole percent C_3H_8, and 20 mole percent nC_4H_{10} at 50 psia and 0°F. Determine the fraction of the stream that is liquid.

Solution:

K-values can be read directly from the De Priester charts at 0°F and 50 psia; we have $K_{CH_4} = 30$, $K_{C_2H_6} = 3.5$, $K_{C_3H_8} = 0.78$, and $K_{nC_4H_{10}} = 0.16$. As previously stated, the first step is to check whether the system contains both vapor and liquid. For this to be possible the K-values must include both values greater than unity and values smaller than unity. Clearly, this set does. The second step is to check the bubblepoint and the dewpoint.

Bubblepoint

$$\sum (z_j K_j) = (0.20)(30) + (0.30)(3.5) + (0.30)(0.78) + (0.20)(0.16)$$
$$= 7.32$$

So the feed is definitely above its bubblepoint.

Dewpoint

$$\sum (z_j/K_j) = (0.20/30) + (0.30/3.5) + (0.30/0.78) + (0.20/0.16)$$
$$= 1.73$$

So the feed is below the dewpoint temperature.

Now we are ready to do the isothermal flash calculation, guessing V/F ratios until equation 7–25 is satisfied.

			Guess V/F = 0.8		*Guess V/F = 0.7*	
Component	z_j	K_j	$(V/F)(K_j - 1)$	x_j	$(V/F)(K_j - 1)$	x_j
C1	0.20	30	23.2	0.0083	20.30	0.0094
C2	0.30	3.5	2.0	0.1000	1.75	0.1091
C3	0.30	0.78	−0.18	0.3659	−0.154	0.3546
C4	0.20	0.16	−0.67	0.6061	−0.588	0.4854
				1.0803		0.9585

The answer will be between the two guesses, probably at about $V/F = 0.74$.

7.4.3 Flash Calculations in Nonideal Systems

If activity coefficients are not equal to unity, flash calculations become more complex. This was not true for bubblepoint calculations because we knew the liquid compositions. Now, however, we do not know them.

Initial guesses of the x_j's must be made so that the activity coefficients (γ_j's) can be calculated. Then a V/F ratio is guessed. New values of the x_j's are calculated from

$$x_j = \sum_{j=1}^{NC} \left\{ \frac{z_j}{1 + \left(\frac{V}{F}\right)\left[\frac{\gamma_j P_j^s}{P} - 1\right]} \right\} \tag{7–34}$$

At each iteration both the ratio V/F and the γ_j's are changing.

7.4.4 Computer Program for Isothermal Flash Calculations

The computer program given in Table 7–3 calculates the vapor and liquid fractions and the compositions of the vapor and liquid phases at 5°C intervals, starting at a given initial temperature and continuing until the system exceeds the dewpoint temperature, for a three-component system consisting of ben-

TABLE 7–3
FORTRAN PROGRAM FOR ISOTHERMAL FLASH CALCULATIONS

```
C     ISOTHERMAL FLASH   -   BTX
C     IDEAL VLE IS ASSUMED
C     INTERVAL HALVING IS USED TO CONVERGE V/F RATIO
C     V/F RATIOS AND LIQUID AND VAPOR COMPOSITIONS ARE CALCULATED
C         OVER A RANGE OF TEMPERATURES AT 760 MM HG PRESSURE.
C
      DIMENSION D(3),C(3),X(3),Y(3),Z(3),PS(3)
      REAL L
      DATA XF,F/.40,.35,.25,100./
      D(1)=-299.1*353.1*ALOG(760./100.)/54.
      D(2)=-324.9*383.6*ALOG(760./100.)/58.7
      D(3)=-354.3*417.4*ALOG(760./100.)/63.1
      C(1)=ALOG(760.)-D(1)/353.1
      C(2)=ALOG(760.)-D(2)/383.6
      C(3)=ALOG(760.)-D(3)/417.4
      P=760.
      VFRAC=.2
      T=95.
      WRITE(6,53)
   53 FORMAT(1X,60H    T      L      V      X1      X2      X3      Y1      Y2
     1      Y3            )
      DO 100 N=1,6
      DVFRAC=.05
      FLAGP=1.
      FLAGM=1.
      LOOP=0
      DO 1 J=1,3
    1 PS(J)=EXP(C(J)+D(J)/(T+273.))
C     CHECK DEW POINT AND BUBBLE POINT
      PDEW=0.
      PBUB=0.
      DO 2 J=1,3
      PBUB=PBUB+Z(J)*PS(J)
    2 PDEW=PDEW+Z(J)/PS(J)
      PDEW=1./PDEW
      IF(P.GT.PBUB) GO TO 3
      IF(P.LT. PDEW) GO TO 5
      GO TO 10
    3 WRITE(6,4)  T
    4 FORMAT(1X,F7.2,20HSUBCOOLED LIQUID        )
      GO TO 100
    5 WRITE(6,6)  T
    6 FORMAT(1X,F7.2,20HSUPERHEATED VAPOR       )
      GO TO 100
   10 LOOP=LOOP+1
      IF(LOOP.GT.50) STOP
      SUMX=0.
      DO 15 J=1,3
      X(J)=Z(J)/(1.+VFRAC*(PS(J)/P-1.))
   15 SUMX=SUMX+X(J)
      IF(ABS(SUMX-1.).LT.  .000001) GO TO 50
```

TABLE 7–3
CON'T.

```
      IF(SUMX-1.) 20,20,30
   20 IF(FLAGP)21,21,22
   21 DVFRAC=DVFRAC*.5
   22 VFRAC=VFRAC+DVFRAC
      FLAGM=-1.
      GO TO 10
   30 IF(FLAGM)31,31,32
   31 DVFRAC=DVFRAC*.5
   32 VFRAC=VFRAC-DVFRAC
      FLAGP=-1.
      GO TO 10
   50 DO 51 J=1,3
   51 Y(J)=X(J)*PS(J)/P
      V=VFRAC*F
      L=F-V
      WRITE(6,52) T,L,V,X,Y
   52 FORMAT(1X,3F7.2,6F7.4)
  100 T=T+5.
      STOP
      END
```

zene, toluene, and o-xylene. Raoult's Law is assumed. The feed flow rate, the feed composition, the vapor-pressure constants for all components, and the flash temperature and pressure are given as input data.

The program first checks to make sure that the system is in the two-phase region. If the system is below the bubblepoint temperature, the program prints out a message to that effect and increments the temperature for a second try. If the system is above the dewpoint temperature, a message indicating this is printed out and the program stops. Interval-halving is used in the flash calculation to converge on the correct V/F ratio. Results for several flash temperatures are given in Table 7–4. Figure 7–6 gives a flowchart of the computer program.

TABLE 7–4
RESULTS FROM PROGRAM IN TABLE 7–3

```
    T       L       V     X1      X2      X3      Y1      Y2      Y3
  95.00SUBCOOLED LIQUID
 100.00  88.87  11.13  .3665   .3610   .2725   .6674   .2624   .0703
 105.00  58.51  41.49  .2750   .3738   .3513   .5763   .3165   .1072
 110.00  32.48  67.52  .2054   .3542   .4405   .4936   .3480   .1584
 115.00   7.72  92.28  .1531   .3110   .5358   .4207   .3533   .2261
 120.00SUPERHEATED VAPOR
```

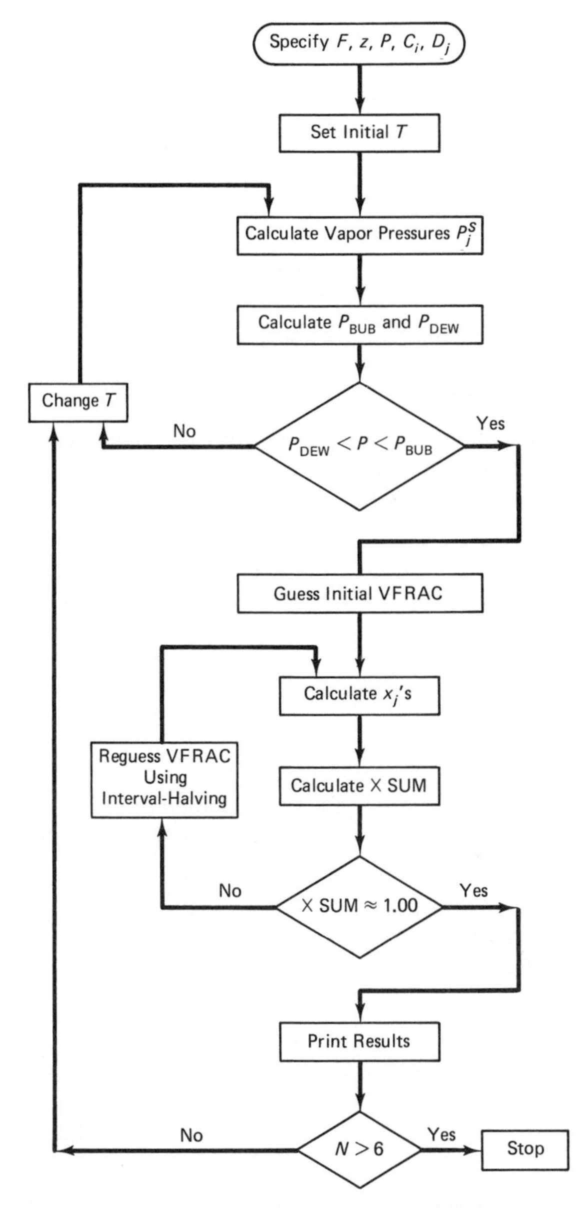

Figure 7–6 Computer flowchart of isothermal flash program.

PROBLEMS

7–1. Vapor pressure data for benzene and toluene are given in the following table.

VAPOR PRESSURES

	100 mm Hg	*760 mm Hg*
Benzene	26.1°C	80.1°C
Toluene	51.9°C	110.6°C

Calculate the equilibrium total pressure P (mm Hg) and vapor composition y (mole fraction benzene) for liquid compositions x of 0.2, 0.4, 0.6, and 0.8 mole fraction benzene at 60°C and 110°C. The system is ideal.

7–2. The liquid composition of a benzene-toluene vapor-liquid system at equilibrium is 0.40 mole fraction benzene. The total system pressure is 760 mm Hg. Calculate the temperature and vapor composition. Assume the system is ideal.

7–3. A liquid stream is a binary mixture of 60 mole percent A and 40 mole percent B. It is flashed into a drum. The temperature in the drum is 140°F. The components A and B form an ideal mixture, and the vapor pressure of each component is given by

$$\ln P_A^S = 11.0 - \frac{4800}{T}$$

$$\ln P_B^S = 10.5 - \frac{5100}{T}$$

where the pressure is in psia and the temperature in °R. If 30 percent of the liquid feed vaporizes, what is the pressure in the flash drum and what is the composition of the vapor stream?

7–4. A three-component mixture of benzene, toluene, and ortho-xylene is separated into three more-or-less pure product streams by using two distillation columns. Component flow rates are given in the figure below.

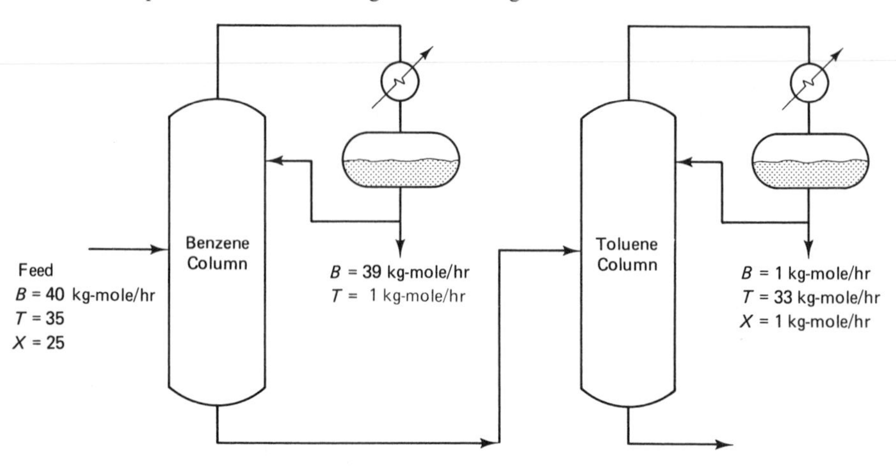

Both columns operate at atmospheric pressure, and overhead vapor is totally condensed. Vapor pressure data for ortho-xylene are 100 mm Hg at 81.3°C and 760 mm Hg at 144.4°C. Vapor pressure data for benzene and toluene are given in problem 7–1. Assume the system is ideal, and calculate

(a) The temperature of the overhead vapor in each column.
(b) The bottom temperature and the reflux drum temperature in each column.
(c) The relative volatility of benzene relative to toluene in the top and in the bottom of the first column.
(d) The relative volatility of toluene relative to ortho-xylene at each end of the second column.

7–5. The overhead composition of a deisobutanizer column is 5 mole percent propane (C_3), 85 mole percent isobutane (iC_4) and 10 mole percent normal butane (nC_4). The overhead vapor must be totally condensed, using cooling-tower water. This means that in the summer the condensed liquid temperature is 120°F, while in the winter it is 70°F. The average is 100°F. Assume ideal VLE. Vapor-pressure data are:

C_3 : 100 psia at 56°F and 188.7 psia at 100°F
iC_4 : 100 psia at 122° F and 71 psia at 100°F
nC_4 : 100 psia at 146°F and 52 psia at 100°F

(a) Calculate the operating pressure of the column at the three temperatures given.
(b) Calculate relative volatilities at the given temperatures. Assume the system is ideal.

7–6. (a) Calculate the operating pressure required to condense the overhead distillate product of a propylene-propane distillation column at summertime conditions of 110°F reflux drum temperature. Distillate product is 90 mole percent propylene (C_3H_6). Assume the system is ideal. Vapor pressure data are:

Propylene	*Propane*
14.7 psia at −53.9°F	14.7 psia at −43.8°F
73.5 psia at +23.4°F	73.5 psia at +34.5°F

$$\ln P^S = 12.75 - \frac{4091.58}{T + 460} \qquad \ln P^S = 12.85 - \frac{4230.17}{T + 460}$$

where P^S is in psia and T is in °F.

(b) Calculate the bottom temperature of the column at the pressure calculated in part (a) if the bottom composition is 15 mole percent propylene.
(c) Calculate relative volatilities at 110°F and at the bottom temperature.
(d) Will the column that makes this separation have many trays and a high reflux ratio, or will it have few trays and a low reflux ratio?

7–7. A conventional distillation column uses steam to boil up vapor in the bottom of the column and cooling water to condense the vapor leaving the top of the column.

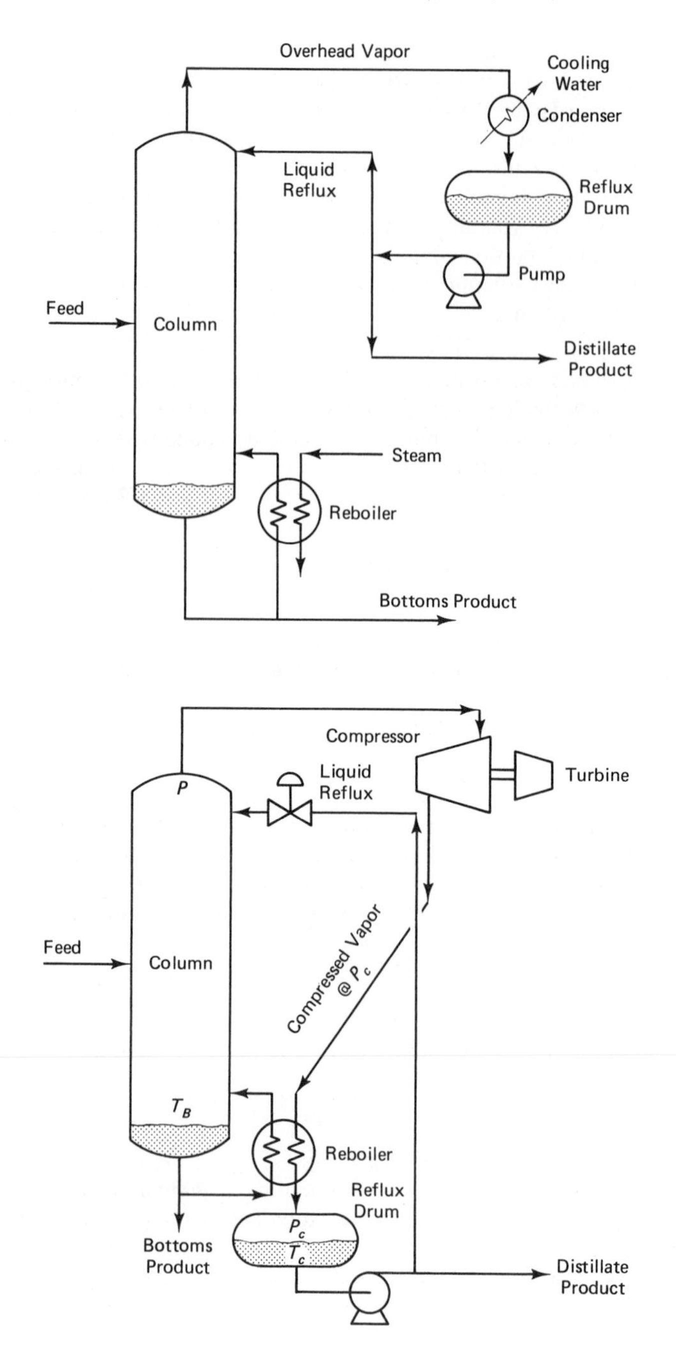
Overhead Vapor
Cooling
Water
Condenser
Liquid
Reflux
Reflux
Drum
Pump
Feed
Column
Distillate
Product
Steam
Reboiler
Bottoms Product
Compressor
Liquid
Reflux
Turbine
P
Feed
Column
Compressed Vapor
@ P_c
T_B
Reboiler
Reflux
Drum
P_c
T_c
Bottoms
Product
Distillate
Product

An alternative configuration (see bottom figure on page 228) uses a compressor to raise the pressure of the overhead vapor so that it can be condensed in a heat exchanger that serves as the reboiler of the same column.

If the propylene-propane separation of problem 7–6 is run in this "vapor re-compression" system, considerably lower pressures can be used than must be used in a conventional column. Assume the column is run at 73.5 psia.

(a) Calculate the bottom temperature T_B and the relative volatility at this temperature if the bottom product is 15 mole percent propylene.

(b) For the reboiler to be of reasonable size, the temperature difference between the hot side (condensing compressed overhead vapor at temperature T_C) and the cold side (boiling bottom liquid at temperature T_B calculated in part (a) above) should be about 15°F, i.e.,

$$T_C = T_B + 15$$

Calculate the discharge pressure P_C of the compressor if the distillate product is 90 mole percent propylene.

7–8. An alkylation reactor in a refinery is operated at 50°F. Liquid in the reactor is 2 mole percent propane, 80 mole percent isobutane, and 18 mole percent normal butane. Assume that the vapor-liquid equilibrium data for the system show it to be ideal. Vapor pressure data are:

	°*F*	*psia*	°*F*	*psia*
Propane C_3	34.5	73.5	100	188.7
Isobutane iC_4	45	29.4	100	71
Normal Butane nC_4	66	29.4	100	52

Vapor is generated in the reactor due to an exothermic chemical reaction and is then compressed and totally condensed at 120°F against cooling-tower water. Liquid from the condensate drum is then flashed back into the reactor. Calculate

(a) The operating pressure of the reactor.

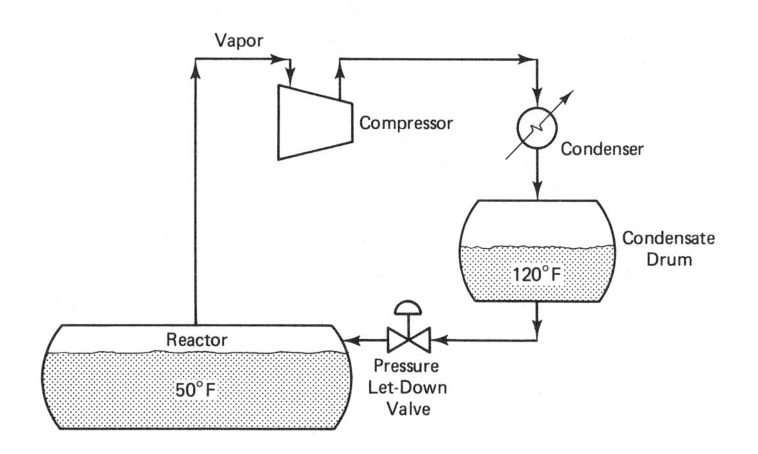

(b) The composition of the vapor boiled off the reactor.
(c) The compressor discharge pressure, i.e., the pressure in the condenser and condensate drum.

7–9. A light-ends stream in a petroleum refinery contains 25 mole percent ethane, 40 mole percent propane, 20 mole percent isobutane, and 15 mole percent n-butane. It is cooled at 100 psia to a temperature of 20°F and is then fed to a separator to disengage the liquid and vapor phases. What fraction of the stream entering the separator is vapor, and what is its composition?

7–10. Write and execute a FORTRAN program that calculates the reflux drum temperature of a depropanizer with a liquid distillate product with the following composition at pressures of 100, 150, 200, and 250 psig.

	Distillate composition (mole fraction)
Ethane	.0524
Propane	.8618
Isobutane	.0419
n-Butane	.0439

Assume that the VLE of the system is ideal. Vapor pressure data are:

	Temperatures (°C)	
	@ 10 atm	*@ 30 atm*
Ethane	−32	10
Propane	26.9	78.7
Isobutane	66.8	120.5
n-Butane	79.5	140.6

Use the interval-halving iteration algorithm.

7–11. Write and execute a FORTRAN program that calculates the base temperature of a depropanizer column whose bottoms product is given in the following table at pressures of 150 and 250 psig. Assume ideal VLE, and use the interval-halving iteration algorithm.

	Bottoms (lb-mole/hr)	*Boiling points (°F) @ psia*		
		14.7	*100*	*500*
$C_3^=$	16.8	−54	44	170
C_3	15.0	−44	56	186
iC_4	126.0	11	122	266
$iC_4^=$	146.7	20	130	280
$nC_4^=$	221.0	27	138	290
nC_4	59.4	31	146	295
iC_5	1.5	82	209	
$C_5^=$	4.2	86	222	
nC_5	0.4	97	224	

7–12. A refinery light-ends stream containing 20 mole percent CH_4, 25 mole percent C_2H_6, 30 mole percent C_3H_8, and 25 mole percent nC_4H_{10} is to be cooled at a constant pressure of 100 psia. Calculate the fraction of the stream that is liquid, and the composition of both the liquid and the vapor phases when the stream has been cooled to 0°F. (*Note:* Be sure to make calculations to ascertain that you are above the dewpoint pressure and below the bubblepoint pressure at 0°F. This will tell you if you do indeed have both phases present, and will allow you to estimate the fraction of liquid to give a reasonable first guess.)

7–13. Using the *K*-values and other data obtained in problem 7–12, calculate activity coefficients for each component in the mix in that problem at 0°F and 100 psia.

7–14. Secondary recovery methods are used to enhance the recovery of petroleum light liquids from an oil field in western Kansas. Natural gas, essentially pure methane (CH_4), is pumped into the well at 600 psia. It reaches the formation at 200°F. The methane returns from the well containing small amounts of n-pentane, n-hexane, and n-heptane. The returning stream is cooled and reduced in pressure to 100 psia and 60°F. The liquid collected from it contains 35 mole percent nC_5, 40 mole percent nC_6, and 25 mole percent nC_7 on a CH_4-free basis. There is also a small amount of CH_4 in the liquid.

(a) How much is that "small" amount of methane?

(b) What is the composition of the uncondensed gases after separation from the liquid?

7–15. Vapor from the top of a distillation column has a composition of 40 mole percent n-hexane, 40 mole percent n-heptane, and 20 mole percent n-octane. The vapor is partially condensed to produce liquid reflux and uncondensed vapor product. If the column and condenser operate at 20 psia, and a reflux ratio of 1.5 is achieved, i.e., the ratio of the liquid formed in the condenser to the vapor leaving the condenser is 1.5, what are the composition and temperature of the liquid reflux? Assume the system is ideal, and that vapor pressure data are as follows:

	1 atm	*10 atm*
n-hexane (C_6H_{14})	68.7°C	102.7°C
n-heptane (C_7H_{16})	98.4°C	166.6°C
n-octane (C_8H_{18})	125°C	235.8°C

7–16. Liquid from a condensate drum is flashed from 47.98 psia to 30 psia. The temperature of the vapor-liquid mixture after the flash is 45°F. The composition of liquid in the condensate drum is 6 mole percent propane, 81.7 mole percent isobutane, and 12.3 mole percent normal butane. The system is ideal.

(a) What are the bubblepoint pressure and dewpoint pressure of the condensate drum liquid at the flash temperature?

(b) Calculate the fraction of vapor formed and the composition of the vapor and liquid phases.

7–17. A binary mixture of isobutane (iC_4) and tertiary butyl alcohol (TBA) is separated in two distillation columns operating in series. Isobutane at a purity of 99.9 mole percent iC_4 is removed overhead as the distillate product from both columns. The bottoms product from the first column is a mixture of iC_4 and TBA and is fed into the second column. The bottoms product from the second column is TBA at a purity of 99.95 mole percent TBA. Base temperatures in both columns are limited to a maximum of 220°F because of safety considerations. (TBA detonates at a temperature of about 250°F.)

The activity coefficients of iC_4 and TBA are 1.2 and 1.7, respectively, under the conditions that exist in the base of the first column. Vapor pressure data are:

iC_4

330 psia @ 220°F

95 psia @ 120°F

32 psia @ 50°F

TBA

32 psia @ 220°F

2.35 psia @ 120°F

0.29 psia @ 50°F

The first column is operated at a pressure such that cooling water can be used in the condenser, giving a reflux drum temperature of 120°F.

(a) What is the minimum concentration of iC_4 that can be tolerated in the bottoms product from the first column?

(b) What is the maximum pressure that can be used in the second column, and what is the reflux drum temperature at this pressure?

7–18. A mixture of 20 mole percent C_2H_6, 30 mole percent C_3H_8, 35 mole percent nC_4H_{10}, and 15 mole percent nC_5H_{12} comes off the top of a light-ends distillation column in a petroleum refinery. The mixture is at least partially condensed to produce reflux. The product from the column is expected to be a vapor. The column operates at 300 psia, and the condenser cools to 120°F. What fraction of the vapor is liquefied for reflux, and what is its composition?

7–19. Develop a computer program that will calculate vapor compositions and temperatures over the complete range of liquid compositions (0–100 percent) at 100 mm Hg total pressure for the binary system normal dodecane–butyl carbitol. Plot your results in the form of a *Txy* diagram. You may use the computer to do the plotting if you desire.

Does an azeotrope occur in this system? Use the interval-halving convergence algorithm. Physical property data are as follows (1 = n-dodecane, 2 = butyl carbitol).

Vapor Pressures

	n-dodecane	*butyl carbitol*
100 mm Hg	146.14°C	159.8
200 mm Hg	167.14	181.2

Activity Coefficients

$$\log_{10} \gamma_1 = \frac{Ax_2^2}{\left(\frac{A}{B}x_1 + x_2\right)^2}$$

$$\log_{10} \gamma_2 = \frac{Bx_1^2}{\left(x_1 + \frac{B}{A}x_2\right)^2}$$

$$A = 0.755$$

$$B = 0.516$$

7–20. The bottoms stream from a distillation column flows to a kettle reboiler where it is partially vaporized, the vapor from this reboiler returning to the column, and the liquid leaving as bottoms product. The stream from the column analyzes 20 mole percent n-hexane, 40 mole percent n-heptane, and 40 mole percent n-octane. The bottoms product from the kettle reboiler flows at a rate of 600 lb-mole/hr. The column itself operates with a V/L ratio of 0.8 in the stripping section. Determine the temperature and composition of the bottoms product. The column and reboiler operate at 40 psia.

7–21. Gmehling and Onken (*VLE Data Collection*, Frankfurt, Germany: Dechema, 1983, Vol. 1, Part 1) give the following formulas for the acetic acid–water system. Acetic acid is component 2 and water is component 1.

Vapor Pressures:

$$\ln P_j^s = A_j - \frac{B_j}{T + C_j}$$

where P_j^s is in mm Hg, and T is in °C, and the constants A_j, B_j, and C_j are given by

$$A_1 = 8.07131 \qquad A_2 = 8.021$$

$$B_1 = 1730.63 \qquad B_2 = 1936.01$$

$$C_1 = 233.426 \qquad C_2 = 258.451$$

Activity Coefficients (Van Laar Equations):

$$\ln \gamma_1 = A_{12}\left(\frac{A_{21}x_2}{A_{12}x_1 + A_{21}x_2}\right)^2$$

$$\ln \gamma_2 = A_{21} \left(\frac{A_{12} x_1}{A_{12} x_1 + A_{21} x_2} \right)^2$$

where $A_{12} = 0.5203$ and $A_{21} = 1.1066$ at 760 mm Hg

Write and execute a computer program that generates data for a *Txy* diagram for the acetic acid–water system at 760 mm Hg. Use the interval-halving convergence algorithm.

7–22. A liquid stream at 130°C from a reactor is a binary mixture of 10 weight percent water and 90 weight percent acetic acid and is flashed into a drum. The temperature in the drum is 120°C. The vapor formed is fed to a heat exchanger, where it is condensed and returned to the flash drum. All the liquid from the flash drum is returned to the reactor. Assume steady-state operation and perfect mixing and phase equilibrium in the flash drum. See problem 7–21 for vapor-liquid equilibrium data.

What is the pressure in the flash drum, and what is the composition of the vapor stream to the waste heat boiler? The molecular weight of acetic acid is 60.

7–23. The binary system consisting of n-hexane (component 1) and ethanol (component 2) is known to be nonideal. Construct the *xy* and *Txy* diagrams for the system at 1 atm. Boiling points for the pure components and activity coefficients for the mixture are given below:

Pressure	*Boiling points (°C)*	
(mm Hg)	*n-Hexane*	*Ethanol*
1,520	93	97.5
760	68.7	78.4
200	31.6	48.4

Wilson Equations:

$$\ln \gamma_1 = -\ln (x_1 + Ax_2) + x_2 \left[\frac{A}{x_1 + Ax_2} - \frac{B}{Bx_1 + x_2} \right]$$

$$\ln \gamma_2 = -\ln (x_2 + Bx_1) - x_1 \left[\frac{A}{x_1 + Ax_2} - \frac{B}{Bx_1 + x_2} \right]$$

where

$$A = 0.281$$

$$B = 0.041$$

Use the Newton-Raphson method to solve any iterative calculations.

7–24. The components A and B form an ideal mixture, and we are interested in obtaining the equilibrium *Txy* diagram at 760 mm Hg. The vapor pressures of the components are given by

$$\ln P_A^s = 19.5564 - \frac{4361.55}{T}$$

$$\ln P_B^s = 20.1735 - \frac{5050.50}{T}$$

where the pressure is in mm Hg and the temperature is in K.

(a) Plot temperature versus liquid composition. Determine the maximum and minimum temperatures.

(b) Complete the plot by calculating two points on the saturated vapor and liquid lines. Do not use any iterations.

7–25. Four alternative process schemes are to be evaluated for separating benzene and toluene. (See the figure on page 236.)

(a) The base case is a single column making essentially pure benzene distillate and essentially pure toluene bottoms, operating at a pressure such that the temperature in the reflux drum is 120°F (water-cooled condenser).

(b) Feed is split between two columns, one running at the same pressure as the base case column (a), the other operated at a high enough pressure so that its overhead vapor can be condensed in the reboiler of the low-pressure column to provide its heat. Each column makes essentially pure benzene distillate and toluene bottoms.

(c) All the feed is fed into a high-pressure column whose overhead vapor reboils the next low-pressure column. One-half of the benzene in the feed is removed as the distillate product in the high-pressure column. The rest of the benzene and all the toluene are taken out of the bottom of the high-pressure column and fed into the low-pressure column. This column operates at the same pressure as the base case column. The rest of the benzene is produced as distillate in this column, and all the toluene is removed as bottoms product.

(d) The process flows are the same as in (c), but the first column is the low-pressure column and the second column is operated at high pressure such that its overhead vapor can reboil the first column.

Assume ideal vapor-liquid equilibrium, negligible pressure drop through the column trays, total condensers, partial reboilers, and a binary system. Feed composition is 60 mole percent benzene. Temperature differentials in the reboiler/condensers are 30°F.

Calculate the base temperatures, reflux drum temperatures, column pressures, and compositions of intermediate feed streams (where applicable) for each of the four alternatives described.

7–26. Propylene and tertiary butyl alcohol (TBA) are separated in a distillation column. The overhead distillate product is 99.99 mole percent propylene. A water-cooled total condenser is used, giving a reflux drum temperature of 120°F.

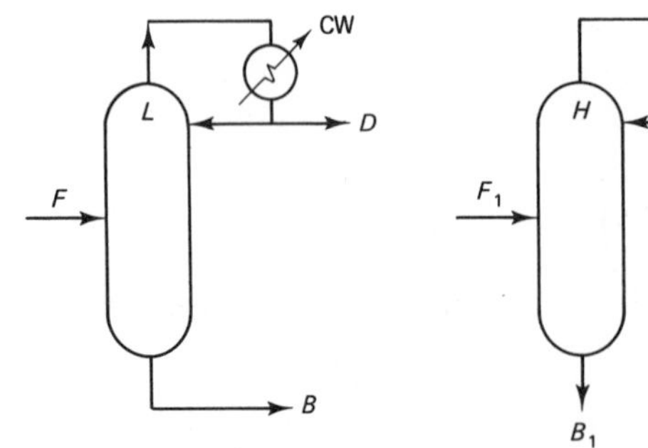

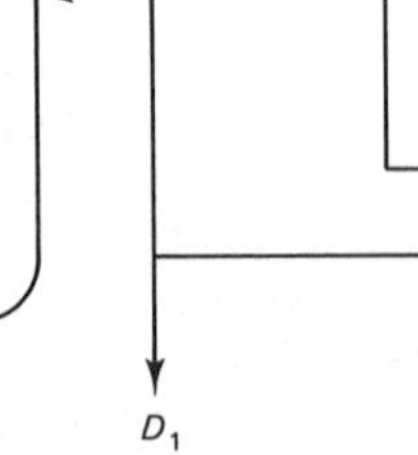

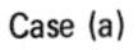

Case (a)

Case (b)

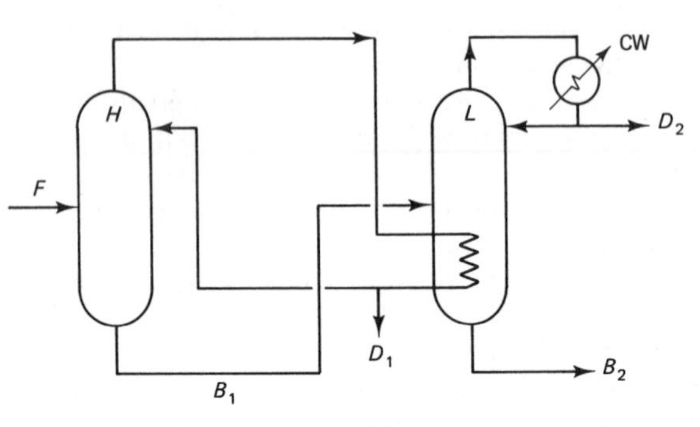

Case (c)

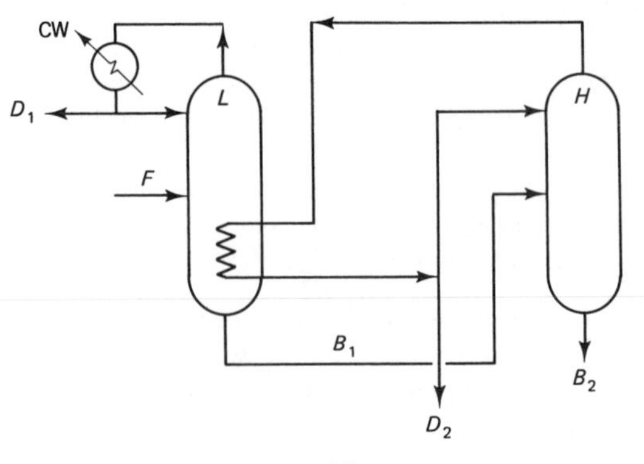

Case (d)

Heat integration schemes for separating benzene and toluene. (a) Base case (b) Feed to two columns at different pressures (c) Feed to high-pressure column (d) Feed to low-pressure column.

Vapor pressure data for propylene and TBA are as follows.

Propylene

297.5 psia at 120°F

839.6 psia at 220°F

TBA

2.35 psia at 120°F

32.0 psia at 220°F

(a) What is the operating pressure in the column?

The base temperature in the column is limited by TBA detonation to 220°F. The activity coefficients of propylene and TBA are given by the equations

$$\ln(\gamma_1) = 1.375\,(x_2)^2$$

$$\ln(\gamma_2) = 4.327\,(x_1)^2$$

where

γ_1 = activity coefficient of propylene

γ_2 = activity coefficient of TBA

x_1 = mole fraction propylene in liquid

x_2 = mole fraction TBA in liquid

(b) What is the composition of the bottoms product if the base temperature is 220°F?

Binary Distillation Column Calculations 8

In this chapter we will study binary distillation, assuming equimolal overflow. Binary distillation is a good example of the application of mass and energy balances. (See also chapter 13.) Several excellent books devoted to the subject are as follows:

1. Matthew Van Winkle, *Distillation*. New York: McGraw-Hill, 1967.
2. Ernest J. Henley and J. D. Seader, *Equilibrium-Stage Separation Operations in Chemical Engineering*. New York: Wiley, 1981.
3. C. Judson King, *Separation Processes*. New York: McGraw-Hill, 1971.
4. Buford D. Smith, *Design of Equilibrium Stage Processes*. New York: McGraw-Hill, 1963.
5. Reinhard Billet, *Distillation Engineering*. New York: Chemical Publishing, 1979.
6. L. M. Rose, *Distillation Design In Practice*. Amsterdam: Elsevier, 1985.

8.1 BASIC CONFIGURATION

The simple single-section countercurrent cascades discussed in chapter 5 involved the simultaneous use of component balances and vapor-liquid equilibrium equations to do tray-to-tray calculations. In this chapter we explore more complex systems of cascades with multiple sections. Most commercial distillation columns have at least two sections, each with different vapor and liquid flow rates. Many columns have more than two sections.

Distillation is a very important industrial separation technique. It is fairly simple and easily understood by beginning students in chemical engineering, in particular in the case of binary systems. In discussing distillation in this chapter, we will limit ourselves to columns where equimolal overflow, i.e., constant liquid and vapor flow rates throughout any one section of the column, can be assumed. The assumption of equimolal overflow gives straight operating lines. In chapter 13 we will consider curved operating lines and non-equimolal overflow.

Figure 8–1 shows a basic distillation column. This "plain vanilla" column has a single feed stream and produces two products: distillate out of the top and bottoms out of the bottom. The feed is introduced onto a tray somewhere in the middle of the column. The location of the feed tray depends on a number of factors that we will discuss later.

Vapor is generated in a heat exchanger (reboiler) in the base of the column and flows up the column. The vapor from the top of the column is then condensed in another heat exchanger (condenser). The liquid from the condenser is collected in a tank called the reflux drum, the purpose of which is to provide liquid surge capacity.

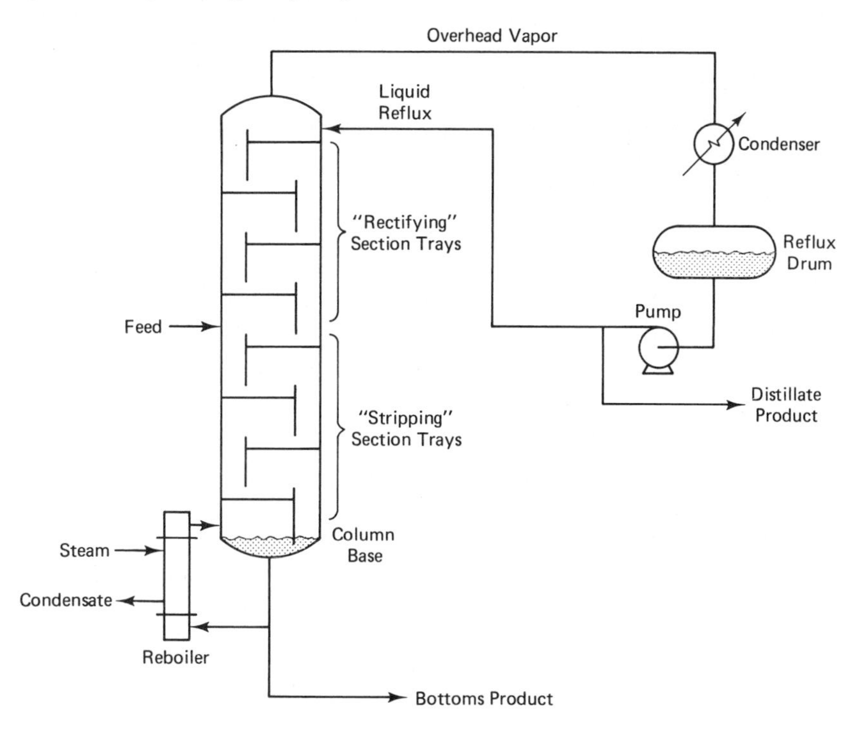

Figure 8–1 Distillation column.

Some of the liquid from the reflux drum is removed as the distillate product, and some of it is sent back to the column to provide liquid flow on the trays. The latter stream is called *reflux*. The countercurrent flow of liquid down and vapor up the column produces the series of equilibrium stages which separate the components. As we will see, the amount of separation, or *fractionation*, depends on the number of trays, the vapor and liquid traffic up and down the column, and the relative volatilities of the components.

One measure of the vapor and liquid flow rates is the *reflux ratio* (RR), which is defined as the ratio of the reflux flow rate R to the distillate flow rate D:

$$\text{Reflux ratio} = \text{RR} = \frac{R}{D} \tag{8-1}$$

The flow rates can be in either molar or weight units per unit time, but molar flow rates are used more frequently. Molar units are needed for consistency with the equimolal overflow assumption.

The basic, simple column has two distinct sections. The one below the feed tray is the *stripping* section, the one above the feed tray the *rectifying* section. These two sections have different liquid and vapor flow rates and therefore different operating lines because of the introduction of the feed stream. Figure 8–2 shows the variables that we will use in treating distillation. Their definition is as follows:

F = feed flow rate (mole/time)

z = feed composition (mole fraction light component)

D = distillate flow rate (mole/time)

x_D = distillate composition (mole fraction light component)

B = bottoms flow rate (mole/time)

x_B = bottoms composition (mole fraction light component)

R = reflux flow rate (mole/time)

$-Q_C$ = heat removal rate in the condenser (energy/time)

Q_R = heat input rate in the reboiler (energy/time)

L_R = liquid flow rate in the rectifying section (mole/time)

L_S = liquid flow rate in the stripping section (mole/time)

V_R = vapor flow rate in the rectifying section (mole/time)

V_S = vapor flow rate in the stripping section (mole/time)

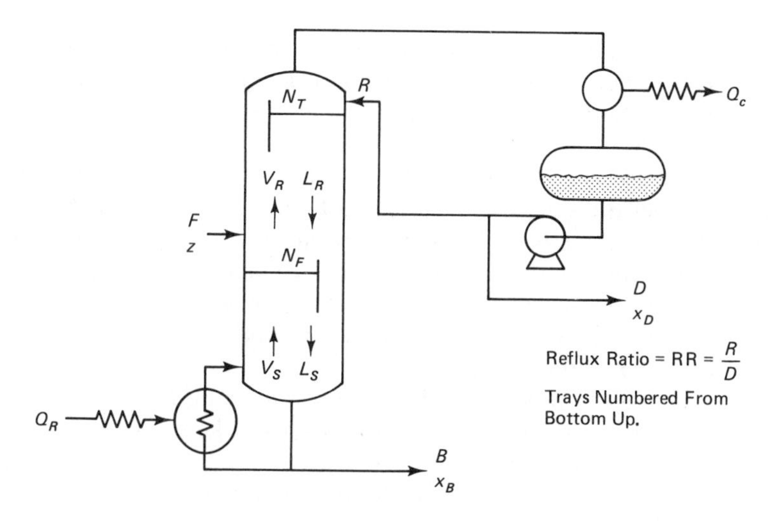

Figure 8–2 Distillation variables.

N_T = total number of trays in the column

N_F = feed tray or number of trays in the stripping section

Note that we start numbering the trays from the bottom of the column: tray 1 is the first tray above the reboiler, tray 2 is the next, etc.

The simple distillation column shown in Figures 8–1 and 8–2 is essentially a stripping column, which we studied in chapter 5, with another section stuck on top. Therefore, we will end up with an operating line for the stripping section that has most of the properties that we found for the operating line in the simple stripping column. In addition, there will be a second operating line for the rectifying section.

8.2 ASSUMPTIONS

We have already mentioned the two vital assumptions we make in our study of distillation in this chapter: binary systems and equimolal overflow. Several other assumptions will often be made when we study a standard column, but they are not critical to the analysis techniques used in this chapter. In fact, they will be relaxed in later sections.

The method of analysis that is used in binary equimolal-overflow distillation systems is called the *McCabe-Thiele* method. It is a very famous and widely used technique.

For a standard column, we make the "USA" (usual simplifying assumptions):

1. *Theoretical stages.* We assume perfect phase equilibrium between the

vapor and liquid streams leaving each tray. A discussion of tray efficiency later in the chapter will show how to handle actual trays, as opposed to theoretical trays, when these phases are not in perfect equilibrium.

2. *Total condenser*. The condenser is called a total condenser if the vapor from the top of the column is totally condensed in the condenser, i.e., the stream leaving the condenser is all liquid. In some columns only a portion of the vapor is condensed, and the stream leaving the condenser is a mixture of vapor and liquid. Such condensers are called partial condensers. We will consider partial condensers later in the chapter.
3. *Partial reboiler*. Most real reboilers operate as partial reboilers. In section 8.10 we will discuss the thermosiphon, kettle, fired, and internal types of reboilers. All of these are normally partial reboilers. In such reboilers, the vapor that leaves the reboiler is in equilibrium with the liquid product that is withdrawn from the column, and only a portion of the liquid that comes into the reboiler from tray 1 is vaporized. A total reboiler would be one in which all of the liquid that enters the reboiler is vaporized. These are rarely seen in industry.

8.3 FEED THERMAL CONDITION

The introduction of the feed into the middle part of the column produces different liquid and vapor flow rates in the two sections of the column. The thermal condition of the feed (how much of it is liquid and how much of it is vapor) dictates how the flow rates of both the liquid and the vapor will vary between the stripping and rectifying sections.

The feed can enter as a saturated liquid at its bubblepoint. In this case the liquid flow rate in the stripping section will be equal to the liquid flow rate in the rectifying section plus the feed flow rate, i.e.,

$$L_S = L_R + F \tag{8–2}$$

and the vapor flow rates in the two sections will be identical.

Alternatively, the feed can enter as a saturated vapor at its dewpoint. In this case the liquid flow rates in the rectifying and stripping sections will be identical, but the vapor flow rate in the rectifying section will be equal to the vapor flow rate in the stripping section plus the feed flow rate, i.e.,

$$V_R = V_S + F \tag{8–3}$$

Finally, the feed can be a mixture of liquid and vapor, or it can be subcooled liquid or superheated vapor. In order to quantify the thermal condition of the feed, it is convenient to define a parameter q given by

$$q = \frac{L_S - L_R}{F} \tag{8-4}$$

That is, q is the fraction of the feed that is liquid.

If the feed is saturated liquid at its bubblepoint temperature T_{BP} (for its composition z and at the pressure of the column), then $q = 1$. If the feed is saturated vapor at its dewpoint temperature T_{DP}, then $q = 0$. For a feed that is between its bubble- and dewpoint temperatures, where it is a mixture of vapor and liquid, q will have a value between 0 and 1.

For subcooled liquid feed at a temperature T_F that is below the bubblepoint temperature of the feed, q will be greater than 1 and is given by

$$q = 1 + c_L \frac{(T_{\text{BP}} - T_F)}{\Delta H_V} \tag{8-5}$$

where

c_L = heat capacity of the liquid feed

ΔH_V = latent heat of vaporization of the feed

We will study more about heat capacities and heats of vaporization in chapter 10. In the meantime, it is sufficient to note that the second term on the right-hand side of equation 8–5 gives the amount of vapor coming up from the stripping section that must condense in order to heat the cold feed up to its bubblepoint temperature.

For superheated vapor feed at a temperature T_F that is above the dewpoint temperature of the feed, q will be negative and is given by

$$q = -c_V \frac{(T_F - T_{\text{DP}})}{\Delta H_V} \tag{8-6}$$

where

c_V = heat capacity of the vapor feed

This equation gives the amount of liquid coming down the column from the rectifying section that must vaporize in order to cool the hot vapor feed down to its dewpoint temperature.

To determine q for mixtures of vapor and liquid, we have to do a flash calculation (see chapter 7). If F_V is the flow rate of the vapor portion of the feed that we calculate from the flash calculation, and F_L is the flow rate of the liquid portion of the feed, then

$$q = \frac{F_L}{F_V + F_L} \tag{8–7}$$

As we will see in section 8.5, the parameter q will be used in the graphical McCabe-Thiele method to draw another straight line called the *q-line* on the xy diagram. The q-line will help us to draw the operating lines for the two sections of the column.

8.4 NUMERICAL PLATE-TO-PLATE CALCULATIONS

Just as we did in chapter 5, we can work our way up the column from plate to plate by alternately using component balances and vapor-liquid equilibrium relationships. In this section we do these calculations numerically by hand. In the next, we will do them graphically. In section 8.11, we will do them on a computer.

8.4.1 Overall Balances

We first write total and component balances around the entire column:

$$F = D + B \tag{8–8}$$

$$Fz = Dx_D + Bx_B \tag{8–9}$$

In the normal design problem, we know the feed flow rate F, the feed thermal condition q, and the feed composition z, as well as the product compositions x_D and x_B. Hence, equations 8–8 and 8–9 can be solved simultaneously to get the distillate and bottoms flow rates:

$$D = \frac{F(z - x_B)}{(x_D - x_B)} \tag{8–10}$$

$$B = F - D \tag{8–11}$$

8.4.2 Internal Liquid and Vapor Flow Rates

Next, we calculate the liquid and vapor flow rates in the two sections of the column. In the normal design problem, the reflux ratio RR $= R/D$ will be known. Thus, the reflux flow rate R can be calculated from

$$R = (\text{RR})(D) = \frac{R}{D}(D) \tag{8–12}$$

If the reflux is saturated liquid at its bubblepoint, the liquid flow rate in the rectifying section of the column will be equal to the reflux flow rate:

$$L_R = R \tag{8-13}$$

The reflux is usually close to its bubblepoint, but not always. Sometimes it is significantly subcooled. In this case the liquid flow rate in the rectifying section will be greater than the reflux flow rate, i.e.,

$$L_R = R\left[1 + c_L \frac{(T_{\text{NT}} - T_R)}{\Delta H_V}\right] \tag{8-14}$$

where

T_{NT} = temperature on the top tray

T_R = temperature of the subcooled reflux

Note the similarity of this equation to equation 8–5.

Once L_R has been calculated, L_S can be determined from equation 8–4. As shown in Figure 8–3, a total mass balance around the top of the column above the feed tray gives

$$V_R = L_R + D \tag{8-15}$$

Similarly, a total mass balance around the bottom section of the column gives

$$L_S = V_S + B \tag{8-16}$$

After this equation is used to calculate V_S, all the internal liquid and vapor flow rates will be known.

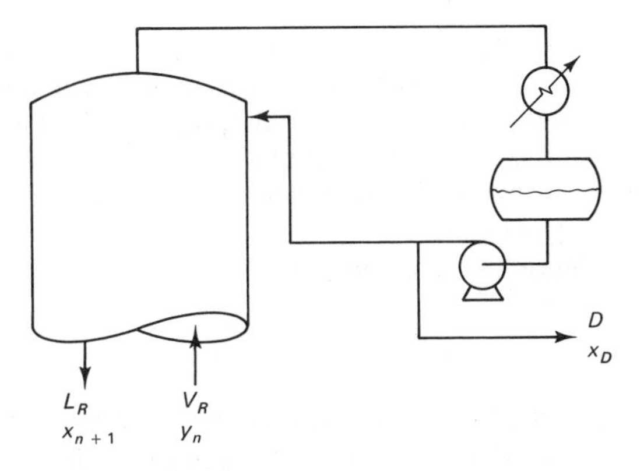

Figure 8–3 Component balance around the top of a distillation column.

8.4.3 Stripping Section Equations

The component and VLE equations describing the stripping section of the column from the base up to the feed tray are exactly the same as those derived for the stripping column in chapter 5. Since only binary systems are being considered, only one component balance is required. Traditionally, the lighter component is used. We have

Reboiler (column base)

$$\text{Component balance: } x_1 L_S = y_B V_S + x_B B \tag{8–17}$$

$$\text{VLE: } y_B = f(x_B) \tag{8–18}$$

Equation 8–17 is used to solve for x_1.

Tray 1

$$\text{Component balance: } x_2 L_S = y_1 V_S + x_B B \tag{8–19}$$

$$\text{VLE: } y_1 = f(x_1) \tag{8–20}$$

Equation 8–19 is used to solve for x_2.

Tray n *in the stripping section*

$$\text{Component balance: } x_{n+1} L_S = y_n V_S + x_B B \tag{8–21}$$

$$\text{VLE: } y_n = f(x_n) \tag{8–22}$$

These tray-to-tray calculations are continued until we get a liquid composition that is near the feed composition. (We will define this more precisely later.) This gives the number of trays in the stripping section.

8.4.4 Rectifying Section Equations

After the stripping section equations are solved, the calculations begin in the rectifying section. The component balances in the rectifying section above the feed tray are similar to those in the stripping section, but they have the additional term involving the feed. Note that the liquid and vapor flow rates in the rectifying section must be used. These are, in general, different from those in the stripping section.

As shown in Figure 8–4, a component balance for tray n in the rectifying section gives

$$x_{n+1} L_R + Fz = y_n V_R + x_B B \tag{8–23}$$

These tray-to-tray calculations are continued until we reach a vapor composition leaving a tray that is equal to or greater than the desired distillate composition x_D. This gives the total number of trays in the column.

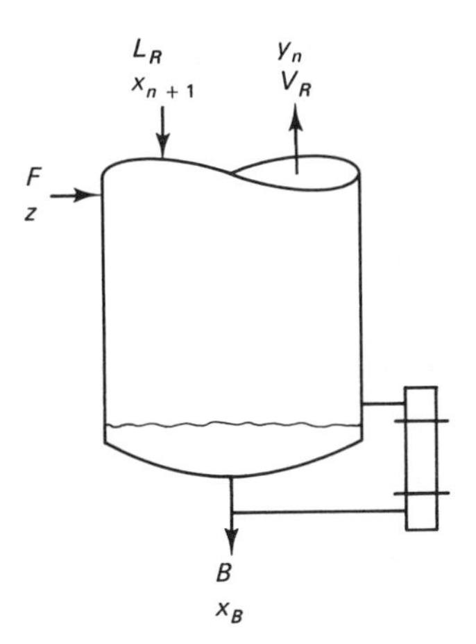

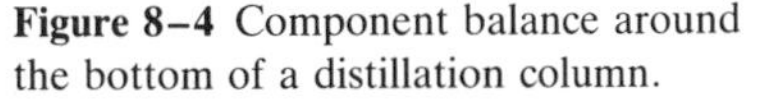

Figure 8–4 Component balance around the bottom of a distillation column.

Example 8.1.

Calculate the number of trays required to produce distillate and bottoms products with compositions $x_D = 0.98$ and $x_B = 0.02$ from 100 kg-mole/hr of saturated liquid feed with composition $z = 0.50$. The reflux ratio is 2.56, and a constant relative volatility of $\alpha = 2$ describes the VLE of the binary system. Equimolal overflow, theoretical trays, saturated reflux, a total condenser, and a partial reboiler can all be assumed.

Solution:

Overall component and mass balances

$$F = D + B = 100$$

$$zF = Dx_D + Bx_B \longrightarrow (0.50)(100) = 0.98\,D + 0.02\,B$$

Solving simultaneously gives

$$B = 50 \text{ kg-mole/hr}$$

$$D = 50 \text{ kg-mole/hr}$$

Internal liquid and vapor flow rates

$$R = L_R = (\text{RR})(D) = (2.56)(50) = 128 \text{ kg-mole/hr}$$

$$L_S = L_R + qF = 128 + (1)(100) = 228 \text{ kg-mole/hr}$$

Note that $q = 1$ was used because the feed was specified as saturated liquid. Finally, we obtain

$$V_R = L_R + D = 128 + 50 = 178 \text{ kg-mole/hr}$$

$$V_S = L_S - B = 228 - 50 = 178 \text{ kg-mole/hr}$$

V_S and V_R are equal in this case, as we would expect, because the feed is saturated liquid. The feed changes the liquid flow rates between the sections, but does not change the vapor flow rate.

Reboiler and column base

$$y_B = \frac{\alpha x_B}{1 + (\alpha - 1)x_B}$$

$$= \frac{(2)(0.02)}{1 + (2 - 1)(0.02)} = 0.0392$$

$$x_1 L_S = y_B V_S + x_B B$$

$$x_1 = \frac{(0.0392)(178) + (0.02)(50)}{228} = 0.035$$

Tray 1

$$y_1 = \frac{\alpha x_1}{1 + (\alpha - 1)x_1}$$

$$= \frac{(2)(0.035)}{1 + (2 - 1)(0.035)} = 0.0676$$

$$x_2 L_S = y_1 V_S + x_B B$$

$$x_2 = \frac{(0.0676)(178) + (0.02)(50)}{228} = 0.0572$$

These calculations are continued up the stripping section until a value of x_{n+1} is reached that is greater than the feed composition $z = 0.50$. We obtain the following results.

Tray	x_n	y_n
0	0.02	0.0392
1	0.035	0.0676
2	0.0572	0.1082
3	0.0888	0.1632
4	0.1318	0.2329
5	0.1862	0.3140
.	.	.
.	.	.
.	.	.
9	0.4339	0.6052
10	0.4769	0.6458
11	0.5086	0.6743

The value of x_{11} calculated using the stripping section component balances is greater than the feed composition z. Therefore, we should put 10 trays in the stripping section and use the rectifying section component balances to get the x values from tray 10 on up the column. Accordingly, the feed tray is $N_F = 10$, and we get

Tray 10

$$x_{11} L_R = y_{10} V_R + x_B B - zF$$

$$x_{11} = \frac{(0.6458)(178) + (0.02)(50) - (0.5)(100)}{128}$$

$$= 0.5152$$

Tray 11

$$y_{11} = \frac{\alpha x_{11}}{1 + (\alpha - 1)x_{11}}$$

$$= \frac{(2)(0.5152)}{1 + (2 - 1)(0.5152)} = 0.6801$$

$$x_{12} L_R = y_{11} V_R + x_B B - zF$$

$$x_{12} = \frac{(0.6801)(178) + (0.02)(50) - (0.5)(100)}{128}$$

$$= 0.5629$$

These calculations continue up the column until a vapor composition is calculated that is greater than x_D. We obtain:

Tray	x_n	y_n
11	0.5152	0.6801
12	0.5629	0.7203
.	.	.
.	.	.
.	.	.
20	0.960753	0.979984
21	0.979977	0.989887

Consequently, our hypothetical column requires 21 trays plus a partial reboiler to achieve the separation. We will do this example by computer in section 8.11.

8.5 GRAPHICAL MCCABE-THIELE METHOD

The equations developed in the last section can be very conveniently and informatively solved graphically on an xy diagram. The method is just like the techniques that we used for the stripping column and the absorber in chapter 5, except that we now deal with two sections.

All the educational advantages of the graphical methods enumerated in chapter 5 apply even more to the treatment of more complex columns. The effects of parameters such as the reflux ratio, relative volatility, feed composition, and feed tray location are easily understood in the picture given by the graphical methods.

8.5.1 Operating Lines

Rectifying Operating Lines Figure 8–3 shows the upper section of the column from tray $n + 1$ on up, including the condenser and reflux drum. A light-component material balance around this section gives

$$y_n V_R = x_{n+1} L_R + x_D D \tag{8–24}$$

Rearranging to put this into the equation of a straight line, we have

$$y_n = \left(\frac{L_R}{V_R}\right) x_{n+1} + \frac{x_D D}{V_R} \tag{8–25}$$

The slope of this *rectifying operating line* (ROL) is the liquid-to-vapor ratio in the rectifying section of the column. Recall that in chapter 5 we said that these operating line slopes were always the L/V ratios. We could do the component balance around the lower part of the column, as shown in Figure 8–4, and get exactly the same result (equation 8–24), remembering that $zF = x_D D + x_B B$.

In many columns, the reflux is saturated liquid, so $L_R = R$. Since $V_R = L_R + D$, the slope of the ROL can be conveniently related to the reflux ratio RR:

$$\text{Slope of ROL} = \frac{L_R}{V_R} = \frac{R}{R + D} = \frac{\text{RR}}{\text{RR} + 1}$$

The intersection of the ROL and the 45° line occurs at the distillate composition x_D. This makes it very convenient to plot the ROL: simply locate x_D on the 45° line, and draw a straight line with slope L_R/V_R.

To prove that the point of intersection is at x_D, let the composition at the point of intersection be x_{int}. On the 45° line, $x = y = x_{\text{int}}$. Substituting into equation 8–24 gives

$$x_{\text{int}} V_R = x_{\text{int}} L_R + x_D D$$

$$x_{\text{int}}(V_R - L_R) = x_D D \tag{8–26}$$

But a total mass balance around the rectifying section yields, of course,

$$V_R = L_R + D \tag{8–27}$$

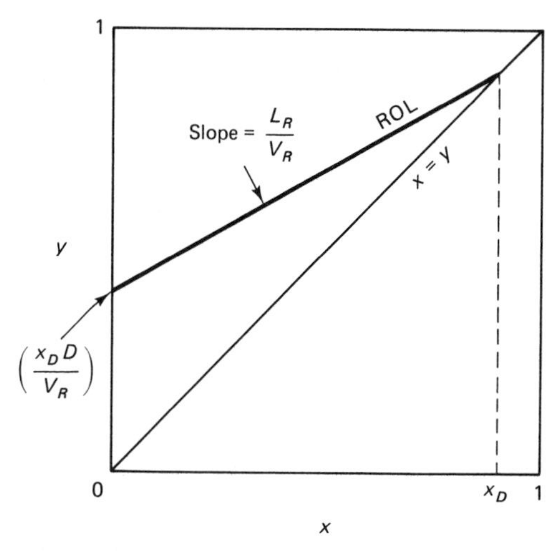

Figure 8–5 Intersection of rectifying section operating line and the 45° line.

Substituting into equation 8–26 gives

$$x_{\text{int}} D = x_D D$$

$$x_{\text{int}} = x_D$$

Thus, the 45° line and the ROL intersect at x_D, as shown in Figure 8–5.

Note that the ROL intersects the vertical axis ($x = 0$) at a value $y = x_D D/V_R$. This is useful for drawing the ROL.

Stripping Operating Line The equations describing the stripping section are identical to those used in chapter 5 for the stripping column. Therefore, the component balance yields a stripping operating line (SOL)

$$y_n = \left(\frac{L_S}{V_S}\right) x_{n+1} + \left(\frac{-x_B B}{V_S}\right) \tag{8–28}$$

The slope of the SOL is the L/V ratio in the stripping section. The SOL intersects the 45° line at x_B, as shown in Figure 8–6.

8.5.2 *q* Line

The intersection of the two operating lines depends on the feed composition z, the feed thermal condition q, and the reflux ratio. We can draw a line on the xy diagram that gives us the loci of possible intersection points of the operating lines for all reflux ratios. This line is called the *q line* and is a function of z and q.

To find a general equation for the q line, let (x_i, y_i) be the location of the

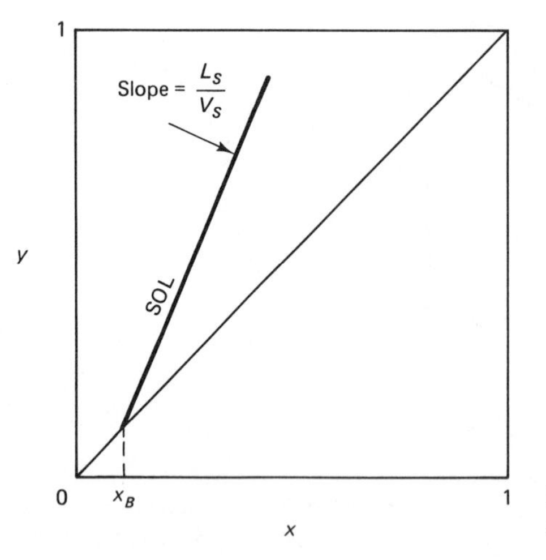

Figure 8–6 Intersection of the stripping section operating line and the 45° line.

point of intersection of the ROL and the SOL. On the ROL, using equation 8–24, x_i and y_i are given by

$$V_R y_i = L_R x_i + x_D D \tag{8–29}$$

On the SOL, using equation 8–28, x_i and y_i are given by

$$V_S y_i = L_S x_i - x_B B \tag{8–30}$$

Subtracting equation 8–30 from equation 8–29 yields

$$(V_R - V_S) y_i = (L_R - L_S) x_i + (x_D D + x_B B) \tag{8–31}$$

Now, from the definition of q given in equation 8–4, q is the fraction of the feed that is liquid. Hence, the fraction of the feed that is vapor is $1 - q$, and the difference between the vapor flow rates in the stripping and rectifying sections is

$$V_R - V_S = (1 - q)F \tag{8–32}$$

Substituting equations 8–4 and 8–32, together with the overall component balance, into equation 8–31 gives

$$(1 - q)F y_i = -qF x_i + zF$$

Rearranging shows that this is the equation of a straight line:

$$y_i = \left(-\frac{q}{1 - q}\right) x_i + \frac{z}{1 - q} \tag{8–33}$$

This is the *q-line* equation, which has a slope of $-q/(1 - q)$ and intersects the 45° line (where $x_i = y_i$) at z.

As shown in Figure 8–7, the q line is easily drawn. It immediately shows

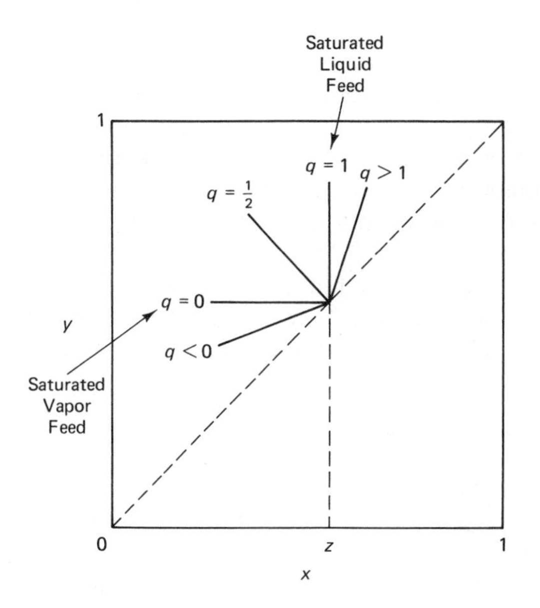

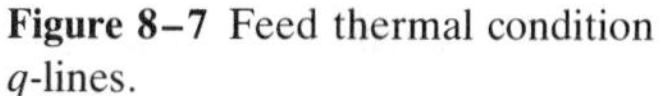
Figure 8–7 Feed thermal condition q-lines.

us the thermal condition of the feed. A vertical line (q-line slope of ∞) corresponds to saturated liquid feed, and a horizontal line (q-line slope of 0) corresponds to a saturated vapor feed.

The intersection of the q line and the stripping operating line is important because it is at that point that we switch from the stripping section to the rectifying section. Accordingly, let the liquid composition at that point be x_{qs} and the vapor composition be y_{qs}, and let us solve the SOL equation (equation 8–28) and the q-line equation (equation 8–33) simultaneously to get x_{qs}. We have

$$y_{qs} = \left(\frac{-q}{1-q} \right) x_{qs} + \frac{z}{1-q}$$

and

$$y_{qs} = \left(\frac{L_S}{V_S} \right) x_{qs} - x_B \frac{B}{V_S}$$

After some algebra, we get

$$x_{qs} = \frac{V_S z + (1-q) x_B B}{L_S (1-q) + V_S q} \tag{8–34}$$

Note that for $q = 1$ (q line vertical, indicating saturated liquid feed), $x_{qs} = z$; and for $q = 0$ (q line horizontal, indicating saturated vapor feed), $y_{qs} = z$.

8.5.3 The Design Problem

There are two basic types of problems in distillation: the *design* problem and the *rating* problem. The design problem is to determine the number of trays to be put into a new column; the rating problem is to determine the performance of an existing column with a fixed number of trays. We will consider the solution of both of these problems using graphical methods in this section and computer solutions in section 8.11.

A step-by-step procedure for graphical solution of the design problem is given below and illustrated in example 8.2. The feed conditions F, z, and q; product compositions x_D and x_B; reflux ratio RR; and vapor-liquid equilibrium curve are given.

1. Draw the VLE curve on an xy diagram.
2. Draw the 45° line.
3. Locate x_B, x_D, and z on the 45° line.
4. Calculate the slope $-q/(1-q)$ of the q line, and draw the q line from the z point on the 45° line.
5. Calculate product flow rates B and D from overall material balances. The equations are $D = F(z - x_B)/(x_D - x_B)$ and $B = F - D$.
6. Calculate the reflux flow rate $R = (\text{RR})(D)$.
7. Calculate the internal liquid and vapor flow rates from the equations $V_R = L_R + D$, $L_S = L_R + qF$, and $V_S = L_S - B$. If the reflux is saturated liquid, $L_R = R$.
8. Calculate the operating line slopes:

$$\text{ROL slope} = \frac{L_R}{V_R} \qquad \text{SOL slope} = \frac{L_S}{V_S}$$

9. Draw the rectifying operating line from the point x_D on the 45° line with a slope of L_R/V_R.
10. Draw the stripping operating line from the point x_B on the 45° line to the intersection of the q line with the rectifying operating line. Check to make sure that the slope of the SOL is what it is supposed to be (L_S/V_S). If it isn't, you have made a mistake somewhere in your calculations or graph. Go back and check your equations and algebra.
11. Start from x_B, and step up the column using the SOL and the equilibrium curve.
12. When a step crosses the intersection of the operating lines (which occurs on the q line), this tray is the "optimum" feed tray N_F, i.e., any other feed tray would require more total trays in the column.
13. Switch to the ROL and continue stepping.
14. When a step reaches x_D on the 45° line, this tray is the top tray in the column, N_T. The design is completed.

Example 8.2.

A binary mixture that is 40 mole percent light component is to be separated into 5 mole percent bottoms product and 90 mole percent distillate product. The feed flow rate is 100 kg-mole/hr, and the feed is 50 percent vapor, 50 percent liquid as it enters the column. The reflux ratio is 2.625. The VLE for the column can be approximated as a constant relative volatility of 2.5. Assume theoretical trays and a partial reboiler, total condenser, and saturated liquid reflux. What is the optimum feed tray location, and how many trays are required?

Solution:

Figure 8–8 shows the McCabe-Thiele diagram for this example.

1. The VLE curve is drawn for $\alpha = 2.5$.
2. The 45° line is drawn.
3. The points $x_B = 0.05$, $x_D = 0.90$, and $z = 0.40$ are located on the 45° line.
4. The slope of the $-q$ line is $q/(1 - q) = -0.5/(1 - 0.5) = -1$. The q line is drawn from the point z on the 45° line with a slope of -1.
5. Product flow rates are calculated:

$$D = \frac{F(z - x_B)}{x_D - x_B} = \frac{(100)(0.4 - 0.05)}{0.9 - 0.05}$$

$$= 41.18 \text{ kg-mole/hr}$$

$$B = F - D = 100 - 41.18 = 58.82 \text{ kg-mole/hr}$$

6. The reflux flow rate R is the reflux ratio times the distillate flow rate:

$$R = (\text{RR})(D) = (2.625)(41.18) = 108.1 \text{ kg-mole/hr}$$

7. Internal flow rates are calculated:

$$L_R = R = 108.1 \text{ kg-mole/hr (since the reflux is saturated liquid)}$$

$$V_R = L_R + D = 108.1 + 41.18 = 149.3 \text{ kg-mole/hr}$$

$$L_S = L_R + qF = 108.1 + (0.5)(100) = 158.1 \text{ kg-mole/hr}$$

$$V_S = L_S - B = 158.1 - 58.82 = 99.3 \text{ kg-mole/hr}$$

8. Slope of ROL $= L_R/V_R = 108.1/149.3 = 0.724$. Slope of SOL $= L_S/V_S = 158.1/99.3 = 1.592$.
9. The ROL is drawn from the point x_D on the 45° line with a slope of 0.724.
10. The SOL is drawn from the intersection of the q line and the ROL to the point x_B on the 45° line. The slope of the SOL is checked to make sure that it is 1.592.
11. The column is stepped up. The first step is the partial reboiler, the next is tray 1, the next tray 2, etc.
12. At tray 5 the intersection of the operating lines is crossed. Therefore, tray 5 is the optimum feed tray.
13. The ROL is stepped up.
14. At tray 9 the vapor composition y_9 is greater than x_D. Therefore, the column must have nine trays plus a partial reboiler to make the separation at the given reflux ratio.

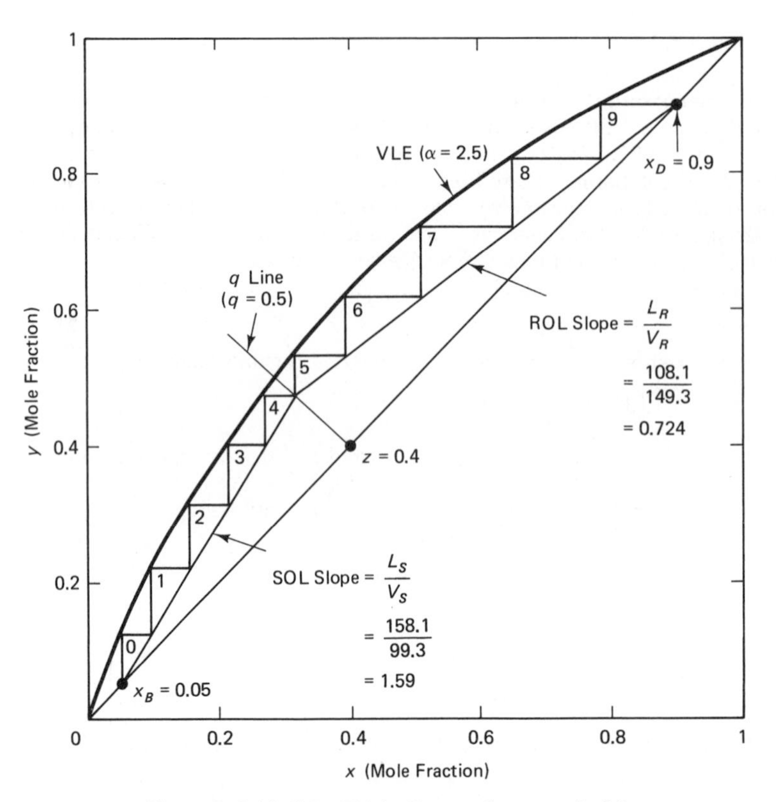

Figure 8–8 McCabe-Thiele diagram for example 8.2.

8.5.4 Rating Problems

Up to this point we have considered only design problems. In a design problem the engineer calculates the size and configuration of equipment needed to do a specified job. In the distillation process the job is to separate a given feed into products of desired purities.

Engineers are also called upon to solve *rating,* or operating, problems. In this situation the equipment is specified, and the engineer is asked to predict how it will perform under different conditions. In the distillation process a typical rating problem is to determine the heat input required to achieve a specified separation in a column with a fixed number of trays as feed composition changes.

There are a number of different types of rating problems. All of them involve iterative (trial-and-error) solutions. A guess is made of one or more unknown variables, and a design solution procedure is carried out using the values guessed. The results are compared with the column specifications, and if they do not match, new guesses are made and the solution is repeated.

One major difference between the design calculation and the operating situation is that the feed to a column will not usually enter the column on the tray where the operating lines intersect, i.e., at the optimum feed tray. Rather, you must stay on the stripping operating line until you have taken as many steps as there are trays in the stripping section of the column. In general, the actual feed tray will not be the optimum feed tray.

Several of the problems at the end of the chapter are representative of rating problems that require iterative solutions. Since the number of calculations grows rapidly in these rating problems, they are ideal candidates for computer solutions. A rating program is presented in section 8.11.

8.6 LIMITING CONDITIONS

There are two limiting conditions that are of great practical importance in distillation: the *minimum reflux ratio* and the *minimum number of trays*. Both are useful in the design of new columns and in the analysis of operating columns. They represent performance limits that the column simply *cannot* exceed.

8.6.1 Minimum Reflux Ratio

Figure 8–9(a) shows how changing the reflux ratio affects the operating lines: the lower the reflux ratio, the closer the operating line moves toward the equilibrium curve, and the larger the number of trays required to make the separation becomes. If the reflux ratio is finally reduced to the point where either operating line intersects or becomes tangent to the VLE curve, an infinite number of steps will be required. The steps get closer and closer to the "pinch point," but they never get past it.

If the shape of the VLE curve is fairly normal, the minimum reflux ratio will occur when the ROL intersects the q line on the VLE curve, as shown in Figure 8–9(a). For some VLE curves, however, the point of intersection or tangency can occur at another location, as illustrated in Figure 8–9(b).

Wherever the intersection occurs, the minimum reflux ratio can be calculated from the slope of the ROL:

$$\text{Slope of ROL} = \frac{L_R}{V_R} = \frac{R}{R + D} = \frac{\text{RR}}{1 + \text{RR}} \tag{8–35}$$

Rearranging to solve for RR in terms of the slope gives

$$\text{RR} = \frac{\text{slope}}{1 - \text{slope}} \tag{8–36}$$

Equation 8–36 holds for any reflux ratio, but if we find the minimum slope where the ROL and VLE curve come together, this slope will give us the minimum reflux ratio.

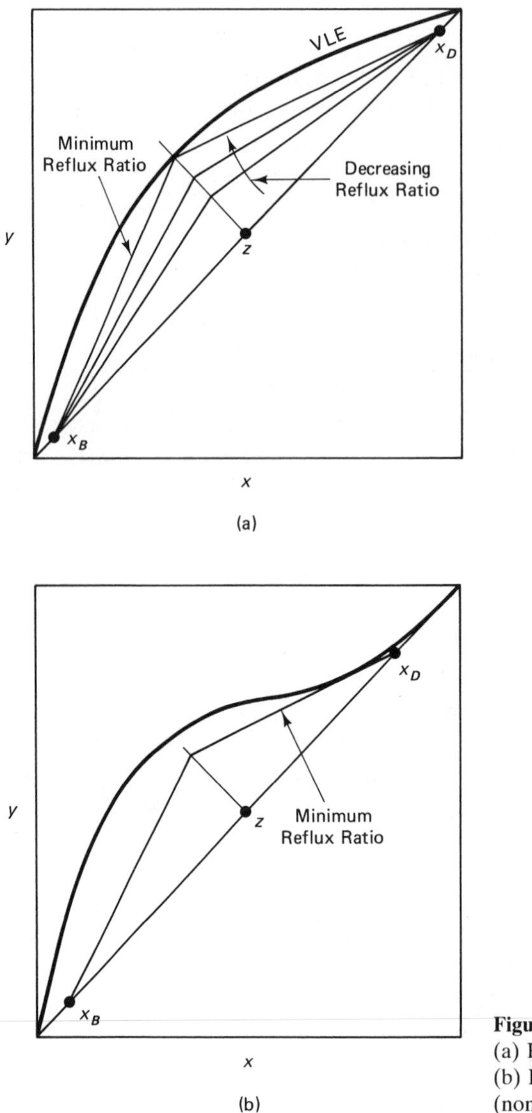

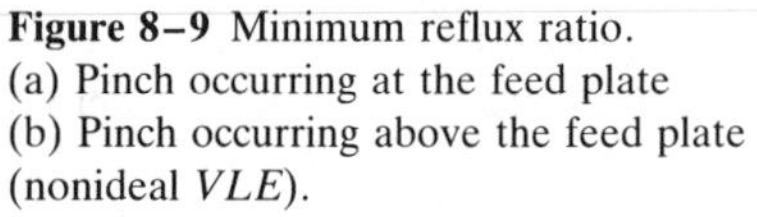

Figure 8–9 Minimum reflux ratio.
(a) Pinch occurring at the feed plate
(b) Pinch occurring above the feed plate (nonideal *VLE*).

The actual reflux ratio must always be greater than the minimum reflux ratio. There is a classical design trade-off between capital costs and energy costs. As we reduce the reflux ratio toward the minimum, the capital investment cost of the column increases because more trays are required and the column must be taller. But the energy consumption decreases as we reduce the reflux ratio because energy is directly related to vapor boilup and vapor rates decrease as RR is decreased. This is evident from the equation

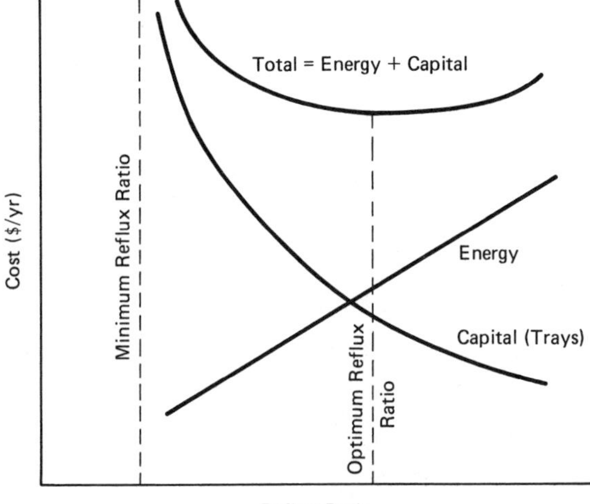

Figure 8–10 Effect of reflux ratio on costs.

$V_R = R + D$: distillate flow rate D is fixed for a given separation, but R decreases as we decrease the reflux ratio.

Consequently, as we decrease the reflux ratio, energy operating costs decrease, but capital investment costs increase. The optimum reflux ratio is the one that minimizes total cost (operating cost plus annual investment depreciation; see Figure 8–10). The optimum ratio depends on the costs of energy and the materials of construction. If the column is made of fairly inexpensive carbon steel, its optimum reflux ratio will be lower than would be the case for a column that is made of platinum or gold. A gold column would have a high reflux ratio, use lots of energy, and be as short as possible.

If energy costs were very low or negligible (as, for example, in some locations in the Middle East where natural gas is simply burned or "flared" to get rid of it), then the optimum reflux ratio would be very high even for a carbon steel column. At the other extreme, when energy costs are very high (as, for example, in Japan,which has to import all of its fuel), the optimum reflux ratio is one that is quite close to the minimum.

Most columns are designed using a reflux ratio that is 1.05 to 1.2 times the minimum reflux ratio. This rule of thumb or "heuristic" is very useful.

Example 8.3.

Calculate the minimum reflux ratio for the system considered in example 8.2.

Solution:

As shown in Figure 8–8, the q line and the VLE line intersect at the coordinates $x = 0.292$ and $y = 0.508$. This point can be found mathematically by solving the equations for the q line and the VLE curve (with constant α) simultaneously.

The slope of the line between x_D on the 45° line and this point of interaction is

$$\text{Minimum ROL slope} = \frac{(0.9 - 0.508)}{(0.9 - 0.292)} = 0.645$$

$$= \frac{RR_{min}}{(1 + RR_{min})}$$

So

$$RR_{min} = 1.815$$

Note that in example 8.2 we used a reflux ratio that was 2.625, or 1.45 times the minimum.

8.6.2 Minimum Number of Trays

If the reflux ratio is increased to a very large number, the operating lines become the 45° line. This *infinite reflux ratio* situation occurs in real life when the column is operated under what are called *total reflux* conditions. Under these conditions, no feed is added to the column and no products are withdrawn, but vapor is boiled up and condensed in the condenser, and all of the liquid is returned as reflux to the column. So the column is just circulating vapor and liquid up and down. Most columns are started up under total reflux conditions.

Under total reflux, the distillate flow rate is zero. Hence, the reflux ratio is infinite for any reflux flow rate. Since the liquid flow rate in the column is the same as the vapor flow rate, the L/V ratio in all sections of the column is 1, and the 45° line is the operating line, as illustrated in Figure 8–11. The steps in moving up the column go between the 45° line and the VLE curve.

The composition in the base of the column under total reflux is x_B, and the composition of the liquid in the reflux drum is x_D. But no products are withdrawn. The number of trays it takes to get from x_B up to x_D under total reflux conditions is called the *minimum number of trays*.

Calculation of the minimum number of trays can be done graphically as shown in Figure 8–11 for any shaped VLE curve. It can also be done analytically, using the Fenske Equation, if constant relative volatility can be assumed. The Fenske Equation gives the number of theoretical stages required under total reflux conditions to move from one set of compositions to another set of compositions. It is applicable to multicomponent systems as well as binary, but it must be kept in mind that it assumes constant relative volatilities and total reflux. It is very useful for getting quick estimates of the size of a column. Typical designs result in columns that have about twice as many trays as the minimum.

The derivation of the Fenske Equation is as follows. Consider two components, one light (L) and the other heavy (H). A component balance for

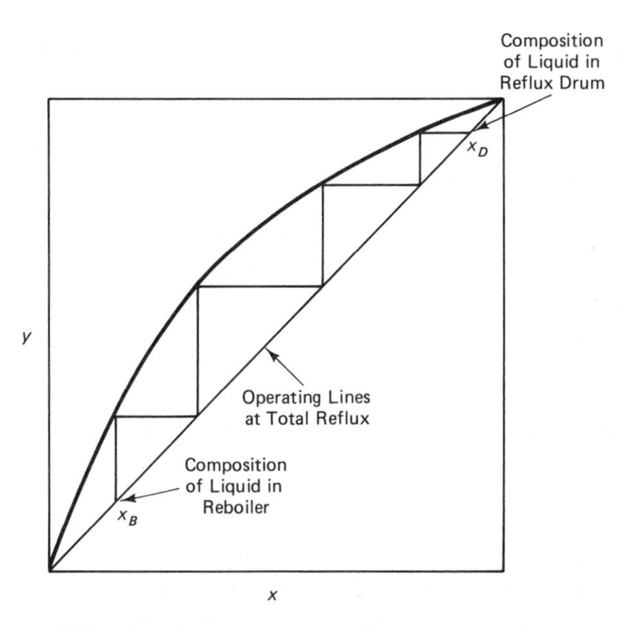

Figure 8–11 Total reflux on a McCabe-Thiele diagram.

the light component around the base of the column gives (see Figure 8–12)

$$L_1 x_{L1} = V_B y_{LB} + B x_{LB} \tag{8–37}$$

where

x_L = mole fraction of light component in liquid

y_L = mole fraction of light component in vapor

Under total reflux conditions, the bottoms product flow rate B is zero. Hence, $L_1 = V_B$, and equation 8–37 reduces to

$$x_{L1} = y_{LB} \tag{8–38}$$

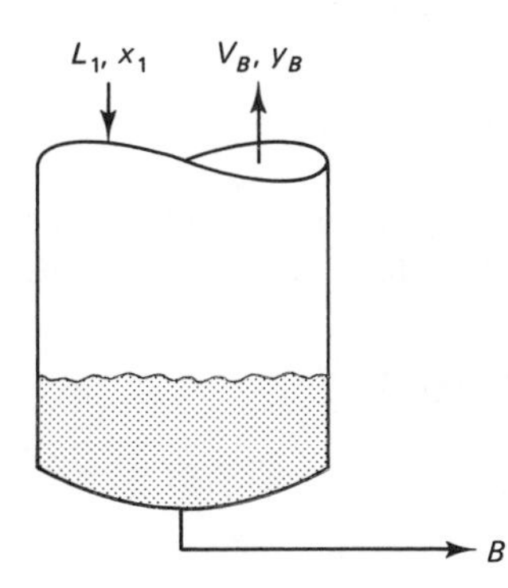

Figure 8–12 Base of column at total reflux.

Likewise a component balance on the heavy component leads to

$$x_{H1} = y_{HB} \tag{8-39}$$

Now we use the definition of the relative volatility α_{LH} between the light and heavy components in the reboiler:

$$\alpha_{LH} = \frac{y_{LB}/x_{LB}}{y_{HB}/x_{HB}} \tag{8-40}$$

Rearranging gives

$$\frac{y_{LB}}{y_{HB}} = \alpha_{LH}\left(\frac{x_{LB}}{x_{HB}}\right) \tag{8-41}$$

Substituting the left sides of equations 8–38 and 8–39 into equation 8–41 gives

$$\frac{x_{L1}}{x_{H1}} = \alpha_{LH}\left(\frac{x_{LB}}{x_{HB}}\right) \tag{8-42}$$

Component balances on tray 1 for the light and heavy components give, under total reflux conditions where $L_2 = V_1$ (note that we are *not* assuming equimolal overflow),

$$x_{L2} = y_{L1} \tag{8-43}$$

$$x_{H2} = y_{H1} \tag{8-44}$$

Applying the relative volatility relationship on tray 1 yields

$$\frac{y_{L1}}{y_{H1}} = \alpha_{LH}\left(\frac{x_{L1}}{x_{H1}}\right) \tag{8-45}$$

Substituting the right sides of equation 8–42, and the left sides of equations 8–43 and 8–44, into equation 8–45 gives

$$\frac{x_{L2}}{x_{H2}} = \alpha_{LH}\alpha_{LH}\left(\frac{x_{LB}}{x_{HB}}\right) \tag{8-46}$$

Continuing this procedure up to the Nth tray yields

$$\frac{x_{LN}}{x_{HN}} = (\alpha_{LH})^{N}\left(\frac{x_{LB}}{x_{HB}}\right) \tag{8-47}$$

Since the condenser is a total condenser, $x_{jD} = y_{jN}$, where j refers to both the light (L) and heavy (H) components. Therefore, we can write the ratio

$$\frac{x_{LD}}{x_{HD}} = \frac{y_{LN}}{y_{HN}} = \alpha_{LH}\left(\frac{x_{LN}}{x_{HN}}\right)$$

$$= (\alpha_{LH})^{(N+1)}\left(\frac{x_{LB}}{x_{HB}}\right) \tag{8-48}$$

Rearranging gives the Fenske Equation:

$$N_{\min} + 1 = \frac{\log\,[(x_{LD}/x_{HD})(x_{HB}/x_{LB})]}{\log\,\alpha_{LH}} \tag{8–49}$$

$N_{\min}$ is the number of theoretical trays in the column. The term "1" comes from the partial reboiler.

Example 8.4.

Calculate the minimum number of trays required to achieve a separation from 5 mole percent bottoms to 90 mole percent distillate in a binary column with a relative volatility of 2.5.

Solution:

The system is the same as the one studied in example 8.2. The Fenske Equation gives

$$N_{\min} + 1 = \log \frac{[(0.9/0.1)\ (0.95/0.05)]}{\log\,(2.5)}$$

$$= 5.61$$

$$N_{\min} = 4.61$$

Recall that the actual number of trays used in the column was 9.

The two limiting conditions discussed above, the minimum number of trays and the minimum reflux ratio, help to illustrate the very important point that there are many, many columns that can be used to make exactly the same separation. This is illustrated in Figure 8–13, where the reflux ratio is

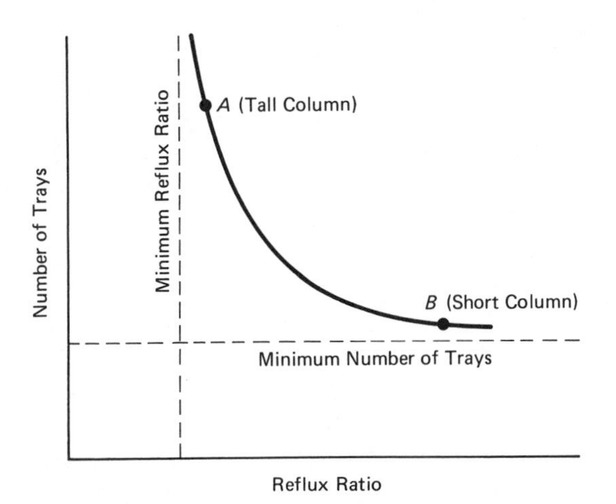

Figure 8–13 Trade-off between trays and reflux ratio to make the same separation.

plotted against the number of trays. This type of curve can be generated for any specified separation and feed.

In the figure, the asymptotes are the limiting conditions. The number of trays goes to infinity as the reflux ratio is reduced toward the minimum, and the reflux ratio goes to infinity as the number of trays is reduced toward the minimum. Any point along the curve represents a column that will make the specified separation, but each column will be different: each will cost different amounts of money to build, and each will consume different amounts of energy. Thus, a column at point A in the figure will be tall and skinny and use little energy, but it will be expensive to build. By contrast, a column at point B will be short and fat (higher vapor rate requires a larger diameter), and it will use lots of energy. Your job as a chemical engineer is to determine which of the infinite number of alternatives is the best. This is classical chemical engineering!

8.7 COMPLEX COLUMNS

So far we have dealt with simple single-feed, two-product columns. However, the McCabe-Thiele method can be very effectively used to analyze more complex configurations. Several of the more common ones will be described below. If you happen to run into a different one from those described, just go back to the fundamental component balances and VLE relationships and derive your own diagrams or computer programs.

8.7.1 Two Feeds

Figure 8–14 shows two feeds that are introduced into a column on two separate feed trays N_{F_1} and N_{F_2}. An xy diagram for the system is also shown. Note that each feed has its own q line.

There are three distinct sections in the column: the top section above the top feed, the middle section between the feeds, and the bottom section below the lower feed. There are also three operating lines, the slope of each of which is the L/V ratio in that section.

8.7.2 Sidestream Columns

Figures 8–15 and 8–16 give sketches of sidestream columns and the corresponding xy diagrams. In Figure 8–15, a liquid sidestream is removed from tray N_S. This reduces the liquid flow rate in the section below the sidestream. The vapor flow rate is unchanged below and above the sidestream. Therefore, the L/V ratio is higher above the sidestream than below it. The operating line changes slope at the sidestream tray.

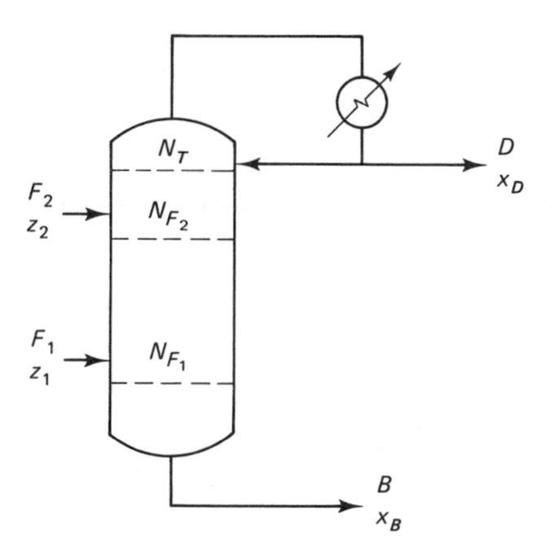

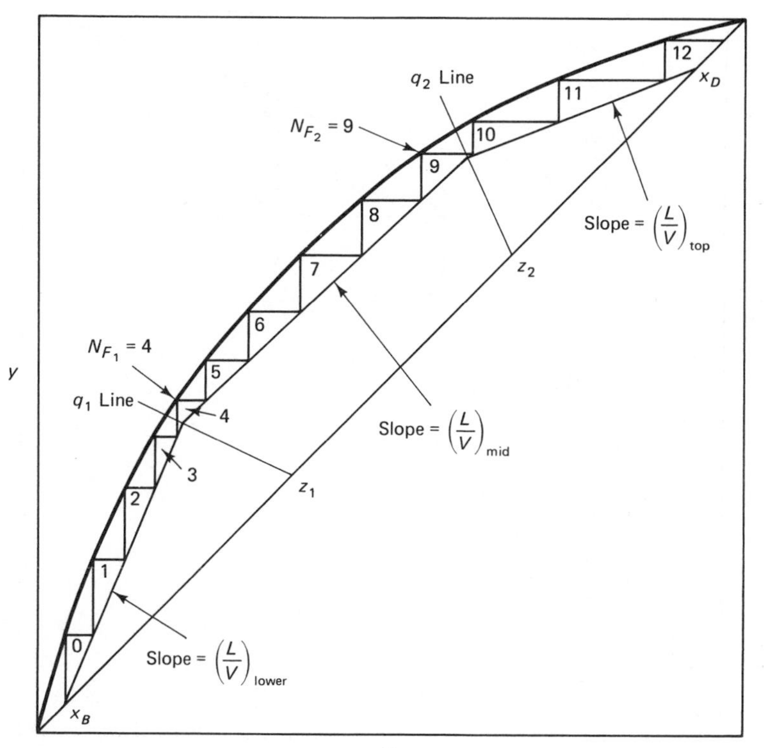

Figure 8–14 Two-feed column.

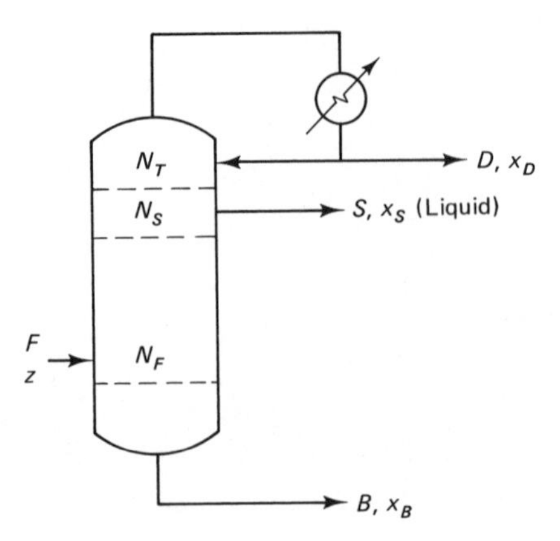

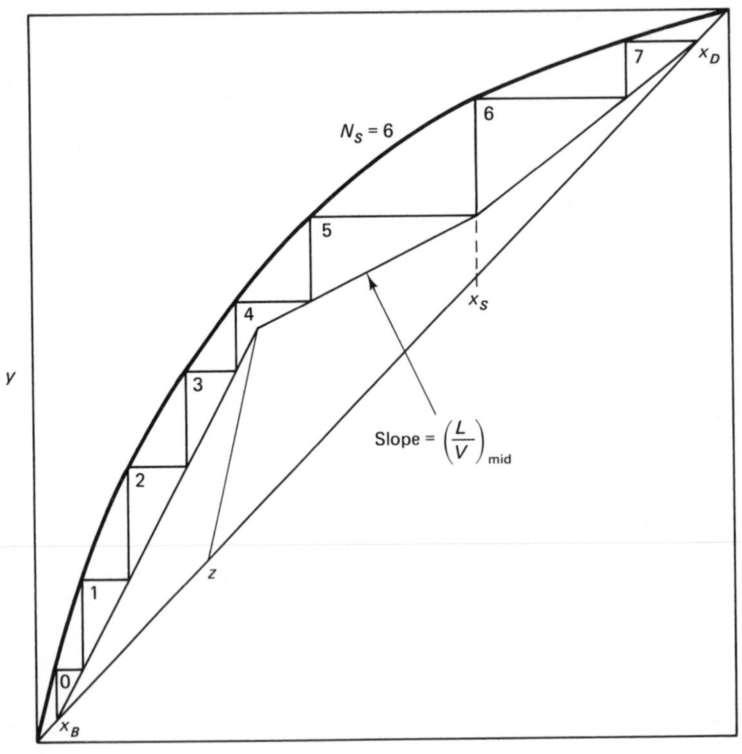

Figure 8–15 Liquid sidestream column.

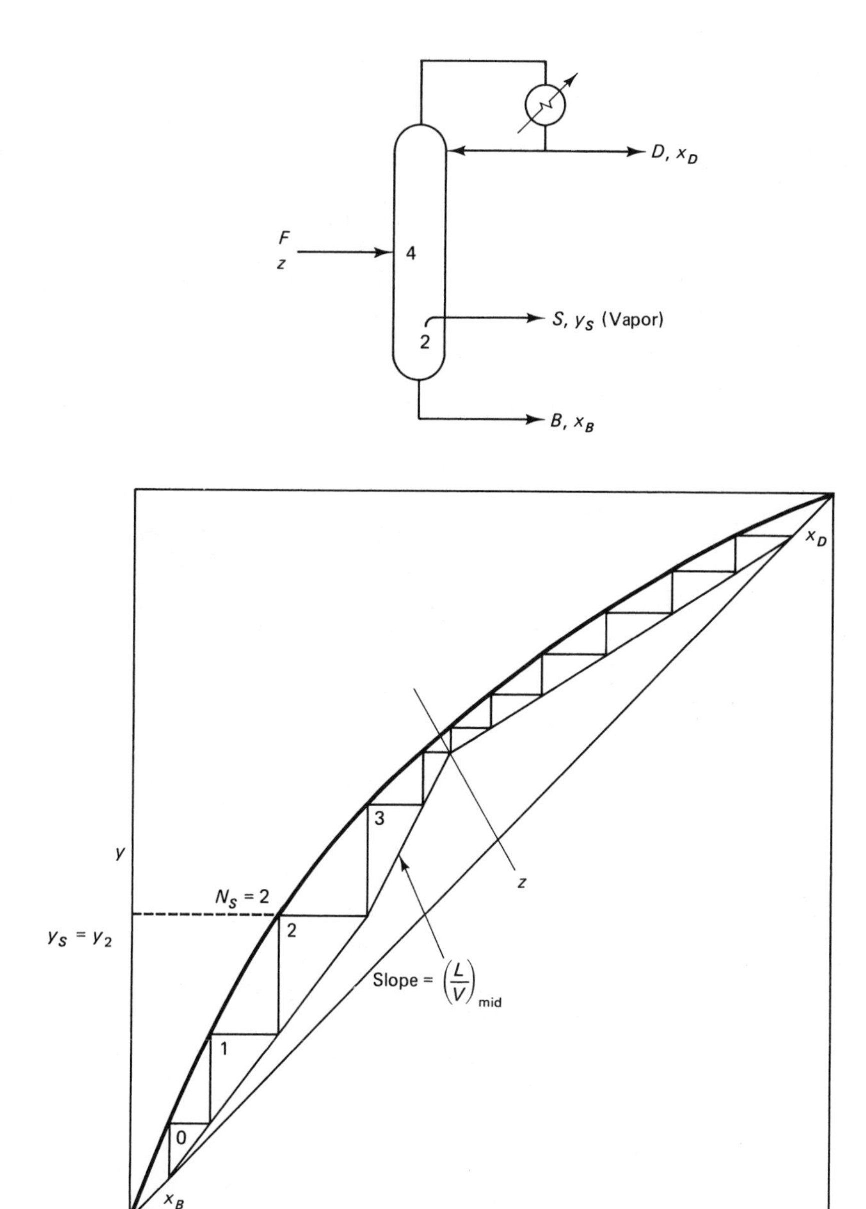

Figure 8–16 Vapor sidestream column.

Figure 8–16 shows a vapor sidestream column. Here the opposite effect is seen: the vapor flow rate is lower above the sidestream than below it, and the liquid flow rate is unchanged.

8.7.3 Other Complex Schemes

Every time material or energy is either added to or removed from the column, the component balances are affected. Therefore, different sections will have different operating lines. For example, Figure 8–17 shows a column with two feeds, one sidestream, and an intermediate reboiler. We looked at a real column in the alkylation process in chapter 3 that was very similar to this.

In the column of Figure 8–17, there are five different sections, each with its own operating line. These operating lines are derived by making component balances around each section. Note that the operating lines in sections I and II have different slopes because the liquid and vapor flow rates are different above and below the intermediate reboiler. Specifically, the vapor and liquid flow rates are higher above the intermediate reboiler than below it. Since the difference between the vapor and liquid flow rates in both sections is B, the bottoms flow rate, the slope in section I will be higher than the slope in section

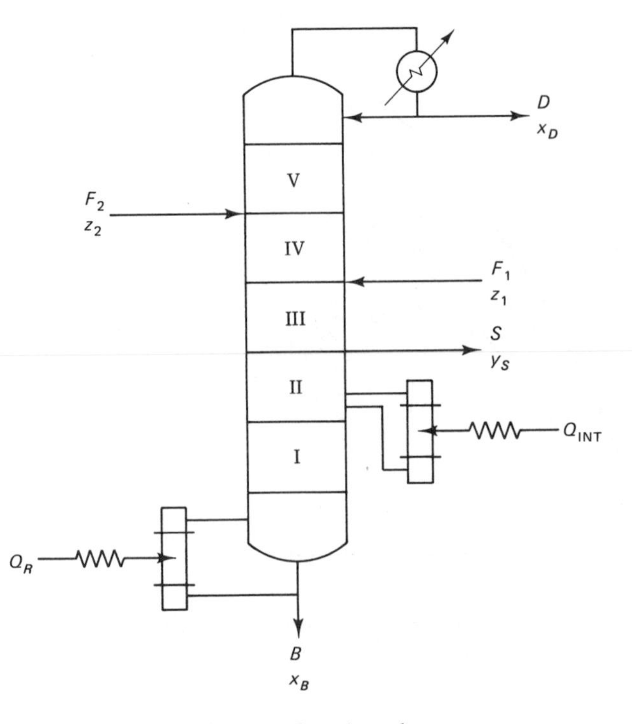

Figure 8–17 Complex column.

II. These slopes are given, respectively, by

$$\left(\frac{L}{V}\right)_{\mathrm{I}} = \frac{L_{\mathrm{I}}}{V_{\mathrm{I}}} = \frac{L_{\mathrm{I}}}{L_{\mathrm{I}} - B} \tag{8–50}$$

and

$$\left(\frac{L}{V}\right)_{\mathrm{II}} = \frac{L_{\mathrm{II}}}{V_{\mathrm{II}}} = \frac{L_{\mathrm{II}}}{L_{\mathrm{II}} - B} \tag{8–51}$$

$(L/V)_{\mathrm{I}}$ is bigger than $(L/V)_{\mathrm{II}}$ because L_{II} is bigger than L_{I}.

However, even if the slopes are different, the two operating lines intersect the 45° line at the bottoms composition x_B since no material (just energy) is added at the intermediate reboiler tray. Problems 8–14 and 8–23 at the end of the chapter provide other intermediate heat removal examples.

8.8 TRAY EFFICIENCY

Real trays seldom achieve perfect phase equilibrium between the vapor and liquid phases leaving them. Therefore, actual trays do not correspond to theoretical trays. Two types of efficiencies are widely used to account for this difference: overall efficiency and Murphree vapor-phase efficiency.

8.8.1 Overall Efficiency

The *overall efficiency* of the trays in a column is defined to be the ratio of the number of theoretical trays required to achieve a certain separation to the number of actual trays performing the same separation, i.e.,

$$\text{Overall efficiency} = (N_T)_{\text{theor}}/(N_T)_{\text{actual}} \tag{8–52}$$

If the VLE data are known, the number of theoretical trays can be calculated by the methods discussed in this book. The number of actual trays is usually determined by experiments in either laboratory or industrial columns; prediction of tray efficiencies from first principles is unreliable.

Typical numbers for overall efficiencies are in the range of 50 to 70 percent. The higher the relative volatility, the lower is the tray efficiency in many systems. Difficult separations with low relative volatilities, like propylene-propane with $\alpha = 1.1$, have tray efficiencies approaching 100 percent. Easy separations with large relative volatilities, like methanol-water, have tray efficiencies around 50 percent.

8.8.2 Murphree Vapor-Phase Tray Efficiency

Instead of looking at the overall column, we can look at the performance of individual trays. The ratio of the actual change in vapor compositions from tray to tray compared to the ideal change is called the *Murphree vapor-phase*

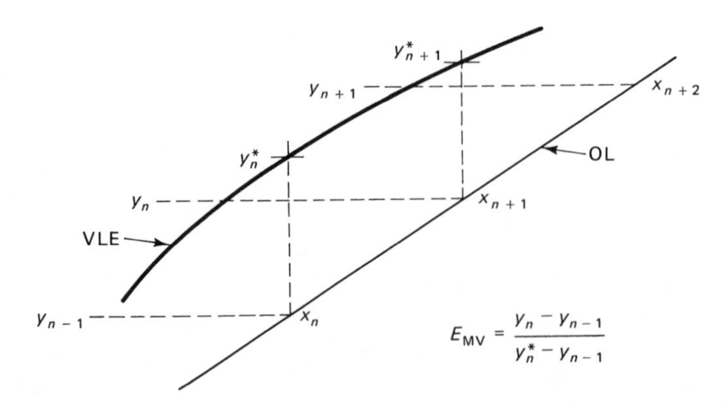

Figure 8–18 Murphree vapor-phase tray efficiency.

tray efficiency and is given by

$$E_{MV} = \frac{y_n - y_{n-1}}{y^*_n - y_{n-1}} \tag{8–53}$$

where

y_n = composition of actual vapor leaving tray n

y_{n-1} = composition of actual vapor entering tray n from tray $n - 1$ below

y^*_n = composition of vapor that would be in equilibrium with the liquid leaving tray n with composition x_n

The graphical representation of E_{MV} is shown in Figure 8–18. The efficiency is the fraction of the vertical height between the operating line and the VLE line that is stepped up. If the trays are 100 percent efficient, you step up all the way; if the efficiency is 50 percent you step up halfway.

Example 8.5.

A ten-tray distillation column is used to separate methanol and water. Feed is introduced into the column on the sixth tray from the bottom. The column has a partial reboiler and a total condenser. A *Txy* diagram for the methanol-water system at atmospheric pressure is given in Figure 8–19. Pressures on all trays throughout the column are essentially atmospheric, and equimolal overflow will be assumed. The molecular weight of methanol is 32, and the density of methanol liquid is 6.5 lb_m/gallon at the temperatures involved.

Tray efficiencies in this real column are not 100 percent. We can assume that they are the same on all trays and in the partial reboiler. Calculate the average Murphree vapor-phase tray efficiency from the operating data given below. The first items that you must calculate are the distillate, bottoms, and feed compositions, and the thermal condition of the feed.

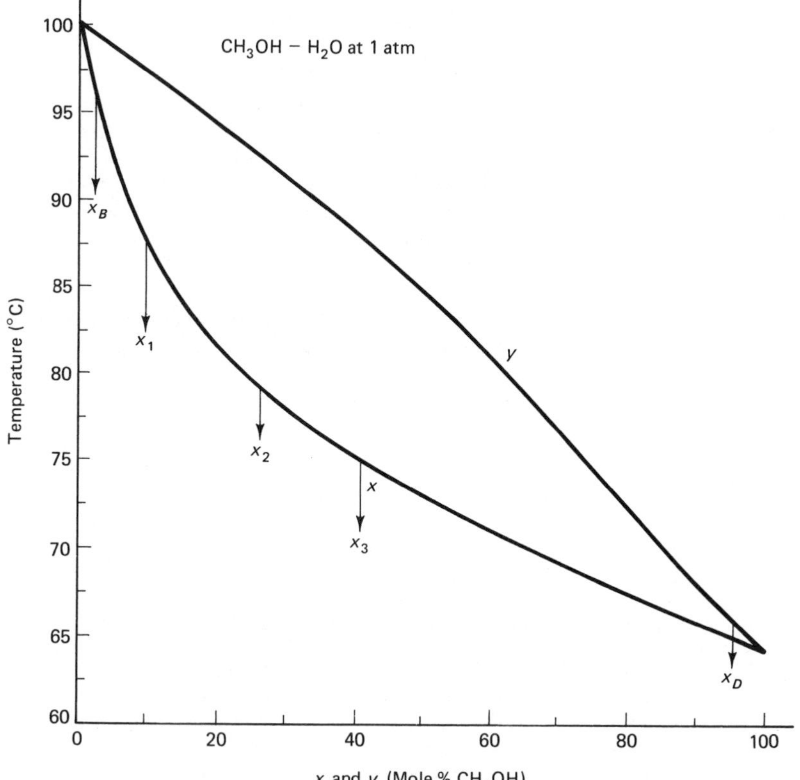

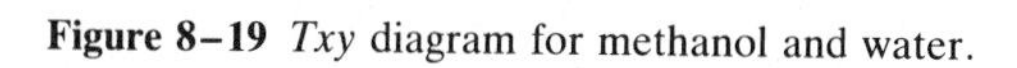
Figure 8–19 *Txy* diagram for methanol and water.

Data

Bottoms flow rate = 23.6 gpm (gallons per minute)
Distillate flow rate = 85.8 gpm
Reflux flow rate = 171.6 gpm

Temperatures (°*C*)

Reflux drum	= 65
Tray 10	= 66
Tray 6	= 69.4
Tray 4	= 71.5
Tray 3	= 75
Tray 2	= 79
Tray 1	= 87.5
Reboiler	= 96
Feed	= 71

Solution:

We can determine product compositions using the temperature data given, since the pressure is 1 atmosphere and the system is binary. The reboiler temperature is 96°C. This corresponds to a composition of 2.5 mole percent methanol (x_B = 0.025). The top tray temperature T_{10} is 66°C. This is the dewpoint temperature of the distillate product, since we have a total condenser. Using the *Txy* diagram gives a vapor composition at 66°C equal to 96 mole percent methanol (x_D = 0.96).

Note that we cannot use the reflux drum temperature to determine x_D because we do not know *a priori* that the liquid in the reflux drum is at its bubblepoint. It might in fact be subcooled. It turns out that in this example the reflux is at its bubblepoint (65°C), but this will not always be true.

Next, we must convert the volumetric flow rates (gpm) of distillate and bottoms into molar flow rates. We have the following data:

DISTILLATE (x_D = 0.96): BASIS = 100 LB-MOLE OF DISTILLATE

	lb-mole	*M*	*lb$_m$*	*lb$_m$/gal*	*gallons*
methanol	96	32	3,072	6.50	472.8
water	4	18	72	8.33	8.6
					481.4

$$(85.8 \text{ gal/min})(100 \text{ lb-mole}/481.4 \text{ gal}) = 17.82 \text{ lb-mole/min of distillate}$$

BOTTOMS (x_B = 0.025): BASIS = 100 LB-MOLE OF BOTTOMS

	lb-mole	*M*	*lb$_m$*	*lb$_m$/gal*	*gallons*
methanol	2.5	32	80	6.50	12.3
water	97.5	18	1,755	8.33	210.7
					223.0

Conversion yields

$$(23.6 \text{ gal/min})(100 \text{ lb-mole}/223 \text{ gal}) = 10.58 \text{ lb-mole/min of bottoms}$$

Now the feed flow rate and composition can be calculated:

$$F = D + B = 17.82 + 10.58 = 28.4 \text{ lb-mole/min}$$

$$z(28.4) = (0.96)(17.82) + (0.025)(10.58)$$

$$z = 0.612$$

Thus, the composition of the feed is 61.2 mole percent methanol. Using the *Txy* diagram at atmospheric pressure, we see that the feed is saturated liquid at its bubblepoint at the feed temperature of 71°C, since it lies on the x curve. Therefore, $q = 1$. If the point (z, T_F) had been in the two-phase region, we could have calculated the feed thermal condition q by first determining the compositions of the liquid and vapor portions of the feed that are in equilibrium with each other from the *Txy* diagram. They would lie at each end of the horizontal line at T_F. Knowing the total feed composition z, the fraction of the feed that is vapor can be calculated.

The average tray efficiency is calculated by drawing in the operating lines and plotting the individual tray liquid compositions from the given tray temperatures. The operating lines are known because we know the reflux ratio (171.6/85.8 = 2) and the q line. We have the following data:

Tray number	*Tray temperature*	*Tray liquid composition*
1	87.5°C	10 mole % methanol
2	79	27
3	75	46
4	71.5	58

Figure 8–20 shows the steps drawn on the McCabe-Thiele diagram. The Murphree vapor-phase efficiencies are about 70 percent. For tray 2, the efficiency is

$$E_{MV} = \frac{y_2 - y_1}{y_2^* - y_1}$$

$$= \frac{0.55 - 0.325}{0.64 - 0.325} = 0.71$$

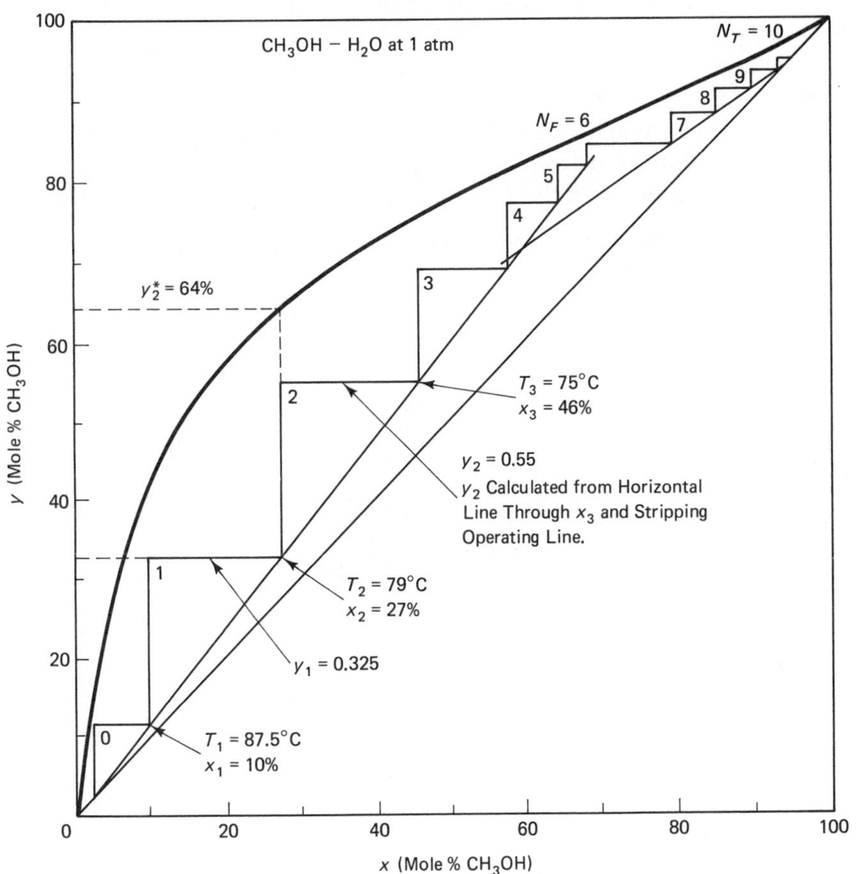

Figure 8–20 Solution to example 8.5 using a McCabe-Thiele diagram.

8.9 EQUIPMENT

In chapter 5 we described briefly some of the hardware that is used in countercurrent vapor-liquid contacting columns. In particular, we discussed columns and types of trays and internals. Here, we give some additional information about columns and some of the other auxiliary pieces of equipment that are used in distillation column systems.

8.9.1 Tray Hydraulics

Figure 8–21 sketches a typical single-liquid-pass tray column. Liquid flows down the downcomer, onto the next tray, across the tray, over an outlet weir, and down into the next downcomer. Vapor flows up through the openings in the trays (bubblecaps, sieves, valves, etc.).

Let us consider the various pressure drops in this fairly complex hydraulic system. The pressure difference between the trays is due to the vapor flowing through the holes in the tray and into the liquid pool on the tray. The vapor pressure drop consists of two terms: the dry hole pressure drop and the liquid hydraulic head on the tray.

The height of liquid in the downcomer must build up to overcome the pressure differential between the trays plus the pressure drop that the liquid produces in its flow down the column. This liquid pressure drop includes

1. The ΔP of flow through the narrow slot that opens onto the tray from the downcomer.

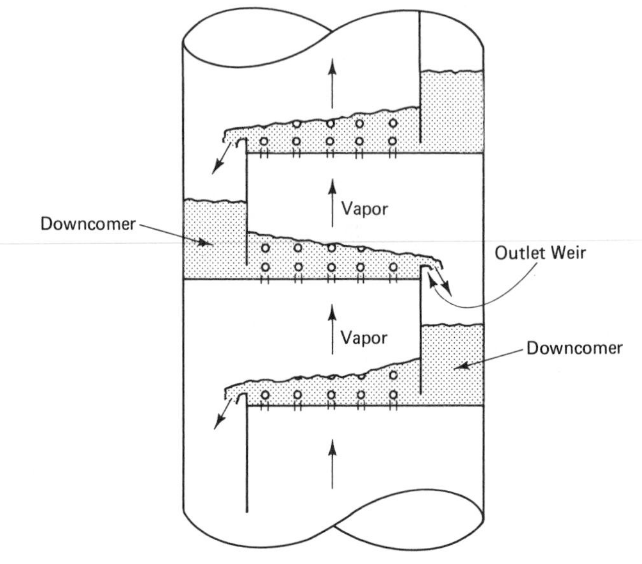

Figure 8–21 Tray hydraulics.

2. The ΔP of flow across the tray (small in sieve trays, but large in bubblecaps). This is called the *gradient*.
3. The height of the liquid over the outlet weir.

The height of the liquid in the downcomer is the sum of all these vapor and liquid pressure drops plus the outlet weir height. If the height of the liquid in the downcomer gets up near the tray spacing, the column floods and separation efficiency becomes very poor. Vapor or liquid flow rates must then be reduced to lower pressure drops.

The size of a column is determined by the number of trays and the vapor throughput. The height of the column is equal to the tray spacing (1 to 3 feet) times the number of actual trays, plus some additional elevation for liquid holdup in the base of the column. The diameter of the column is set by the maximum superficial vapor velocity that can be handled in the column.

Example 8.6.

A column has to handle 4,833 lb-mole/hr of vapor. The maximum superficial vapor velocity up the column is 3 ft/sec. Assume perfect gases, a molecular weight of 32, a temperature of 154°F, and a pressure of 2 psig. What must be the diameter of the column?

Solution:

The vapor density is calculated from the Perfect Gas Law:

$$\frac{MP}{RT} = \frac{(32)(14.7 + 2)(144)}{(1{,}545)(154 + 460)}$$

$$= 0.081 \text{ lb}_m/\text{ft}^3$$

The vapor volumetric flow rate is

$$(4{,}833 \text{ lb-mole/hr})(32 \text{ lb}_m/\text{lb-mole})(\text{ft}^3/0.081 \text{ lb}_m)(\text{hr}/3{,}600 \text{ sec}) = 530 \text{ ft}^3/\text{sec}$$

The required cross-sectional area of the column must be equal to the volumetric throughput divided by the maximum superficial vapor velocity:

$$\frac{(530 \text{ ft}^3/\text{sec})}{(3 \text{ ft/sec})} = 176.5 \text{ ft}^2$$

The area of the column is

$$\pi \frac{D^2}{4} = 176.5 \text{ ft}^2$$

where D is the diameter we seek. So

$$D = 15 \text{ ft}$$

8.9.2 Condensers

A large variety of condensers and coolants are used on distillation columns. Cooling water from a river, ocean, or cooling tower is the most frequently used coolant, but refrigeration is used on some columns operating at low temperature, and some column overhead condensers are used to generate steam for use elsewhere in the plant. The overhead vapor from a high-temperature column can be used to provide heat in the reboiler of a lower temperature column. Some of the problems at the end of the chapter give illustrations of these heat-integrated distillation columns.

Figure 8–22 sketches several condensers. Tube-in-shell heat exchangers are frequently used. The coolant can be on either the shell side or the tube side. The condensing process vapors are usually put on the tube side if (1) the pressure in the system is high, since only the tubes have to withstand the high pressure, or (2) the material is corrosive or tends to plug the tubes, since it is much easier to clean the inside of tubes than to clean the shell side.

Air-cooled condensers are used in some columns. They have the advantage of not requiring makeup water that would be needed in a cooling tower system. However, not quite as low temperatures can be obtained in the reflux drum with air as with water (130°F versus 120°F respectively). This causes column

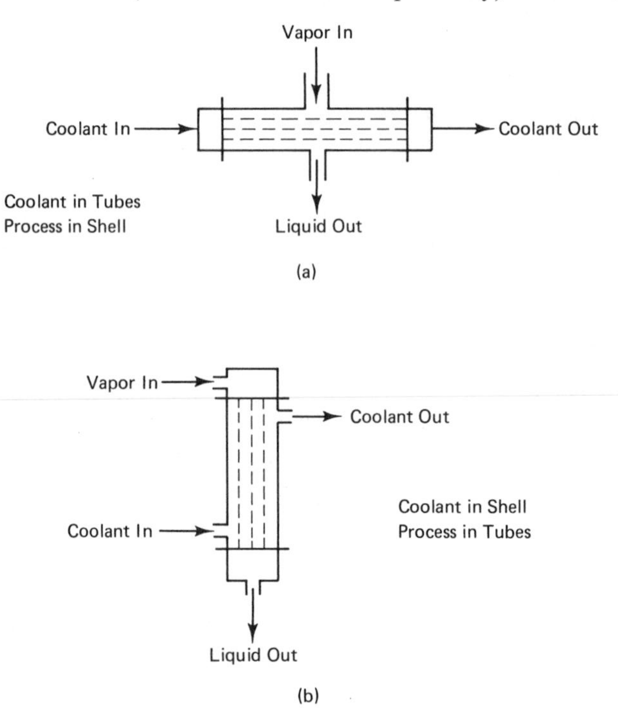

Figure 8–22 Condensers. (a) Horizontal (b) Vertical (c) Air coolers (d) Spray condenser (direct contact) (e) Gravity-flow condenser.

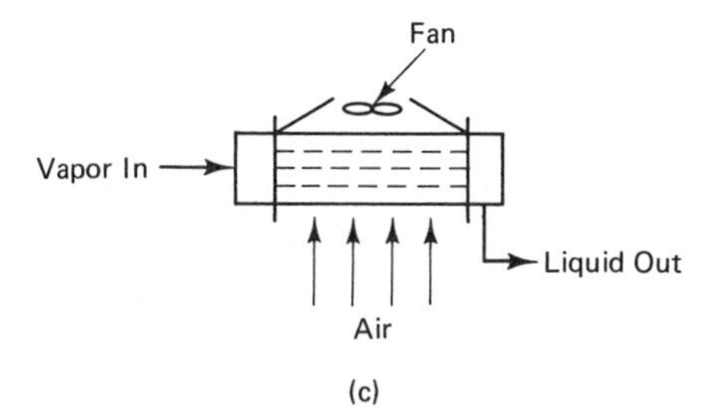

(c)

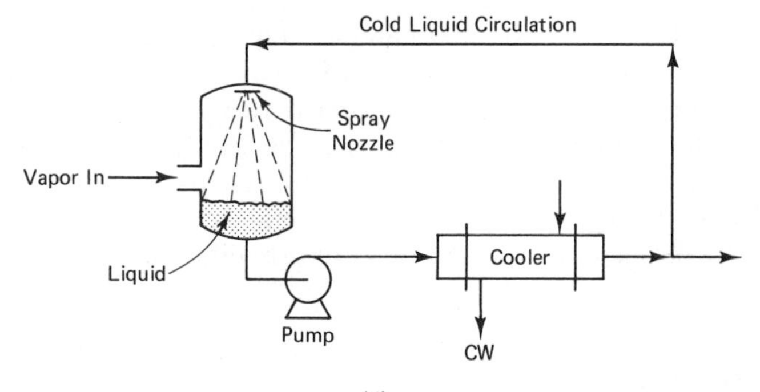

(d)

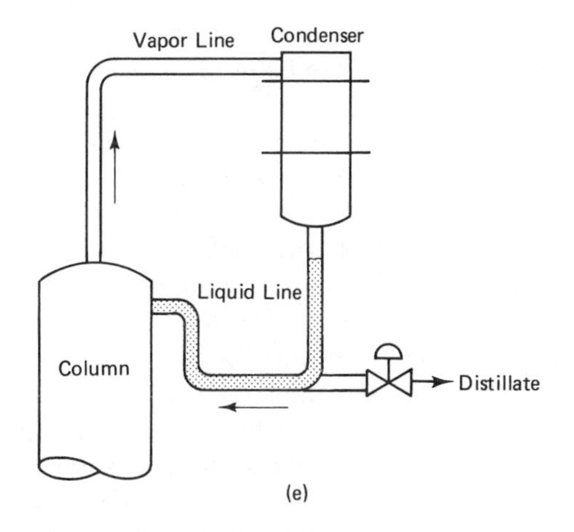

(e)

Figure 8–22 Con't.

pressures to be somewhat higher, which can increase energy consumption in some chemical systems since relative volatilities tend to decrease as pressure increases.

Spray condensers are used in high-vacuum columns to eliminate undesirable pressure drop through the overhead vapor piping and condenser. Liquid is circulated around through a cooler at a high flow rate and is sprayed into the vapor space in the condenser. The subcooled liquid condenses the vapor; the sensible heat change in the liquid is used to remove the latent heat of condensation of the vapor. Spray condensers are usually used in situations where the overhead temperature is fairly high, so that the large amount of subcooling required in the cooled liquid can be achieved.

Most condensers are mounted above the reflux drum on the second level of a platform, perhaps 20 feet above ground level. The reflux drum is mounted on the first level, perhaps 10 feet off the ground. The pumps are located at ground level. The reflux is pumped up to the top of the column.

Some condensers are mounted in or above the top of the column so that the reflux can flow back into the column without using pumps. These *gravity flow* systems are useful when very corrosive, toxic, or hazardous material is being handled and pumping is to be avoided because of possible leaks and explosions. The hydraulics of these gravity flow condensers have to be carefully considered to ensure that the desired flow rate of reflux can be achieved despite changes in vapor flow rates to the condenser. The liquid has to have enough hydraulic head to overcome the pressure difference between the condenser and the column.

8.9.3 Reboilers

The hot material that is used in reboilers depends on the temperature level required in the base of the column. For very low temperatures, warm water or some other warm stream can be used. A warm feed stream is used in some cryogenic columns.

Low pressure (perhaps 15 psig) steam is the next higher temperature heat source. For higher temperatures, steam at progressively higher pressure levels (50, 150, and 300 psig) is used. For temperatures higher than about 400°F, steam cannot be used economically because pressures become too high. Some high-boiling material like Dow-Therm is sometimes used. In other high-temperature systems, a fired reboiler is employed. Process liquid from the base of the column is pumped through a furnace in which fuel gas or oil is burned.

Several types of reboilers are sketched in Figure 8–23. Thermosiphon and kettle reboilers are probably the most common. Internal reboilers save capital investment, but require a complete column shutdown to clean or repair the reboiler. Kettle or thermosiphon reboilers are often installed as two separate units that operate in parallel. This permits column operation to continue when one of the reboilers must be taken off line for cleaning or repair.

Forced circulation reboilers are used when the material in the base of the column has a tendency to coke or polymerize on the hot tubes in the reboiler. Circulating a large liquid stream through the reboiler keeps film temperatures low and reduces fouling. Remember that the reboiler has the highest temperature in the distillation column, so it is usually the place where fouling and polymerization occur.

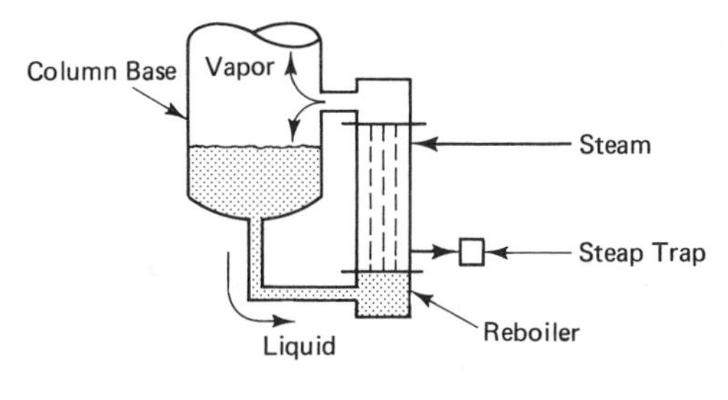

(a)

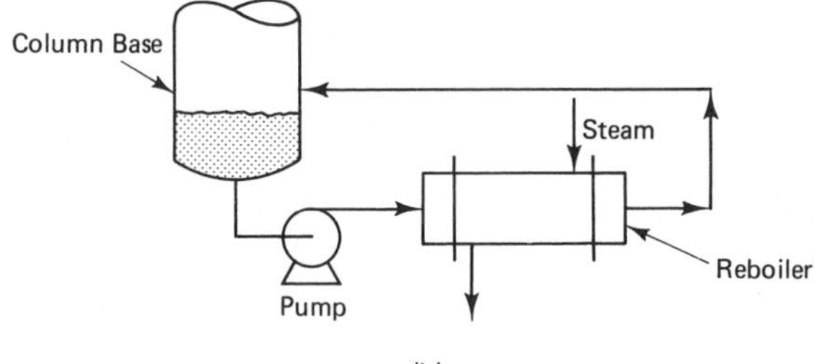

(b)

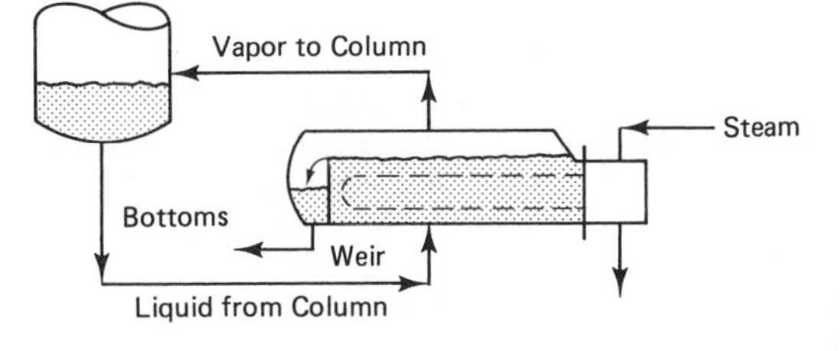

(c)

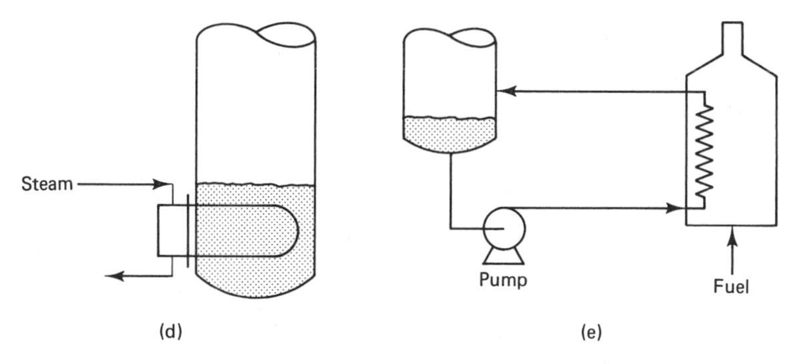

(d) (e)

Figure 8–23 Reboilers. (a) Thermosiphon: circulation through tubes due to density differences (b) Forced circulation (c) Kettle reboiler (d) Internal reboiler (e) Fired reboiler (furnace).

8.10 SELECTION OF OPERATING PRESSURE

8.10.1 Cooling Limitations

The pressure in a distillation column is very often established by the bubblepoint pressure of the distillate product at a temperature that is about 20° higher than the available cooling water temperature. The reason for this is that we usually want the lowest pressure possible so that the relative volatility is as high as possible. We could go to even lower pressures by using refrigerant in the condenser, but this is seldom done because the cost of heat removal to cooling water or air is much less (by a factor of 10) than to refrigerants.

Normal summertime maximum cooling water temperatures are about 90°F (32°C) in most locations around the world. Therefore, most reflux drum temperatures are designed for 110 to 120°F. In the wintertime, lower temperatures can be achieved, so many columns operate at lower pressures when the weather is cooler. This saves energy in those systems where relative volatility increases as pressure decreases.

Example 8.7.

A deisobutanizer column has a distillate product coming from a total condenser that is 5 mole percent propane, 85 mole percent isobutane, and 10 mole percent n-butane. The bottoms product from the column is 21 mole percent isobutane and 79 mole percent n-butane. The pressure drop from the reflux drum to the base of the column is 2 psi. Assuming ideal (Raoult's Law) VLE, calculate the pressures in the reflux drum and in the column base if the reflux drum temperature is 120°F.

Solution:

We have the following data:

	x_{Dj}	P_j^s at 120°F	$x_{Dj}P_j^s$
C_3	0.05	240 psia	12
iC_4	0.85	97	82
nC_4	0.10	71	7
			101 psia

The bubblepoint pressure calculation gives a reflux drum pressure of 101 psia.

Note that the pressure in the base of the column is 101 + 2 = 103 psia. An iterative bubblepoint temperature calculation, using the bottoms liquid composition, yields a base temperature of 142°F at 103 psia.

8.10.2 Base Temperature Limitations

Some distillation columns have their pressures set by temperature limitations in the base of the column. Some problems at the end of chapter 7 illustrated this point. If thermally sensitive materials are present in the bottoms product, the base temperature may have to be kept below some maximum temperature limit. Operation above this temperature would result in rapid reboiler fouling, excessive product degradation, excessive corrosion rates, or explosions. Many distillation systems have this type of restriction.

If the temperature and composition of the base are fixed, then the pressure in the base is completely specified, and the column operating pressure is accordingly fixed. The temperature that must be attained in the reflux drum at this pressure (less the pressure drop through the column) can then be calculated since the composition of the distillate is known. If this reflux drum temperature is lower than 100°F, some type of refrigeration system will be required. Since cooling costs will be a significant part of energy costs, you may have to pay for your energy twice: once to put it into the reboiler, and again to remove it in the condenser.

8.11 COMPUTER PROGRAMS FOR DESIGN AND RATING

The digital computer is a very useful tool in analyzing distillation columns. In this section we give two programs that are simple extensions of the program discussed in chapter 5 for single sections. For both programs, we put the rectifying section on top of the stripping section.

8.11.1 Design Program

The FORTRAN program that solves the design problem is given in Table 8–1. The feed composition, feed flow rate, feed thermal condition, relative volatility, distillate composition, bottoms composition, and reflux ratio are given in a DATA statement. The assumptions made are equimolal overflow, constant relative volatility, a total condenser, a partial reboiler, theoretical stages, and saturated liquid reflux.

The program calculates the distillate and bottoms flow rates, the liquid and vapor flow rates in the stripping and rectifying sections, and the point of intersection of the SOL with the q line. Then it steps up the column, using the component balance equations for the stripping section, until the optimum feed tray is reached. The program then switches to the component balances for the rectifying section and steps up the column until the distillate com-

TABLE 8–1

FORTRAN BINARY DISTILLATION DESIGN PROGRAM

```
C    BINARY DISTILLATION DESIGN PROGRAM
C    GIVEN XB,XD AND REFLUX RATIO,
C       THE TOTAL NUMBER OF TRAYS AND OPTIMUM FEED TRAY
C       ARE CALCULATED.
C
C CONSTANT RELATIVE VOLATILITY AND EQUIMOLAL OVERFLOW
      DIMENSION X(100),Y(100)
      REAL LS,LR
C     RELATIVE VOLATILITY FUNCTION
      EQUIL(XX)=ALPHA*XX/(1.+(ALPHA-1.)*XX)
      DATA ALPHA,XB,XD,Z,F,Q,RR/2.,.02,.98,.5,100.,1.,2.56/
      D=F*(Z-XB)/(XD-XB)
      B=F-D
      LR=RR*D
      VR=D+LR
      VS=VR-(1.-Q)*F
      LS=LR+Q*F
C   INTERCEPT Q-LINE AND OPERATING LINES
      XINT=(VS*ZF+(1.-Q)*B*XB)/((1.-Q)*LS+VS*Q)
      YB=EQUIL(XB)
      X(1)=(VS*YB+B*XB)/LS
      Y(1)=EQUIL(X(1))
      DO 10 N=2,100
      X(N)=(VS*Y(N-1)+B*XB)/LS
      IF(X(N).GT. XINT) GO TO 15
   10 Y(N)=EQUIL(X(N))
      WRITE(6,11)
   11 FORMAT(1X,21HDESIGN LOOP STRIPPING )
      STOP
   15 NF=N-1
      NFP1=NF+1
      DO 20 N=NFP1,100
      X(N)=(VR*Y(N-1)+B*XB-F*Z)/LR
      Y(N)=EQUIL(X(N))
      IF(Y(N).GT. XD) GO TO 25
   20 CONTINUE
      WRITE(6,21)
   21 FORMAT(1X,22HDESIGN LOOP RECTIFYING )
      STOP
   25 NT=N
      WRITE(6,30) ALPHA,XB,XD,Z,F,Q,RR
   30 FORMAT(1X,10HDATA GIVEN,/,1X,8H ALPHA =,F6.2,5H XB =,F8.5,5H XD =
     1,F8.5,5H Z =,F8.5,4H F =,F10.4,/,1X,4HQ = ,F6.3,5H RR =,F6.3)
      WRITE(6,31)
   31 FORMAT(1X,18HCALCULATED RESULTS ,/,1X,20H  N        X          Y   )
      N=0
      WRITE(6,40) N,XB,YB
   40 FORMAT(1X,I2,3X,2F10.6)
      DO 41 N=1,NT
   41 WRITE(6,40) N,X(N),Y(N)
      WRITE(6,42) NF
   42 FORMAT(1X,15HFEED PLATE =   ,I2)
      STOP
      END
```

position is reached. Figure 8–24 gives a flow diagram of the logic in the program.

Table 8–2 lists the output of the program, including liquid and vapor compositions on all trays. Note that this is the same problem that was solved by hand in example 8.1.

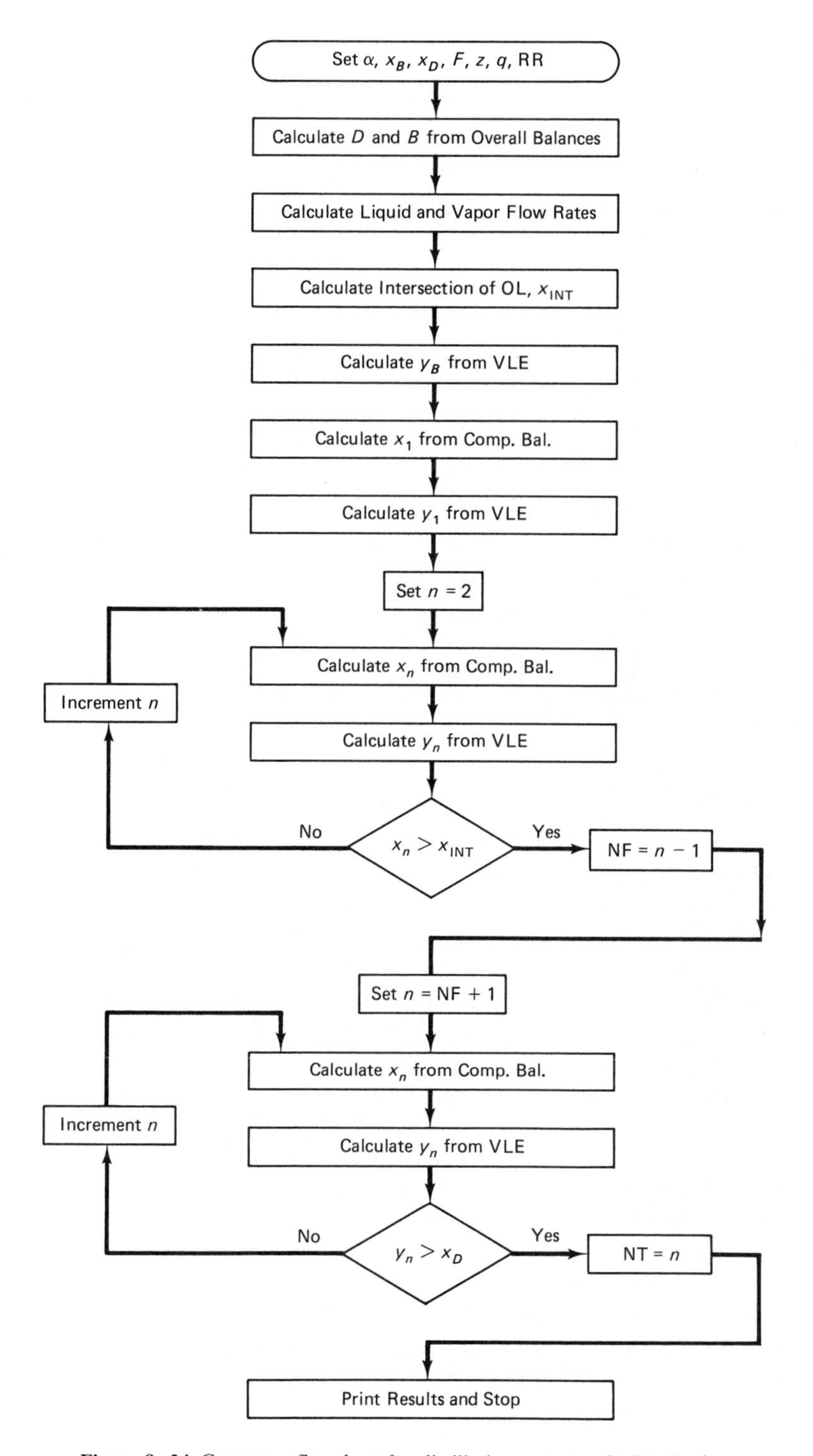

Figure 8–24 Computer flowchart for distillation program (Table 8–1).

TABLE 8–2
DATA AND OUTPUT FOR BINARY DISTILLATION DESIGN PROGRAM (TABLE 1)

```
DATA GIVEN
 ALPHA =   2.00 XB =   .02000 XD =   .98000 Z =   .50000 F
Q =  1.000 RR = 2.560

CALCULATED RESULTS
  N        X          Y
 0       .020000    .039216
 1       .035002    .067636
 2       .057190    .108192
 3       .088851    .163202
 4       .131798    .232900
 5       .186212    .313960
 6       .249495    .399354
 7       .316162    .480430
 8       .379459    .550156
 9       .433894    .605197
10       .476864    .645779
11       .515224    .680063
12       .562900    .720328
13       .618894    .764588
14       .680443    .809838
15       .743368    .852795
16       .803106    .890803
17       .855960    .922391
18       .899887    .947306
19       .934535    .966160
20       .960753    .979984
21       .979978    .989888
 FEED PLATE =    10
```

8.11.2 Rating Program

A FORTRAN program that solves a rating problem is given in Table 8–3. The system used is methanol-water separation at a pressure of 17 psia. Nonideal VLE is used in the subroutine that does the bubblepoint calculations. The van Laar equations are used for activity coefficients (GAM1 and GAM2), and a two-constant Antoine equation is used for vapor pressures (PS1 and PS2). Newton-Raphson iteration is used to converge on temperature.

The rating program finds the value of the reflux flow rate R that is needed to achieve product purities of $x_B = 0.001$ and $x_D = 0.999$ in a 40-tray column with feed introduced on tray 15. Feed composition, feed flow rate, and feed thermal condition are given. Equimolal overflow, saturated liquid reflux, a total condenser, a partial reboiler, and 40 theoretical trays are assumed.

After an initial guess of R is made, the internal liquid and vapor flow rates are calculated. Then, starting at x_B, the program steps 15 trays up the column using the stripping section component balances. The rectifying component balances are used for the next 25 trays.

If the correct value of R has been guessed, the composition of the vapor leaving tray 40 will be equal to x_D. If the difference between these two compositions is not sufficiently small (10^{-5}), a new guess of R is made.

TABLE 8–3
FORTRAN BINARY DISTILLATION RATING PROGRAM

```
C    BINARY DISTILLATION "RATING" PROGRAM

C     FINDS R AND V REQUIRED TO GIVE SPECIFIED
C       XB AND XD IN A GIVEN COLUMN (NT AND NF FIXED)
C   SPECIFIC SYSTEM IS   METHANOL/WATER
C     NON-IDEAL VLE
C     EQUIMOLAL OVERFLOW
      DIMENSION X(40),Y(40),T(40)
      COMMON D1,C1,D2,C2,AV,BV
      REAL LR,LS
      DATA P,NT,NF,F/17.,40,15,2300./
      DATA XB,XD,Z,Q/.001,.999,.80,1./
      AV=.85
      BV=.24
      D1=-337.5*373.*ALOG(2600./760.)/35.5
      C1=ALOG(760.)-D1/337.5
      D2=-332.4*373.*ALOG(760./145.4)/40.6
      C2=ALOG(760.)-D2/373.
      R=1200.
      DR=200.
      LOOP=0
      FLAGM=1.
      FLAGP=1.
C    INITIAL GUESS OF TRAY TEMPERATURES
      TB=100.
      DO 10 N=1,40
   10 T(N)=80.
      CALL EQUIL(XB,TB,YB,P)
      D=F*(Z-XB)/(XD-XB)
      B=F-D
  100 VR=D+R
      LR=R
      VS=VR-(1.-Q)*F
      LS=LR+Q*F
      LOOP=LOOP+1
      IF(LOOP.GT.50) STOP
      X(1)=(B*XB+VS*YB)/LS
      CALL EQUIL(X(1),T(1),Y(1),P)
      DO 20 N=2,NF
      X(N)=(B*XB+VS*Y(N-1))/LS
   20 CALL EQUIL(X(N),T(N),Y(N),P)
      X(NF+1)=(VR*Y(NF)+B*XB-F*ZF)/LR
      IF(X(NF+1).LT.X(NF)) GO TO 30
      CALL EQUIL(X(NF+1),T(NF+1),Y(NF+1),P)
      NFP2=NF+2
      DO 25 N=NFP2,NT
      X(N)=(VR*Y(N-1)+B*XB-F*Z)/LR
      IF(X(N).GT.XD) GO TO 40
   25 CALL EQUIL(X(N),T(N),Y(N),P)
      IF(ABS(Y(NT)-XD) .LT. .00001) GO TO 90
      IF(Y(NT)-XD) 30,30,40
   30 IF(FLAGP.LT.0.) DR=DR/2.
      R=R+DR
      FLAGM=-1.
      GO TO 100
   40 IF(FLAGM.LT.0.) DR=DR/2.
      R=R-DR
      FLAGP=-1.
      GO TO 100
   90 WRITE(6,91)
```

TABLE 8-3 CON'T.

```
  91 FORMAT(/,1X,50H XB       XD        Z          VS      R       F
     WRITE(6,92) XD,XB,Z,VS,R,F,Q
  92 FORMAT(1X,3F7.4,3F7.1,F6.3)
     WRITE(6,93)
  93 FORMAT(1X,45H   N          X          Y          T
     N=0
     WRITE(6,96) N,XB,YB,TB
     DO 95 N=1,NT
  95 WRITE(6,96)N,X(N),Y(N),T(N)
  96 FORMAT(1X,I3,3X,2F10.7,2F10.2)
     STOP
     END
C
     SUBROUTINE EQUIL (X,T,Y,PSIA)
C   BUBBLE POINT CALCULATION FOR METHANOL-WATER USING NEWTON-I
     COMMON D1,C1,D2,C2,AV,BV
     A=AV
     B=BV
     X1=X
     X2=1.-X
C USING VAN LAAR EQUATIONS FOR ACTIVITY COEFFICIENTS
     GAM1=EXP(A*X2*X2/((A*X1/B+X2)**2))
     GAM2=EXP(B*X1*X1/((X1+B*X2/A)**2))
     P=PSIA*760./14.7
     LOOP=0
  10 LOOP=LOOP+1
     IF(LOOP.GT.50) STOP
     PS1=EXP(C1+D1/(T+273.))
     PS2=EXP(C2+D2/(T+273.))
     PCALC=X1*PS1*GAM1+X2*PS2*GAM2
     IF(ABS(P-PCALC) .LT. P/10000.) GO TO 50
     F=PCALC-P
     DF=(-D1*PS1*X1*GAM1-D2*PS2*X2*GAM2)/((T+273.)**2)
     T=T-F/DF
     GO TO 10
  50 Y=X1*PS1*GAM1/P
     RETURN
     END
```

Interval-halving is used as the convergence algorithm. Figure 8–25 gives a flow diagram of the program.

Table 8–4 gives the output of the program for a feed composition of 80 mole percent methanol. Table 8–5 gives the output for a feed composition of 60 mole percent. Notice that the reflux flow rate has to change from 1,071.1 kg-mole/hr to 995.6 kg-mole/hr when the feed composition changes from 80 to 60 mole percent methanol. The column produces products that have exactly the same compositions in both cases, although, naturally, the distillate and bottoms flow rates are different for the two feed compositions. Notice also that more vapor boilup (energy) is required with a feed composition of 80 mole percent methanol than with a feed composition of 60 mole percent, i.e., V_S is bigger for $z = 0.8$ than for $z = 0.6$.

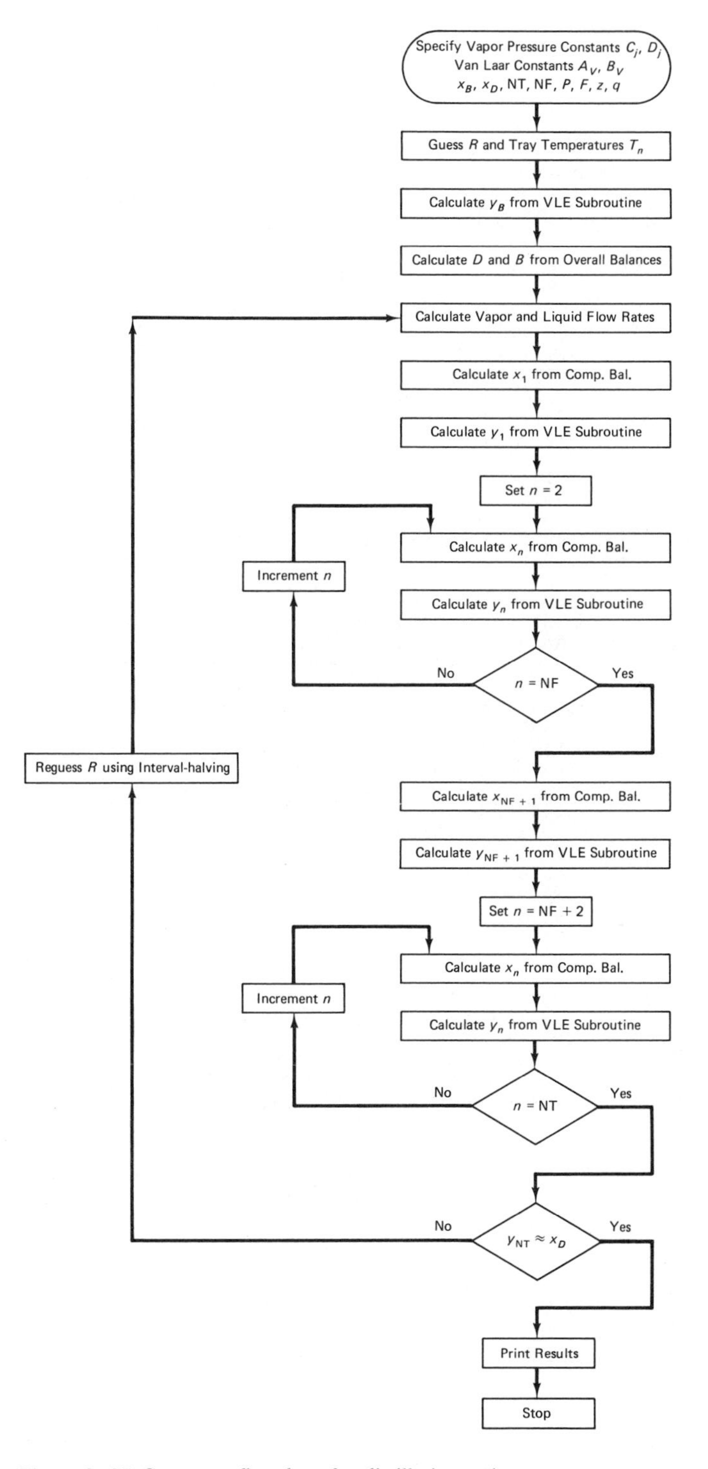

Figure 8–25 Computer flowchart for distillation rating program.

TABLE 8–4

DATA AND OUTPUT FOR BINARY DISTILLATION RATING PROGRAM (TABLE 8-3) WITH FEED COMPOSITION OF 80 MOLE PERCENT

XB	XD	Z	VS	R	F	Q
.9990	.0010	.8000	2912.5	1071.1	2300.0	1.000

N	X	Y	T
0	.0010000	.0077546	103.86
1	.0068357	.0496068	102.81
2	.0429941	.2252972	98.07
3	.1947829	.5156832	88.93
4	.4456636	.7384936	80.73
5	.6381620	.8536575	75.84
6	.7376585	.9014393	73.59
7	.7789399	.9194046	72.70
8	.7944612	.9258611	72.37
9	.8000392	.9282657	72.26
10	.8021167	.9290488	72.21
11	.8027933	.9293180	72.20
12	.8030259	.9293875	72.19
13	.8030859	.9295165	72.19
14	.8031973	.9294947	72.19
15	.8031785	.9294754	72.19
16	.8099515	.9322295	72.05
17	.8174404	.9352449	71.89
18	.8256395	.9385107	71.72
19	.8345198	.9420064	71.54
20	.8440252	.9457013	71.35
21	.8540722	.9495547	71.15
22	.8645502	.9535174	70.93
23	.8753254	.9575338	70.72
24	.8862468	.9615451	70.50
25	.8971541	.9654924	70.29
26	.9078874	.9693201	70.08
27	.9182957	.9729794	69.88
28	.9282458	.9764300	69.69
29	.9376286	.9796418	69.51
30	.9463620	.9825953	69.34
31	.9543931	.9852810	69.19
32	.9616958	.9876983	69.05
33	.9682687	.9898539	68.93
34	.9741303	.9917604	68.82
35	.9793144	.9934343	68.72
36	.9838660	.9948944	68.64
37	.9878363	.9961610	68.56
38	.9912802	.9972542	68.50
39	.9942529	.9981938	68.45
40	.9968079	.9989985	68.40

TABLE 8–5

DATA AND OUTPUT FOR BINARY DISTILLATION RATING PROGRAM (TABLE 8–3) WITH FEED COMPOSITION OF 60 MOLE PERCENT

XB	XD	Z	VS	R	F	Q
.9990	.0010	.6000	2376.0	995.6	2300.0	1.000

N	X	Y	T
0	.0010000	.0077546	103.86
1	.0058699	.0430564	102.98
2	.0313217	.1799601	99.35
3	.1300262	.4267602	91.90
4	.3079636	.6310791	84.84
5	.4552729	.7450541	80.47
6	.5374463	.7975953	78.30
7	.5753273	.8196558	77.35
8	.5912324	.8285505	76.96
9	.5976453	.8320134	76.80
10	.6001420	.8334277	76.74
11	.6011617	.8339771	76.72
12	.6015577	.8342143	76.71
13	.6017288	.8343159	76.71
14	.6018021	.8343662	76.71
15	.6018383	.8343969	76.71
16	.6061584	.8366513	76.60
17	.6115387	.8395603	76.47
18	.6184813	.8433578	76.31
19	.6275445	.8481390	76.09
20	.6389554	.8540108	75.82
21	.6529690	.8611615	75.49
22	.6700349	.8696767	75.10
23	.6903572	.8795484	74.64
24	.7139170	.8906401	74.11
25	.7403885	.9026687	73.53
26	.7690959	.9152167	72.91
27	.7990428	.9277811	72.28
28	.8290290	.9398500	71.65
29	.8578326	.9509833	71.07
30	.8844033	.9608720	70.54
31	.9080039	.9693612	70.08
32	.9282640	.9764360	69.69
33	.9451489	.9821869	69.36
34	.9588740	.9867669	69.10
35	.9698044	.9903547	68.90
36	.9783672	.9931435	68.74
37	.9850230	.9952642	68.62
38	.9900841	.9968750	68.52
39	.9939285	.9980915	68.45
40	.9968317	.9990060	68.40

PROBLEMS

8–1. The following equilibrium data are available for the benzene-toluene system at 1 atmosphere total pressure:

	mole fraction benzene	
T°C	*x*	*y*
80.02	1.0000	1.0000
84.0	0.8227	0.9223
88.0	0.6589	0.8297
92.0	0.5077	0.7201
96.0	0.3760	0.5957
100.0	0.2560	0.4528
104.0	0.1547	0.3043
103.0	0.0581	0.1278
110.4	0.00	0.00

A distillation column is to be used to separate 100 kg-mole/minute of feed containing 50 mole percent benzene into a distillate containing 96 mole percent benzene and a bottoms product containing 4 mole percent benzene.

(a) If the feed is a saturated liquid, and a reflux ratio of 3.0 is used, what is the number of theoretical trays required in the column, and on which tray should the feed enter?

(b) For the same reflux ratio, but using vapor feed, what will be the theoretical tray requirement and the feed tray location?

(c) Using a vapor feed with a reflux ratio of 6.0, what will be the number of theoretical trays required and the feed plate location?

(d) For the conditions of (c), what will be the vapor boilup required in the reboiler?

For all of these calculations, use a total condenser and a liquid bottoms product taken from the reboiler. Assume that the reboiler acts as one theoretical stage, but no separation occurs in the condenser.

8–2. 100 kg-mole/minute of saturated vapor feed (composition 60 mole percent) is introduced onto tray 5 of a 13-tray distillation column which has a total condenser and a partial reboiler. The relative volatility is constant at 2. Equimolal overflow, saturated liquid reflux, and 100 percent efficient trays can be assumed. The bottoms composition is 4 mole percent, and the distillate composition is 94 mole percent.

(a) What is the reflux ratio?

(b) Calculate the liquid and vapor rates in the stripping and rectifying sections.

(c) Draw a composition-versus-tray-number diagram.

8–3. A depropanizer column is fed 784 kg-mole/hour of saturated liquid feed with a composition of 28.4 mole percent propane and 71.6 mole percent isobutane. The column has a total condenser and a partial reboiler. Assume equimolal overflow, a reflux drum temperature of 120°F, and a column base temperature of 200°F.

(a) If the system is ideal, calculate the relative volatilities of propane and isobutane at 120°F and 200°F.

(b) Draw an xy diagram using an average of the two relative volatilities calculated above.

(c) Calculate the number of trays N_T required to make an overhead distillate product that is 96 mole percent propane and a bottoms product that is 5.8 mole percent propane when the reflux ratio is 2.774. What is the optimum feed tray N_F?

(d) Calculate N_T and the optimum N_F for the same product purities and reflux ratio as above when the feed composition z increases to 55 mole percent propane.

(e) Calculate N_T and the optimum N_F for the same product purities and vapor boilup as found in part (c) when the feed composition is 55 mole percent propane.

8–4. Calculate the number of theoretical trays required to achieve the separations indicated below under total reflux conditions. Assume constant relative volatility and a partial reboiler.

(a) $x_D = 0.95$, $x_B = 0.05$, $\alpha = 2$
(b) $x_D = 0.99$, $x_B = 0.01$, $\alpha = 2$
(c) $x_D = 0.95$, $x_B = 0.05$, $\alpha = 1.2$

8–5. The depropanizer of problem 8–3 ($z = 0.284$, $q = 1$, $x_B = 0.058$, $x_D = 0.95$, $\alpha = 2.4$) has been found to operate with a 75 percent Murphree vapor-phase tray efficiency. How many actual trays are required to make the specified separation with a reflux ratio of 2.774? What is the feed tray?

8–6. Two streams feed binary fluids into a distillation column where the two components are separated into a distillate containing 97 mole percent A and a bottoms of 4 mole percent A. Operating requirements are:

System relative volatility = 2.0.
Feed F: 75 mole percent A at 150 lb-mole/hr. Feed is a saturated liquid.
Feed G: 35 mole percent A at 100 lb-mole/hr and 50 mole percent vaporized.
Reboiler: partial, bottoms taken as saturated liquid.
Condenser: partial, distillate taken as saturated vapor.
Reflux ratio: 1.6.

Determine the number of ideal stages required and the location of each feed point, assuming constant molar overflow.

8–7. A binary, equimolal-overflow, constant relative volatility mixture with $\alpha = 2$ is to be distilled in a 100 percent tray efficiency distillation column with a partial reboiler and a total condenser. A 3 mole percent bottoms product and a 95 mole percent distillate are to be produced from 8 lb-mole/minute of 35 mole percent feed which is 50 percent vaporized. Heat input to the reboiler is 4.324 $\times 10^6$ Btu/hr. The molecular weight of the vapor from the reboiler is 73 lb_m/lb-mole and its heat of vaporization is 85 Btu/lb_m. Reflux is saturated liquid.

How many trays are required to make the separation, and what is the optimum feed tray?

8–8. A process consists of an exothermic reactor followed by a heat exchanger (cooler) and distillation column:

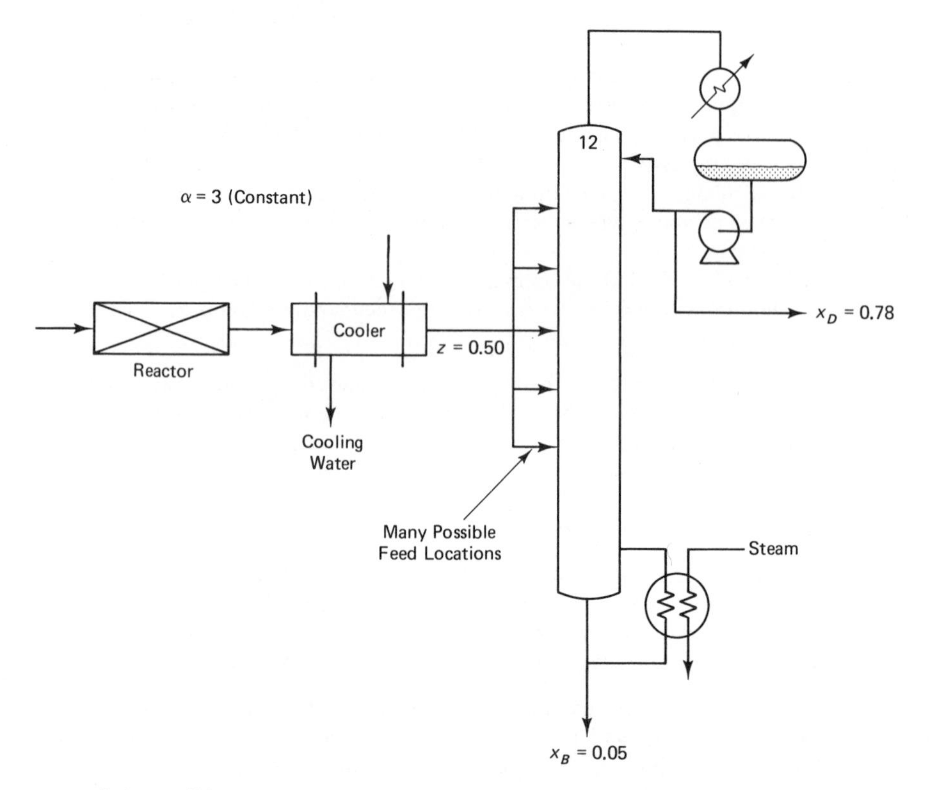

It is possible to vary the feed from a subcooled liquid to a superheated vapor.

Plot, qualitatively, (1) the optimum feed plate location N_F as a function of the feed q-value, and (2) the reflux ratio and the heat input to the reboiler as functions of feed q-value. No exact calculations need be made. You may wish to consult a qualitative McCabe-Thiele diagram, however. If you do, include this in your solution.

(a) What is the optimum process design?

(b) Does the optimum amount of feed preheat depend on the relative costs of heating and cooling? Explain.

8–9. Calculate bottoms and distillate compositions in a distillation column operating under total reflux conditions when the amount of light component impurity in the bottoms is the same as the amount of heavy component impurity in the distillate. The system is binary with a constant relative volatility of 2. The column has a partial reboiler, 20 theoretical trays, and a total condenser.

8–10. A binary, constant relative volatility ($\alpha = 2$) distillation column, with 13 theoretical trays, partial reboiler, total condenser, and feed on tray 8, produces

162.2 kg-mole/hr of 95 mole percent distillate and 254.8 kg-mole/hr of 5 mole percent bottoms when the reflux ratio is 5.25. (The *xy* diagram is shown below.)

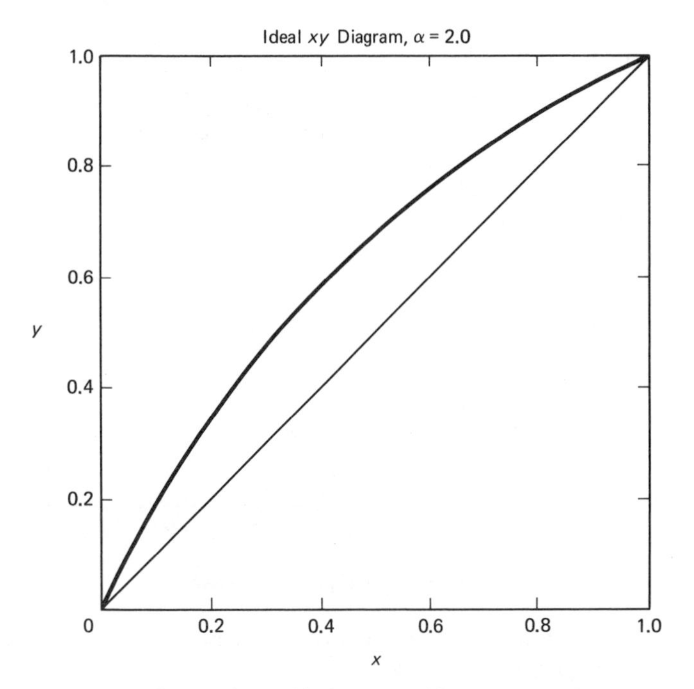

xy diagram for an ideal system with a constant = 2.

What are the feed rate, feed composition, and feed thermal condition?

8–11. A distillation column is used to separate the binary system that corresponds to the nonideal *xy* diagram shown on page 294. The column feed is 40 mole percent light component (A) and is 50 percent vapor.

(a) If the distillate is 70 mole percent A with a bottoms of 4 mole percent A, what is the minimum reflux ratio?

(b) If the distillate composition is increased to 80 mole percent A, with the bottoms composition held at 4 mole percent A, what is the minimum reflux ratio?

8–12. Oxygen and nitrogen are produced by the fractional distillation of air at 1 atm pressure (101.3 k N/m^2). The feed (79 mole percent nitrogen, 21 mole percent oxygen) is 75 percent liquid. The composition of the distillate product, *a vapor withdrawn from the condenser*, is 98 mole percent nitrogen. The bottoms product, a liquid withdrawn from the bottom plate before the reboiler, has a composition of 3 mole percent nitrogen.

(a) What percentage of the oxygen in the feed is recovered in the bottoms product?

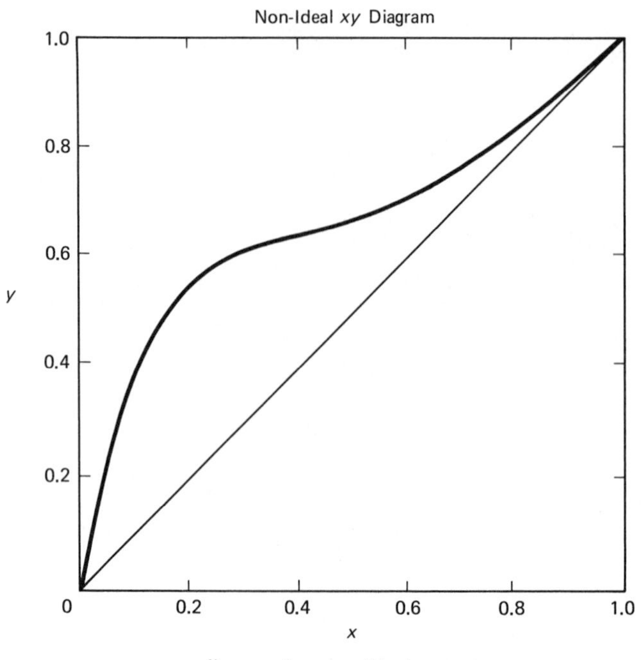

xy diagram for a nonideal system.

(b) What are the minimum reflux ratio and the minimum number of equilibrium stages required for this separation?

(c) What reflux ratio would be required if a column with nine theoretical trays were used?

(d) For 100 kg-mole/hr of feed, determine the liquid and vapor flow rates in both the rectifying section and the stripping section for the column of part (c).

(e) What is the composition of the reboiler liquid? Of the condenser liquid?

VAPOR-LIQUID EQUILIBRIUM DATA FOR N_2/O_2 SYSTEM, 1 ATM

Mole % N_2 in liquid x	*Mole % N_2 in vapor* y	*Mole % N_2 in liquid* x	*Mole % N_2 in vapor* y
3.85	13.97	33.80	67.95
8.02	26.10	40.47	73.74
12.40	36.60	47.83	78.95
17.05	46.00	56.62	84.35
22.20	54.20	66.65	88.95
27.73	61.60	78.40	93.50
		91.90	97.70

Normal Boiling Point of N_2 = −320.5°F

Normal Boiling Point of O_2 = −297.3°F

8–13. Design a distillation column to separate a 50 liquid vol. percent propane and 50 liquid vol. percent isobutane mixture into a 6 liquid vol. percent propane liquid bottoms product and a 96 liquid vol. percent propane vapor product leaving a partial condenser. A partial reboiler (thermosiphon) is used. The feed to this depropanizer is 20,000 barrels per day (42 gal per barrel), entering the column at 140°F. Since cooling water is to be used in the condenser, the design reflux drum temperature is 110° F. Physical property data are as follows.

Vapor pressure	
propane:	10 atm @ 26.9°C
	20 atm @ 58.1°C
isobutane:	5 atm @ 39°C
	10 atm @ 66.8°C
Heat capacities: 0.5	

Specific gravity, liquid	
propane:	0.5
isobutane:	0.6

Latent heat of vaporization	
propane =	147 Btu/lb_m
isobutane =	141 Btu/lb_m

Molecular weight	
propane =	44.1
isobutane =	58.1

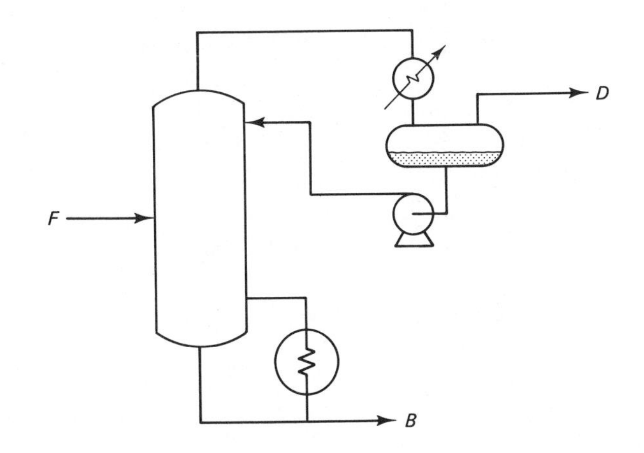

Assume equimolal overflow, 100 percent tray efficiencies, ideal vapor-liquid equilibrium, and constant relative volatility.

(a) Calculate the operating pressure of the column.

(b) Calculate the minimum reflux ratio.

(c) Calculate the number of trays N_T and the feed tray location N_F if an actual reflux ratio of 1.1 times the minimum is used.

(d) What is the composition of the liquid reflux?

(e) Calculate the reboiler and condenser duties (Btu/hr).

(f) Calculate the diameter of the column if the maximum F-factor is 3, where

$$F\text{-factor} \equiv V\sqrt{\rho_V}$$

for vapor velocity V in ft/sec and vapor density ρ_V in lb_m/ft^3.

8–14. Heat is frequently removed from a distillation column in a "pumparound" circulating stream located in the rectifying section of the column. Removing heat condenses some of the vapor coming up the column at that point. Therefore, the liquid and vapor rates below the pumparound tray are larger than the liquid and vapor rates above the pumparound tray.

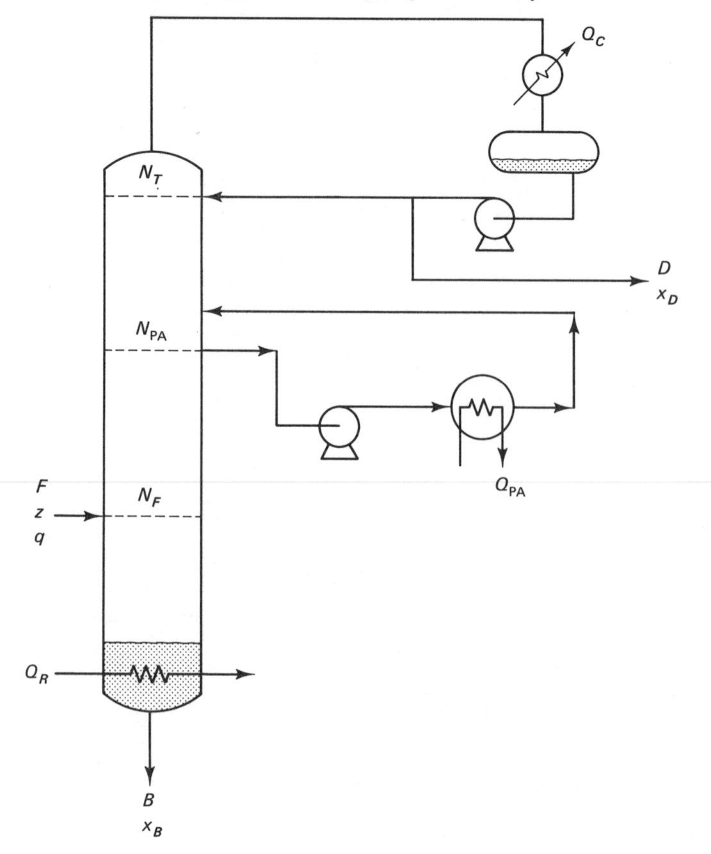

The reason for using this intermediate heat removal is that temperatures at this point in the column are higher than at the top of the column. Therefore, heat can be removed at a higher temperature, which is more valuable than at a lower temperature. For example, it may be possible to preheat the column feed using this higher temperature heat source, whereas the heat removed in the overhead condenser may have to be rejected to cooling water.

Design a distillation column (calculate the total number of trays, the optimum feed tray location, and the location of the pumparound tray) which has the following specifications:

x_B = 0.05 mole fraction more volatile component

x_D = 0.95 " " " " "

z = 0.50 " " " " "

Feed thermal condition: $q = 0.5$
Partial reboiler and total condenser
Feed rate = 100 kg-mole/minute
Reflux ratio = 1.5
Equimolal overflow and saturated liquid reflux
Constant relative volatility $\alpha = 2$
Composition of liquid at the pumparound tray should be approximately 0.70
Heat removed in the pumparound is 170 percent of the heat removed in the overhead condenser

8–15. Two hundred kg-mole/hr of saturated liquid feed containing 20 mole percent ethanol and 80 mole percent water is fed into a stripping column which operates at 760 mm Hg. Assume that the relative volatility of ethanol out of water is 2.0 at 760 mm Hg. The bottoms product from the stripper contains 2 mole percent ethanol.

The vapor from the top of the stripper is used to reboil a second distillation column operating at 250 mm Hg where the relative volatility of ethanol out of water is 2.5. After the stripper overhead is condensed, it is fed into this second column. Since it is saturated liquid at 760 mm Hg, 10 percent of it flashes when it is fed into the column at 250 mm Hg.

The second column produces a distillate product with composition 90 mole percent ethanol and bottoms products with composition 2 mole percent ethanol.

The bottoms product flow rates from the two columns are equal. Assume equimolal overflow, theoretical trays, saturated reflux to the second column, partial reboilers, and total condensers.

(a) What is the composition of the overhead vapor from the first column?
(b) How many trays are required in the stripper?
(c) Calculate the minimum reflux ratio in the second column.

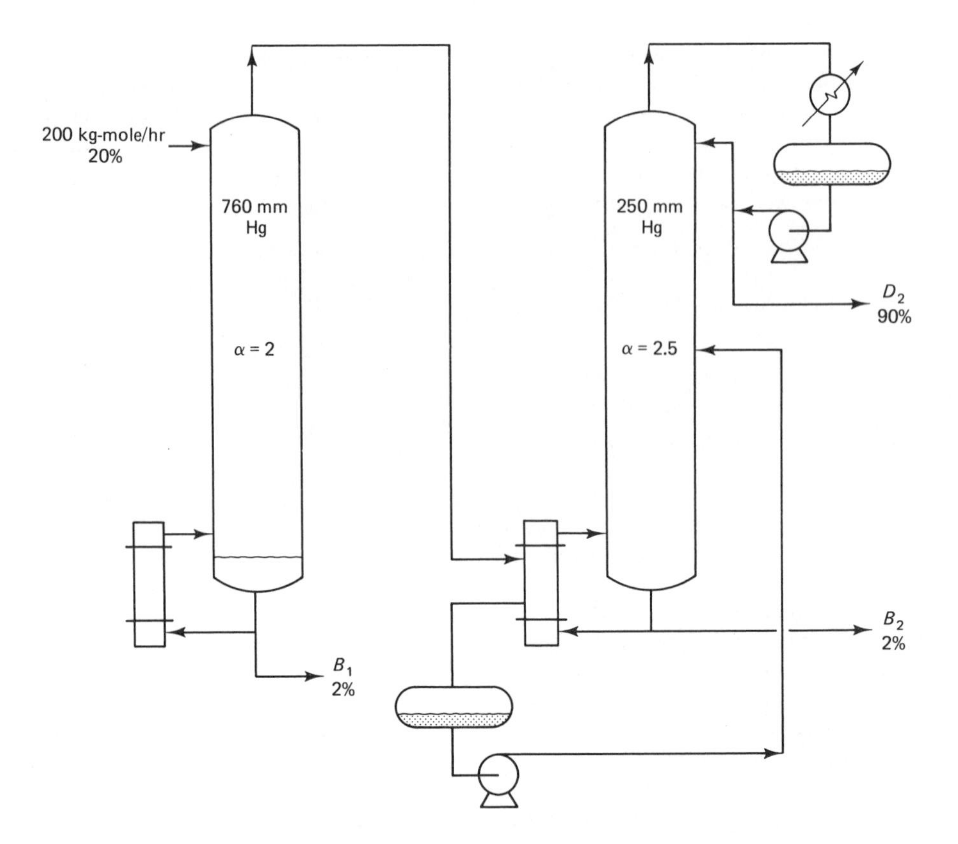

(d) Using 1.2 times the minimum reflux ratio, calculate the vapor rate in the stripping section of the second column and the number of trays.

(e) Is there enough vapor from the stripper to supply the heat input requirements of the second column?

8–16. A distillation column with a partial reboiler and a partial condenser is to be designed to distill a methanol-water mixture. Two light products will be formed: the distillate, a saturated vapor taken from the condenser, which operates at 66°C; and a saturated liquid sidestream with an 80 mole percent methanol composition. The flow rate of the sidestream is equal to the distillate flow, and 98 percent of the methanol in the feed is recovered in the light streams. The feed is 60 mole percent methanol and 28.6 percent liquid at 600 kg-mole/hr, fed on an optimum feed tray.

(a) What is the minimum reflux ratio, and how does the sidestream affect it?

(b) What are the compositions and flow rates of the three product streams?

(c) If a reflux ratio of 2.5 is used, how many stages are needed and what are the feed and sidestream trays?

The *Txy* diagram for this column is on p. 304.

8–17. Steady-state data from a 24-tray distillation column separating methanol and water at atmospheric pressure are as follows:

Feed Rate	1.5 gpm
Reflux Ratio	1.8
Feed Composition	40 mole % methanol
Distillate Composition	98 " " "
Bottoms Composition	6 " " "
Feed Thermal Condition	$q = 1.083$

Feed is introduced on the 10th tray. The column has a total condenser and a partial reboiler.

Assume equimolal overflow. VLE data for the methanol-water system are:

Vapor pressures (mm Hg)

CH_3OH	760 @ 64.5°C
	2,600 @ 100°C
H_2O	760 @ 100°C
	145.4 @ 59.4

Activity coefficients (van Laar)

$$\ln \gamma_1 = \frac{Ax_2^2}{\left[\frac{A}{B}x_1 + x_2\right]^2}$$

$$\ln \gamma_2 = \frac{Bx_1^2}{\left[x_1 + \frac{B}{A}x_2\right]^2}$$

where

γ_1 = activity coefficient of CH_3OH

γ_2 = activity coefficient of H_2O

x_1 = mole fraction CH_3OH

x_2 = mole fraction H_2O

$A = 0.85$

$B = 0.46$

Use a Newton-Raphson iteration algorithm for bubblepoint calculations, and write and execute a computer program in FORTRAN that calculates the average tray efficiency (Murphree vapor efficiency) of the column.

8–18. Write and execute a digital computer FORTRAN program that calculates the number of theoretical trays required in a C_4-splitter operating at pressures of 40, 60, 80, and 100 psia. Two feed streams (300 kg-mole/hr of 80 mole percent iC_4 and 100 kg-mole/hr of 50 mole percent iC_4) are to be separated into a 5 mole percent iC_4 bottoms product and a 98 mole percent iC_4 distillate product. Assume that the iC_4/nC_4 binary system is ideal, and that the column has a partial reboiler and a total condenser. The thermal conditions of the two feeds are $q = 1.2$ for the 50 percent feed and $q = 0.8$ for the 80 percent feed. The reflux ratio should be 1.1 times the minimum reflux ratio. Use Newton-Raphson convergence in your bubblepoint subroutine. Vapor pressure data are:

	2 atm	*10 atm*
iC_3	7.5°C	66.8°C
nC_4	18.8	79.5

8–19. Write a FORTRAN computer program that will determine the number of trays and the feed tray location for a binary, equimolal overflow, constant relative volatility distillation column. Your program should handle any thermal condition of the feed from subcooled liquid to superheated vapor. Parameters given are x_D, x_B, and z. The actual-to-minimum reflux ratio is 1.1. The reboiler is a partial reboiler, but the condenser is a total condenser.

8–20. A distillation column is to be designed to separate two feeds: stream G, consisting of 1,200 lb-mole/hr of a saturated vapor containing 70 mole percent A in B; and stream F, consisting of 800 lb-mole/hr of a saturated liquid containing 40 mole percent A in B. A distillate of 96 mole percent A and a bottoms of 4 mole percent A are desired. The relative volatility of A in B is 3.90, equimolar overflow is expected in the system, and the column will operate at a reflux ratio of 0.85.

(a) If the column is to have a partial reboiler and a partial condenser, determine the stage number on which stream F should be fed, the stage number on which stream G should be fed, and the total number of stages in the column. Assume that the stages operate at 100 percent efficiency.

(b) Determine the vapor boilup in the reboiler in lb-mole/hr.

8–21. A petroleum paraffins separation column is fed with a mixture of 15 mole percent propane (C_3H_8), 30 mole percent n-butane (nC_4H_{10}), 30 mole percent n-pentane (nC_5H_{12}), and 25 mole percent n-hexane (nC_6H_{14}). This feed is flashed adiabatically to a column pressure of 100 psia from an upstream pressure of 1,000 psia. After flash the temperature is 200°F. What is the q-value of this feed? Was the temperature before the flash greater than, equal to, or less than 200°F?

8–22. A binary mixture of isobutane and normal butane is fed into a chemical reactor. Some of the iC_4 reacts to form a high-boiling product which is removed from the reactor in a stream that is pure product. The light components (unreacted iC_4 and nC_4) flow from the reactor at a rate of 1,000 kg-mole/hr and with an iC_4 concentration of 80 mole percent.

A fraction of the iC_4/nC_4 stream from the reactor is fed to a C_4 splitter (a distillation column which removes nC_4 from the system). The amount of iC_4 that can be lost out the bottom with the nC_4 purge is 1 mole of iC_4 for every 10 moles of nC_4. The relative volatility of iC_4 out of nC_4 in the C_4 splitter is constant at 1.4. The feed to the C_4 splitter is a saturated liquid, and the column has a total condenser and a partial reboiler.

The remaining portion of the iC_4/nC_4 stream from the reactor is bypassed around the C_4 splitter and fed back into the reactor. The total feed to the reactor consists of (1) the distillate product from the C_4 splitter, (2) the portion of the iC_4/nC_4 stream from the reactor that is bypassed around the C_4 splitter, and (3) a fresh makeup iC_4 feed stream (200 kg-mole/hr of 95 mole percent iC_4, 5 mole percent nC_4). The C_4 splitter is operated at 1.2 times the minimum reflux ratio for any distillate composition.

For the C_4 splitter, the higher the feed rate, the lower the distillate composition can be to achieve the specified reactor effluent concentration of 80 mole percent iC_4. This reduces the reflux ratio. However, the higher the feed rate to the C_4 splitter, the more distillate product must be produced, which increases the required vapor boilup. Thus, there is an optimum feed rate F to the C_4 splitter (and a corresponding optimum distillate composition x_D) which minimizes the energy input in the reboiler of the C_4 splitter, i.e., minimizes the total overhead vapor rate.

(a) Calculate the flow rate and composition of the C_4 splitter bottoms product.
(b) Calculate the flow rate and composition of the total reactor feed stream.
(c) Derive a relationship between the distillate composition x_D of the C_4 splitter and the feed rate F to the C_4 splitter.
(d) What are the optimum values of F and x_D?

8–23. A distillation column uses vapor recompression to provide most of the heat required to generate vapor in the lower section of the column. Overhead vapor is compressed and then condensed in an intermediate reboiler in the stripping section of the column. A steam reboiler at the base of the column provides 53.56 percent of the total heat input to the column. (See figure on page 302.)

The system is binary with constant relative volatility ($\alpha = 2$), equimolal overflow, a partial reboiler, a total condenser, and ideal trays. The two saturated liquid feeds are introduced into the column at two different feed plates:

$$F_1 = 73.93 \text{ kg-mole/hr} \qquad z_1 = 0.40 \text{ mole fraction}$$

$$F_2 = 59.02 \text{ kg-mole/hr} \qquad z_2 = 0.70 \text{ mole fraction}$$

Distillate composition is 0.93 mole fraction; bottoms composition is 0.05 mole fraction. The column operates with an internal reflux ratio of 1.3. The composition of the liquid at the intermediate reboiler is approximately 20 mole percent.

Design the column described, i.e., calculate the total number of trays required, the optimum feed tray locations, and the tray location where the intermediate reboiler is located.

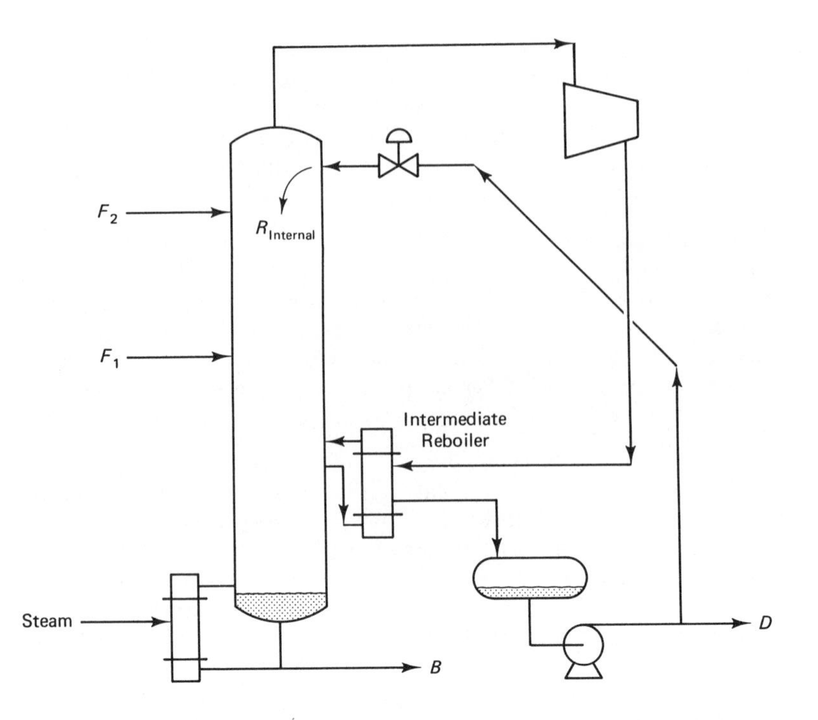

8–24. A heat-integrated two-column distillation system is fed 100 kg-mole/hr of saturated liquid feed with a composition of 70 mole percent methanol and 30 mole percent water. The first column is operated at a high pressure where the relative volatility of methanol out of water is 2. The overhead vapor from the high-pressure column is condensed in the reboiler of the second column, which is operated at low pressure. The relative volatility in the low pressure column is 3. Since the system has equimolal overflow, the vapor rate in the stripping section of the low-pressure column is equal to the vapor rate in the rectifying section of the high-pressure column.

As shown in the sketch on page 303, the bottoms product from the high-pressure column is fed into the low-pressure column. Distillate product purities on both columns are 95 mole percent methanol. The bottoms composition in the high-pressure column is 54 mole percent methanol. In the low-pressure column, bottoms composition is 5 mole percent methanol. Both columns have a partial reboiler, a total condenser, and 100 percent efficient trays. The reflux ratio in the high-pressure column is 1.38.

When the bottoms from the high-pressure column are flashed into the low-pressure column, 20 percent (on a molar basis) of the stream vaporizes. How many trays are required in each column, and what are the optimum feed tray locations?

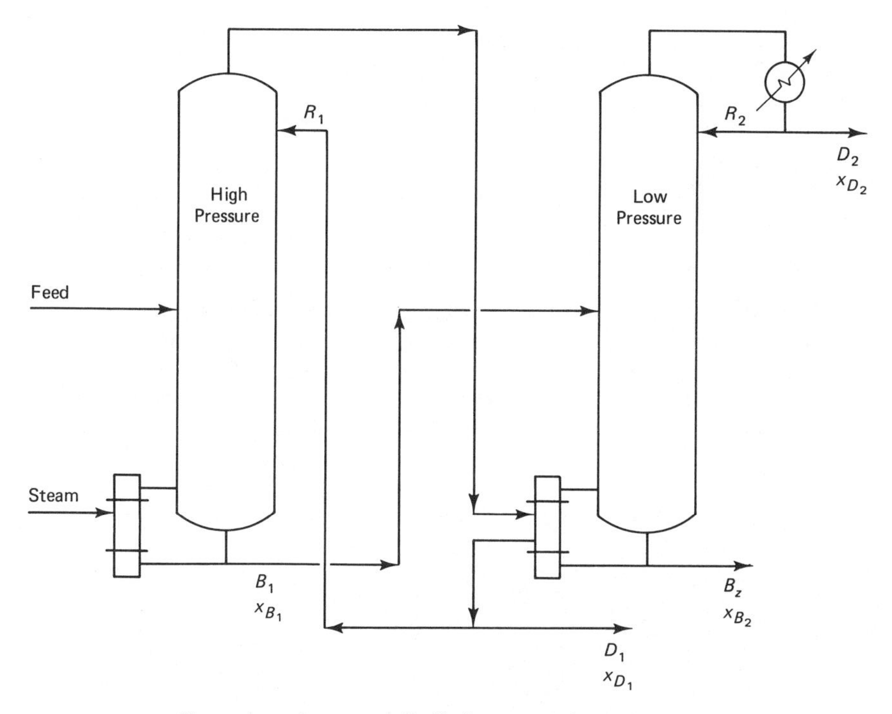

Two-column integrated distillation system for problem 8–24.

8–25. Write a FORTRAN computer program to design several C_2 splitter distillation columns. Saturated liquid feed (1,000 kg-mole/hr) which is 60 mole percent ethylene and 40 mole percent ethane is fed into a column which has a partial reboiler, a total condenser, 100 percent tray efficiency, equimolal overflow, and saturated liquid reflux. Two different operating pressures are to be explored: 200 psia and 500 psia. Two different product purity levels are also to be studied:

(a) $x_B = 5$ mole % $C_2^=$, $x_D = 95$ mole % $C_2^=$

(b) $x_B = 0.5$ mole % $C_2^=$, $x_D = 99.5$ mole % $C_2^=$

Your program should calculate the minimum reflux ratio, set the actual reflux ratio at 1.1 times the minimum, and then calculate the total number of trays and the optimum feed tray for each of the four cases. Assume ideal VLE relationships and the following vapor pressure data:

Ethylene, $C_2^=$	515 psia @ 20°F
	214 psia @ −40°F
Ethane, C_2	500 psia @ 60°F
	160 psia @ −20°F

Use the Newton-Raphson iteration algorithm in your bubblepoint subroutine. Also, calculate the total annual cost of each of the four columns as follows:

$$\text{Total Cost} = \text{Energy Cost} + \text{Capital Cost}$$

Assume that each tray costs about \$2,000, and capital investment must be depreciated over a five-year period.

Energy cost is approximately \$5 per million Btu added in the reboiler. The heat of vaporization of ethane and ethylene is about 160 Btu/lb_m.

8–26. A distillation column is used to separate methanol and water. Feed is introduced into the column on the optimum feed tray. The column has a partial reboiler and a partial condenser. *Txy* and *xy* diagrams for methanol (CH_3OH) and water at atmospheric pressure are given in the following figures. Both vapor and liquid products are withdrawn from the reflux drum.

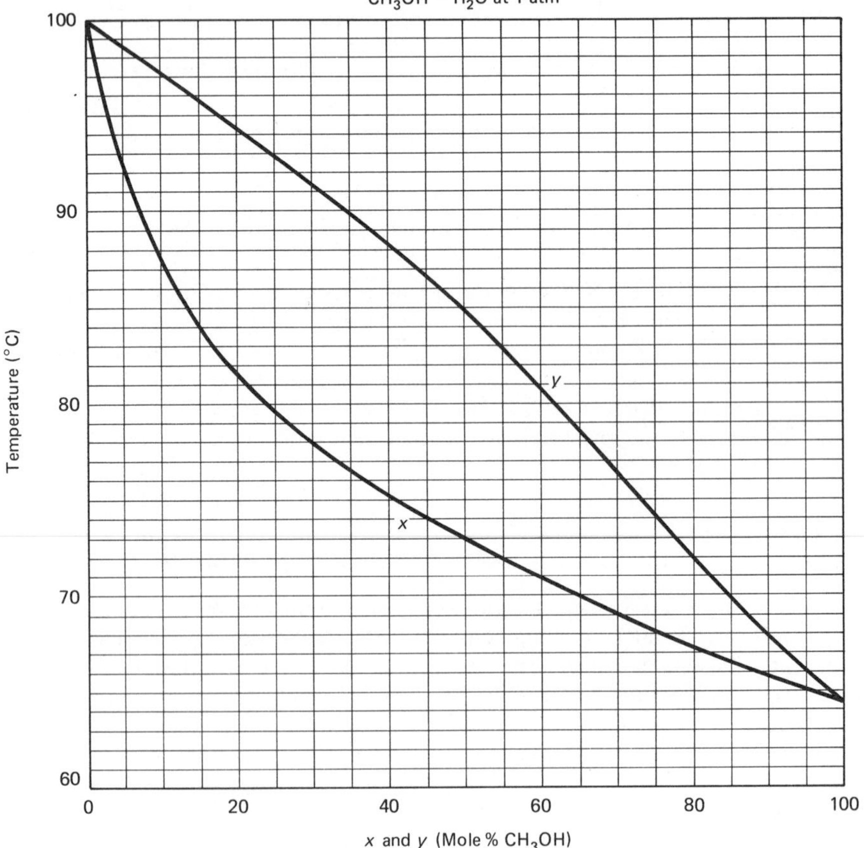

Txy diagram for methanol-water system, $P = 1$ atm.

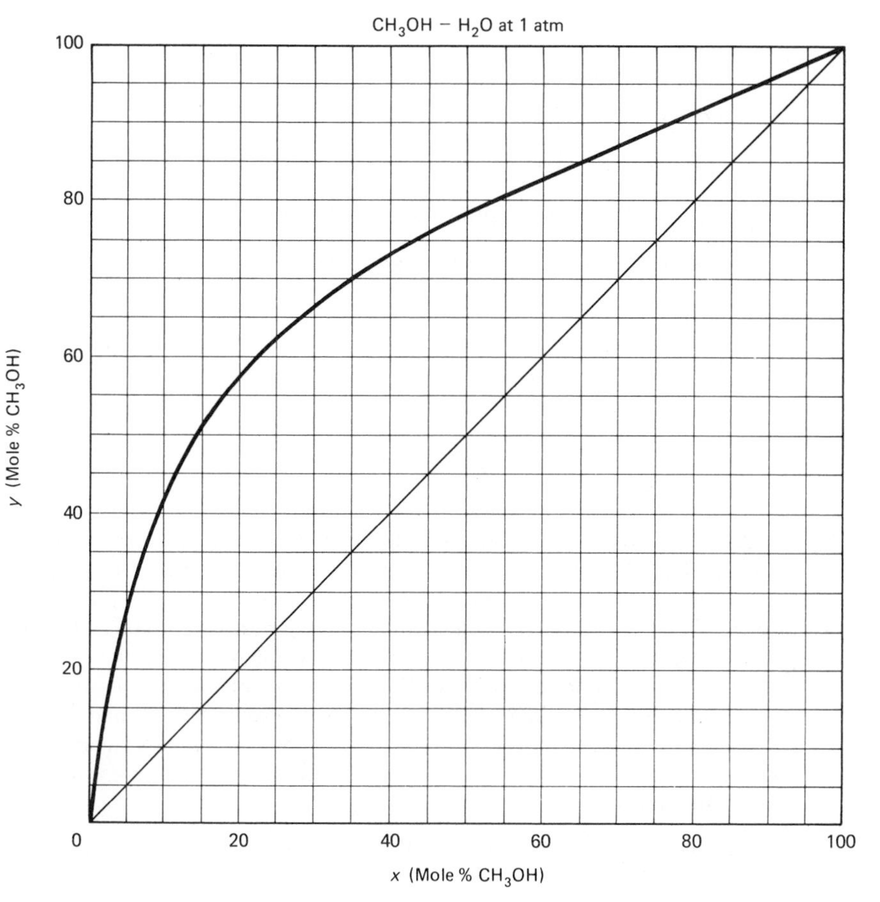

xy diagram for methanol-water system, $P = 1$ atm.

Assume that pressures throughout the column are 1 atmosphere. Also, assume equimolal overflow. The specific gravity of methanol is 0.78, and its molecular weight is 32.

Tray efficiencies are 100 percent. From the operating data given on page 306, calculate

(a) The composition of the liquid and vapor products from the reflux drum
(b) The bottoms and feed compositions
(c) The feed thermal condition
(d) The number of trays in the column

DATA

gpm = gallons per minute

SCFM = standard cubic feet per minute (@32°F, 1 atm)

Bottoms Flow Rate = 60 gpm

Liquid Product from Reflux Drum = 36.6 gpm

Vapor Product from Reflux Drum = 4908 SCFM

Reflux Flow Rate = 81.7 gpm

Temperature (°C)

Reflux Drum = 68

Reboiler = 93

Feed = 80

8–27. Using the VLE data from problem 7–26, together with the data given below, write a FORTRAN program to do the tray-to-tray design calculations for three acetic acid dehydration columns operating at pressures of 250, 760, and 2,053 mm Hg. Feed is saturated liquid (30,000 kg/hr) with composition 40 mole percent water, 60 mole percent acetic acid. The column has a partial reboiler, a total condenser, 100 percent tray efficiency, equimolal overflow, and saturated liquid reflux. Distillate and bottoms compositions are 80 and 4 mole percent water, respectively. Use the Newton-Raphson iteration algorithm in your bubblepoint subroutine. Set the actual reflux ratio at 1.2 times the minimum reflux ratio.

Also, calculate the diameter of each of the columns, assuming a maximum "*F*-Factor" of 2, where

$$F\text{-Factor} = V\sqrt{\rho_V}$$

in which

$$V = \text{vapor velocity (ft/sec)}$$

$$\rho_V = \text{vapor density } (\text{lb}_m/\text{ft}^3)$$

Additional VLE Data (Van Laar constants) are as follows:

P (mm Hg)	A_{12}	A_{21}
250	0.5930	1.2465
2,053	0.4486	1.1407

8–28. Compare the energy consumption (vapor rate from the reboiler) of two alternative configurations of distillation columns to separate a binary 50 mole percent saturated liquid mixture into products that are 95 mole percent and 5 mole percent light component. Assume ideal trays, a partial reboiler, a total con-

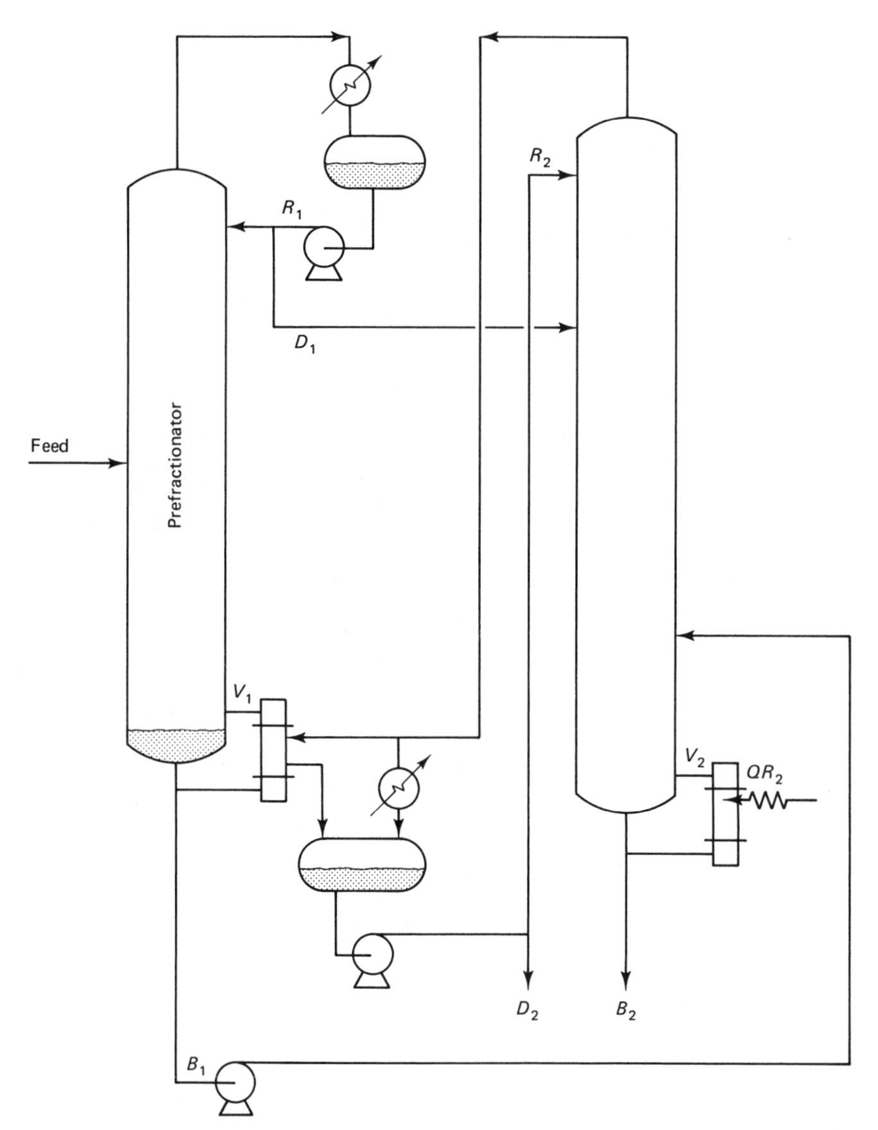

denser, equimolal overflow, constant relative volatility, and saturated liquid feeds and reflux.

(a) In scheme 1, a single conventional column is used, operating at low pressure with a relative volatility of 3. Calculate the minimum reflux ratio and the vapor boilup if the actual reflux ratio is 1.1 times the minimum. Sketch a McCabe-Thiele diagram of this column.

(b) In scheme 2, a two-column heat-integrated prefractionator system is used as sketched above.

The first column operates at low pressure ($\alpha = 3$) and produces an 85 mole percent distillate product and a 25 mole percent bottoms product. These are both fed into a second column at different feed tray locations. Distillate and bottoms products from the second column are 95 and 5 mole percent, respectively.

The second column operates at high pressure (giving a relative volatility of only 2), so that its overhead vapor can be used to reboil the first column (heat integration).

(a) Sketch McCabe-Thiele diagrams for each of these columns.

(b) Calculate the minimum reflux ratio for each column and the vapor rate from each reboiler if the actual reflux ratios are set at 1.1 times the minimum.

8–29. A binary mixture with relative volatility 2.5, containing 60 mole percent of light component A, enters a column as 80 percent vapor. This stream enters the column, which has nine ideal trays, on the fourth tray. The column operates at a reflux ratio of 1.5, and has a partial reboiler and a total condenser. If you operate the column to produce a 90 mole percent A distillate product, what bottoms product would you expect? What would be the boilup (vapor rate) from the reboiler?

8–30. 100 kg-mole/hr of a binary mixture of 30 mole percent n-hexane and 70 mole percent ethanol is to be separated in a two-column system to overcome the limitations resulting from the formation of a minimum-boiling azeotrope. The first column operates at atmospheric pressure, producing a 98 mole percent ethanol bottoms product and a distillate product that is 3 mole percent less than the azeotrope, which is about 68 mole percent n-hexane.

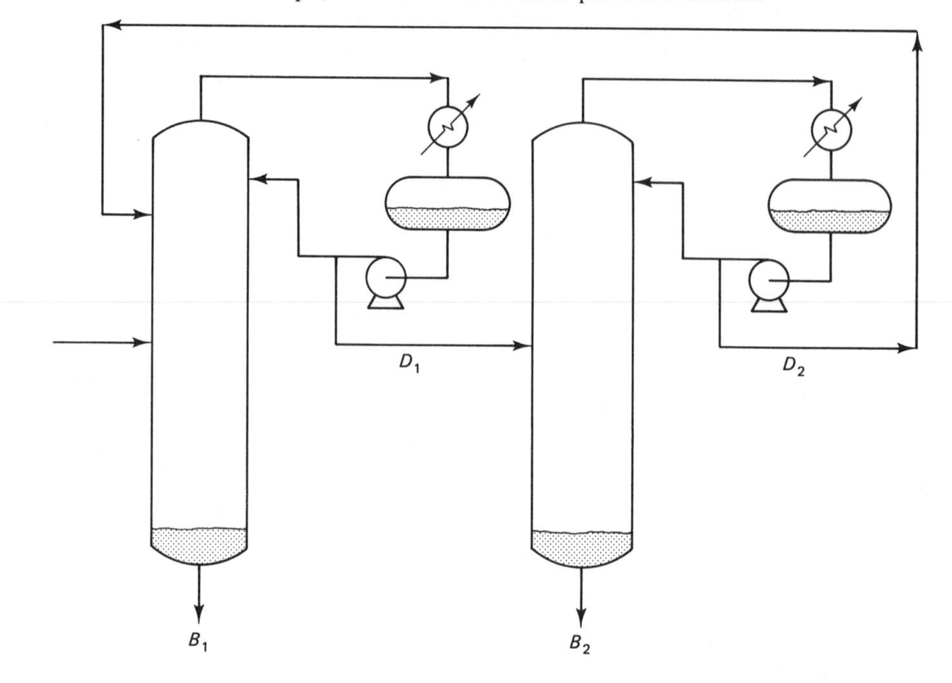

The distillate is fed into a second column operating at 10 atmospheres. The bottoms product from the high-pressure column is 99 mole percent n-hexane. The distillate product from the high-pressure column is 3 mole percent higher than the azeotrope at 10 atm, which is about 47 mole percent n-hexane. It is recycled back as a second feed to the low-pressure column.

The reflux ratio is 0.206 in the first column and 0.235 in the second column. The thermal condition of the fresh feed to the low-pressure column is saturated liquid. The distillate from the high-pressure column and the distillate from the low-pressure column are saturated liquids. Reflux streams in both columns are also saturated liquids. Assume partial reboilers, total condensers, and equimolal overflow, and use the vapor pressures and Wilson Equation parameters given in problem 7–28 in a subroutine that calculates y and T given x, P, and a guess of T. The Murphree vapor phase tray efficiency in both columns is 50 percent.

(a) By graphical methods, determine the number of actual trays and optimum feed trays in each column.

(b) Write a FORTRAN program that designs both columns.

(c) Change the compositions of the two distillates to be only 2 mole percent from their respective azeotropes, and repeat the calculations using your program from (b).

(d) How does energy consumption depend on the compositions of the distillate products compared to the azeotropic conditions?

8–31. 350 kg-mole/hr of a binary mixture that is 60 mole percent light component is fed into a distillation column. The feed is 50 percent vapor. The column produces a distillate product with 92.5 mole percent light component and a bottoms product with 5 mole percent light component. Assume equimolal overflow, 100 percent tray efficiency, constant relative volatility ($\alpha = 2.5$; see the graph on page 311), a partial reboiler, and a total condenser.

Vapor recompression is used to conserve energy. (See figure on page 310.) The overhead vapor from the column is compressed to a high enough pressure so that it can be condensed in the reboiler. The portion of the vapor that is not needed to reboil the column is condensed in a condenser. Since the vapor from the compressor is superheated, every mole that is condensed in the reboiler produces 1.1 moles of vapor in the column.

Since the reflux drum is at a higher pressure than the column, 20 percent of the reflux liquid flashes as it enters the column. Therefore, the liquid rate in the rectifying section is not equal to the external reflux flow rate. Use graphical solution methods to calculate

(a) The minimum liquid rate in the rectifying section.

(b) The liquid and vapor rates in the rectifying and stripping sections of the column if the actual liquid rate in the rectifying section is 1.2 times the minimum.

(c) The number of trays required.

(d) The flow rate of the external reflux.

(e) The total vapor flow rate through the compressor.

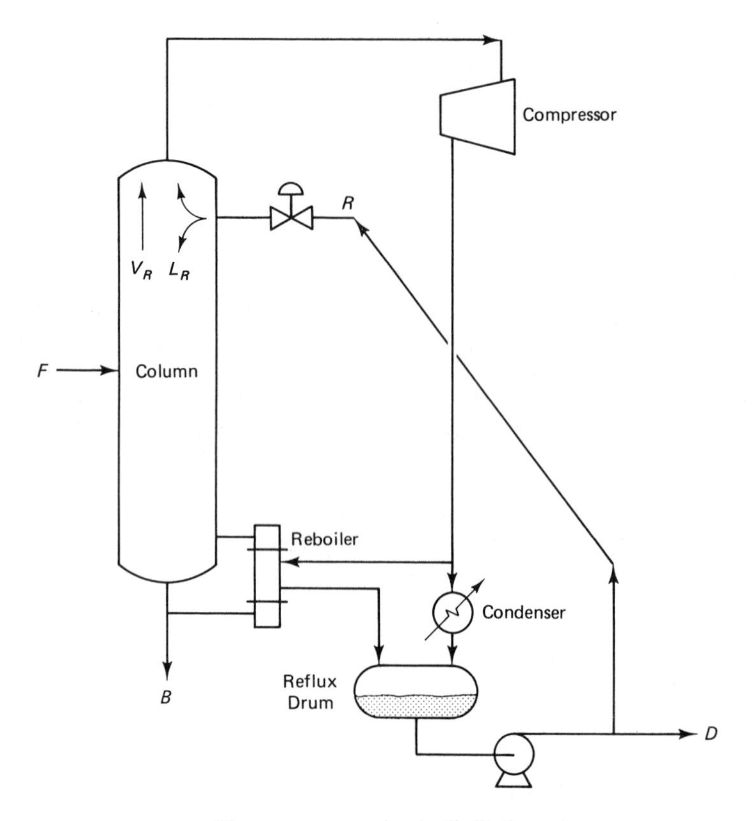

Vapor recompression in distillation.

(f) The portion of the hot vapor from the compressor that is used in the reboiler to vaporize the necessary vapor in the stripping section of the column.
(g) The flow rate of vapor to the condenser.

8–32. A methanol-water solution containing 50 mole percent methanol is fed at 1 atm and with 70 percent of the material in the vapor phase into a distillation column. The column has a total condenser and a total reboiler, thus producing a liquid distillate and a vapor bottoms product. Ninety percent of the methanol fed in is collected in the distillate product at 95 mole percent methanol. The column operates at 1.2 times the minimum vapor boilup from the reboiler. How many ideal stages are required for the separation?

Actually, the column operates with a Murphree vapor efficiency of 70 percent. How many actual trays should be specified for the column?

8–33. The figure on page 312 gives the *xy* diagram for a binary heterogeneous azeotrope that is to be separated in a distillation column. Feed flow rate is 400 kg-

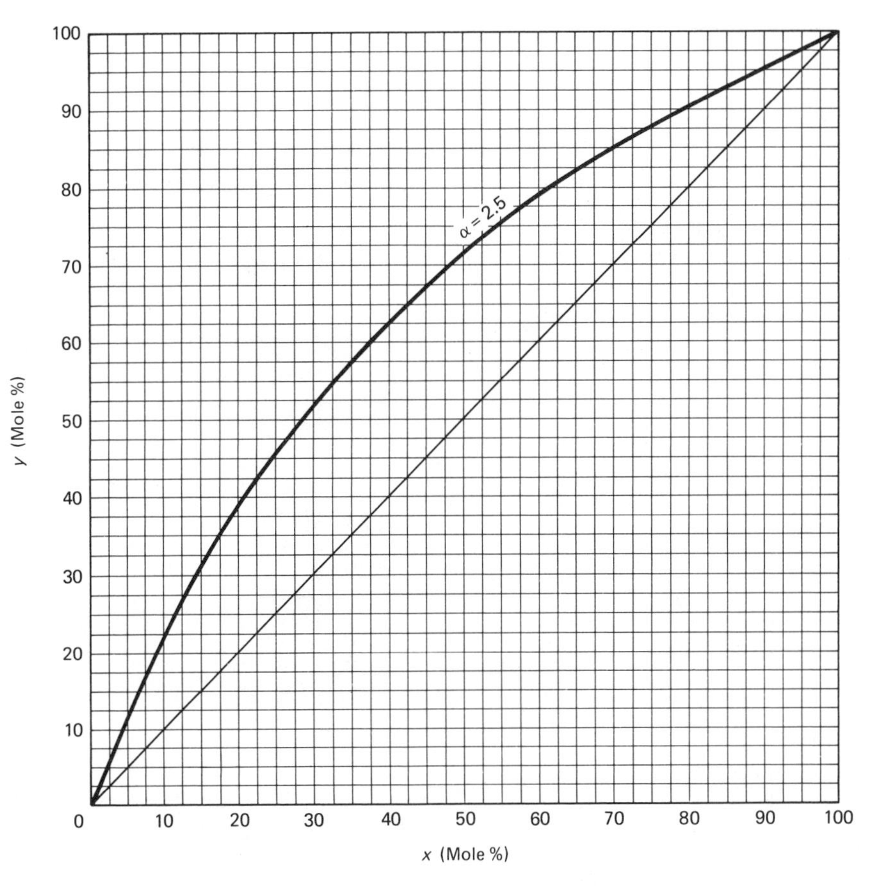

xy diagram for an ideal system with $\alpha = 2.5$.

mole/hr. Feed composition is 25 mole percent light component. The feed is 50 percent vapor as it enters the column.

The vapor from the top of the column is totally condensed and forms two liquid phases with compositions $x_\alpha = 50$ mole percent and $x_\beta = 90$ mole percent. The β liquid phase is removed as distillate product. All of the α liquid phase is refluxed back to the column. The bottoms product composition is 2.5 mole percent. A reflux ratio of 3.83 is used.

Assume theoretical trays, partial reboiler, equimolal overflow and saturated liquid reflux.

(a) Calculate the flow rates of the bottom and distillate products.
(b) Calculate the composition of the vapor leaving the top of the column and entering the condenser.
(c) Determine the number of trays required.

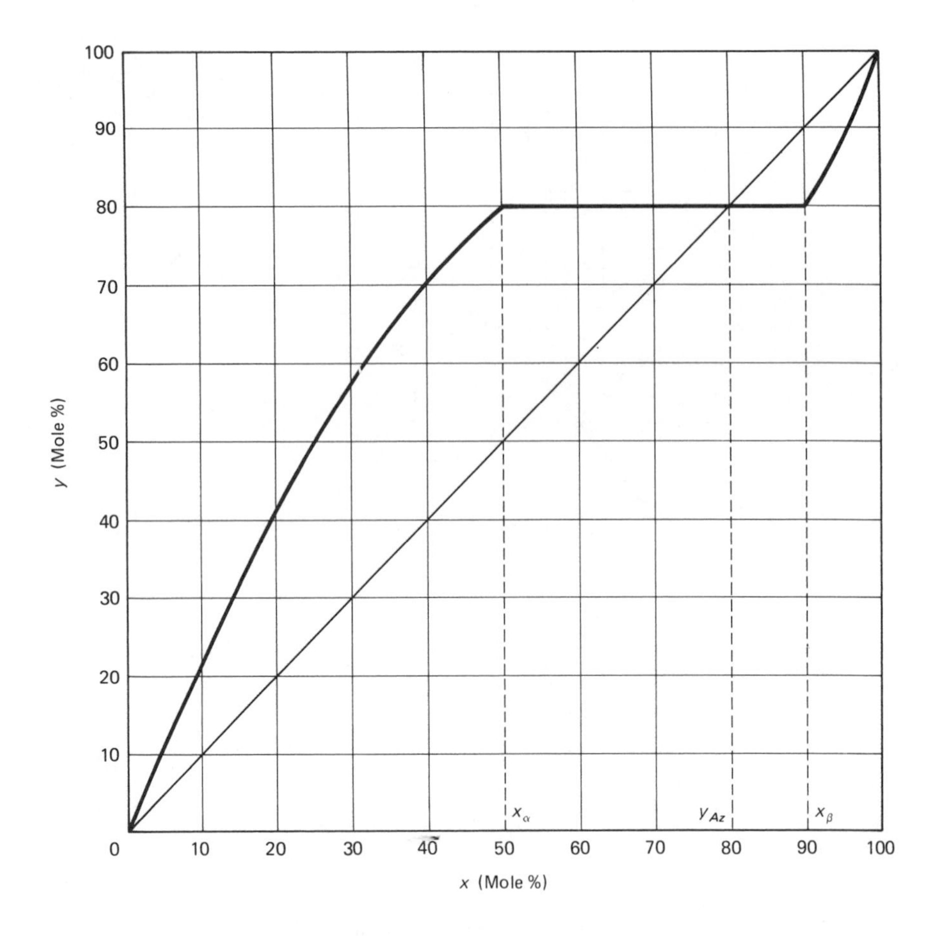

8–34. The figure on page 313 shows a distillation system in which the bottoms (B_1) from the high-pressure distillation column is flashed into a drum. The amount of vapor flashed in the drum is 20 percent of B_1. The relative volatility of the binary system is 2 in the high-pressure distillation column and is 3 in the low-pressure flash drum and stripper.

The liquid from the flash drum is fed to the top of a stripping column. The bottoms from the stripper is 2.5 mole percent light component. The vapor from the top of the stripper (V_2) is combined with the vapor from the flash drum (V_F) to give a total vapor stream of composition $y_{\text{TOT}} = 48$ mole percent light component. This vapor is compressed and fed into the high-pressure distillation column as a saturated vapor feed.

The feed to the system is 100 kg-mole/hr of saturated liquid with a composition 70 mole percent light component. Assume equimolal overflow, theoretical trays, partial reboilers, total condenser, constant relative volatilities (see figure on page 314) and saturated liquid reflux.

The composition of the distillate product from the high-pressure column is 95 mole percent light component. This column has a reflux ratio of 1.5.

(a) Calculate the flow rates D_1 and B_2.

(b) Calculate the flow rates B_1 and V_{TOT}.

(c) Calculate the compositions of the liquid and vapor streams from the flash drum (x_F and y_F).

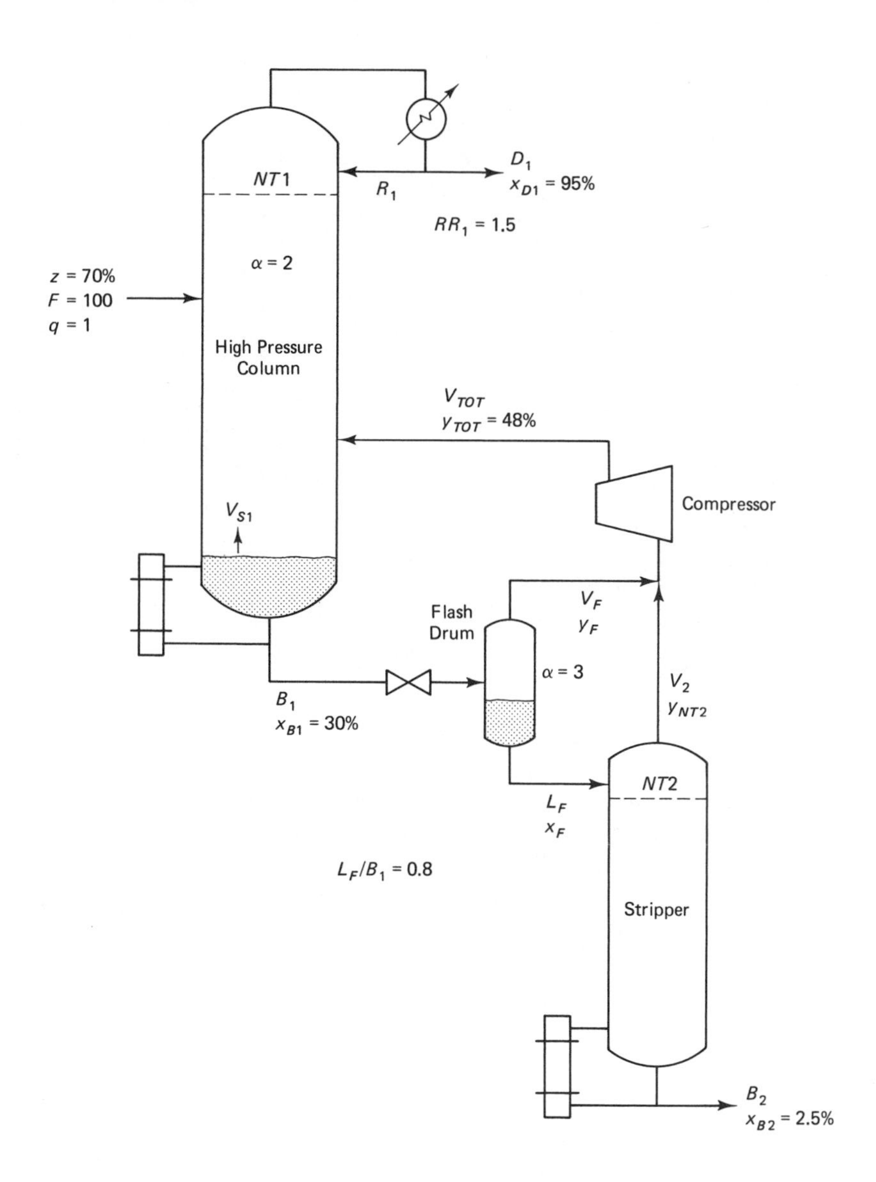

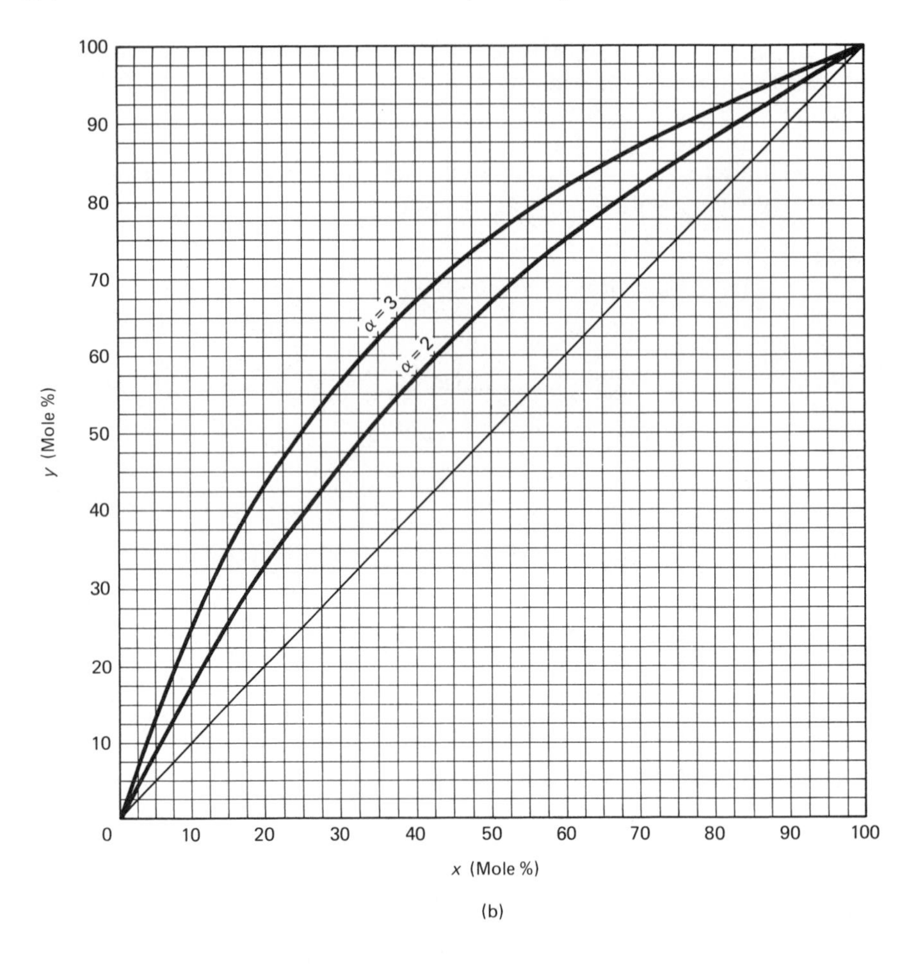

(b)

(d) Calculate the vapor flow rate in the stripper (V_2) and the composition of the vapor leaving the top of the stripper (y_{NT2}).

(e) Use the graphical McCabe-Thiele method to determine the number of trays in the stripper ($NT2$) and in the column ($NT1$).

9 Liquid-Liquid Extraction

9.1 INTRODUCTION

We study liquid-liquid extraction at this point because it is another important process that can be analyzed primarily by using component balances and phase equilibrium relationships. The phase equilibrium involves liquid-liquid equilibrium (LLE) instead of vapor-liquid equilibrium. Liquid-liquid extraction almost always involves at least three components, so we will have to use the ternary phase diagrams introduced in chapter 6, as well as keep track of the three components by means of two component balances and one total mass balance. A ternary system is more complex than the binary distillation systems that we have studied so far; but the basic approach to its analysis is exactly the same: the simultaneous use of phase equilibrium and component balances.

Most liquid-liquid extraction columns run essentially isothermally. Since the heat effects of transferring one component from one liquid phase into another liquid phase are usually small, energy balances are seldom needed to describe liquid-liquid extraction.

9.1.1 Basic Process

Liquid-liquid extraction involves the mixing of a liquid solvent with a liquid process stream in order to achieve a separation between the components in the process stream. The key problem is to find a solvent that dissolves one component but not the other. A typical and industrially important example is the separation of aromatics from paraffins using a tetraethylene glycol solvent as shown in Figure 9–1. "Lean" solvent (an aqueous solution of tetraethylene glycol) is fed into the top of a liquid-liquid extraction column. Organic feed, a mixture of paraffins and aromatics, is fed into the base of the column. The lighter (lower density) organic liquid rises upward in the column, contacting the falling heavier (more dense) solvent liquid. Since aromatics dissolve more readily in the solvent than do paraffins, the organic stream leaving the top of the extraction column (called the *raffinate* stream) has less aromatic material in it than was in the organic feed.

The "fat" solvent leaving the bottom of the column contains quite a bit of dissolved aromatics (and also a little of the paraffins). This stream is fed into a distillation stripping column where the aromatics are separated from the solvent. The solvent is then recycled back to the extractor.

9.1.2 References

We will discuss only simple liquid-liquid extraction systems here. For a more in-depth treatment of the subject, see the following texts:

1. Robert E. Treybal, *Mass Transfer Operations*, 2nd Edition. New York: McGraw-Hill, 1968.
2. Ernest J. Henley and J. D. Seader, *Equilibrium-Stage Separation Operations in Chemical Engineering*. New York: Wiley, 1981.

An excellent compilation of liquid-liquid phase equilibrium data is given

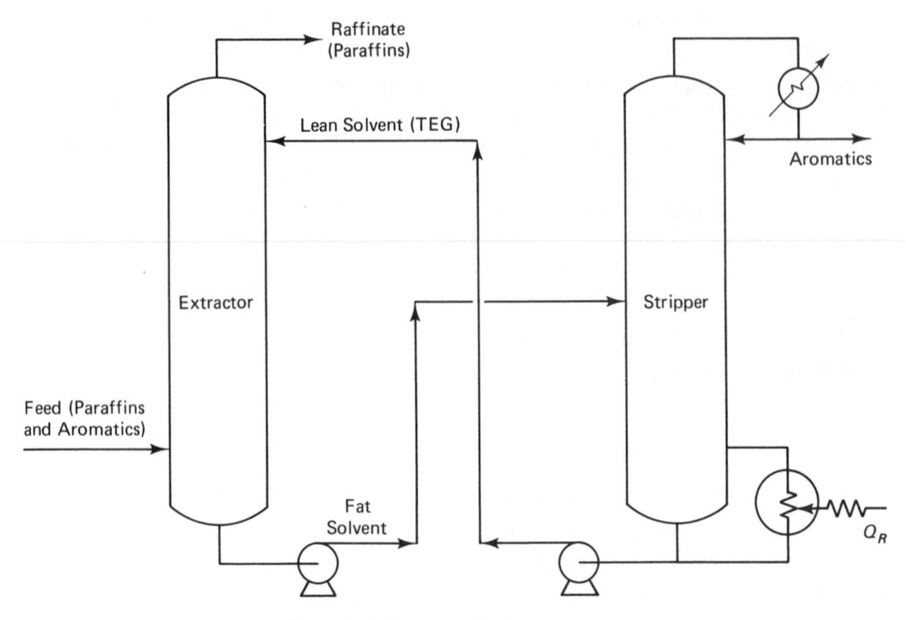

Figure 9–1 Extractor/stripper process.

in Sorenson and Arlt, *LLE Data Collection Series*, Frankfurt: DECHEMA, 1980.

9.1.3 Applicability

Liquid-liquid extraction is used to separate components in situations where

1. Relative volatilities are quite close to unity ($\alpha < 1.1$), making distillation very costly. (Distillation requires tall towers due to the existence of many trays, and high energy consumption because of high reflux ratios.)
2. Thermally sensitive components will not permit high enough temperatures to produce a vapor-liquid system at reasonable pressures (pressures greater than 10–50 mm Hg).

An example of the first case is the aromatic-paraffin separation, e.g., a mixture of benzene and cyclohexane. The normal boiling points of these organics are 80.1°C and 80.7°C, respectively, making their separation by distillation impractical. However, benzene is much more soluble in a solvent such as tetraethylene glycol, so liquid-liquid extraction can be used to separate the two.

One of the problems that liquid-liquid extraction presents is the addition of another component (the solvent). Since separations are never perfect, small amounts of solvent will be lost in the raffinate stream from the extractor and in the product stream from the stripper. This can be a problem if the solvent is expensive or if its presence in the other streams is undesirable.

9.2 EQUIPMENT

A variety of mechanical devices are used in liquid-liquid extraction. The simplest is a mixer/settler, or *decanter*, in which the two liquid phases are separated. Plate towers, packed towers, and mechanically agitated mixers (rotating disk contactors) are also used. In all of these devices, the number of stages tends to be much smaller than in distillation columns. This is due to the larger settling times required for liquid-liquid separation because of the small density differences between the liquid phases.

The devices can be operated with either phase (light or heavy) as the continuous phase. The dispersed phase is the liquid phase which forms the droplets. The design of the internals and of the control system naturally depends on this choice. Note that sometimes the solvent liquid is the heavy phase. In this context, "light" and "heavy" refer to density, not to volatility, as was the case in distillation. (See Figure 9–2.)

Liquid-liquid extraction columns are sometimes operated in a *pulsed* mode, i.e., the flow rate of one or both of the input streams is alternated between zero and some high value. Alternatively, the trays in the column are some-

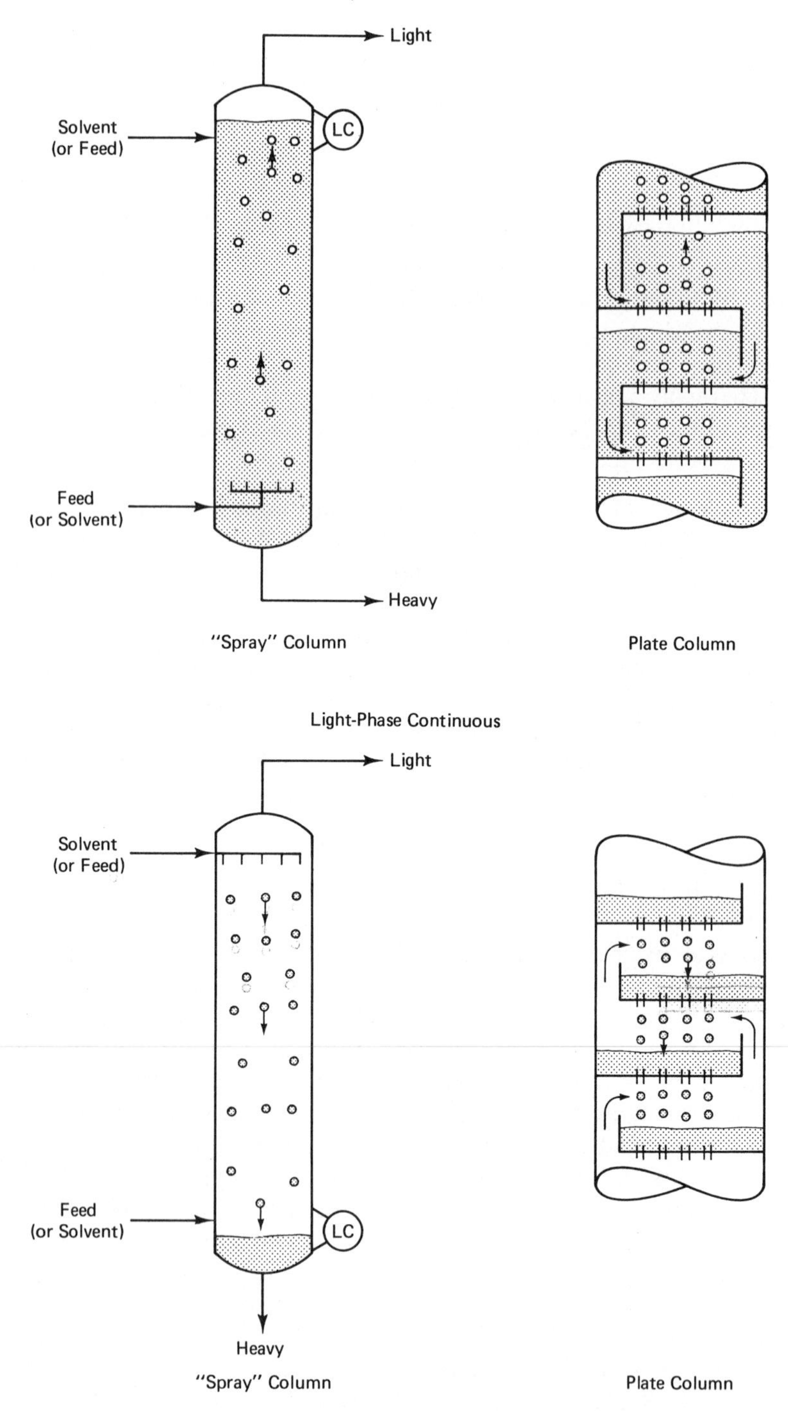

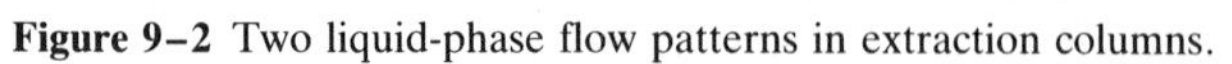

Figure 9–2 Two liquid-phase flow patterns in extraction columns.

times mechanically moved in a periodic manner. The idea is to achieve alternating flow rates of light phase up and heavy phase down the column. This periodic operation can sometimes produce much higher effective tray efficiencies than does continuous operation. Efficiencies in the 150–200 percent region have been reported.

9.3 GRAPHICAL MIXING RULES

Suppose we take two streams that contain three components and mix them together. Let one of these streams be stream A with flow rate F_A (kg/hr) and composition x_1^A, x_2^A, and x_3^A (weight fractions of components 1, 2, and 3), and let the other be stream F_B with corresponding composition x_1^B, x_2^B, and x_3^B. The mixed stream leaving the mixer will have a flow rate F_M and composition x_1^M, x_2^M, and x_3^M. A flow diagram is as follows:

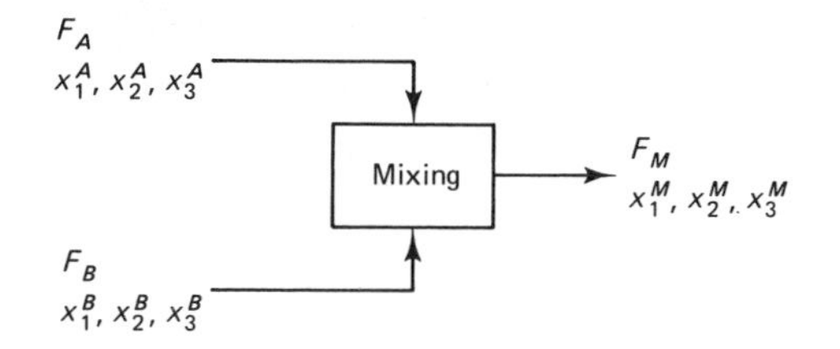

We want to prove that we can easily determine the location of the mixture composition on a graph. Since there are three components, only two coordinates are needed to completely specify the composition of any stream. We can use either right or equilateral triangular plots.

On a right-triangular plot (see Figure 9–3(a)) let us locate point A with coordinates (x_1^A, x_2^A) and point B with coordinates (x_1^B, x_2^B). The point M with coordinates x_1^M and x_2^M representing the mixture will lie someplace on the graph. We wish to show that M in fact lies on a straight line joining the A and B points. Accordingly, let M lie at an arbitrary location in Figure 9–3(b). If we can show that the angles α and β in the figure are equal, then M must lie on a straight line between A and B.

Now, the total mass balance for the system is

$$F_M = F_A + F_B \tag{9–1}$$

and the independent component balances for components 1 and 2 are

$$x_1^M F_M = x_1^A F_A + x_1^B F_B \tag{9–2}$$

and

$$x_2^M F_M = x_2^A F_A + x_2^B F_B \tag{9–3}$$

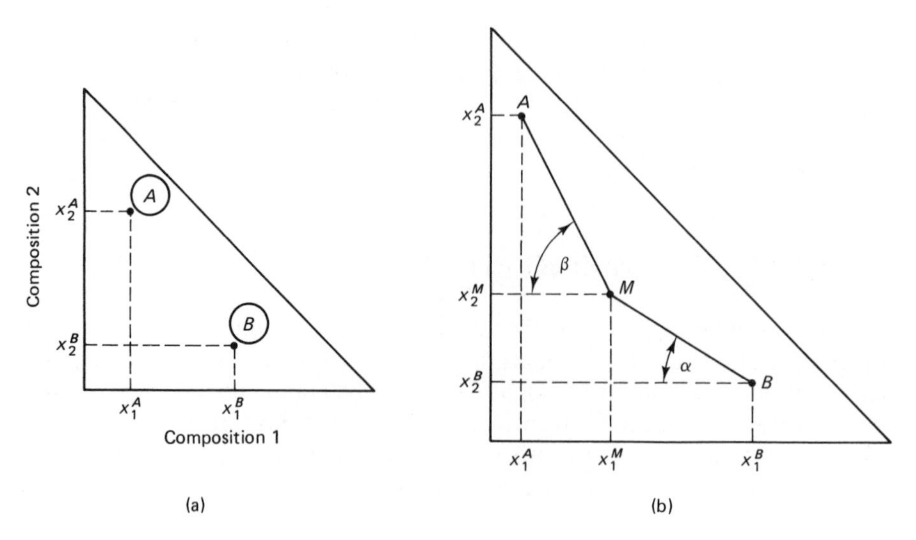

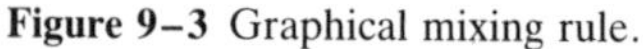

Figure 9–3 Graphical mixing rule.

Combining gives

$$x_1^M(F_A + F_B) = x_1^A F_A + x_1^B F_B$$

$$x_2^M(F_A + F_B) = x_2^A F_A + x_2^B F_B$$

Rearranging these two equations, we obtain

$$F_A(x_1^M - x_1^A) = F_B(x_1^B - x_1^M)$$

$$F_A(x_2^M - x_2^A) = F_B(x_2^B - x_2^M)$$

Solving for the ratio F_A/F_B, we have

$$\frac{F_A}{F_B} = \frac{x_1^B - x_1^M}{x_1^M - x_1^A} = \frac{x_2^B - x_2^M}{x_2^M - x_2^A} \tag{9–4}$$

or, by means of a little algebra,

$$\frac{x_2^A - x_2^M}{x_1^A - x_1^M} = \frac{x_2^M - x_2^B}{x_1^M - x_1^B} \tag{9–5}$$

These two ratios are the tangents of the angles α and β; hence, tan α = tan β. Therefore, α = β, and we have proven that the line AMB is a straight line.

The coordinates of the point M can be solved for analytically by using equations 9–1, 9–2, and 9–3. Alternatively, M can be located graphically by using the *inverse lever-arm rule*, which says that the distance from the point A to the point M divided by the distance from the point M to the point B is equal to the ratio F_B/F_A. Equation 9–4 shows this relationship in terms of the abscissae. The same ratio holds for the ordinates since the triangles shown in Figure 9–3(b) are similar.

9.4 CHOICE OF SOLVENT

Finding the best solvent is the most critical aspect of developing a liquid-liquid extraction process. The solvent should have a high *selectivity* for the extracted solute. The selectivity β of a solvent is similar to relative volatility and is given by

$$\beta = \frac{x_1^S/x_2^S}{x_1^R/x_2^R} \tag{9-6}$$

where

x_1^S = weight fraction of component 1 (solute) in the solvent phase

x_2^S = weight fraction of component 2 in the solvent phase

x_1^R = weight fraction of component 1 (solute) in the raffinate phase

x_2^R = weight fraction of component 2 in the raffinate phase

Selectivity is thus the ratio of the compositions in the solvent phase of the two components being separated to the ratio of the compositions in the raffinate phase of the same components. If the ratio is high, it means that the solvent preferentially soaks up the solute. Note that β goes to unity at the plait point.

Of course, the solvent should be easily recoverable from the extracted solute. Usually, this is done by distillation. The solvent and raffinate phases should have as large a density difference as possible to help settling.

It is also desirable to find a solvent that is inexpensive, nontoxic, noncorrosive, and without safety or pollution problems. Satisfying these last properties simultaneously is rarely possible.

9.5 SINGLE-STAGE CALCULATIONS

Consider the one-stage liquid-liquid extractor sketched in Figure 9–4. Component 1 is the solute, component 2 is the other component in the feed that we are trying to separate from component 1, and component 3 is the solvent. The feed stream has a flow rate F (mass/time) and composition z_1, z_2, and z_3 (weight fractions). In the normal situation, the process feed contains no solvent, so that $z_3 = 0$. We need to specify only two weight fractions in this ternary system.

A solvent stream with flow rate S_0 (mass/time) and composition $x_1^{S_0}$, $x_2^{S_0}$,

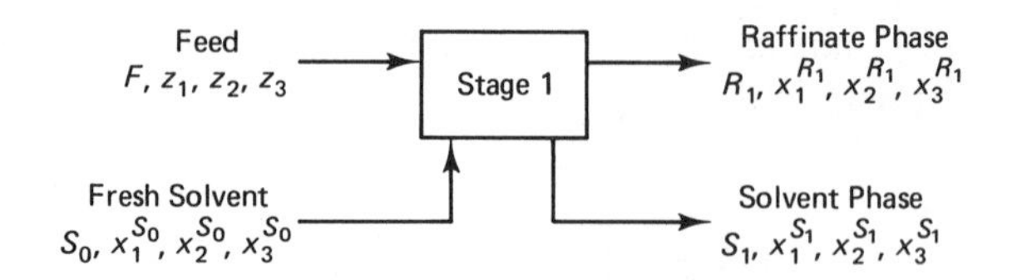

Figure 9–4 Single-stage extractor.

and $x_3^{S_0}$ is introduced. If the solvent is essentially pure, as is often the case, $x_3^{S_0} = 1$ and $x_1^{S_0} = x_2^{S_0} = 0$.

After mixing and achieving phase equilibrium, the two liquid phases are allowed to separate (due to their difference in density). The raffinate-rich liquid phase is removed at a flow rate R_1 and has a composition $x_1^{R_1}$, $x_2^{R_1}$, and $x_3^{R_1}$. The solvent-rich liquid phase is removed at a flow rate S_1 and has a composition $x_1^{S_1}$, $x_2^{S_1}$, and $x_3^{S_1}$. The total mass balance at this stage is

$$F + S_0 = S_1 + R_1 \tag{9–7}$$

and a component balance on the jth component yields

$$Fz_j + S_0 x_j^{S_0} = S_1 x_j^{S_1} + R_1 x_j^{R_1} \tag{9–8}$$

Now we define the parameter

$$M = F + S_0 = S_1 + R_1 \tag{9–9}$$

Mathematically, we are mixing together the F and S_0 streams to give a fictitious total stream M. However, from equation 9–7, M is also equal to $S_1 + R_1$. From the graphical mixing rule, the point M must lie on the straight line joining F and S_0. (See Figure 9–5.) It also must lie on a straight line joining S_1 and R_1. (See Figure 9–6.)

The coordinates x_1^M and x_3^M of the point M can be calculated from the material balance equations. We have three components, so three independent equations can be used. Typically, we use the total mass balance, the solute (component 1) balance and the solvent (component 3) balance.

$$\text{Total: } M = F + S_0 \tag{9–10}$$

$$\text{Solute component: } x_1^M M = Fz_1 + S_0 x_1^{S_0} \tag{9–11}$$

$$\text{Solvent component: } x_3^M M = Fz_3 + S_0 x_3^{S_0} \tag{9–12}$$

Since the two liquid phases leaving the system are in phase equilibrium, the points R_1 and S_1 must be connected by an LLE tie-line as discussed in chapter 6 and illustrated in Figure 9–6. Given the process feed and fresh solvent flow rates and compositions, the procedure for determining the compositions and flow rates of the two liquid streams leaving the system is as follows:

1. Calculate M from equation 9–10.

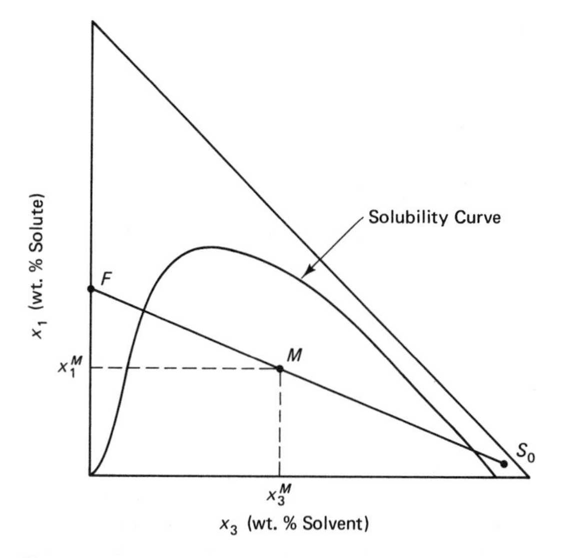

Figure 9–5 Mix point M of feed F and fresh solvent S_0.

2. Calculate x_1^M and x_3^M from equations 9–11 and 9–12.
3. Locate the point M using x_1^M and x_3^M.
4. Find the LLE tie-line that passes through the point M.
5. The points at the two ends of the LLE tie-line give the compositions of the two phases leaving the system: the raffinate-rich phase with

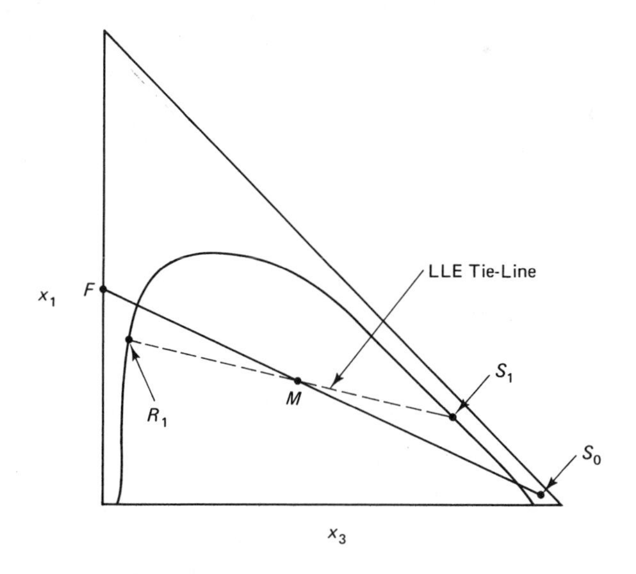

Figure 9–6 Splitting the mixture of F and S_0 into raffinate-rich phase R_1 and solvent-rich phase S_1 at ends of a tie-line.

composition $x_1^{R_1}$ and $x_3^{R_1}$, and the solvent-rich phase with composition $x_1^{S_1}$ and $x_3^{S_1}$.

6. Calculate the flow rates S_1 and R_1 by solving equations 9–7 and 9–8 simultaneously:

$$Fz_1 + S_0x_1^{S_0} = S_1x_1^{S_1} + (F + S_0 - S_1)x_1^{R_1}$$

$$S_1 = \frac{F(z_1 - x_1^{R_1}) + S_0(x_1^{S_0} - x_1^{R_1})}{x_1^{S_1} - x_1^{R_1}} \quad (9\text{–}13)$$

$$R_1 = F + S_0 - S_1 \quad (9\text{–}14)$$

Example 9.1.

An organic stream, with composition 30 weight percent acetone and 70 weight percent methyl isobutyl ketone and flow rate 10,000 kg/hr, is mixed with a pure water solvent with flow rate 5,000 kg/hr. What are the compositions and flow rates of the two liquid phases leaving a single-stage liquid-liquid extractor operating at 25°C?

Solution:

The ternary LLE diagram for the methyl isobutyl ketone–acetone–water system at 25°C is given in Figure 9–7. First, the point M is located; acetone is the solute (component 1), and water is the solvent (component 3). We have, for the system,

$$z_1 = 0.30$$

$$z_2 = 0.70$$

$$z_3 = 0$$

$$x_1^{S_0} = 0$$

$$x_2^{S_0} = 0$$

$$x_3^{S_0} = 1.0$$

$$M = F + S_0 = 10{,}000 + 5{,}000 = 15{,}000 \text{ kg/hr}$$

$$x_1^M = \frac{Fz_1 + S_0x_1^{S_0}}{M}$$

$$= \frac{(10{,}000)(0.30) + (5{,}000)(0)}{15{,}000} = 0.20$$

$$x_3^M = \frac{Fz_3 + S_0x_3^{S_0}}{M}$$

$$= \frac{(10{,}000)(0) + (5{,}000)(1)}{15{,}000} = 0.333$$

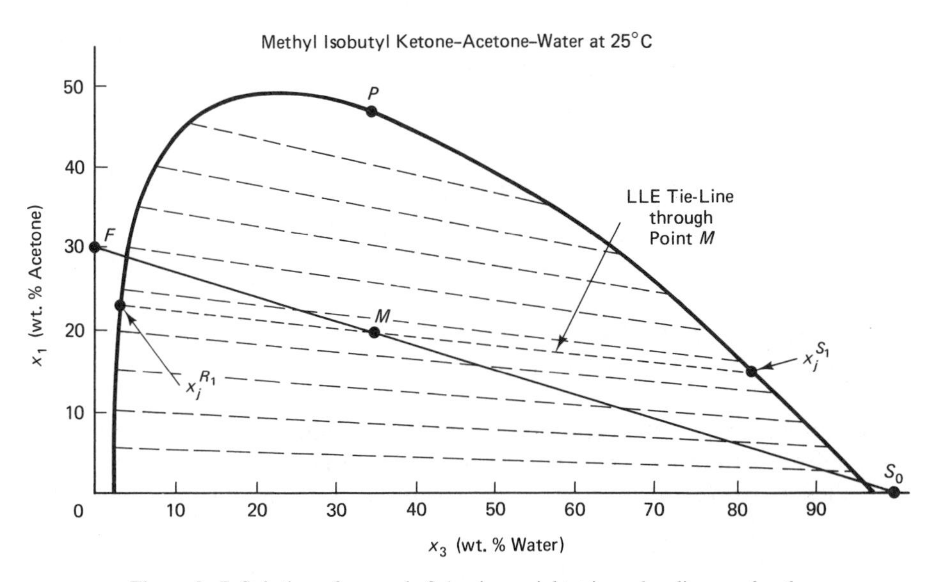

Figure 9–7 Solution of example 9.1 using a right triangular diagram for the methyl isobutyl ketone-acetone-water system at 25°C.

The point M is thus located at (0.333, 0.20), as shown in Figure 9–7. Note that M lies on the straight line connecting the points F and S_0, one-third of the way from the point F on the straight line, as predicted by the inverse lever-arm rule (equation 9–4):

$$\frac{(5{,}000 \text{ kg/hr water solvent})}{(5{,}000 + 10{,}000 \text{ kg/hr feed})} = 0.333$$

The LLE tie-line that passes through M is shown in the figure. An approximate tie-line (also shown) can be sketched by drawing a line through M that is roughly parallel to the adjacent tie-lines already drawn on the diagram.

A more rigorous approach involves a trial-and-error solution for the tie-line through M, using the LLE conjugate line. A guess is made of the end of the tie-line on the raffinate phase solubility curve. The LLE conjugate line is used to determine the other end of the LLE tie-line on the solvent phase solubility curve. If the tie-line goes through the point M, the compositions of the water solvent phase and organic raffinate phase have been found. We have

Solvent phase

$$x_1^{S_1} = 14 \text{ wt.\% acetone}$$

$$x_3^{S_1} = 83 \text{ wt.\% water}$$

Raffinate phase

$$x_1^{R_1} = 23.5 \text{ wt.\% acetone}$$

$$x_3^{R_1} = 4.0 \text{ wt.\% water}$$

A component balance on acetone (equation 9–13) gives the flow rate of the two liquid phases leaving the single-stage extractor:

$$S_1 = \frac{(10{,}000)(0.30 - 0.235) + (5{,}000)(0 - 0.235)}{0.14 - 0.235}$$

$$= 5{,}526 \text{ kg/hr solvent phase}$$

$$R_1 = 15{,}000 - 5{,}526 = 9{,}474 \text{ kg/hr raffinate phase}$$

9.6 MULTIPLE STAGES WITH CROSSFLOW OF SOLVENT

If the process liquid stream from the first stage is fed into a second extractor and mixed with more fresh solvent, as shown in Figure 9–8, we have what is called *cross-flow extraction*. In cross-flow extraction, the procedure de-

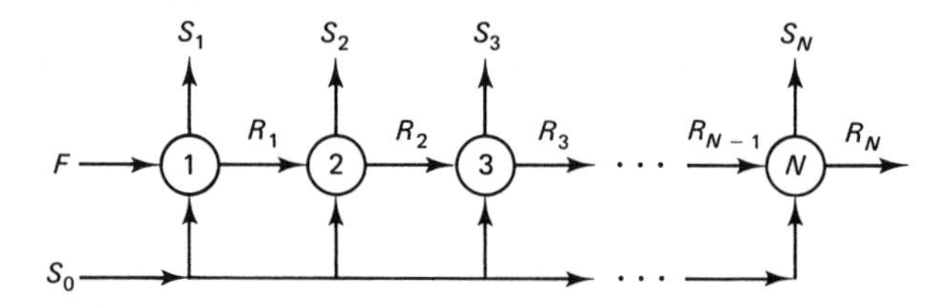

Figure 9–8 Multistage extractor with cross-flow of solvent.

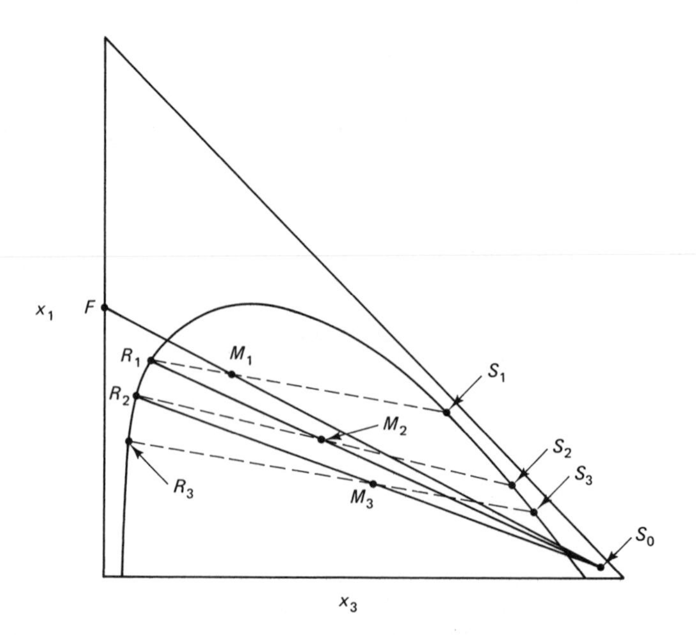

Figure 9–9 Mix points for cross-flow extractor.

scribed in the previous section for a single stage is simply repeated again for each stage, using the raffinate phase from the upstream stage as the feed to each stage. The graphical solution is sketched in Figure 9–9 for a three-stage cross-flow extractor. For each stage, the point M is calculated from the flow rates of the feed and fresh solvent streams to that particular stage.

9.7 MULTISTAGE COUNTERCURRENT EXTRACTION

Multistage countercurrent extraction is the most commonly encountered liquid-liquid extraction process. In it, the raffinate and solvent streams travel countercurrent to each other through N stages. The flow rate of the raffinate leaving the last stage (tray N) is R_N. (See Figure 9–10.) The flow rate of fresh solvent to tray N is S_0. (The notation "S_{N+1}" could also be used for this stream.)

Mass and component balances around the entire cascade give

$$F + S_0 = R_N + S_1 \tag{9–15}$$

$$z_1 F + x_1^{S_0} S_0 = x_1^{R_N} R_N + x_1^{S_1} S_1 \tag{9–16}$$

$$z_3 F + x_3^{S_0} S_0 = x_3^{R_N} R_N + x_3^{S_1} S_1 \tag{9–17}$$

We next define a pseudo flow rate M and pseudo compositions x_1^M and x_3^M:

$$M = F + S_0 \tag{9–18}$$

$$x_1^M M = z_1 F + x_1^{S_0} S_0 \tag{9–19}$$

$$x_3^M M = z_3 F + x_3^{S_0} S_0 \tag{9–20}$$

If the fresh solvent flow rate S_0 and composition are given along with the feed flow rate F and composition, we can locate the point M on the straight line connecting the points F and S_0. (See Figure 9–11.) From equations 9–15 through 9–17, it is clear that M must also lie on the straight line connecting S_1 and R_N, as shown in the figure.

In the typical design problem, the point R_N will be given, i.e., the concentration of the raffinate phase leaving the final stage will be specified so as to recover the desired amount of the solute from the feed. Obviously, in any real system we cannot recover all of it.

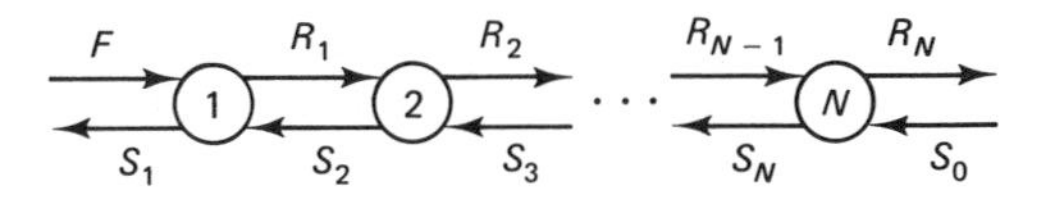

Figure 9–10 Multistage countercurrent extractor.

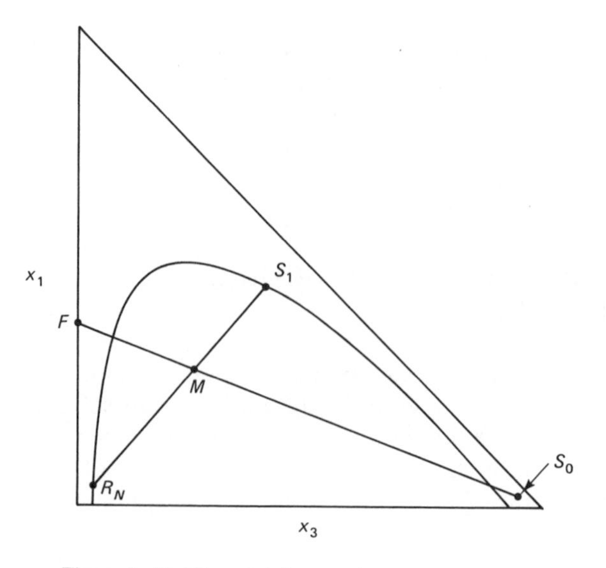

Figure 9–11 Mix point for countercurrent extractor.

Since the points R_N and M are known, a straight line can be drawn to the solubility curve to determine the composition of the $x_1^{S_1}$ and $x_3^{S_1}$ of the S_1 stream. Then equations 9–15 and 9–16 can be used to solve for the flow rates R_N and S_1.

Since S_1 and R_1 are in phase equilibrium, we can use an LLE tie-line to determine the point R_1 on the solubility curve. (See Figure 9–12.) Component and mass balances around the first stage can then be used to calculate the flow rate S_2 and compositions $x_1^{S_2}$ and $x_3^{S_2}$ of the solvent entering stage 1 from stage 2:

$$F + S_2 = R_1 + S_1 \tag{9–21}$$

$$z_1 F + x_1^{S_2} S_2 = x_1^{R_1} R_1 + x_1^{S_1} S_1 \tag{9–22}$$

$$z_3 F + x_3^{S_2} S_2 = x_3^{R_1} R_1 + x_3^{S_1} S_1 \tag{9–23}$$

Once S_2 is known, R_2 can be found at the other end of the LLE tie-line.

This computational procedure can be repeated from stage to stage until enough stages have been used to produce a process stream that meets or exceeds the specifications on R_N, the final raffinate phase leaving the unit. Thus, the number of stages required to achieve the desired separation for a given solvent flow rate has been determined.

These calculations can all be done graphically with relative ease. The advantages of the graphical procedure are the same as those found in the graphical McCabe-Thiele method used in binary distillation. The effects of various design and operating parameters can be clearly seen on the graph, giving the engineer considerable insight as to what makes the process tick.

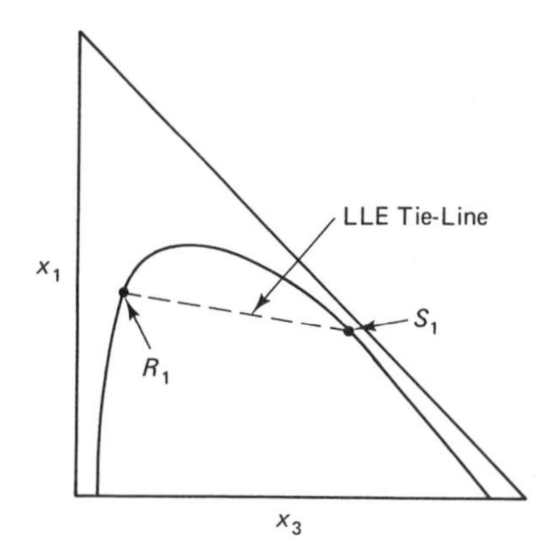

Figure 9–12 Use LLE tie-line to get R_1 composition from S_1.

To develop such a graphical procedure, it is convenient to define another pseudo flow rate Δ and pseudo compositions x_1^Δ and x_3^Δ. This fictitious Δ stream serves the same purpose as the operating lines did in binary distillation: it lets us calculate the compositions of "passing" streams in the column. In effect, it provides a graphical way to solve the three mass balances that describe this ternary system simultaneously:

$$\Delta = R_N - S_0 \tag{9–24}$$

$$x_1^\Delta \Delta = x_1^{R_N} R_N - x_1^{S_0} S_0 \tag{9–25}$$

$$x_3^\Delta \Delta = x_3^{R_N} R_N - x_3^{S_0} S_0 \tag{9–26}$$

The Δ stream can be thought of as the "net" outward flow of material from the last (Nth) stage and from each of the other stages. The pseudo composition x_j^Δ is entirely fictitious and, therefore, can be less than zero or greater than unity.

Using equations 9–15 and 9–24, we see that

$$\Delta = R_N - S_0 = F - S_1 \tag{9–27}$$

Therefore, the Δ point must lie on two straight lines, one through the points R_N and S_0 and the other through the points F and S_1. Δ can lie either to the left or to the right of the phase diagram, as illustrated in Figure 9–13. Δ lies to the left of the diagram when the LLE tie-lines have positive slopes; it lies to the right when the tie-lines have negative slopes.

The mass balances for the first stage (equations 9–21 through 9–23), the definition of Δ (equations 9–24 through 9–26), and equation 9–27 can be combined to give

$$\Delta = R_N - S_0 = F - S_1 = R_1 - S_2 \tag{9–28}$$

Therefore, the two streams R_1 and S_2 that pass each other between the first and second stages are related to each other by the Δ point. Hence, if we

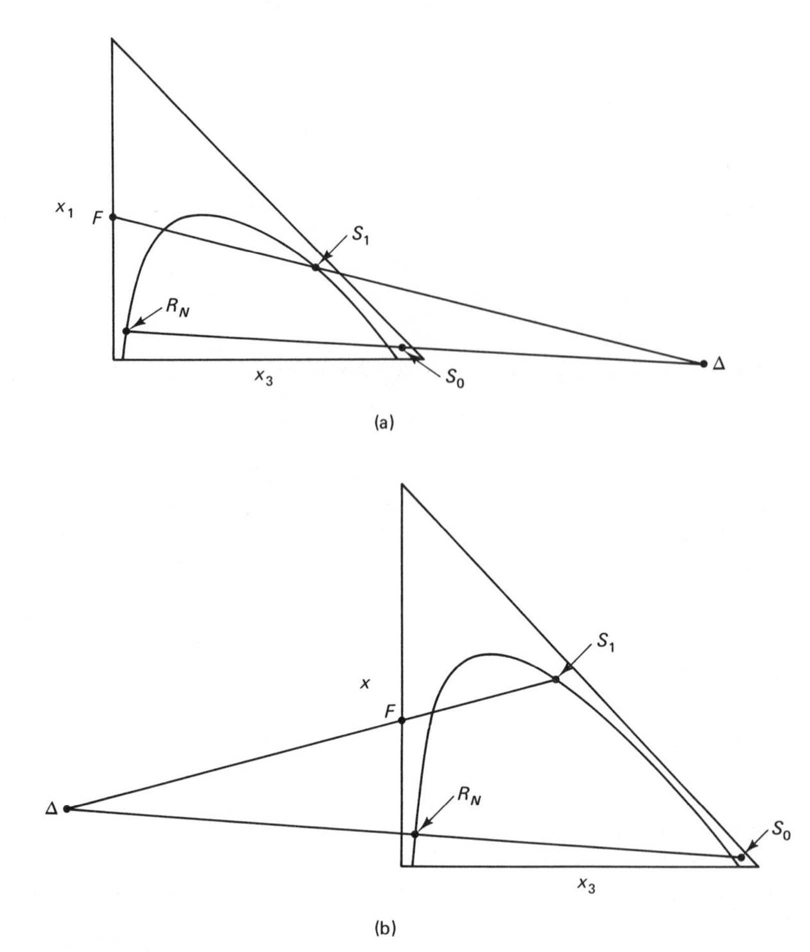

Figure 9–13 Drawing the Δ point. (a) Δ lies to the right (b) Δ lies to the left.

know R_1, we can determine S_2 by using the straight line that connects R_1 and Δ.

The preceding development shows that the Δ point serves the same function as the operating line in a McCabe-Thiele diagram, relating compositions of passing streams. Thus, we can step from stage to stage by alternately using Δ and the LLE tie-lines as illustrated in Figure 9–14 for a three-stage column. The lines drawn to the Δ point can be considered as operating lines representing component balances.

Example 9.2.

Figure 9–15 shows a liquid-liquid ternary phase diagram for isopropyl alcohol (IPA), toluene, and water at 25°C. Feed flow rate is 100 kg/hr, and feed composition

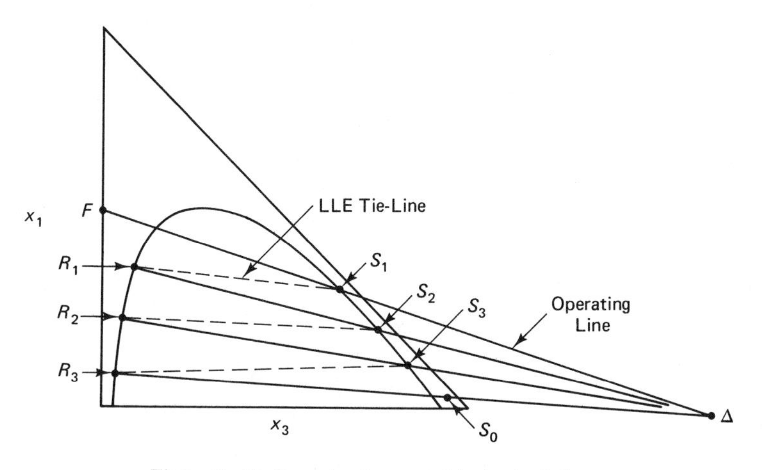

Figure 9–14 Stage-to-stage graphical calculations.

is 40 weight percent IPA and 60 weight percent toluene. Fresh solvent is pure water at a flow rate of 100 kg/hr. Determine the number of equilibrium stages required to produce a raffinate stream that contains 3 weight percent IPA.

Solution:

In this system, water is the solvent and IPA is the solute that we wish to extract from the toluene-IPA feed mixture. The raffinate stream leaving the extractor will be mostly toluene. Figure 9–15 gives the graphical solution.

First the points F and S_0 are plotted, and a straight line is drawn joining them. Then the point M is determined. Since the flow rates of F and S_0 are both 100 kg/hr, M will lie halfway between F and S_0 on the straight line.

Next a straight line is drawn from the point R_N up through M. The intersection of this straight line with the solubility curve gives us the point S_1. The composition of the solvent phase leaving the first stage is read off the graph: 27.5 weight percent IPA and 70 weight percent water.

Two straight lines are drawn to locate the Δ point. The first line passes through F and S_1, the second through R_N and S_0. The intersection of these two lines occurs at Δ.

An LLE tie-line is then followed from S_1 down to R_1 on the solubility curve. This line represents the first stage. The composition of the raffinate phase leaving the first stage is 21 weight percent IPA and 2 weight percent water.

Next an "operating" straight line is drawn from R_1 to Δ. This line intersects the solubility curve at S_2. The composition of the solvent phase leaving the second stage is then read off the graph: 12 weight percent IPA and 86 weight percent water.

Finally, an LLE tie-line is drawn from S_2 to R_2. Since the raffinate phase from stage 2 contains about 3 weight percent IPA, we have achieved the required separation in two stages.

Notice that the solvent came into the top of the column (since it is the heavier liquid phase) as pure water. The "fat" solvent leaving stage 1 contained 27.5 weight percent IPA. The organic feed was introduced into the bottom of the column (since it is the lighter liquid phase) and contained 40 weight percent IPA and 60 weight percent toluene. The organic raffinate stream leaving the top of the column from

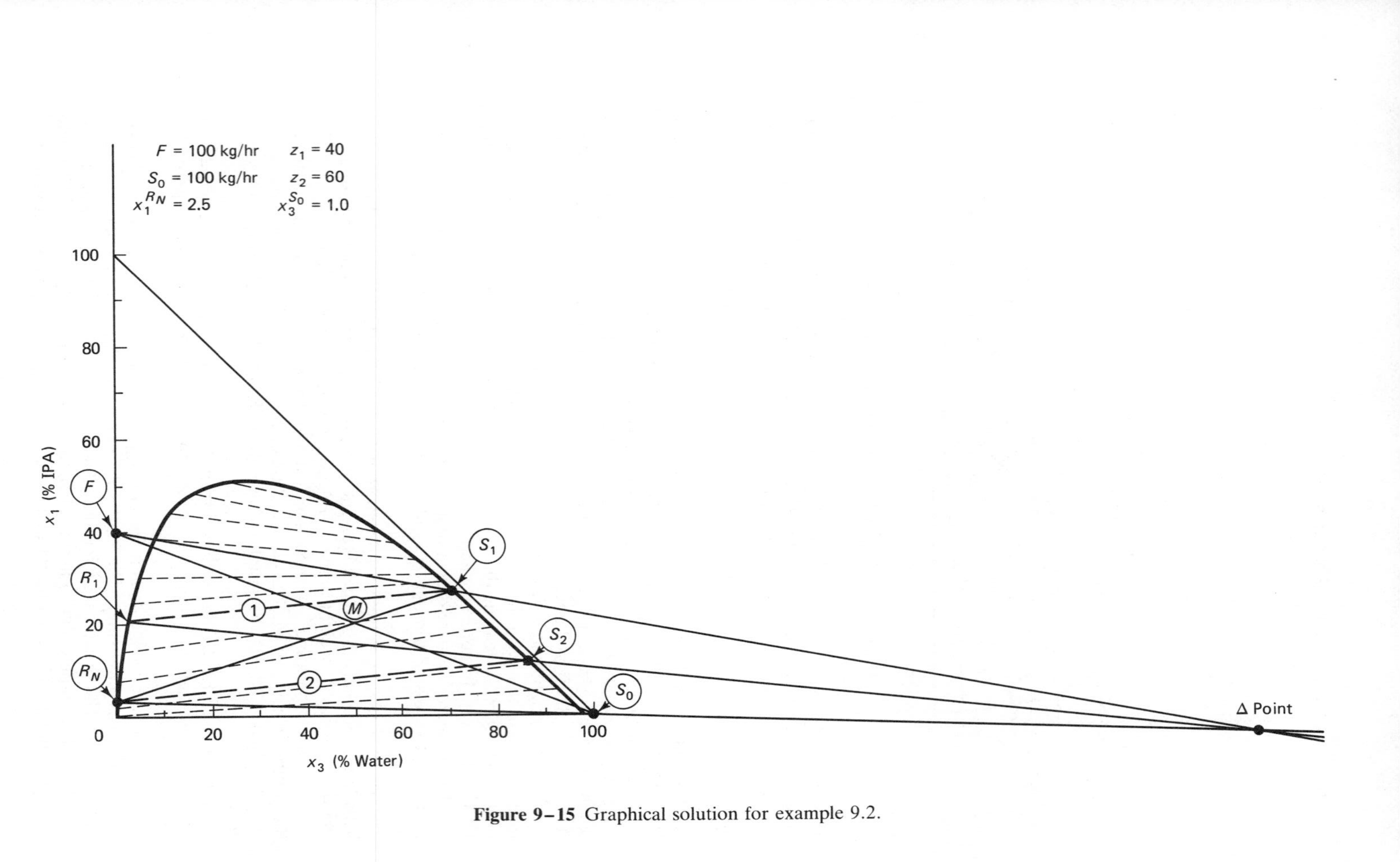

Figure 9–15 Graphical solution for example 9.2.

stage 2, being mostly toluene, contained only 3 weight percent IPA. So we have extracted most of the IPA from the original feed.

9.8 MINIMUM SOLVENT RATE

In the preceding discussion we assumed that the fresh solvent flow rate S_0 was given. However, in the normal design situation the solvent rate must be specified by the designer. Usually, the smaller the solvent flow rate the lower the energy costs because less solvent must be circulated between the extractor and the stripping column and the concentration of solute in the "fat solvent" from the extractor is higher. Hence, separation costs are reduced.

However, the smaller the solvent flow the higher the capital costs, because more stages are required in the liquid-liquid extractor. These engineering trade-offs between the number of stages and the solvent flow rate are analogous to the situation in distillation, where trade-offs must be made between the reflux ratio and the number of trays.

A commonly used heuristic in distillation discussed in chapter 8 was setting the reflux ratio to 1.2 times the minimum reflux ratio. In liquid-liquid extraction, there is similarly a minimum solvent flow rate. Once this minimum is found, the actual solvent flow rate is set somewhat higher. Since liquid-liquid extraction stages are usually more expensive than distillation trays, the ratio of actual to minimum solvent flow rates is usually higher in extraction (typically 2).

The minimum solvent flow rate S_0^{min} occurs when an LLE tie-line is coincident with an "operating line" drawn to the Δ point. The minimum solvent rate is usually found by extending the LLE tie-line that passes through the feed point.The intersection of this straight line with the straight line through the points S_0 and R_N gives the minimum Δ, Δ_{min}. As shown in Figure 9–16, Δ_{min} can lie either to the left or to the right of the phase diagram.

In some unusual LLE systems, there are other tie-lines (not through the feed point) that intersect the S_0–R_N line farther to the left. The extensions

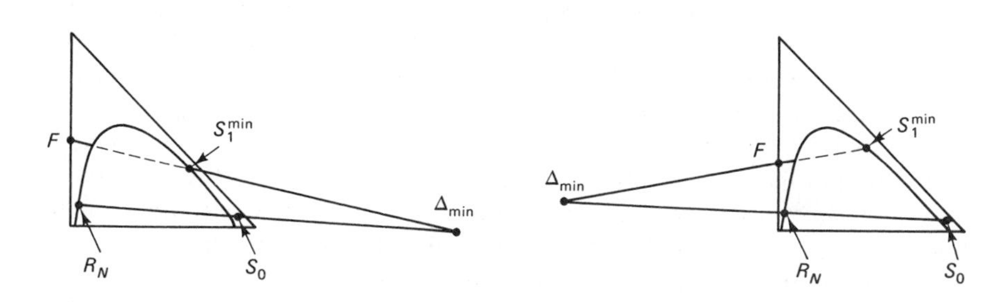

Figure 9–16 Minimum solvent flow rate.

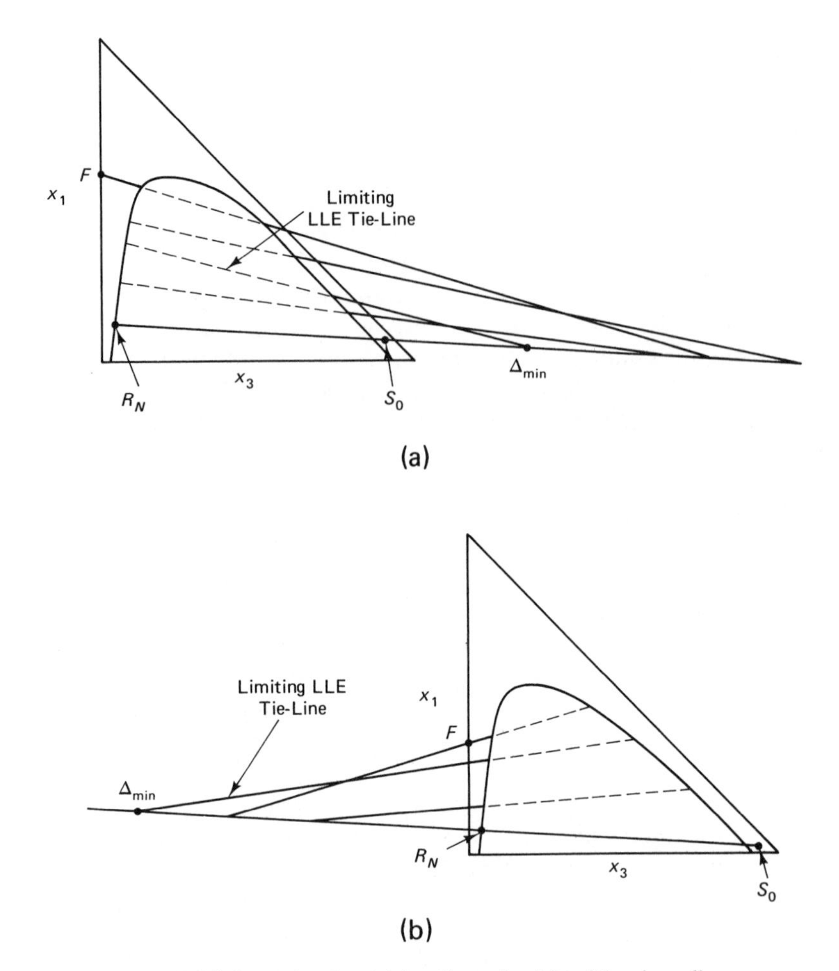

Figure 9–17 Minimum Δ point. (a) Δ_{min} lies to the right of the phase diagram (b) Δ_{min} lies to the left of the phase diagram.

of the tie-lines that occur farthest to the *left* dictate the point Δ_{min}. Figure 9–17 illustrates this situation for two cases: when Δ_{min} lies to the right of the diagram, and when it lies to the left. For both cases, the leftmost intersection of a tie-line with the line through the points R_N and S_0 gives Δ_{min}.

The point S_1^{min} is found on the straight line from the feed point to Δ_{min}, and a straight line is drawn from it to the point R_N which has been specified. The point M^{min} is then given by the intersection of this line with the straight line connecting the points F and S_0. S_0^{min} can then be calculated from equations 9–18 and 9–19 using the known compositions x_1^{min} and x_3^{min} of M^{min}:

$$S_0^{min} = F \frac{z_1 - x_1^{min}}{x_1^{min} - x_1^{S_0}} \tag{9–29}$$

9.9 DESIGN PROCEDURE

The foregoing graphical construction may sound somewhat complicated on first reading. However, it is fairly straightforward once you get the basic idea. It parallels the construction used for a single stripping or rectifying column in binary distillation. Let us summarize the specific sequence of steps to design a countercurrent liquid-liquid extractor for a ternary system.

1. Parameters specified.
   (a) Feed flow rate and compositions (F, z_1, z_3).
   (b) Composition of solute in the raffinate stream leaving the last stage ($x_1^{R_N}$).
   (c) LLE data for the ternary system.
   (d) Economic ratio of actual to minimum solvent flow rates ($S_0/S_0^{\min}$).
   (e) Composition of fresh solvent ($x_1^{S_0}$, $x_3^{S_0}$).
2. Calculate the minimum solvent flow rate $S_0^{\min}$.
   (a) Extend the tie-line through the point F.
   (b) Draw a line through the points S_0 and R_N.
   (c) Locate $S_1^{\min}$ on the tie-line through F.
   (d) Draw a line from $S_1^{\min}$ to R_N.
   (e) The intersection of this line with a line through F and S_0 gives the point $M^{\min}$.
   (f) Read off the composition ($x_1^{\min}$) at $M^{\min}$.
   (g) Calculate $S_0^{\min}$ from equation 9–29.
3. Set the actual fresh solvent flow rate S_0.
4. Calculate the point M from equations 9–18 and 9–19.
5. Draw a line from R_N through M to find S_1 on the solubility curve.
6. Locate the Δ point by drawing two lines, one through S_0 and R_N, and the other through S_1 and F.
7. Step off stages, alternately using the LLE tie-lines and the Δ point, as follows:

$$S_1 \xrightarrow{\text{LLE}} R_1 \xrightarrow{\Delta} S_2 \xrightarrow{\text{LLE}} R_2 \xrightarrow{\Delta} S_3 \longrightarrow \ldots$$

Example 9.3.

Design a countercurrent liquid-liquid extractor to separate acetone from methyl isobutyl ketone (MIK) using water as a solvent. The column is to operate at 25°C.

The organic feed rate is 10,000 kg/hr, and the composition is 45 weight percent acetone and 55 weight percent ketone. The fresh solvent is pure water, and twice the minimum solvent rate is to be used. The concentration of acetone in the organic raffinate leaving the top of the extractor is to be 2.5 weight percent acetone.

Solution:

In this system, acetone is the solute to be separated from MIK using water as the solvent. Thus, component 1 is acetone and component 3 is water.

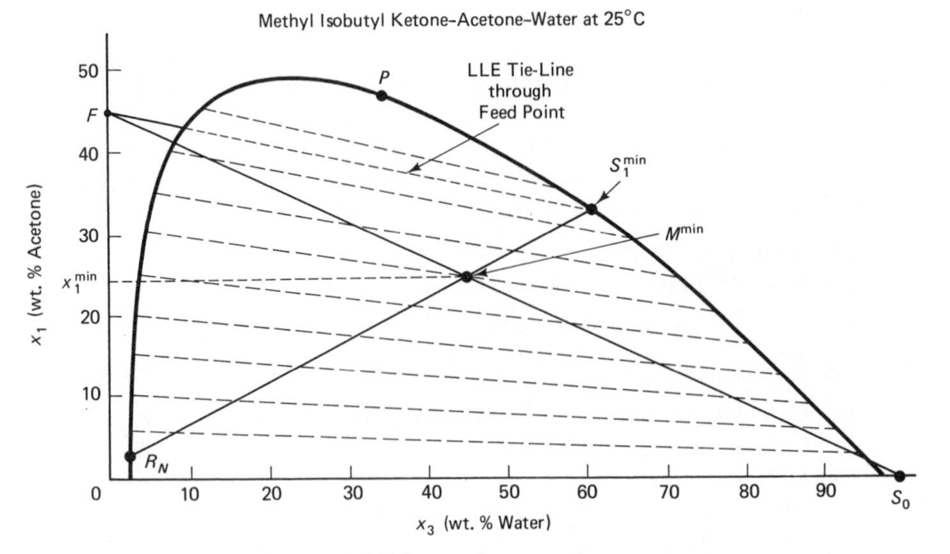

Figure 9–18 Minimum solvent rate for example 9.3.

Minimum Solvent Rate The point R_N is plotted on the solubility curve at $x_1^{R_N} = 2.5$ weight percent acetone. (See Figure 9–18.) The point S_0 is plotted at the right lower corner of the diagram, since $x_3^{S_0} = 100$ weight percent water. The point F is plotted on the vertical axis at $z_1 = 45$ weight percent acetone.

The LLE tie-line that passes through F is used to determine the points Δ_{min} and S_1^{min}. A straight line is drawn from R_N to S_1^{min}, and another from F to S_0. The intersection of these lines is the point M^{min}, giving a composition $x_1^{min} = 24$ weight percent acetone. The minimum solvent flow rate is calculated from equation 9–29:

$$S_0^{min} = F\frac{z_1 - x_1^{min}}{x_1^{min} - x_1^{S_0}}$$

$$= \frac{(10{,}000)(0.45 - 0.24)}{(0.24 - 0)}$$

$$= 8{,}750 \text{ kg/hr minimum solvent flow rate}$$

The actual solvent rate is twice the minimum:

$$S_0 = 2\,(8750) = 17{,}500 \text{ kg/hr}$$

Calculate the point M

$$M = F + S_0 = 10{,}000 + 17{,}500 = 27{,}500 \text{ kg/hr}$$

$$x_1^M = \frac{(Fz_1 + S_0\, x_1^{S_0})}{M}$$

$$= \frac{[(10{,}000)(45) + (17{,}500)(0)]}{27{,}500}$$

$$= 16.4 \text{ weight percent acetone}$$

Determine S_1 Draw a straight line from R_N through M to the solubility curve. The composition of the solvent phase on stage 1 is found from Figure 9–19 to be $x_1^{S_1}$ = 19 weight percent acetone.

The flow rate of S_1 can be calculated from a total mass balance and an acetone component balance around the entire system:

$$F + S_0 = R_N + S_1$$

$$z_1 F + x_1^{S_0} S_0 = x_1^{R_N} R_N + x_1^{S_1} S_1$$

$$10{,}000 + 17{,}500 = R_N + S_1$$

$$(45)(10{,}000) + (0)(17{,}500) = (2.5) R_N + (19) S_1$$

$$S_1 = 23{,}100 \text{ kg/hr}$$

$$R_N = 4{,}400 \text{ kg/hr}$$

Note that the original organic feed rate was 10,000 kg/hr, but the organic raffinate leaving the unit is only 4,400 kg/hr. The recovery of acetone is found as follows:

$$\text{Acetone in Feed} = (10{,}000)(0.45) = 4{,}500 \text{ kg/hr acetone}$$

$$\text{Acetone in Raffinate } (R_N) = (4{,}400)(0.025) = 110 \text{ kg/hr acetone}$$

$$\text{Acetone in Fat Solvent } (S_1) = (23{,}100)(0.19)$$

$$= 4{,}390 \text{ kg/hr acetone recovered}$$

$$\text{Recovery of acetone} = 4{,}390/4{,}500 = 0.976$$

Thus, 97.6 percent of the acetone fed to the unit is captured by the solvent.

Step Off Stages As shown in Figure 9–19, LLE tie-lines and the Δ point are used alternately to step off five stages. On the fifth stage, the concentration of the raffinate phase has been reduced to the desired value of 2.5 weight percent acetone. Thus, five equilibrium stages are required to achieve the separation at a solvent rate twice the minimum.

The problems discussed have been design problems, i.e., problems concerned with determining the number of stages of a system. Just as in distillation, however, there are also many interesting rating problems in liquid-liquid extraction. For example, we may want to find the flow rate of fresh solvent required to attain a specified recovery of solute in a column with a fixed number of stages. These rating problems are usually solved iteratively. For the previous example, a fresh solvent flow rate would be guessed, and the number of stages in the actual column would be stepped off to see if the desired recovery was achieved. If not, a new value of fresh solvent flow rate would be guessed. Some of the problems at the end of the chapter are rating problems.

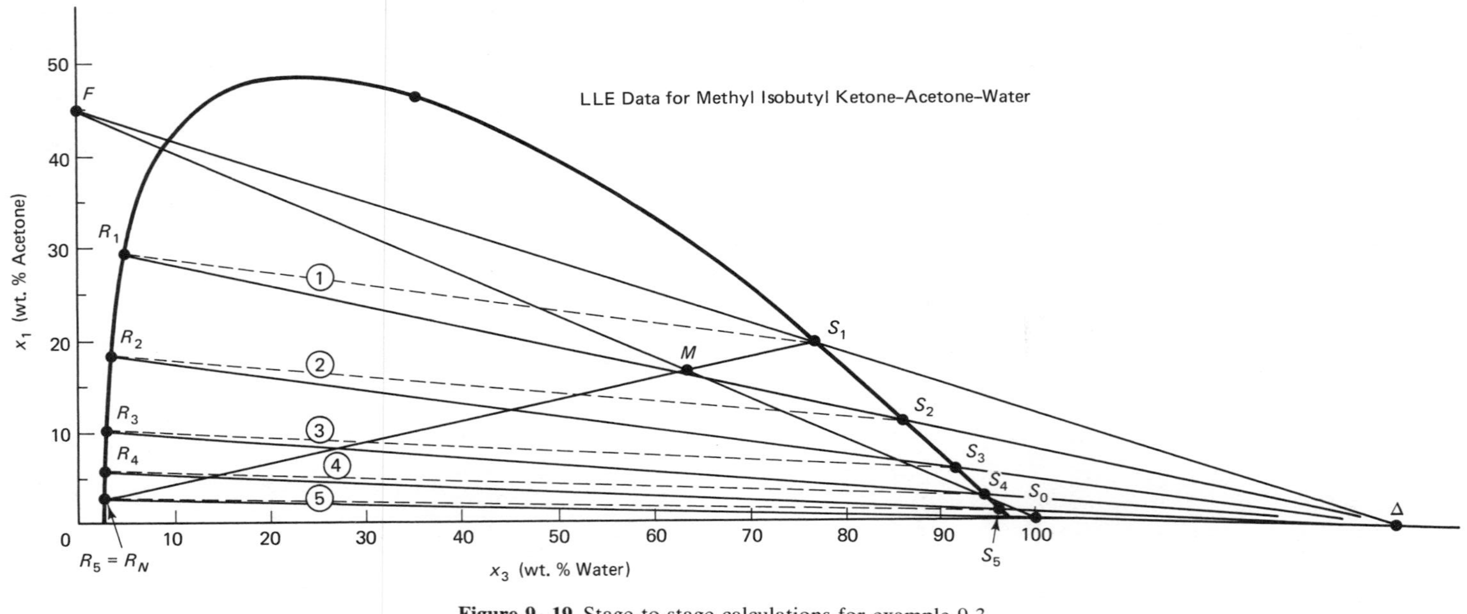

Figure 9–19 Stage-to-stage calculations for example 9.3.

9.10 COMPUTER SOLUTIONS

The stage-to-stage calculations can be solved numerically on a computer. Table 9–1 gives a FORTRAN program that determines the number of stages required to achieve a specified raffinate purity $x_1^{R_N}$, given the feed and fresh solvent flow rates and compositions, and the LLE data.

The main problem in doing the calculations on the computer is expressing the liquid-liquid phase equilibrium data in equation form. The fundamental thermodynamic relationships tell us that the chemical potentials of each component must be the same in all phases. That is,

$$\mu_j^\alpha = \mu_j^\beta \quad j = 1, 2, 3, \ldots, \mathrm{NC} \tag{9–30}$$

The chemical potential of component j in the β liquid phase is usually given by

$$\mu_j^\beta = x_j^\beta \, P_j^S \, \gamma_j^\beta \tag{9–31}$$

where

μ_j^β = chemical potential of component j in the β liquid phase

x_j^β = mole fraction of component j in the β liquid phase

P_j^S = vapor pressure of component j

γ_j^β = liquid phase activity coefficient of component j in the β liquid phase

Equation 9–31 can be written for each component and for each liquid phase. Suppose we have two phases, α and β. Then, from equation 9–30,

$$x_j^\alpha \, P_j^S \, \gamma_j^\alpha = x_j^\beta \, P_j^S \, \gamma_j^\beta \tag{9–32}$$

which, since the vapor pressures are the same in both phases, reduces to

$$x_j^\alpha \, \gamma_j^\alpha = x_j^\beta \, \gamma_j^\beta \tag{9–33}$$

If we have equations (such as van Laar or NRTL) that give the dependence of the activity coefficients on composition, we can solve for the unknown compositions in both phases. For example, in a ternary system there would be four unknown compositions—x_1^S, x_3^S, x_1^R, and x_3^R—and three LLE equations:

$$x_1^S \, \gamma_1^S = x_1^R \, \gamma_1^R \tag{9–35}$$

$$x_2^S \, \gamma_2^S = x_2^R \, \gamma_2^R \tag{9–36}$$

$$x_3^S \, \gamma_3^S = x_3^R \, \gamma_3^R \tag{9–37}$$

Remember that the activity coefficients are known functions of the x_j's and that the sum of the x_j's must be unity in each phase.

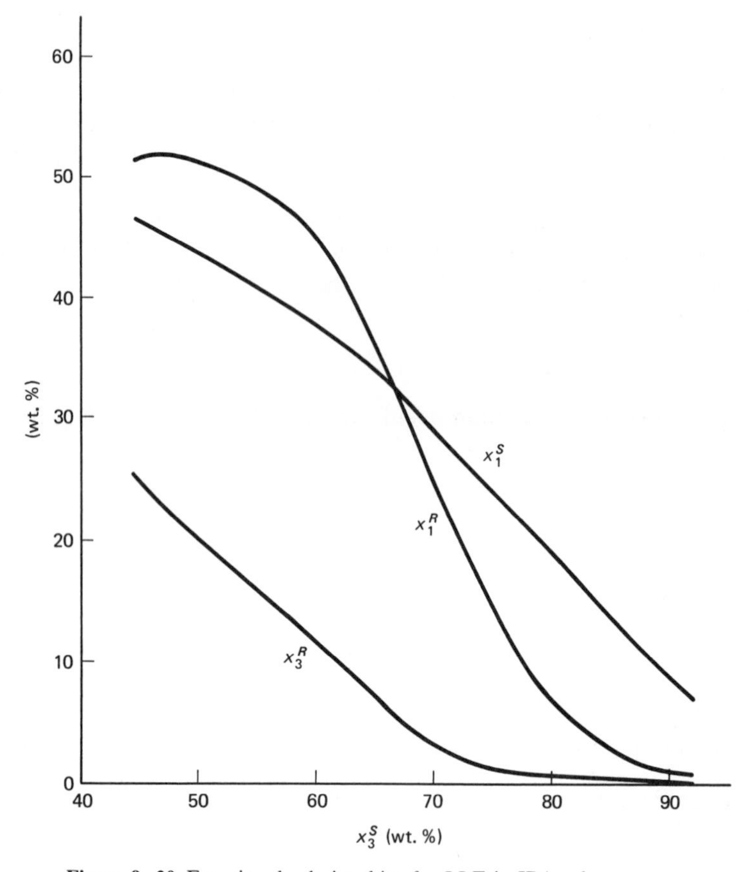

Figure 9–20 Functional relationships for LLE in IPA-toluene-water.

Now if we specify any one composition, we should be able to calculate all of the three remaining compositions because we have three unknowns and three equations. We do this in our graphical solutions by using the two solubility curves (one for the raffinate-rich phase and one for the solvent-rich phase) and the LLE conjugate line that tells us where the two ends of the LLE tie-lines are.

The numerical calculations can get somewhat involved in the numerical activity-coefficient approach. In addition, the computed results from the equations for the γ's sometimes don't fit the experimental data too well. Therefore, we will take a simple curve-fitting approach to the LLE calculations.

Solubility and tie-line data are used in the LLE and FR subroutines. Straight-line approximations to the data are used to interpolate in between the data points. The solvent composition x_3^S in the solvent-rich phase is specified, and the remaining three compositions x_1^S, x_1^R, and x_3^R are calculated.

Figure 9–20 gives the three curves that are used for the IPA-toluene-water example.

The subroutine LLE calculates the x_1^S, x_1^R, and x_3^R for a given value of x_3^S. The subroutine FR calculates x_3^R for a given value of x_1^R. We need this solubility curve relationship in order to solve the overall material balances for R_N and S_1.

The program in Table 9–1 is a design program: it determines the number

TABLE 9–1

LIQUID-LIQUID EXTRACTION COLUMN DESIGN PROGRAM

```
C
C************************************
C LIQUID-LIQUID EXTRACTION COLUMN DESIGN PROGRAM
C************************************
C    GIVEN: FEED FLOWRATE AND COMPOSITION
C           FRESH SOLVENT FLOWRATE AND COMPOSITION (SO AND XSO)
C           SOLUTE CONCENTRATION IN RAFFINATE PRODUCT (XRN1)
C
C    PROGRAM CALCULATES:
C           NUMBER OF EQUILIBRIUM STAGES
C           FLOWRATES AND COMPOSITIONS OF SOLVENT PHASE ON EACH
C                  STAGE ( S(N) AND XS(N,J) )
C           FLOWRATES AND COMPOSITIONS OF RAFFINATE PHASE ON
C                  EACH STAGE ( R(N) AND XR(N,J) )
C
C    LLE PHASE EQUILIBRIUM:
C           DATA POINTS FROM TWO SOLUBILITY LINES AND CONJUGATE
C                 LINE ARE FED INTO LLE AND FR SUBROUTINES
C           STRAIGHT-LINE INTERPOLATION BETWEEN DATA POINTS
C
      PROGRAM EXTRACT
      DIMENSION R(20),S(20),XR(20,3),XS(20,3),XS0(3),Z(3)
      DATA F,Z/100.,40.,60.,0./
      DATA SO,XS0/100.,0.,0.,100./
C CHANGE THE FRESH SOLVENT FLOWRATE
      DO 500 K=1,4
      XRN1=3.
C GUESS A VALUE FOR SOLVENT COMPOSITION OF S1
      XS(1,3)=70.
      LOOP=0
      CALL FR(XRN1,XRN3)
C OVERALL CALCULATIONS TO GET S1 AND RN
   10 CALL LLE(XS(1,3),XS(1,1),XR(1,3),XR(1,1))
      S(1)=(F*(Z(1)-XRN1)+SO*(XS0(1)-XRN1))/(XS(1,1)-XRN1)
      RN=F+SO-S(1)
      XCALC=(F*Z(3)+SO*XS0(3)-RN*XRN3)/S(1)
    2 FORMAT(11X,3F9.2)
      IF(ABS(XCALC-XS(1,3)).LT..01)GO TO 20
      LOOP=LOOP+1
      IF(LOOP.GT.50)THEN
      WRITE(6,*)' OVERALL LOOP'
      STOP
      ENDIF
C REGUESS XS(1,3)
      XS(1,3)=XS(1,3)+0.2*(XCALC-XS(1,3))
      GO TO 10
   20 CONTINUE
      WRITE(6,199)SO
  199 FORMAT(/,'  FRESH SOLVENT FLOWRATE = ',F8.2)
      WRITE(6,*)'     F       Z1       Z2       Z3'
```

TABLE 9-1 CON'T.

```
      WRITE(6,201)F,Z
      WRITE(6,*)'  S0       XS01     XS02       XS03'
  201 FORMAT(1X,4F8.2)
      WRITE(6,201)S0,XS0
C STAGE CALCULATIONS TO GET S(N+1) AND R(N)
      DO 100 N=1,19
C GUESS XS(N+1,3)
      XS(N+1,3)=(XS(N,3)+99.)/2.
      IF(XS(N+1,3).GT.98.9) XS(N+1,3)=98.9
      LOOP=0
   50 CALL LLE(XS(N+1,3),XS(N+1,1),XR(N+1,3),XR(N+1,1))
      R(N)=(F*(Z(3)-XS(N+1,3))+S(1)*(XS(N+1,3)-XS(1,3)))/(XR(N,3)-
     + XS(N+1,3))
      S(N+1)=S(1)+R(N)-F
      IF(LOOP.GT.30)WRITE(6,701)R(N),S(N+1)
  701 FORMAT('  R(N) AND S(N+1)    ',2F10.2)
      XCALC=(S(1)*XS(1,1)+R(N)*XR(N,1)-F*Z(1))/S(N+1)
      IF(LOOP.GT.30)WRITE(6,700)XCALC,XS(N+1,1)
  700 FORMAT('  XCALC AND XS(N+1,1)    ',2F10.3)
      IF(ABS(XCALC-XS(N+1,1)).LT..01)GO TO 90
      LOOP=LOOP+1
      IF(LOOP.GT.50) THEN
      WRITE(6,*)' STAGE N LOOP'
      STOP
      ENDIF
      IF(XCALC.LT.0.) XCALC=0.
C REGUESS XS(N+1,3)
      XS(N+1,3)=XS(N+1,3)-0.5*(XCALC-XS(N+1,1))
      IF(XS(N+1,3).GT.98.9)XS(N+1,3)=98.9
      GO TO 50
   90 CONTINUE
      XS(N,2)=100.-XS(N,1)-XS(N,3)
      XR(N,2)=100.-XR(N,1)-XR(N,3)
      WRITE(6,*)' N     S      XS1      XS2      XS3'
      WRITE(6,202) N,S(N),(XS(N,J),J=1,3)
      WRITE(6,*)'       R      XR1      XR2      XR3'
      WRITE(6,203) R(N),(XR(N,J),J=1,3)
      IF(XR(N+1,1).LT.XRN1)GO TO 400
  202 FORMAT(1X,I2,3X,8F6.1)
  203 FORMAT(6X,8F6.1)
  100 CONTINUE
      WRITE(6,*)' TOO MANY TRAYS'
      STOP
  400 NT=N+1
      XS(NT,2)=100.-XS(NT,1)-XS(NT,3)
      XR(NT,2)=100.-XR(NT,1)-XR(NT,3)
      WRITE(6,*)' N     S      XS1      XS2      XS3'
      WRITE(6,202)NT,S(NT),(XS(NT,J),J=1,3)
      R(NT)=F+S0-S(1)
      WRITE(6,*)'       R      XR1      XR2      XR3'
      WRITE(6,203)R(NT),(XR(NT,J),J=1,3)
  450 XRN1=XRN1-1.
  500 S0=S0-10.
      STOP
      END
C
      SUBROUTINE LLE(XX,YY1,YY2,YY3)
C XX IS XS(3), Y1 IS XS(1), Y2 IS XR(3), Y3 IS XR(1)
      DIMENSION X(14),Y1(14),Y2(14),Y3(14)
      DATA X/99.,93.12,87.54,79.95,75.34,69.98,67.35,64.2,59.64,55.75,
     + 52.32,48.78,46.8,44./
      DATA Y1/1.,6.87,12.16,19.77,24.14,29.,31.4,34.34,37.4,39.92,42.1,
     + 44.13,44.73,46.50/
```

TABLE 9-1 CON'T.

```
      DATA Y2/0.,0.,.09,.71,1.32,3.2,5.03,7.9,11.52,15.08,17.45,21.84,
     + 22.55,25.24/
      DATA Y3/0.0,0.89,2.27,7.45,14.34,24.9,30.62,37.9,44.23,48.57,
     + 49.95,51.,52.05,51.2/
      DO 10 N=1,13
      IF(XX.LT.X(N))GO TO 10
      YY1=Y1(N-1)+(XX-X(N-1))*(Y1(N)-Y1(N-1))/(X(N)-X(N-1))
      YY2=Y2(N-1)+(XX-X(N-1))*(Y2(N)-Y2(N-1))/(X(N)-X(N-1))
      YY3=Y3(N-1)+(XX-X(N-1))*(Y3(N)-Y3(N-1))/(X(N)-X(N-1))
      RETURN
   10 CONTINUE
      WRITE(6,*)' XX TOO BIG'
      STOP
      END
C
      SUBROUTINE FR(XR1,XR3)
      DIMENSION X(13),Y(13)
      DATA X/0.89,2.27,7.45,14.34,24.9,30.62,37.9,44.23,48.57,
     + 49.95,51.,52.05,51.2/
      DATA Y/0.,.09,.71,1.32,3.2,5.03,7.9,11.52,15.08,17.45,
     + 21.84,22.55,25.24/
      DO 10 N=1,12
      IF(XR1.GT.X(N)) GO TO 10
      XR3=Y(N)+(XR1-X(N))*(Y(N+1)-Y(N))/(X(N+1)-X(N))
      RETURN
   10 CONTINUE
      WRITE(6,*)' FR LOOP'
      STOP
      END
```

of stages in the column. Overall balances are done first, by means of an iterative algorithm, to calculate the flow rate and composition of the solvent phase leaving stage 1 (S_1). A value of $x_3^{S_1}$ is guessed. Subroutine LLE is used to calculate all of the compositions of the two phases in stage 1 that correspond to this guessed composition. Then the overall total mass balance and the overall solute component balance (component 1) are solved simultaneously to get the flow rates S_1 and R_N. A new value of $x_3^{S_1}$ is calculated from the overall solvent component balance (component 3). If the calculated and guessed values are not close enough, a new guess is made by using the difference between the guessed and calculated values (a modified direct-substitution iteration).

After the overall-balance loop converges, a similar procedure is used with balances around stage 1 to get the flow rate and composition of the solvent phase coming from stage 2. These calculations are repeated from stage to stage until the composition of the raffinate phase is reduced to the specified level. Figure 9–21 gives a flow diagram of the logic and the program.

Table 9–2 gives output results from the program for several values of fresh solvent flow rate. The number of stages increases as the fresh solvent flow rate is reduced. The specific system studied is the same as used in example 9–2: isopropyl alcohol–toluene–water.

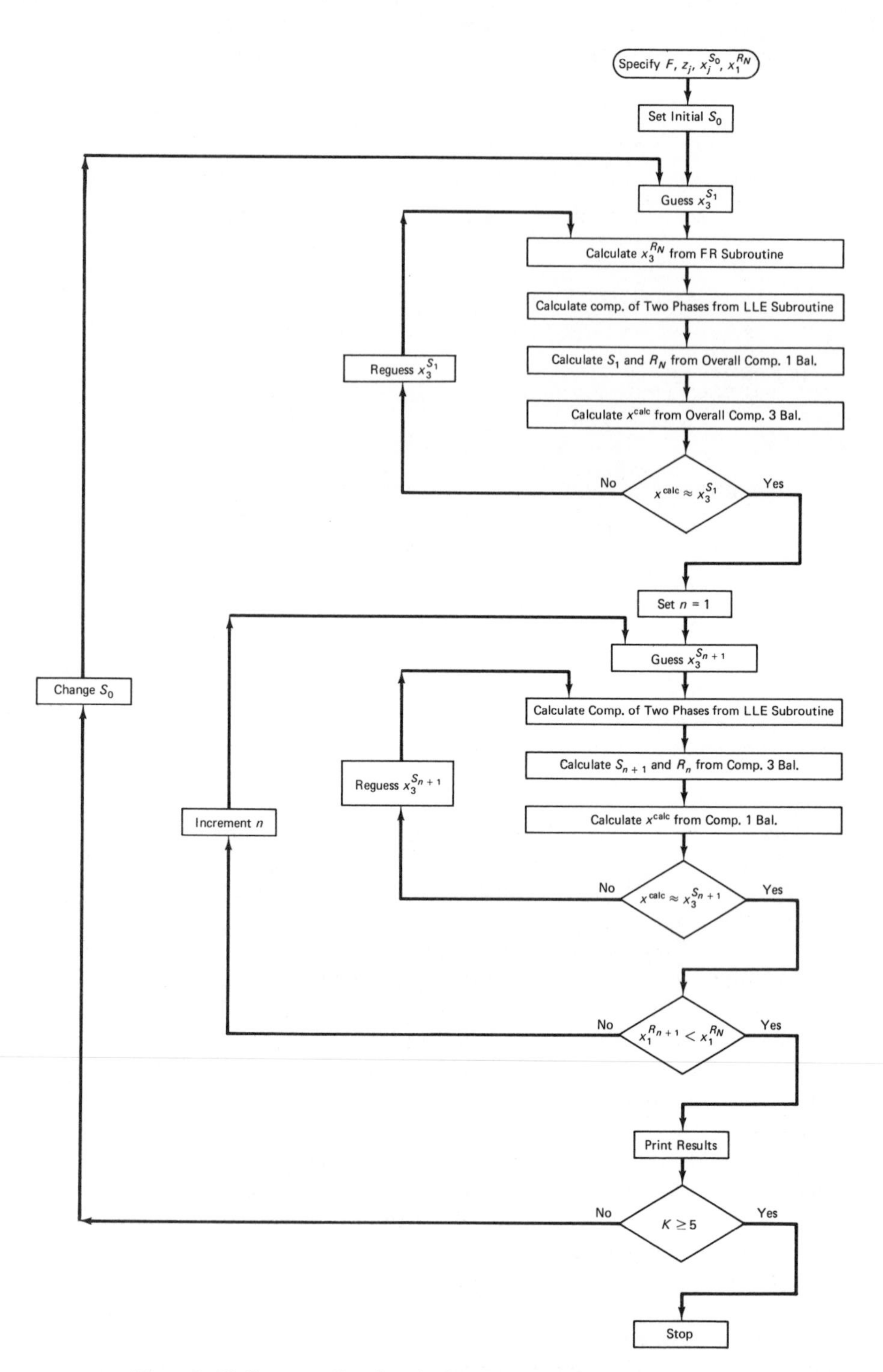

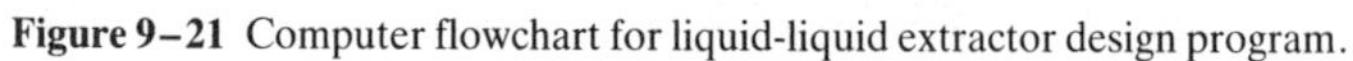

Figure 9–21 Computer flowchart for liquid-liquid extractor design program.

TABLE 9–2
OUTPUT FROM PROGRAM IN TABLE 9–1

```
(A)   FRESH SOLVENT FLOWRATE =    100.00

       F         Z1         Z2         Z3
    100.00     40.00      60.00       .00

    SO        XSO1       XSO2        XSO3
    100.00      .00        .00     100.00

  N      S        XS1        XS2        XS3
  1      139.2   27.4        .9    71.7
         R        XR1        XR2        XR3
          77.9   21.5      75.9     2.6

  N      S        XS1        XS2        XS3
  2      117.1   12.8        .3    86.9
         R        XR1        XR2        XR3
          60.8    2.7      97.2      .1

(B)   FRESH SOLVENT FLOWRATE =     90.00

       F         Z1         Z2         Z3
    100.00     40.00      60.00       .00

    SO        XSO1       XSO2        XSO3
     90.00      .00        .00     100.00

  N      S        XS1        XS2        XS3
  1      129.4   29.5       1.1    69.4
         R        XR1        XR2        XR3
          83.8   26.1      70.3     3.6

  N      S        XS1        XS2        XS3
  2      113.2   17.7        .3    82.0
         R        XR1        XR2        XR3
          62.8    6.1      93.4      .5

  N      S        XS1        XS2        XS3
  3       92.1    2.2        .0    97.8
         R        XR1        XR2        XR3
          60.6     .2      99.8      .0

(C)   FRESH SOLVENT FLOWRATE =     80.00

       F         Z1         Z2         Z3
    100.00     40.00      60.00       .00

    SO        XSO1       XSO2        XSO3
     80.00      .00        .00     100.00

  N      S        XS1        XS2        XS3
  1      119.5   31.9       1.3    66.8
         R        XR1        XR2        XR3
          94.7   32.0      62.5     5.6

  N      S        XS1        XS2        XS3
  2      114.2   24.9        .6    74.5
         R        XR1        XR2        XR3
          71.2   16.0      82.4     1.6
```

TABLE 9-2 CON'T.

```
N     S         XS1        XS2        XS3
3      90.7    10.6         .2    89.2
      R         XR1        XR2        XR3
       60.5       1.9    98.1         .1

(D) FRESH SOLVENT FLOWRATE =      70.00

        F          Z1          Z2          Z3
     100.00     40.00      60.00        .00

      SO        XS01        XS02        XS03
       70.00        .00         .00    100.00

N     S         XS1        XS2        XS3
1     109.9     34.8       1.7    63.6
      R         XR1        XR2        XR3
      113.5     38.8      52.8      8.4

N     S         XS1        XS2        XS3
2     123.4     34.2       1.5    64.3
      R         XR1        XR2        XR3
      109.5     37.6      54.6      7.8

N     S         XS1        XS2        XS3
3     119.3     33.0       1.4    65.6
      R         XR1        XR2        XR3
      100.9     34.6      58.9      6.6

N     S         XS1        XS2        XS3
4     110.7     29.8       1.1    69.1
      R         XR1        XR2        XR3
       84.6     26.9      69.2      3.8

N     S         XS1        XS2        XS3
5      94.4     22.2        .4    77.4
      R         XR1        XR2        XR3
       66.4     11.3      87.7      1.0

N     S         XS1        XS2        XS3
6      76.2      7.4        .0    92.5
      R         XR1        XR2        XR3
       60.1      1.0      99.0       .0
```

PROBLEMS

9–1. A single-stage liquid-liquid extractor operating at 25°C is fed 10,000 lb_m/hr of a 40 weight percent ethylene glycol, 60 weight percent water mixture. Calculate the compositions and flow rates of the raffinate-rich and solvent-rich streams leaving the extractor if the flow rate of pure furfural solvent is

(a) 3,000 lb_m/hr

(b) 6,000 lb_m/hr

Also, calculate the recovery of glycol in the solvent-rich phase. The LLE phase diagram is given on page 347.

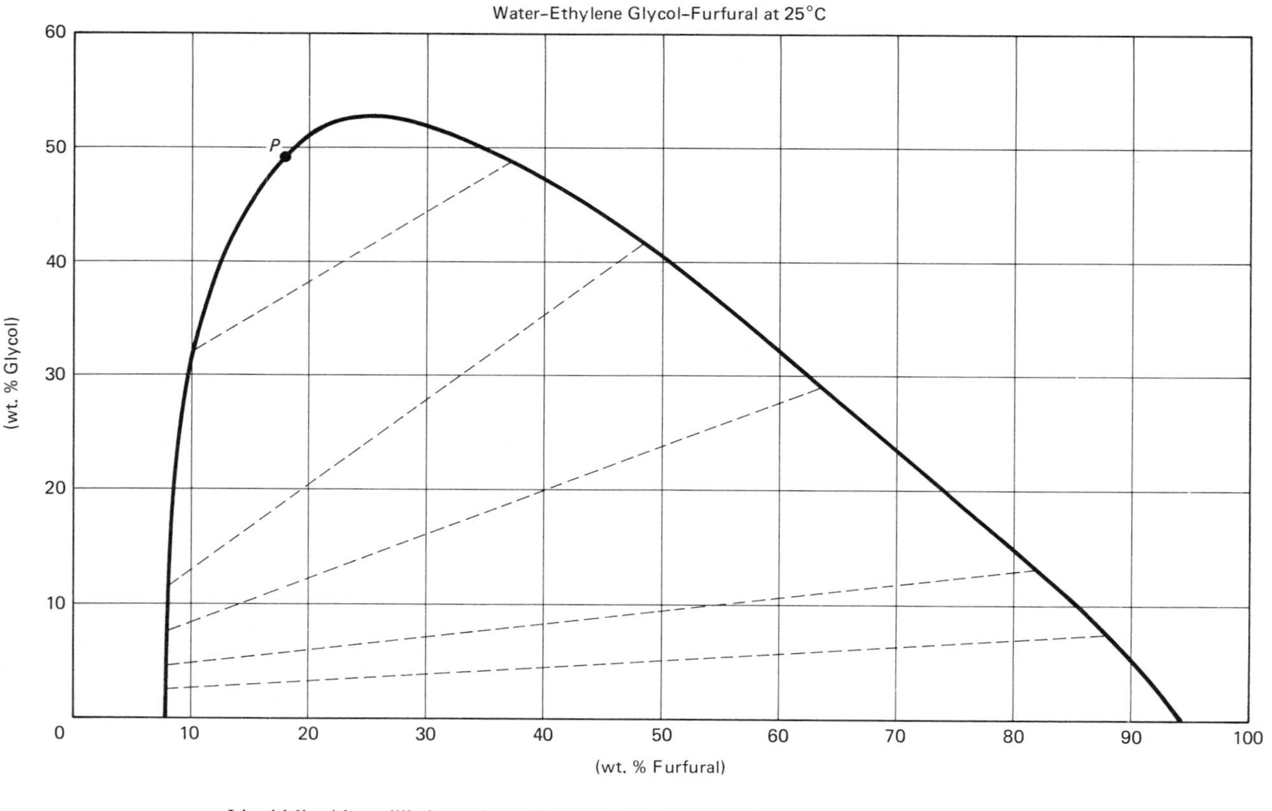

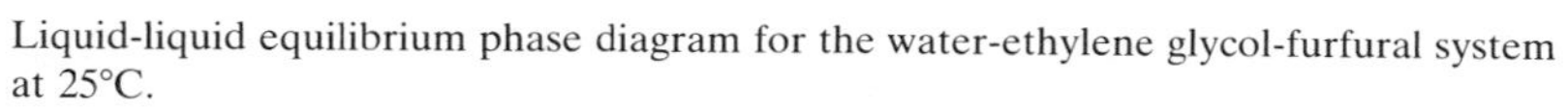
Liquid-liquid equilibrium phase diagram for the water-ethylene glycol-furfural system at 25°C.

9–2. A chemical plant produces 10,000 lb_m/hr of a mixture that contains 60 weight percent toluene and 40 weight percent isopropyl alcohol. It is desired to utilize cross-current extraction at 25°C to remove the alcohol, using pure water as a solvent. A mass of water equal to 80 percent of the weight of the feed stream to each stage is fed to each stage. How many stages will be needed to reduce the alcohol concentration in the raffinate to 5 weight percent? If all of the solvent streams are mixed together, what is the alcohol concentration in water? Equilibrium data are as follows.

Water layer wt. %			*Toluene layer wt. %*		
Alcohol	*Water*	*Toluene*	*Water*	*Alcohol*	*Toluene*
6.87	93.12	0.01	0.0	0.89	99.14
12.16	87.54	0.3	0.09	2.27	97.64
19.77	79.95	0.28	0.71	7.45	91.84
24.14	75.34	0.42	1.32	14.34	84.30
29.00	69.98	1.02	3.2	24.90	71.90
31.40	67.35	1.25	5.03	30.62	64.35
34.34	64.20	1.46	7.9	37.90	54.20
37.40	59.64	2.96	11.52	44.23	44.35
39.92	55.75	4.33	15.08	48.57	36.35
42.10	52.32	5.58	17.45	49.95	32.60
44.13	48.78	7.09	21.84	51.00	27.16
44.73	46.80	8.47	22.55	52.05	26.40
46.50	44.00	9.50	25.24	51.20	23.56

9–3. Calculate the number of theoretical stages required to extract acetone from 5,000 kg/hr of a binary mixture of 35 weight percent acetone and 65 weight percent methyl isobutyl ketone in a countercurrent liquid-liquid extraction column at 25°C. The solvent is water containing 2 weight percent ketone. The concentration of acetone in the raffinate stream leaving the column is to be 2 weight percent. The ratio of the actual solvent rate to the minimum is 1.7.

9–4. Acetone is to be recovered from an acetone–methyl isobutyl ketone mixture in a liquid-liquid extractor using pure water as a solvent. Feed is 17,000 lb_m/hr and 40 weight percent acetone. The concentration of acetone in the raffinate is 5 weight percent. The extractor contains six theoretical stages.

(a) How much solvent must be used?

(b) What is the ratio of actual to minimum solvent flow rate?

(c) What is the percent recovery of acetone?

9–5. We wish to recover 89.6 percent of the glycol fed to a liquid-liquid extraction column operating at 25°C. Feed is 35,000 kg/hr of 30 weight percent ethylene glycol, 70 weight percent water. Solvent is 8,750 kg/hr of pure furfural. What are the compositions of the raffinate and rich solvent leaving the column? How many stages are required?

9–6. A liquid-liquid extraction column, operating at 25°C, is fed 20,000 lb_m/hr of a mixture of 40 weight percent acetone, 55 weight percent methyl isobutyl ketone, and 5 weight percent water. A solvent stream fed to the top of the column contains 95 weight percent water, 2.5 weight percent acetone, and 2.5 weight percent ketone.

The raffinate-rich product stream from the top of the extractor has a composition 89.5 weight percent ketone, 7.5 weight percent acetone, and 3 weight percent water. The solvent-rich product stream from the bottom of the column contains 81 weight percent water, 16 weight percent acetone, and 3 weight percent ketone.

Calculate the solvent flow rate, the number of theoretical stages in the extractor, and the amount of acetone recovered.

9–7. A three-stage countercurrent liquid-liquid extractor is used to remove acetone from an organic feed that contains acetone and methyl isobutyl ketone. An aqueous solvent is used which contains 2 weight percent acetone and 98 weight percent water. The fat solvent stream leaving the bottom of the extractor has a concentration of 15 weight percent acetone. The organic raffinate stream leaving the top of the extractor contains 5 weight percent acetone. A ternary phase equilibrium diagram for acetone-ketone-water is shown on page 350.

(a) What is the feed composition?
(b) What is the ratio of fresh solvent to feed flow rates?
(c) What is the percent recovery of acetone?

9–8. A 50 weight percent glycol, 50 weight percent water mixture is pumped into a two-stage cross-flow liquid-liquid extractor. The solvent is water-free furfural containing 4 weight percent glycol and is fed into each stage at 55 weight percent of the flow rate of the feed stream to that stage. The feed rate is 10,000 lb_m/hr.

(a) What is the weight percent recovery of glycol in each stage? Overall?
(b) What is the composition of the raffinate leaving stage 2?
(c) If the same *total* solvent flow rate is used for a *countercurrent* extraction of the same feed using two stages, what is the weight percent recovery?

9–9. Fish oil is to be extracted from the pulp into trichlorethylene (TCE) using three batch extractions. Originally, TCE and fish pulp containing 30 weight percent oil are mixed together on an equal mass basis. The solids are removed from the TCE-oil mixture by centrifugation, and they contain 30 weight percent liquid phase after centrifugation. On the next two extractions TCE and the wet pulp are mixed together on equal mass bases. After each mix step, separation occurs by centrifugation, and in each case the centrifuged solids contain 30 weight percent liquid phase. If each extraction solubilizes 60 percent of the oils present in the fish pulp,

(a) What is the fraction of the original fish oil recovered in the liquid phases removed from the three centrifugation processes?
(b) What are the compositions and amounts of liquid phase removed from each stage?

9–10. A countercurrent liquid-liquid extractor separates ethylene glycol from water

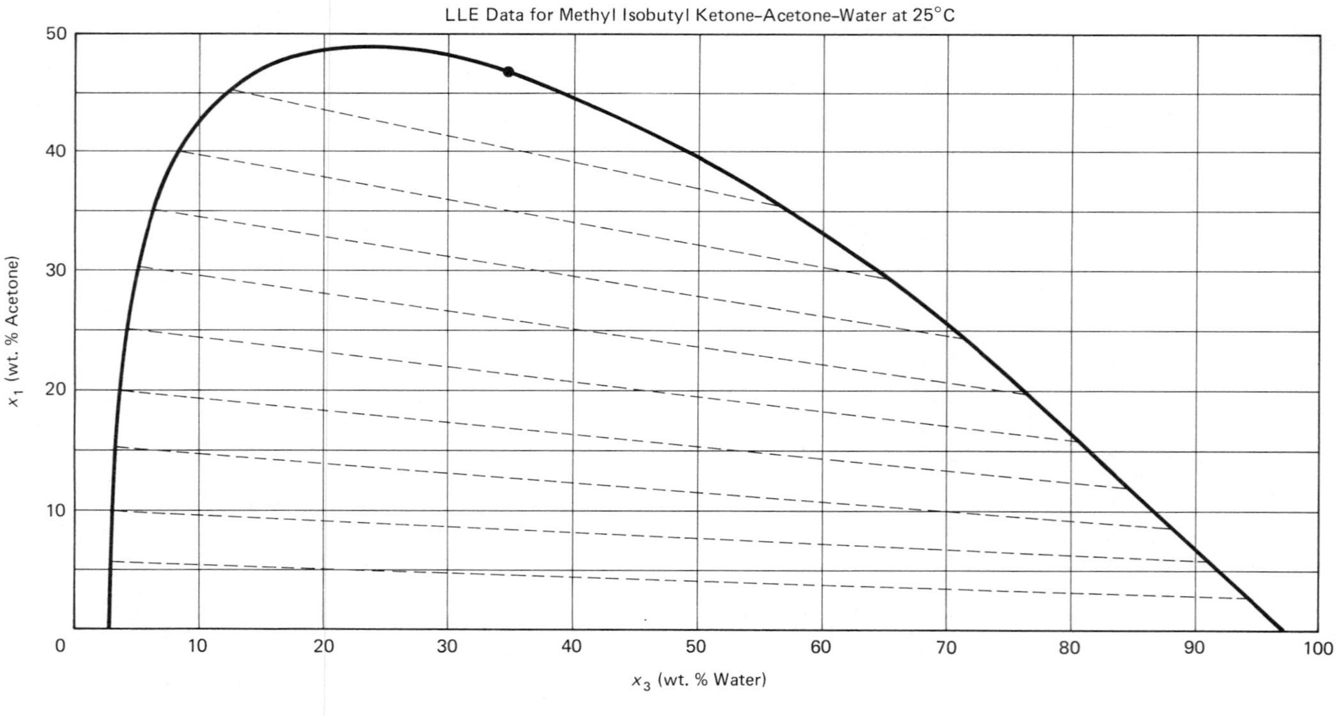

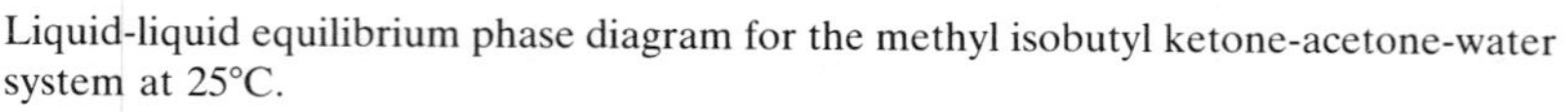

Liquid-liquid equilibrium phase diagram for the methyl isobutyl ketone-acetone-water system at 25°C.

using a pure furfural solvent. The feed is 4,000 kg/hr of a binary mixture of glycol and water (25 weight percent glycol). The fat solvent stream leaving the extractor flows at a rate of 2,050 kg/hr and contains 45 weight percent glycol. The figure in problem 9–1 gives the ternary LLE diagram for the glycol/water/furfural system. Determine

(a) The composition and flow rate of the raffinate stream leaving the extractor.
(b) The flow rate of the pure furfural solvent fed to the extractor.
(c) The number of stages in the extractor.

9–11. A liquid-liquid extraction column removes ethylene glycol from water using a furfural solvent. The fat solvent stream leaving the column contains 48 weight percent glycol and has a flow rate of 8,800 kg/hr.

The raffinate stream leaving the column has a flow rate of 10,500 kg/hr and contains 2 weight percent glycol. The fresh solvent is pure furfural. The feed contains no furfural.

(a) Calculate the feed composition.
(b) Calculate the flow rates of the feed and the fresh solvent.
(c) Calculate the number of ideal stages in the column. The figure in problem 9–1 gives the LLE data for the ethylene glycol/water/furfural system at 25°C.

Energy and Thermal Properties **10**

10.1 FIRST LAW OF THERMODYNAMICS

The First Law of Thermodynamics says that energy is conserved. It is based on centuries of experience in dealing with quantities such as heat, work, kinetic energy, potential energy, and chemical energy.

Any conservation relationship (conservation of mass, energy, money, or whatever) is written as

$$\text{Input} - \text{Output} = \text{Accumulation} \tag{10–1}$$

To use this relationship, we must specify carefully what part of the universe we are considering. That is, we must define a *system*, exactly as we did in dealing with mass balances in chapter 4. This system can be a whole plant or any small part of a plant. Everything outside the system is called the *surroundings*. The separation is a barrier (or envelope) through which energy may flow as either heat or work and through which material may flow, both in and out of the system.

10.1.1 Closed Systems

If the system has a barrier which is impervious to mass, we call it a *closed system*. Note that this barrier is *not* impervious to energy, i.e., work and heat can be transferred across it. For a closed system, the conservation of energy can be expressed as

$$\Delta K + \Delta\Phi + \Delta U = Q_C - W_C \tag{10–2}$$

where

U = internal energy of the material in the system, which is a result of the random motion of its molecules and their chemical nature.

K = kinetic energy of the material in the system, which is the result of the motion (velocity) of the system.

Φ = potential energy of the material in the system, which is the result of its position in a gravity field.

W_C = work done *by* the closed system on the surroundings.

Q_C = heat transferred *into* the closed system from the surroundings.

"Δ" is the standard mathematical notation for the change in any quantity.

The sign convention used in chemical engineering for W_C is different from that used by chemists: work is *positive* when the system is *doing work* on the surroundings. Since we are usually pumping or compressing something in a process, work terms are usually negative, i.e., the surroundings are doing work on the process.

The sign convention on Q_C is the same as the chemists use: heat transfer is *positive* if heat is *added* to the system.

A brief discussion of the units of equation 10–2 is in order. A variety of units are used, depending on the problem and the person solving it. Chemists tend to use units of energy per unit mass (kcal/kg) or total energy (kcal). Engineers tend to use units of energy per unit time (Btu/hr or Joules/sec). Whatever units are used, there is one vitally important factor to remember when using equation 10–2: *Each and every term in the equation must have the same units*. Thus, you cannot express heat transfer (Q_C) in Btu/hr and the change in internal energy in kcal/kg. Similarly, work cannot be expressed in ft-lb_F/sec or in horsepower, and heat transfer in kcal. You must be consistent!

Work and heat are both forms of energy. Heat is energy in transit as a result of a temperature driving force; work is energy in transit as the result of any other driving force. The reason for distinguishing between heat and work is that work is a higher quality of energy than is heat. According to the Second Law of Thermodynamics, work can be converted entirely into heat, but heat cannot be converted entirely into work.

The three forms of energy given on the left side of equation 10–2 are all *state functions*, i.e., those properties of the system that can be defined in terms of measurable system properties (temperature, pressure, composition, etc.) at any point. These state functions are independent of the path that is taken to get to the state. For example, if we specify the temperature and the pressure of some pure water vapor (steam) contained in a vessel, we know all of its other properties—its density, internal energy, etc. Or, if we know the internal energy and the pressure of the steam, we can determine its temperature, density, etc.

If the internal energy at temperature T_1 and pressure P_1 is U_1, and if it is U_2 at T_2 and P_2, the change ΔU in internal energy in moving from (T_1,P_1) to (T_2,P_2) is

$$\Delta U = U_2 - U_1 \tag{10–3}$$

This change is independent of the path that is taken to get from (T_1,P_1) to (T_2,P_2). For example, suppose we have a process in which steam is first compressed at a constant temperature T_1 up to a pressure P_2, and then is heated up to a temperature T_2 at a constant pressure P_2. Each of these steps would involve both heat transfer (Q_C) and work (W_C). Now suppose we have another two-step process in which the steam is first heated to T_2 at constant pressure P_1 and then compressed to P_2 at constant temperature T_2. These steps would involve different amounts of heat transfer and work than were used in the first process. However, the final net change in the internal energy of the steam in the two processes (neglecting kinetic and potential energy changes) would be exactly the same:

$$\Delta U = Q_C - W_C \tag{10–4}$$

In other words, the differences between the Q_C's and the W_C's for the two foregoing processes must be exactly the same, despite the fact that the Q_C's and W_C's are themselves different in the processes. (You will learn much more about state functions in thermodynamics courses.)

As indicated in the preceding example, heat and work are not state functions; their values depend on the path that is taken. Consequently, they are referred to as *path functions*. They are the integrated area under the curve showing property changes from the initial state to the final state. For pressure/volume changes, work can be expressed as the integral of force times distance. Specifically, in going from position s_1 to position s_2 against an opposing force F, the amount of work done is

$$W_C = \int_{s1}^{s2} F\, ds \tag{10–5}$$

Since force is equal to pressure P times area A,

$$W_C = \int_{s1}^{s2} PA\, ds \tag{10–6}$$

But $A\ ds = dV$, or change in volume. So

$$W_C = \int_{V_1}^{V_2} P\, dV \tag{10–7}$$

Heat is also a path function, often shown as

$$Q_C = \int_{T_1}^{T_2} C\,dT \tag{10–8}$$

where C is the heat capacity of the material in the system. Heat capacities are expressed in either mass or molar units—energy per unit mass per degree change in temperature, or energy per mole per degree change in temperature.

The amount of heat transferred is a function of the path taken in going from T_1 to T_2 in the process. The path is usually indicated by the subscript used on the heat capacity. For example, if the path is at constant pressure, C_p would be used; if the path is at constant volume, C_v would be used.

10.1.2 Open Systems

Chemical engineers are usually interested in what happens to a chemical plant or process that has mass flows in and out of the system. Consider, for example, the following diagram:

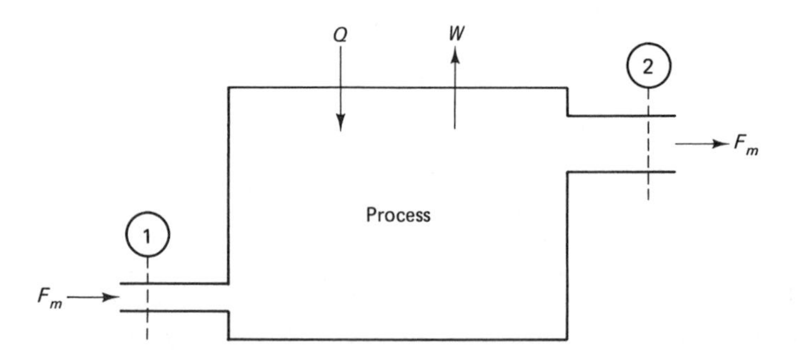

Mass enters the process at point 1 and leaves at point 2. Since we are assuming steady-state operation, the mass flow rate F_m into the system at point 1 must be exactly equal to the mass flow rate out the system at point 2.

Q is the rate of heat transferred from the surroundings into the process. W is the work done by the process on the surroundings. Both have units of energy per unit time. Q is usually given in Btu/hr or Joules/sec. Work *must* have exactly the same units, so remember to convert the work expressed as horse-power or kilowatts into Btu/hr or Joules/sec by using the appropriate conversion factors. (See Appendix A.1.)

Let us apply equation 10–1 to an open system operating at steady state with a flow rate of material into and out of the system equal to F_m. We have

$$\text{IN} = \text{OUT}$$

$$F_m(U_1 + K_1 + \Phi_1 + P_1V_1) + Q = F_m(U_2 + K_2 + \Phi_2 + P_2V_2) + W \tag{10–10}$$

where

F_m = mass flow rate in and out of the system (mass/time)

U_1 = internal energy of the material entering the system in the feed (energy/mass)

U_2 = internal energy of the material leaving the system in the outflowing stream (energy/mass)

K_1 = kinetic energy of the material entering (energy/mass)

K_2 = kinetic energy of the material leaving (energy/mass)

Φ_1 = potential energy of the material entering (energy/mass)

Φ_2 = potential energy of the material leaving (energy/mass)

P_1 = pressure at the point where the feed enters (force/area)

P_2 = pressure at the point of exit of material from the system (force/area)

V_1 = specific volume of the material entering (volume/mass)

V_2 = specific volume of the material leaving (volume/mass)

Q = heat transfer rate into the process (energy/time)

W = work done by the process on the surroundings (energy/time)

The two "*PV*" terms come from the work that must be done as the mass is pushed into and out of the process at points 1 and 2. Since the internal energy *U*, kinetic energy *K*, and potential energy Φ are expressed in units of energy per unit mass, the "*PV*" terms must also have the same units. Thus, if *P* is in lb_F/in^2 and *V* is in ft^3/lb_m, a conversion factor of 778 ft-lb_F/Btu must be used to get Btu/lb_m:

$$(lb_F/in^2)(144\ in^2/ft^2)(ft^3/lb_m)(Btu/778\ ft\text{-}lb_F) = Btu/lb_m$$

The *Q* and *W* terms in equation 10–10, describing an open system, are different than the Q_C and W_C terms used for the closed system, described by equation 10–4. It can be shown that $W = -\int V dP$. You will learn more about these distinctions in your thermodynamics courses.

Since the "*U*" and "*PV*" terms are both energy quantities and state functions, it is convenient to combine them into another useful energy quantity called *enthalpy*, given by

$$H = U + PV \qquad (10\text{–}11)$$

Substituting into equation 10–10 gives

$$F_m\,[(H_2 - H_1) + (K_2 - K_1) + (\Phi_2 - \Phi_1)] = Q - W \qquad (10\text{–}12)$$

In most chemical processes, kinetic energy and potential energy changes are insignificant. Equation 10–12 will then reduce to

$$\boxed{\Delta H = Q - W} \tag{10–13}$$

where

$$\Delta H = F_m(H_2 - H_1) \tag{10–14}$$

ΔH is the difference in enthalpy between the outflowing and the inflowing streams. The "Δ" means "OUT − IN."

If a chemical reaction occurs inside a process, the enthalpy of the outlet fluid system will include the difference in enthalpy resulting from the reaction.

Equation 10–13 is the form of energy balance (the First Law of Thermodynamics) that is the most widely used in chemical engineering. Remember that it includes the important assumptions of steady-state operation and negligible kinetic and potential energy differences between the inflowing and outflowing streams.

For a process with a number N of feed streams entering it at flow rates F_n with enthalpies h_n^F and with a number M of product streams leaving it at flow rates P_m with enthalpies h_m^P, equation 10–13 becomes

$$\sum_{m=1}^{M} P_m h_m^P - \sum_{n=1}^{N} F_n h_n^F = Q - W \tag{10–15}$$

We will work many problems in chapter 11 using equation 10–15.

10.2 HEAT EFFECTS

There are three major types of heat effects:

1. *Sensible heat* is associated with simple temperature changes. No phase changes (liquid to vapor or liquid to solid) and no chemical reactions are considered.
2. *Latent heat* is the heat effect associated with a phase change (heat of vaporization or heat of fusion).
3. *Heats of reaction* involve energy produced or consumed by chemical reactions. This heat is negative for exothermic reactions and positive for endothermic reactions.

Each of these effects changes the enthalpy of the system and is involved in the calculation of enthalpy changes. For gases, pressure also affects enthalpy.

10.2.1 Sensible Heat Effects

Sensible heat is directly related to an increase or decrease in temperature and is path dependent. If the path is at constant pressure, the change in sensible heat is

$$\Delta H = \int_{T_1}^{T_2} C_p dT \tag{10–16}$$

where

C_p = heat capacity at constant pressure (cal/g-mole K)

ΔH = enthalpy change (cal/g-mole)

If the path is at constant volume, C_v = heat capacity at constant volume is used in place of C_p.

For an ideal gas, the relationship between heat capacities is

$$C_p - C_v = R \tag{10–17}$$

where R is the perfect gas constant (1.99 Btu/lb-mole °R or 1.99 cal/g-mole K).

Heat capacities are approximately constant over small temperature ranges, particularly for liquids and solids. In general, however, temperature-dependent heat capacities must be used. Tables 10–1 and 10–2 give parameters of quadratic equations that determine values of C_p for several gases in the ideal state. The equations in Table 10–1 have the form

$$C_p = a + bT + cT^2 \tag{10–18}$$

while those in Table 10–2 have the form

$$C_p = a + bT + \frac{c}{T^2} \tag{10–19}$$

Thus, Table 10–1 uses T^2, while Table 10–2 uses T^{-2}. In both cases temperatures must be in Kelvin. C_p is in units of cal/g-mole K or Btu/lb-mole °R.

The use of temperature-dependent heat capacities is illustrated in the next two examples.

Example 10.1.

Methane gas is to be heated from 300°F to 600°F in a heat exchanger. The flow rate of the gas is 32,000 lb_m/hr. Assuming constant-pressure operation and assuming that methane is an ideal gas, how much heat must be transferred into the gas to raise its temperature?

Solution:

Let us consider the system as the gas-side of the heat exchanger. Since there is no work being done, $W = 0$, and equation 10–13 reduces to

$$\Delta H = Q = F_m (H_2 - H_1) = F_m \int_{T_1}^{T_2} C_p \, dt \tag{10–20}$$

TABLE 10–1
MOLAR HEAT CAPACITIES OF ORGANIC GASES IN THE IDEAL-GAS STATE*

Constants for the Equation $C_p = a + bT + cT^2$ where T is in K and C_p is in Btu/lb-mole °F or cal/g-mole °C
T from 298 to 1,500 K

Compound	*Formula*	*a*	*b × 10³*	*c × 10⁶*
Normal paraffins:				
Methane	CH_4	3.381	18.044	−4.300
Ethane	C_2H_6	2.247	38.201	−11.049
Propane	C_3H_8	2.410	57.195	−17.533
n-Butane	C_4H_{10}	3.844	73.350	−22.655
n-Pentane	C_5H_{12}	4.895	90.133	−28.039
n-Hexane	C_6H_{14}	6.011	106.746	−33.363
n-Heptane	C_7H_{16}	7.094	123.447	−38.719
n-Octane	C_8H_{18}	8.136	140.217	−44.127
Increment per C atom above C_8		1.097	16.667	−5.338
Normal monoolefins (1-alkenes)				
Ethylene	C_2H_4	2.830	28.601	−8.726
Propylene	C_3H_6	3.253	45.116	−13.740
1-Butene	C_4H_8	3.909	62.848	−19.617
1-Pentene	C_5H_{10}	5.347	78.990	−24.733
1-Hexene	C_6H_{12}	6.399	95.752	−30.116
1-Heptene	C_7H_{14}	7.488	112.440	−35.462
1-Octene	C_8H_{16}	8.592	129.076	−40.775
Increment per C atom above C_8		1.097	16.667	−5.338
Miscellaneous materials:				
Acetaldehyde**	C_2H_4O	3.364	35.772	−12.263
Acetylene	C_2H_2	7.331	12.622	−3.889
Benzene	C_6H_6	−0.409	77.621	−26.429
1,3-Butadiene	C_4H_6	5.432	53.224	−17.649
Cyclohexane	C_6H_{12}	−7.701	125.675	−41.584
Ethanol	C_2H_6O	6.990	39.741	−11.926
Methanol	CH_4O	4.394	24.274	−6.855
Toluene	C_7H_8	0.576	93.493	−31.227

*Selected from H. M. Spencer, *Ind. Eng. Chem.* (1948), Vol. 40, p. 2,152.
**298 to 1,000 K.

where

$$F_m = \text{flow rate in g-mole/hr}$$

$$C_p = \text{heat capacity in cal/g-mole K}$$

Table 10–1 gives the parameters for determining C_p for methane:

$$C_p = 3.381 + 18.044 \times 10^{-3}\, T - 4.3 \times 10^{-6}\, T^2 \qquad (10\text{–}21)$$

TABLE 10–2
MOLAR HEAT CAPACITIES OF INORGANIC GASES IN THE IDEAL-GAS STATE

Constants for the Equation $C_p = a + bT + cT^{-2}$ where T is in K and C_p is in Btu/lb-mole °F or cal/g-mole °C

Compound	*Formula*	*Range (K)*	*a*	*b × 10³*	*c × 10⁻⁵*
Ammonia	NH_3	298–1,800	7.11	6.00	−0.37
Bromine	Br_2	298–3,000	8.92	0.12	−0.30
Carbon Monoxide	CO	298–2,500	6.79	0.98	−0.11
Carbon Dioxide	CO_2	298–2,500	10.57	2.10	−2.06
Carbon Disulfide	CS_2	298–1,800	12.45	1.60	−1.80
Chlorine	Cl_2	298–3,000	8.85	0.16	−0.68
Hydrogen	H_2	298–3,000	6.52	0.78	+0.12
Hydrogen Sulfide	H_2S	298–2,300	7.81	2.96	−0.46
Hydrogen Chloride	HCl	298–2,000	6.72	1.24	+0.30
Hydrogen Cyanide	HCN	298–2,500	9.41	2.70	−1.44
Nitrogen	N_2	298–3,000	6.83	0.90	−0.12
Nitrous Oxide	N_2O	298–2,000	10.92	2.06	−2.04
Nitric Oxide	NO	298–2,500	7.03	0.92	−0.14
Nitrogen Dioxide	NO_2	298–2,000	10.07	2.28	−1.67
Nitrogen Tetroxide	N_2O_4	298–1,000	20.05	9.50	−3.56
Oxygen	O_2	298–3,000	7.16	1.00	−0.40
Sulfur Dioxide	SO_2	298–2,000	11.04	1.88	−1.84
Sulfur Trioxide	SO_3	298–1,500	13.90	6.10	−3.22
Water	H_2O	298–2,750	7.30	2.46	0.00

Converting °F into K gives 300°F = 422 K and 600°F = 589 K, and we have

$$Q = F_m \int_{T_1}^{T_2} (a + bT + cT^2)\, dT$$

$$= F_m \left[a(T_2 - T_1) + \frac{b(T_2^2 - T_1^2)}{2} + \frac{c(T_2^3 - T_1^3)}{3} \right]$$

Since

$$F_m = (32{,}000\ \text{lb}_\text{m}/\text{hr})(454\ \text{g/lb}_\text{m})(\text{g-mole}/16\ \text{gram})$$

$$= 908{,}000\ \text{g-mole/hr}$$

it follows that

$$Q = (908{,}000)\left[(3.381)(589 - 422) + \frac{(18.044 \times 10^{-3})(589^2 - 422^2)}{2} - \frac{(4.3 \times 10^{-6})(589^3 - 422^3)}{3}\right]$$

$$= (908{,}000)(564.6 + 153.2 - 185.1) = 1.727 \times 10^9\ \text{cal/hr}$$

$$= (1.727 \times 10^9\ \text{cal/hr})(\text{Btu}/252\ \text{cal}) = 6.86 \times 10^6\ \text{Btu/hr}$$

Sometimes it is faster to use a mean heat capacity instead of doing the integration. Figures 10–1 and 10–2 give values for mean heat capacities of several gases. A mean heat capacity is simply an average heat capacity defined in terms of an integral. Formally,

$$C_p^{\text{mean}} = \frac{\int_{T_1}^{T_2} C_{p(T)}\, dT}{T_2 - T_1} \qquad (10\text{–}22)$$

where C_p^{mean} is the mean heat capacity between T_1 and T_2. Note the difference between C_p and C_p^{mean}: the former is the heat capacity at a given temperature, whereas the latter is an average value of heat capacity that applies over a range of temperatures. Mean heat capacities make the calculations a little easier by avoiding the need to integrate.

The mean heat capacities given in Figures 10–1 and 10–2 are used to find the change in enthalpy in heating any of the gases shown from 77°F (25°C) up to the temperature given on the abscissa. The enthalpy change for any other temperature range T_1 up to T_2 is found by using a two-step approach. First the gas is cooled from T_1 down to 77°F. The enthalpy change generated thereby is called ΔH_1 and is calculated using the mean heat capacity read off the chart at T_1. Then the gas is heated up from 77°F to T_2, and the enthalpy change produced thereby is called ΔH_2 and is calculated using the mean heat capacity read off the chart at T_2. Since enthalpy is a state function, it doesn't matter which path we take to get from T_1 to T_2. Therefore the enthalpy change is the sum of ΔH_1 and ΔH_2.

$$\Delta H = \Delta H_1 + \Delta H_2 \qquad (10\text{–}23)$$

The following example illustrates the procedure.

Example 10.2.

Do the problem given in example 10.1 using mean heat capacities.

Solution:

Figure 10–1 gives mean heat capacity values for methane (CH_4). For cooling from 300°F down to 77°F, the mean heat capacity is 9.2 Btu/lb-mole °F. For heating up from 77°F to 600°F, the mean heat capacity is 10.4 Btu/lb-mole °F. The molar flow rate of the gas is

$$(32{,}000\ \text{lb}_\text{m}/\text{hr})(\text{lb-mole}/16\ \text{lb}_\text{m}) = 2{,}000\ \text{lb-mole/hr}$$

So

$$\Delta H_1 = (2{,}000)(9.2)(77 - 300) = (2{,}000)(-2{,}052)$$

$$\Delta H_2 = (2{,}000)(10.4)(600 - 77) = (2{,}000)(5{,}439)$$

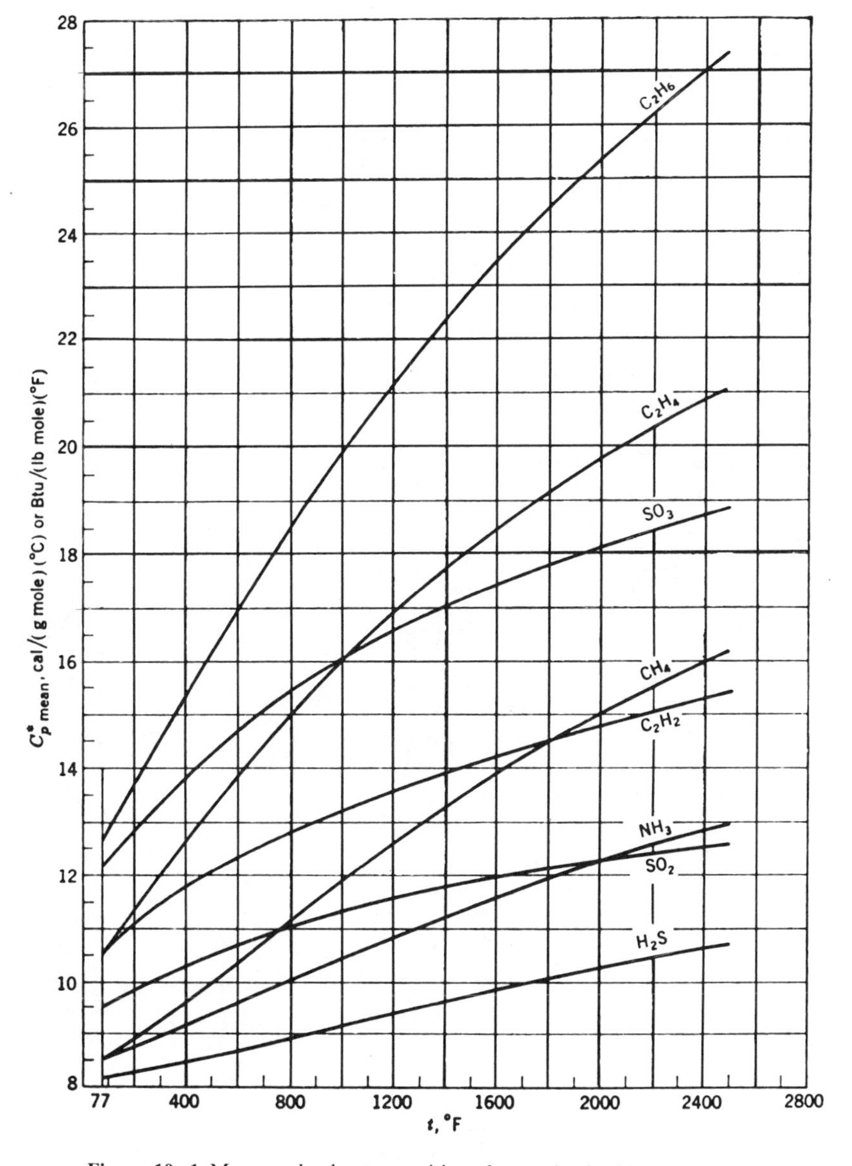

Figure 10–1 Mean molar heat capacities of gases in the ideal state. Base temperature, 77°F. *Based mainly on data from D. D. Wagman (ed.)*, Selected Values of Chemical Thermodynamic Properties, *Natl. Bur. Stand. Circ.* 500, 1952. *From J. M. Smith and H. C. Van Ness,* Introduction to Chemical Engineering Thermodynamics, Second Edition. *Copyright 1959 McGraw-Hill Book Co. Reprinted by permission.*

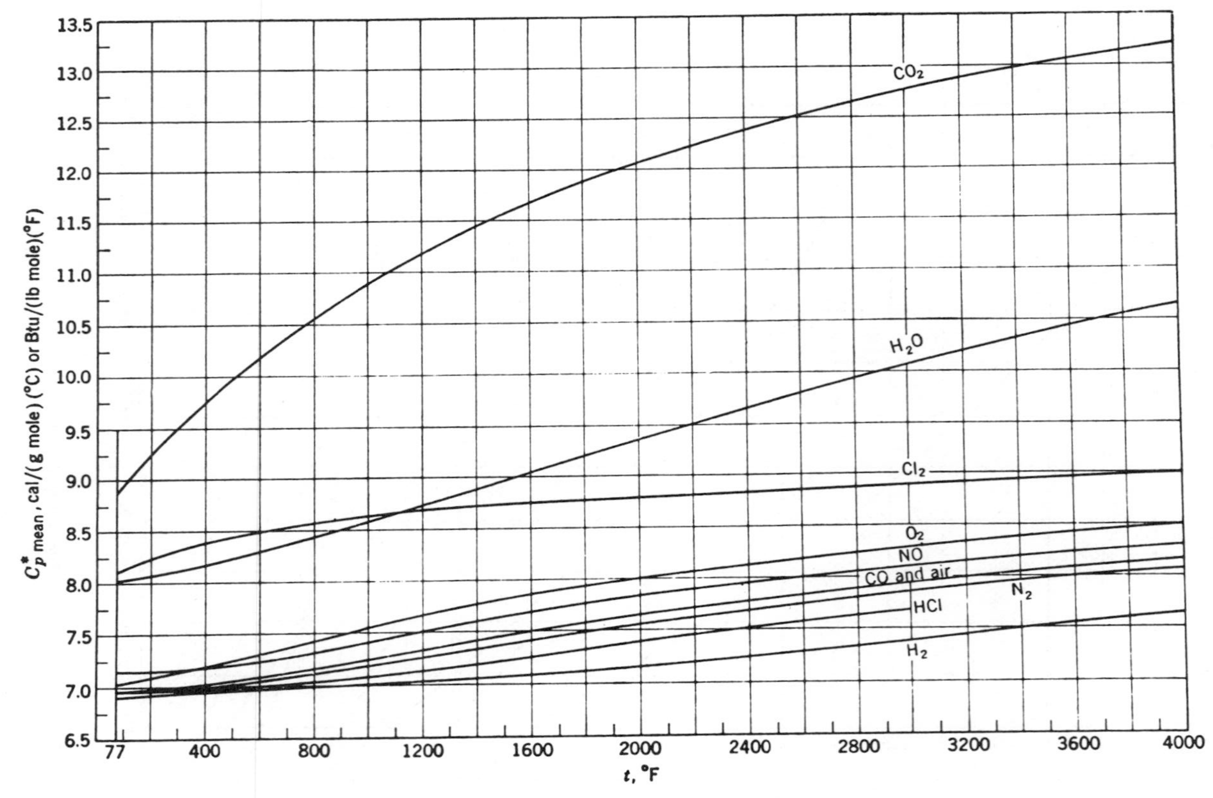

Figure 10–2 Mean molar heat capacities of gases in the ideal-gas state. Base temperature, 77°F. *Based mainly on data from D. D. Wagman (ed.)*, Selected Values of Chemical Thermodynamic Properties, *Natl. Bur. Stand. Circ.* 500, 1952. *From J. M. Smith and H. C. Van Ness,* Introduction to Chemical Engineering Thermodynamics, Second Edition. *Copyright 1959 McGraw-Hill Book Co. Reprinted by permission.*

and

$$\Delta H = \Delta H_1 + \Delta H_2$$
$$= (2{,}000)(-2{,}052 + 5{,}439) = 6.78 \times 10^6 \text{ Btu/hr}$$

Notice that this is approximately the same answer we got in example 10.1. The slight difference is due to the inaccuracy in reading the graphs for the mean heat capacities.

For a mixture of gases, C_p^{mean} is calculated from the mole fractions y_j of each component in the gas and the individual molar heat capacities $(C_p^{\text{mean}})_j$:

$$C_p^{\text{mean}} = \sum_{j=1}^{NC} (C_p^{\text{mean}})_j y_j \tag{10–24}$$

Example 10.3.

A gas mixture containing 10 mole percent CO_2, 6 mole percent H_2O, 15 mole percent O_2, and 69 mole percent N_2 is to be cooled from 3,000°F to 1,200°F. How much heat must be removed from 100 lb-moles of the mixture?

Solution:

Figure 10–2 is used to determine mean heat capacities for the gas species present for the temperature intervals 3,000°F to 77°F and 77°F to 1,200°F:

		Mean heat capacity C_p^{mean}	
Compound	*Mole %*	*3,000–77°F*	*77–1,200°F*
CO_2	10	12.75	11.2
H_2O	6	10.1	8.75
O_2	15	8.3	7.7
N_2	69	7.9	7.25

$$\Delta H = \Delta H_1 + \Delta H_2 = \Delta H_{3,000-77} + \Delta H_{77-1,200}$$

$$\Delta H_1 = (100 \text{ lb-mole})[(0.1)(12.75) + (0.06)(10.1) + (0.15)(8.3) + (0.69)(7.9)][77 - 3{,}000]$$
$$= -2.508 \times 10^6 \text{ Btu}$$

$$\Delta H_2 = (100 \text{ lb-mole})[(0.1)(11.2) + (0.06)(8.75) + (0.15)(7.7) + (0.69)(7.25)][1200 - 77]$$
$$= 0.876 \times 10^6 \text{ Btu}$$

$$\Delta H = -1.632 \times 10^6 \text{ Btu}$$

The mean heat capacity of the gas mixture for the temperature range from 1,200°F

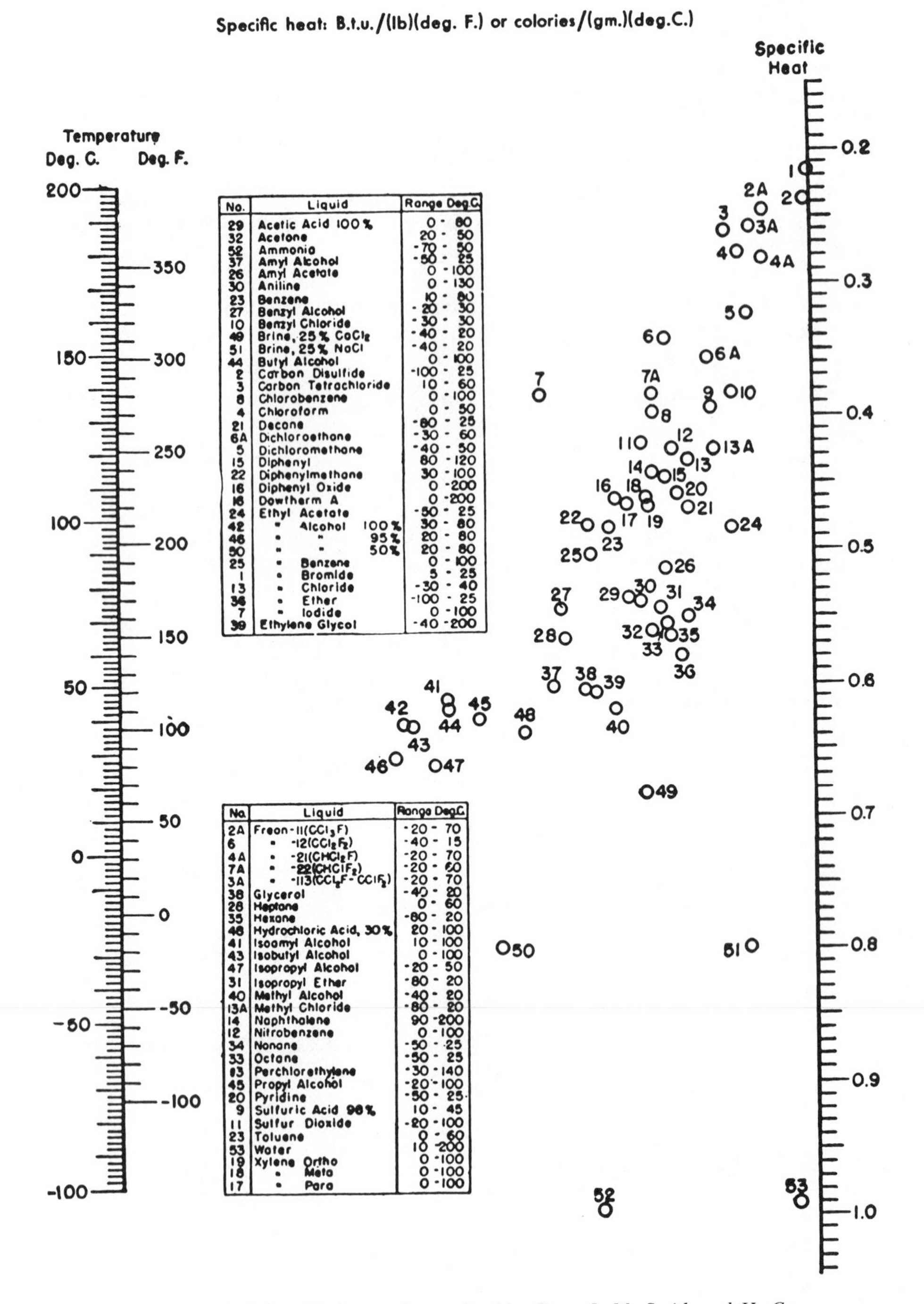

No.	Liquid	Range Deg.C.
29	Acetic Acid 100%	0 - 80
32	Acetone	20 - 50
52	Ammonia	-70 - 50
37	Amyl Alcohol	-50 - 25
26	Amyl Acetate	0 - 100
30	Aniline	0 - 130
23	Benzene	10 - 80
27	Benzyl Alcohol	-20 - 30
10	Benzyl Chloride	-30 - 30
49	Brine, 25% $CaCl_2$	-40 - 20
51	Brine, 25% NaCl	-40 - 20
44	Butyl Alcohol	0 - 100
2	Carbon Disulfide	-100 - 25
3	Carbon Tetrachloride	10 - 60
8	Chlorobenzene	0 - 100
4	Chloroform	0 - 50
21	Decane	-80 - 25
6A	Dichloroethane	-30 - 60
5	Dichloromethane	-40 - 50
15	Diphenyl	80 - 120
22	Diphenylmethane	30 - 100
16	Diphenyl Oxide	0 - 200
16	Dowtherm A	0 - 200
24	Ethyl Acetate	-50 - 25
42	" Alcohol 100%	30 - 80
46	" " 95%	20 - 80
50	" " 50%	20 - 80
25	" Benzene	0 - 100
1	" Bromide	5 - 25
13	" Chloride	-30 - 40
36	" Ether	-100 - 25
7	" Iodide	0 - 100
39	Ethylene Glycol	-40 - 200

No.	Liquid	Range Deg.C.
2A	Freon-11(CCl_3F)	-20 - 70
6	" -12(CCl_2F_2)	-40 - 15
4A	" -21($CHCl_2F$)	-20 - 70
7A	" -22($CHClF_2$)	-20 - 60
3A	" -113(CCl_2F-$CClF_2$)	-20 - 70
38	Glycerol	-40 - 20
28	Heptane	0 - 60
35	Hexane	-80 - 20
48	Hydrochloric Acid, 30%	20 - 100
41	Isoamyl Alcohol	10 - 100
43	Isobutyl Alcohol	0 - 100
47	Isopropyl Alcohol	-20 - 50
31	Isopropyl Ether	-80 - 20
40	Methyl Alcohol	-40 - 20
13A	Methyl Chloride	-80 - 20
14	Naphthalene	90 - 200
12	Nitrobenzene	0 - 100
34	Nonane	-50 - 25
33	Octane	-50 - 25
3	Perchlorethylene	-30 - 140
45	Propyl Alcohol	-20 - 100
20	Pyridine	-50 - 25
9	Sulfuric Acid 98%	10 - 45
11	Sulfur Dioxide	-20 - 100
23	Toluene	0 - 60
53	Water	10 - 200
19	Xylene Ortho	0 - 100
18	" Meta	0 - 100
17	" Para	0 - 100

Figure 10–3 Specific heats of pure liquids. *From J. M. Smith and H. C. Van Ness,* Introduction to Chemical Engineering Thermodynamics, Second Edition. *Copyright 1959 McGraw-Hill Book Co. Reprinted by permission.*

to 3,000°F is thus

$$\frac{(1{,}632{,}000\ \text{Btu})}{[(100\ \text{lb-mole})(1{,}800°\text{F})]} = 9.07\ \text{Btu/lb-mole °F}$$

Heat capacities of liquids and solids are usually obtained directly from measurement. Figure 10–3 is a nomograph from which heat capacities of liquids can be obtained. A straightedge pivoted at the point given for the liquid of interest connects temperature on the left to a specific heat on the right. The specific heat of a material is its heat capacity divided by the heat capacity of some reference material. Usually this reference material is water, which has a heat capacity of 1 (cal/g °C or Btu/lb_m °F).

Table 10–3 lists some values of heat capacities of various solids. Although only approximate, the values may be used over the indicated temperature ranges.

10.2.2 Latent Heat

Latent heat is associated with phase changes. Converting liquid into a vapor is a phase change that consumes energy; converting vapor into a liquid gives off energy. Figure 10–4 gives latent heats of vaporization for some compounds. Latent heats of vaporization for water can be found in Appendix A.2 in the Steam Tables.

The Clausius-Clapyron equation

$$\frac{dP}{dT} = \frac{\Delta H_v}{T\,\Delta V} \tag{10–25}$$

can also be used to estimate latent heats. In this equation, dP/dT is the slope of the phase boundary (for vaporization it is the slope of the vapor pressure curve). ΔH_v is the latent heat, and ΔV is the change in volume accompanying the phase change. (For vaporization, $\Delta V = V_V - V_L$.) For a liquid-vapor phase change at low pressure, the Perfect Gas Law

$$V_V = \frac{RT}{P}$$

can be used for the vapor. Neglecting the volume of the liquid (i.e., setting $V_L = 0$), equation 10–25 gives

$$\frac{dP}{dT} = \frac{\Delta H_v}{RT^2/P} \tag{10–26}$$

$$\frac{dP}{P} = \left(\frac{\Delta H_v}{R}\right) dT/T^2 \tag{10–27}$$

$$\ln\left(\frac{P_2}{P_1}\right) = \left(\frac{\Delta H_v}{R}\right)\left(\frac{1}{T_1} - \frac{1}{T_2}\right) \tag{10–28}$$

TABLE 10–3
SPECIFIC HEATS OF MISCELLANEOUS MATERIALS*

Material	*Specific heats, cal/g°C*
Alumina	0.2 (100°C); 0.274 (1,500°C)
Aluminum	0.099 (0–100°C)
Asbestos	0.25
Asphalt	0.22
Bakelite	0.3 to 0.4
Brickwork	0.2
Carbon	0.168 (26° to 76°C)
	0.314 (40° to 892°C)
	0.387 (56° to 1,450°C)
Carbon steel	0.107 (0–100°C)
Cellulose	0.32
Cement	0.186
Charcoal (wood)	0.242
Clay	0.224
Coal	0.26 to 0.37
Copper alloys	0.090
Diamond	0.147
Gasoline	0.53
Glass	0.16 to 0.20
Graphite	0.165 (26°–76°C); 0.390 (56°–1,450°C)
Inconel	0.109
Kerosene	0.47
Limestone	0.217
Magnesia	0.234 (100°C); 0.188 (1,500°C)
Nickel (pure)	0.11
Porcelain	0.186 (100°C); 0.324 (1,100°C)
Pyrites (iron)	0.136 (15° to 98°C)
Rubber (vulcanized)	0.415
Sand	0.191
Silk	0.33
Silver and gold	0.056
Stainless steel	0.12
Turpentine	0.42 (18°C)
Wood	0.45 to 0.65
Wool	0.325
Zinc	0.034

*Data selected from *Chemical Engineering Handbook, 4th Ed.*, R. H. Perry, C. H. Chilton, S. D. Kirkpatrick, Eds., McGraw-Hill Book Co., (1963) by permission.

The numbers 1 and 2 refer to two points on the phase boundary. Equation 10–28 can be used to estimate the latent heat of vaporization from vapor pressure data. It is satisfactory only at low pressures, however, because of the two assumptions made in deriving it.

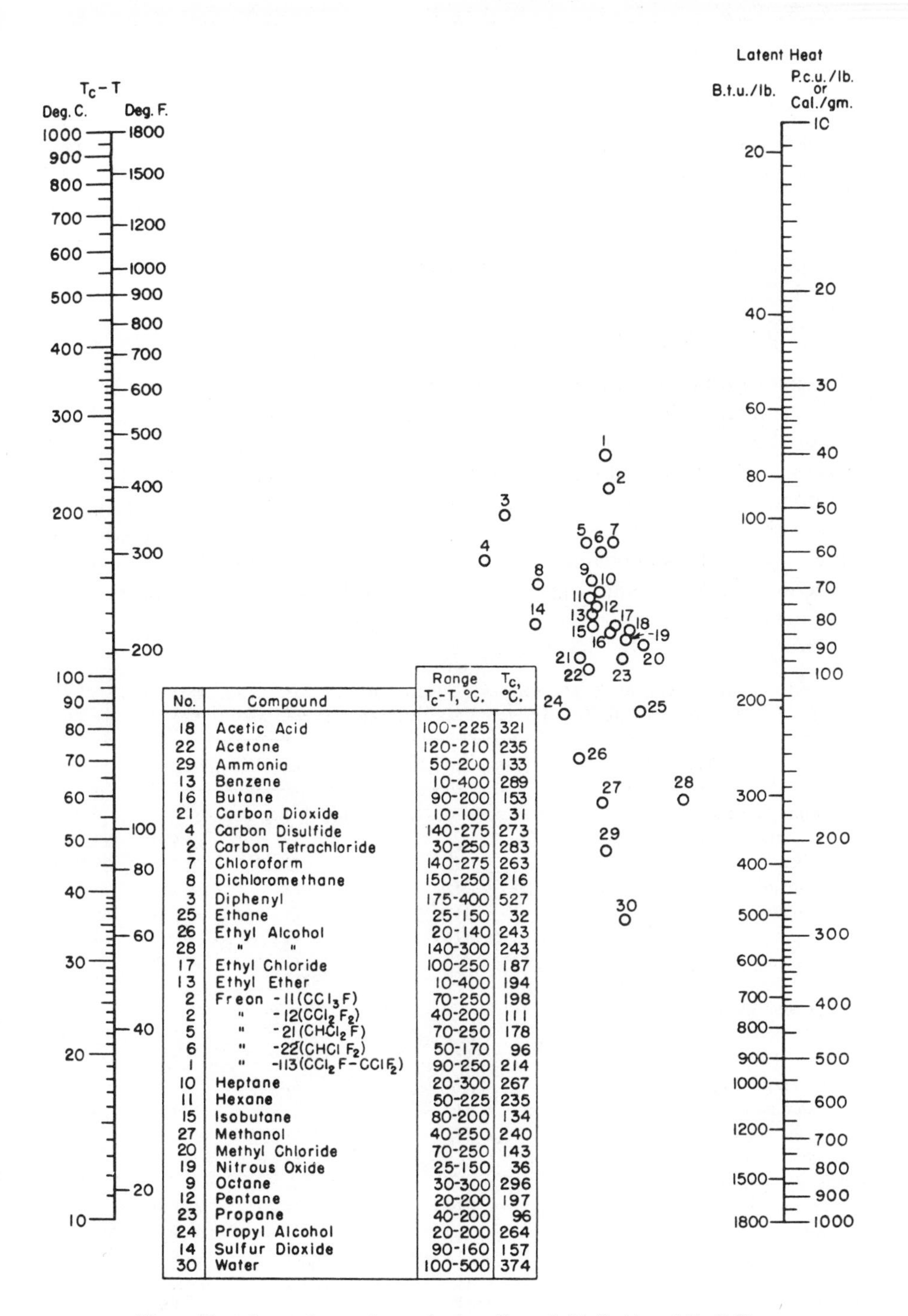

No.	Compound	Range T_c-T, °C.	T_c, °C.
18	Acetic Acid	100-225	321
22	Acetone	120-210	235
29	Ammonia	50-200	133
13	Benzene	10-400	289
16	Butane	90-200	153
21	Carbon Dioxide	10-100	31
4	Carbon Disulfide	140-275	273
2	Carbon Tetrachloride	30-250	283
7	Chloroform	140-275	263
8	Dichloromethane	150-250	216
3	Diphenyl	175-400	527
25	Ethane	25-150	32
26	Ethyl Alcohol	20-140	243
28	" "	140-300	243
17	Ethyl Chloride	100-250	187
13	Ethyl Ether	10-400	194
2	Freon -11(CCl_3F)	70-250	198
2	" -12(CCl_2F_2)	40-200	111
5	" -21($CHCl_2F$)	70-250	178
6	" -22($CHClF_2$)	50-170	96
1	" -113($CCl_2F-CClF_2$)	90-250	214
10	Heptane	20-300	267
11	Hexane	50-225	235
15	Isobutane	80-200	134
27	Methanol	40-250	240
20	Methyl Chloride	70-250	143
19	Nitrous Oxide	25-150	36
9	Octane	30-300	296
12	Pentane	20-200	197
23	Propane	40-200	96
24	Propyl Alcohol	20-200	264
14	Sulfur Dioxide	90-160	157
30	Water	100-500	374

Figure 10–4 Latent heats of vaporization. *From J. M. Smith and H. C. Van Ness,* Introduction to Chemical Engineering Thermodynamics, Second Edition. *Copyright 1959 McGraw-Hill Book Co. Reprinted by permission.*

10.2.3 Heat of Reaction

Energy is released or consumed whenever a chemical reaction occurs. This energy change is the heat of reaction and is given by

$$\Delta H_R = H_{\text{Products}} - H_{\text{Reactants}}$$

Heats of reactions can be calculated from the information tabulated in Table 10–4. The *heat of formation* of a chemical compound is defined as

$$\Delta H_F = H_{\text{compound}} - \Sigma H_{\text{elements forming the compound}} \qquad (10\text{–}29)$$

Heats of formation of pure elements (e.g., H_2, N_2, O_2, and C) are zero.

The heats of formation given in Table 10–4 are for a temperature of 25°C. Thus, heats of reaction can be calculated at 25°C. However, few reactions occur at this temperature. To calculate the heat of reaction at a different temperature, the sensible heat effects of heating the reactants and products up to the reaction temperature must be accounted for. We will do some of these calculations in chapter 11. For now, we calculate some simple heats of reaction at 25°C. In many highly exothermic reactions, the sensible heat effects are negligible.

TABLE 10–4
STANDARD HEATS OF FORMATION AND COMBUSTION AT 25°C IN CAL/G-MOLE. [FOR COMBUSTION REACTIONS THE PRODUCTS ARE H_2O(L) AND CO_2(g).]

Substance	*Formula*	*State*	ΔH_F	$-\Delta H_C$
Normal paraffins:				
Methane	CH_4	G	−17,889	212,800
Ethane	C_2H_6	G	−20,236	372,820
Propane	C_3H_8	G	−24,820	530,600
n-Butane	C_4H_{10}	G	−30,150	687,640
n-Pentane	C_5H_{12}	G	−35,000	845,160
n-Hexane	C_6H_{14}	G	−39,960	1,002,570
Increment per C atom above C_6		G	−4,925	157,440
Normal monoolefins (1-alkenes):				
Ethylene	C_2H_4	G	12,496	337,150
Propylene	C_3H_6	G	4,879	491,990
1-Butene	C_4H_8	G	−30	649,380
1-Pentene	C_5H_{10}	G	−5,000	806,700
1-Hexene	C_6H_{12}	G	−9,960	964,240
Increment per C atom above C_6		G	−4,925	157,440
Miscellaneous organic compounds:				
Acetaldehyde	C_2H_4O	G	−39,760	
Acetic acid	$C_2H_4O_2$	L	−116,400	
Acetylene	C_2H_2	G	54,194	310,620
Benzene	C_6H_6	G	19,820	789,080
Benzene	C_6H_6	L	11,720	780,980
1,3-Butadiene	C_4H_6	G	26,330	607,490
Cyclohexane	C_6H_{12}	G	−29,430	944,770
Cyclohexane	C_6H_{12}	L	−37,340	936,860

TABLE 10–4 (CON'T.)

Ethanol	C_2H_6O	G	−56,030	
Ethanol	C_2H_6O	L	−66,200	
Ethylbenzene	C_8H_{10}	G	7,120	1,101,130
Ethylene glycol	$C_2H_6O_2$	L	−108,580	
Ethylene oxide	C_2H_4O	G	−12,190	
Methanol	CH_4O	G	−48,050	
Methanol	CH_4O	L	−57,110	
Methylcyclohexane	C_7H_{14}	G	−36,990	1,099,580
Methylcyclohexane	C_7H_{14}	L	−45,450	1,091,130
Styrene	C_8H_8	G	35,220	1,060,900
Toluene	C_7H_8	G	11,950	943,580
Toluene	C_7H_8	L	2,870	934,500
Miscellaneous inorganic compounds:				
Ammonia	NH_3	G	−11,040	
Calcium carbide	CaC_2	S	−15,000	
Calcium carbonate	$CaCO_3$	S	−288,450	
Calcium chloride	$CaCl_2$	S	−190,000	
Calcium chloride	$CaCl_2 \cdot 6H_2O$	S	−623,150	
Calcium hydroxide	$Ca(OH)_2$	S	−235,800	
Calcium oxide	CaO	S	−151,900	
Carbon	C	*	....	94,051
Carbon dioxide	CO_2	G	−94,051	
Carbon monoxide	CO	G	−26,416	67,636
Hydrochloric acid	HCl	G	−22,064	
Hydrogen	H_2	G	....	68,317
Hydrogen sulfide	H_2S	G	−4,815	
Iron oxide	FeO	S	−64,300	
Iron oxide	Fe_3O_4	S	−267,000	
Iron oxide	Fe_2O_3	S	−196,500	
Iron sulfide	FeS_2	S	−42,520	
Lithium chloride	LiCl	S	−97,700	
Lithium chloride	$LiCl \cdot H_2O$	S	−170,310	
Lithium chloride	$LiCl \cdot 2H_2O$	S	−242,100	
Lithium chloride	$LiCl \cdot 3H_2O$	S	−313,500	
Nitric acid	HNO_3	L	−41,404	
Nitrogen oxides	NO	G	21,570	
	NO_2	G	7,930	
	N_2O	G	19,513	
	N_2O_4	G	2,190	
Sodium carbonate	Na_2CO_3	S	−270,300	
Sodium carbonate	$Na_2CO_3 \cdot 10H_2O$	S	−975,600	
Sodium chloride	NaCl	S	−98,232	
Sodium hydroxide	NaOH	S	−101,990	
Sulfur dioxide	SO_2	G	−70,960	
Sulfur trioxide	SO_3	G	−94,450	
Sulfur trioxide	SO_3	L	−104,800	
Sulfuric acid	H_2SO_4	L	−193,910	
Water	H_2O	G	−57,798	
Water	H_2O	L	−68,317	

*Graphite

Example 10.4.

Calculate the heat of reaction at 25°C for the formation of ammonia gas from hydrogen and nitrogen gases. The reaction equation is

$$N_{2(g)} + 3H_{2(g)} \longrightarrow 2NH_{3(g)}$$

Solution:

The heat of reaction is equal to twice the heat of formation of ammonia (because of the stoichiometric coefficient 2) minus three times the heat of formation of hydrogen minus the heat of formation of nitrogen. That is,

$$\Delta H_R = 2(\Delta H_F \text{ of } NH_3) - 3(\Delta H_F \text{ of } H_2) - (\Delta H_F \text{ of } N_2) \qquad (10\text{–}30)$$

Since the heats of formation of elements are zero, the heat of this reaction is just equal to $2(\Delta H_F$ of $NH_3)$. From Table 10–4, then, we have

$$\Delta H_R = 2(-11{,}040) \text{ cal/g-mole}$$

$$= -22{,}080 \text{ calories for two g-mole of ammonia}$$

Clearly, this reaction is exothermic (the heat of reaction is negative), so heat is given off. Thus, the reaction of one g-mole of gaseous nitrogen and three g-mole of gaseous hydrogen to produce two g-mole of ammonia gas will generate 22,080 calories of heat at 25°C.

Example 10.5.

Calculate the heat of reaction for the formation of methanol gas from carbon dioxide and hydrogen gases at 25°C.

Solution:

The balanced reaction is

$$CO_{(g)} + 2H_{2(g)} \longrightarrow CH_3OH_{(g)}$$

The heat of reaction is

$$\Delta H_R = (\Delta H_F)_{CH_3OH} - (\Delta H_F)_{CO} - 2\,(\Delta H_F)_{H_2}$$

$$= -48{,}050 - (-26{,}416) - 2(0) \qquad (10\text{–}31)$$

$$= -21{,}634 \text{ cal/g-mole of methanol produced}$$

Table 10–4 also gives the heat of combustion of several compounds. The heat of combustion, which is related to the heat of formation, is the enthalpy change for the reaction of the specified compound with oxygen to form $CO_{2(g)}$, $H_2O_{(L)}$, $NO_{2(g)}$, and $SO_{2(g)}$ at 25°C. For example, the heat of combustion of liquid benzene is the enthalpy change for the reaction

$$C_6H_{6(L)} + \left(\frac{15}{2}\right) O_{2(g)} \longrightarrow 6CO_{2(g)} + 3H_2O_{(L)} \qquad (10\text{–}32)$$

Table 10–4 lists the heat of combustion of benzene as −780,980 cal/g-mole of benzene. Note that the rightmost column of the table gives the negative of the heat of combustion.

10.3 ENTHALPY DATA

In order to be able to do energy balances, we need to be able to calculate the enthalpies of the various streams entering and leaving a process. These enthalpies are strong functions of temperature and composition. Pressure affects the enthalpy of gases somewhat, but has almost no effect on liquid and solid enthalpies at normal processing pressures.

Enthalpy data are given in various forms in the literature. Sometimes we use just heat capacity and latent heat of vaporization data to calculate enthalpy. Sometimes the data are given in graphical form (enthalpy versus composition). This is often the case in binary systems, since temperature, pressure, and only one composition specify the enthalpy. We will use graphs called *Hxy* diagrams extensively in the next three chapters.

In using enthalpy, we always deal with enthalpy *differences*. Therefore, a convenient reference state (typically, saturated liquid at 0°C and 1 atmosphere) is assigned a value of zero enthalpy.

10.3.1 Pure Components

Figure 10–5 sketches a typical enthalpy diagram for a single pure component. The enthalpy of saturated liquid and the enthalpy of saturated vapor are plotted as functions of temperature. Enthalpies are usually given in energy per mass units (e.g., Btu/lb_m or kcal/kg), but sometimes molar units are used (e.g., kg/kg-mole).

Both vapor and liquid enthalpies increase with temperature. Pressure increases as well, because the vapor pressure of a pure component increases with temperature. At any temperature, the difference between the saturated

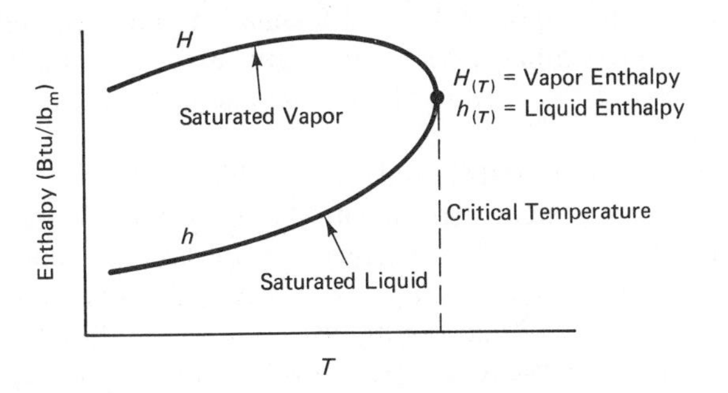

Figure 10–5 Enthalpies of vapor and liquid for a pure component.

vapor and liquid enthalpy lines is the heat of vaporization at that temperature. At the critical temperature, where two phases cease to exist, the two curves come together and the heat of vaporization is zero.

We will use the variable h for liquid enthalpy and the variable H for vapor enthalpy. These pure component enthalpy curves can often be accurately approximated by straight lines over a limited range of temperatures. The straight-line equations are

$$H = C_V (T - T_0) + \Delta H_v \quad (10\text{–}33)$$

and

$$h = C_L (T - T_0) \quad (10\text{–}34)$$

where

H = enthalpy of saturated vapor (Btu/lb$_m$ or cal/g)

h = enthalpy of saturated liquid (Btu/lb$_m$ or cal/g)

C_V = heat capacity of the vapor (Btu/lb$_m$ °F or cal/g °C)

C_L = heat capacity of the liquid (Btu/lb$_m$ °F or cal/g °C)

ΔH_v = heat of vaporization at reference temperature T_0 (Btu/lb$_m$ or cal/g)

T = temperature (°F or °C)

T_0 = reference temperature (°F or °C)

In addition to enthalpy-versus-temperature diagrams, there are several other types of plots that are used to give thermodynamic data. Pressure-versus-enthalpy (P-H) plots and entropy-versus-enthalpy (S-H) plots are commonly used. You will learn more about these in courses in thermodynamics.

10.3.2 Steam Tables

One of the most important components in process and power systems is water. Steam tables are a detailed compilation of the thermodynamic properties of water in its vapor and liquid forms. The basis (or reference state) for enthalpy is zero enthalpy for liquid water at 32°F and 0.08854 psia (the triple-point pressure). A portion of the steam tables is given in Appendix A.2 in English units.

If water is saturated (at its bubblepoint, which is the same as its dewpoint since it is a pure component), specification of only one variable (temperature or pressure) fixes all physical properties of both the liquid and vapor phases, e.g., enthalpy, density, and entropy. Table A.2–1 gives temperature, pressure, liquid enthalpy, and vapor enthalpy data for saturated steam.

If the steam is superheated vapor, two variables (typically, temperature and pressure) must be specified to define the properties. Table A.2–2 gives the enthalpy of superheated steam vapor as a function of temperature and pressure. Interpolation between listed data is usually done linearly.

Example 10.6.

Superheated steam at 75 psia and 350°F flows at a rate of 50,000 lb_m/hr through a valve into the shell side of a reboiler on a distillation column. The pressure in the shell is 20 psia. The steam condensate (liquid water) leaves as saturated liquid through a steam trap. How much heat is transferred in the reboiler?

Solution:

We choose as our system the shell side of the reboiler. Assuming steady-state operation and negligible kinetic and potential energy effects, we have, from equation 10–13,

$$\Delta H = Q - W$$

There is no work done, so $W = 0$. Hence, all we need to do is find the enthalpies of the superheated steam entering and the saturated liquid condensate leaving. From the steam tables in Appendix A.2,

$$\text{Superheated vapor } H_{75\text{ psia, }350°\text{F}} = 1{,}205.3\text{ Btu/lb}_m$$

$$\text{Saturated liquid } h_{20\text{psia}} = 196.16\text{ Btu/lb}_m$$

$$Q = \Delta H = (\text{Steam flow rate})\,(H_{out} - H_{in})$$

$$= (50{,}000\text{ lb}_m/\text{hr})(196.16 - 1{,}205.3)(\text{Btu/lb}_m)$$

$$= -\ 50.46 \times 10^6\text{ Btu/hr}$$

That Q is negative means that heat is transferred out of the system, which is the shell side of the reboiler. This makes good sense, of course, because the steam is heating the process material in the bottom of the column.

10.3.3 *Hxy* Diagrams

We can use *Hxy* diagrams for binary systems. Figure 10–6 gives a sketch of two typical *Hxy* diagrams showing isobaric (constant-pressure) data—i.e., the enthalpies, compositions, and temperatures apply for the pressure given. In both diagrams, the top saturated-vapor curve $H_{(y)}$ shows how vapor enthalpy varies with vapor composition y. The bottom saturated-liquid curve $h_{(x)}$ shows how liquid enthalpy varies with liquid composition x. The region above the H curve is where superheated vapor exists at the given pressure, and the region below the h curve is where subcooled liquid exists at the given pressure. The region between the curves is the two-phase region, where both liquid and vapor phases coexist.

Two other important items are also shown on the diagrams. First, lines of constant temperature are given. In Figure 10–6(a), the vertical line at $x = 0$ corresponds to pure heavy component, so the temperature T_H along

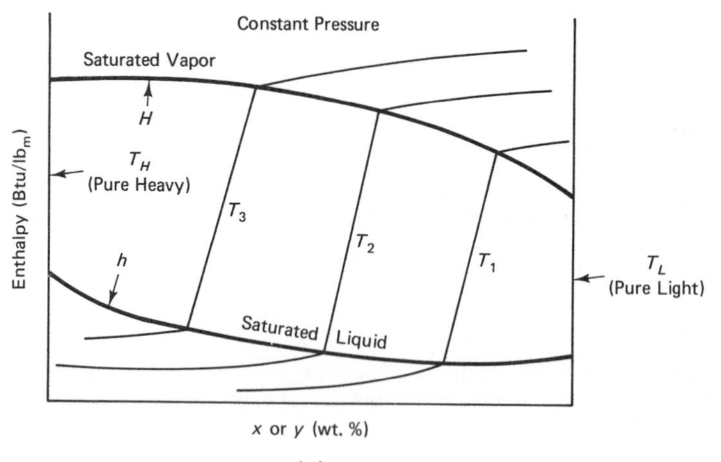

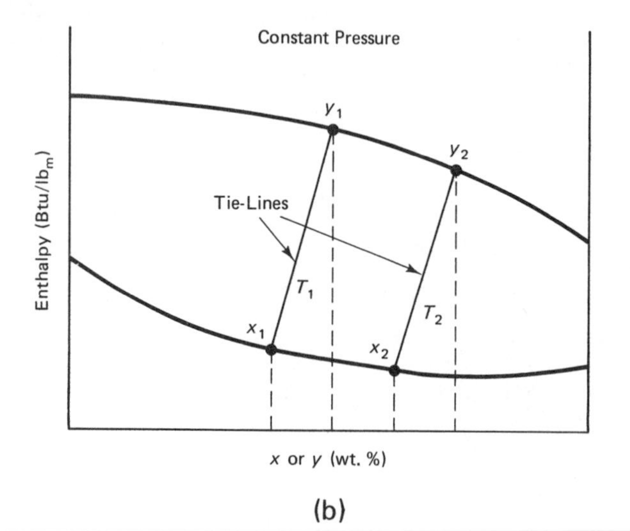

Figure 10–6 *Hxy* diagram. (a) Constant temperature lines (b) VLE tie-lines.

this line is the boiling-point temperature of the heavy component at the specified pressure. Similarly, the vertical line at $x = 1$ corresponds to pure light component, so the temperature T_L along this line is the boiling-point temperature of the light component at the specified pressure. The lower end of a straight line in the two-phase region between the H and h curves intersects the h curve at a composition x which is the composition of the liquid that would exist at the specified temperature and pressure in a two-phase system. The top end of the same line intersects the H curve at a composition y which is the composition of the vapor that is in equilibrium with the liquid at the other end of the constant-temperature straight line.

Second, vapor-liquid phase equilibrium data are given in Figure 10–6(b). The constant temperature lines in the two-phase region are called *VLE tie-lines*. These are exactly the same as the LLE tie-lines we examined in liquid-liquid extraction. From the two VLE tie-lines shown in the figure, a liquid with composition x_1 would be in equilibrium at temperature T_1 with a vapor with composition y_1. Similarly, the two phases x_2 and y_2 are in equilibrium at temperature T_2. Remember, these data are for the specific binary chemical system in question at the specified constant pressure. Figure 10–7 gives *Hxy* data for ethanol and water at two pressures.

These *Hxy* diagrams contain a lot of information. They give us both enthalpy and VLE data. Figure 10–8 shows how VLE tie-lines of an *Hxy* diagram are related to an *xy* diagram. The diagrams are less cluttered if the VLE is displayed on a separate *xy* diagram, but then two figures are required.

An alternative to putting in a lot of VLE tie-lines is to plot a VLE *conjugate* line on the *Hxy* diagram which can be used to locate any VLE tie-line. This conjugate line involves exactly the same idea that we used in chapters 6 and 9 for liquid-liquid equilibrium data. Figure 10–9 shows the VLE conjugate line (the dashed line) for the methanol-water system at atmospheric pressure. To find the composition y that is in equilibrium with a given composition x, you simply draw a vertical line at x up to the VLE conjugate line. Then, from this point of intersection, you draw a horizontal line over to the right until it intersects the saturated-vapor curve H. The composition y at this intersection is in equilibrium with the composition x on the saturated-liquid curve h. Conjugate lines make it easy to construct tie-lines, so we will use them extensively in chapter 13 when we combine component balances, energy balances, and vapor-liquid equilibrium relationships to analyze distillation columns with nonequimolal overflow.

Example 10.7.

Grandpa McCoy is producing his famous "Liquid Lightning" moonshine in a still that produces 1 gallon/hr. The density of the moonshine liquid is 7.5 lb_m/gallon. The temperature of the vapor coming off the top of the still is 175°F. The pressure is atmospheric. The temperature of the liquid leaving the total condenser is 100°F. How much heat must be removed in the old Model T radiator that Grandpa uses as his overhead condenser?

Solution:

We use the *Hxy* diagram for ethanol and water (Figure 10–7) to read off the enthalpies of the saturated vapor at 175°F and 1 atmosphere and the subcooled liquid at 100°F. We have

$$H_{175} = 560 \text{ Btu/lb}_m \text{ on the saturated vapor line}$$

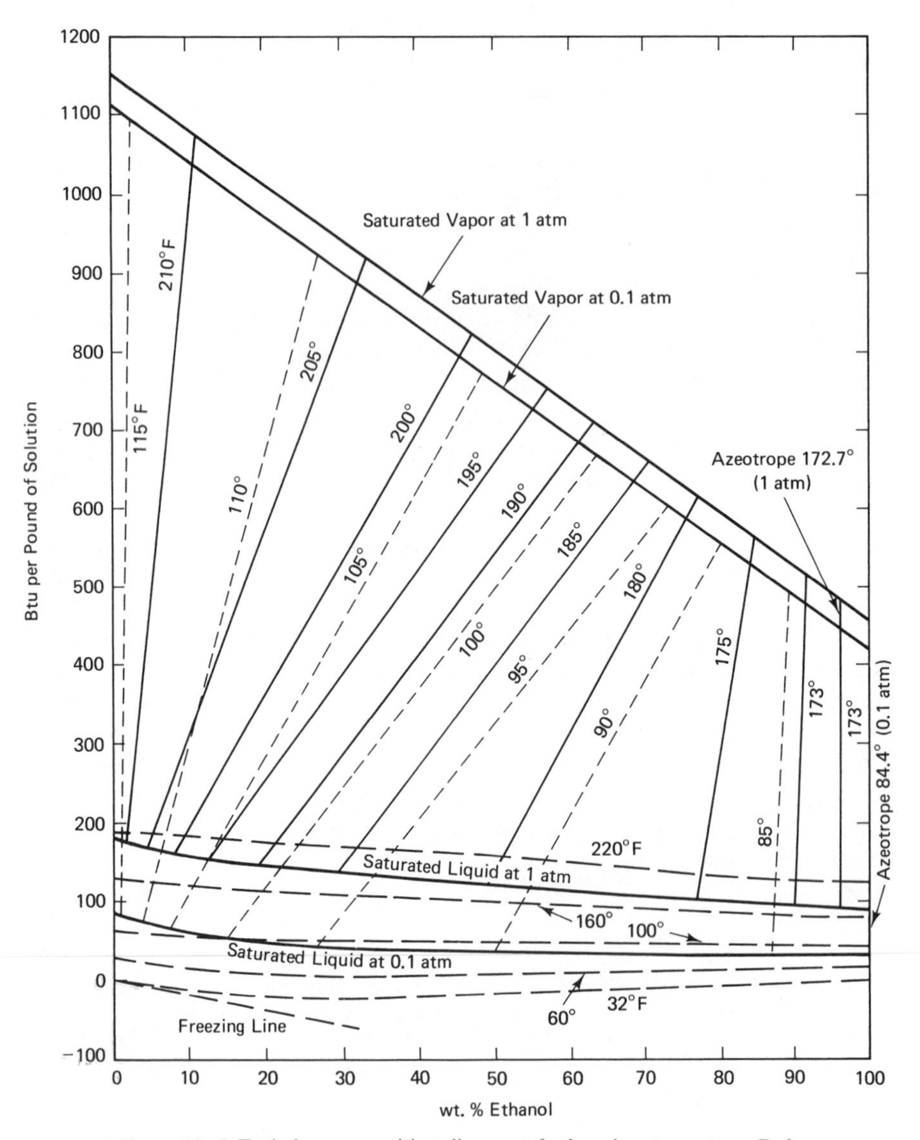

Figure 10–7 Enthalpy-composition diagram of ethanol-water system. Reference states, pure liquid alcohol and pure liquid water, each at 32°F saturated. *From A. O. Hougen, K. M. Watson, R. A. Rogatz*, Chemical Process Principles, Part I, Material and Energy Balances, 2nd Ed. *Copyright © 1954 John Wiley & Sons, Inc. Reprinted by permission.*

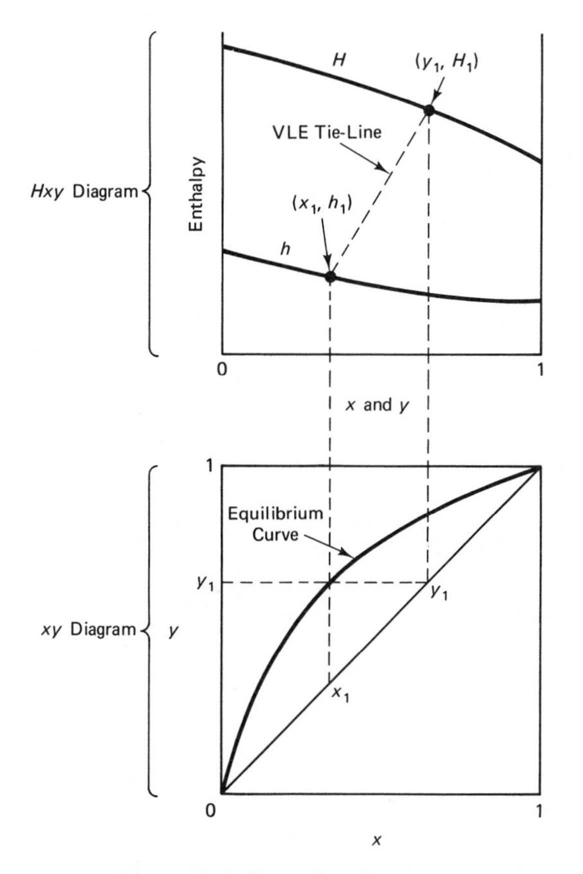

Figure 10–8 *Hxy* and *xy* diagrams.

The composition of the moonshine can be read off the *Hxy* diagram also. At 175°F and 1 atmosphere, the vapor composition is 84 weight percent ethanol.

The vapor is totally condensed, so the liquid has the same composition. At $x = 84$ weight percent and at a temperature of 100°F (note that we are below the saturated liquid line, so the liquid is subcooled), the enthalpy is 40 Btu/lb_m. Consequently, we obtain

$$\Delta H = Q - W$$

$$(1 \text{ gallon/hr})(7.5 \text{ lb}_m/\text{gallon})(560 - 40)(\text{Btu/lb}_m) = Q - 0$$

$$Q = 3{,}900 \text{ Btu/hr}$$

10.3.4 Multicomponent Systems

For systems with more than two components, and for computer solutions, we need equations that give us vapor and liquid enthalpies as functions of composition and temperature. In many chemical systems, when heat of mixing

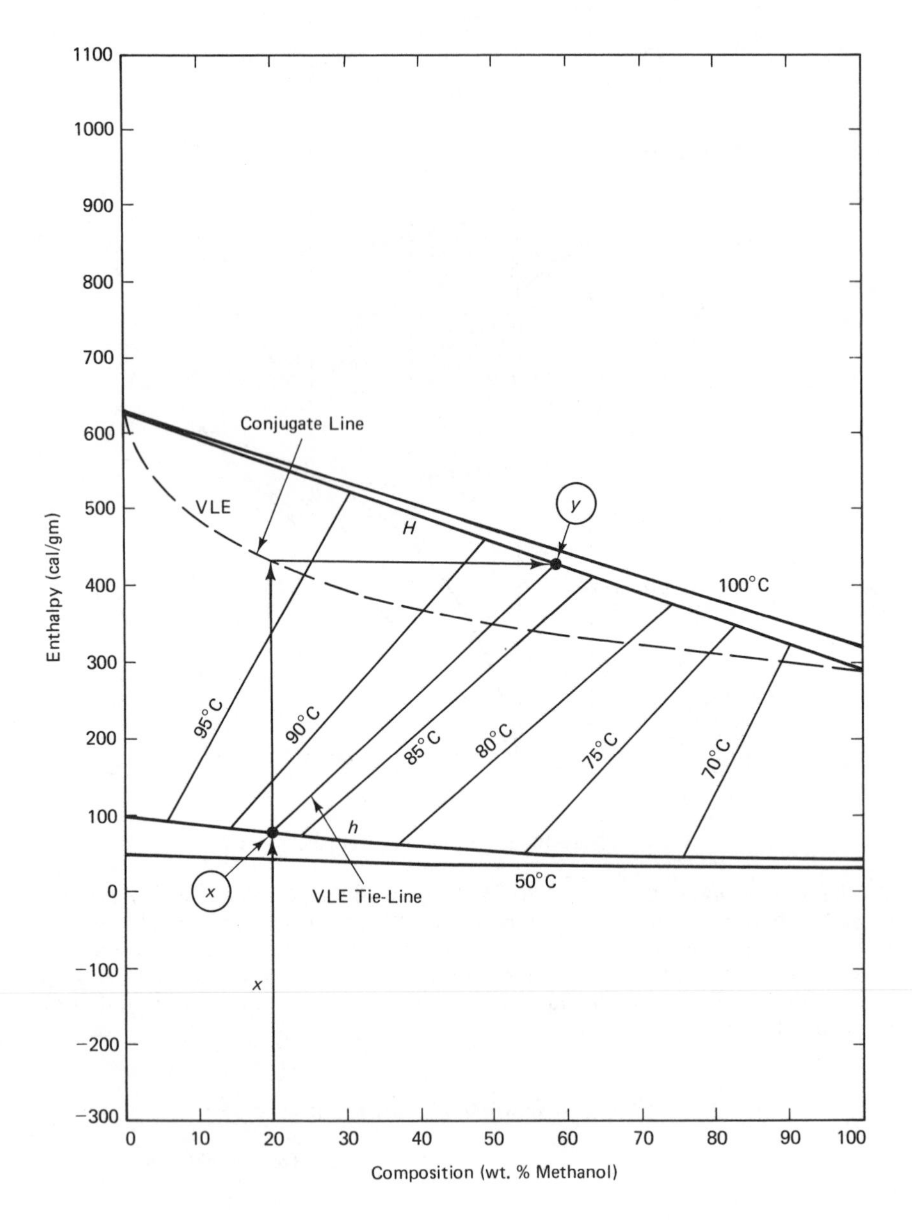

Figure 10–9 *Hxy* diagram, methanol and water @ 760 mm Hg.

effects are small, the total enthalpy can be calculated by simply averaging the enthalpies of the individual pure components on the basis of composition. The equations are

$$H = \sum_{j=1}^{NC} y_j H_j \tag{10–35}$$

$$h = \sum_{j=1}^{NC} x_j h_j \tag{10–36}$$

where H_j and h_j are respectively the vapor and liquid enthalpies of pure component j at a given temperature. Simple relationships like equations 10–33 and 10–34 might then be used to calculate H_j and h_j. If mole fraction compositions are given, molar enthalpies (Btu/lb-mole or cal/g-mole) must be used; if weight fraction compositions are used, enthalpies must be in Btu/lb_m or cal/g.

PROBLEMS

10–1. Synthesis gas containing 63.6 mole percent H_2, 9.1 mole percent CO_2, and 27.3 mole percent CO is cooled from 800°F to 77°F. How much heat must be removed if the gas flow rate to the cooler is 700 kg-mole/hr?
 (a) Use mean heat capacities.
 (b) Use heat capacities as functions of temperature, and integrate.

10–2. 200,000 lb_m/hr of condensate (liquid water) at 300°F is fed into the Dow-Corning wood-fired boiler in Midland, Michigan. The super-heated steam leaving the boiler is at 1,275 psia and 900°F.
 (a) How much heat must be transferred into the water in the boiler?
 (b) If 500 tons per day of wood chips are fed into the boiler with a heat of combustion of 8,500 Btu/lb_m, what is the thermal efficiency of the boiler? Use the formula

$$\text{Efficiency} = \frac{\text{Heat Transferred to Water}}{\text{Heat Produced by Combustion}}$$

10–3. Heat is removed from a 5,000 gal chemical reactor by feeding cooling water through a jacket that surrounds the vessel. The heat removal rate is 2.8×10^6 Btu/hr. The cooling water inlet and exit temperatures are 70°F and 99.3°F, respectively. What is the cooling water flow rate in gpm?

10–4. A gas compressor is driven by a 200 hp motor. During startup, gas is circulated through a bypass cooler from the compressor discharge to the compressor suction. How much heat must be removed in the cooler under these conditions? (See figure on page 382.)

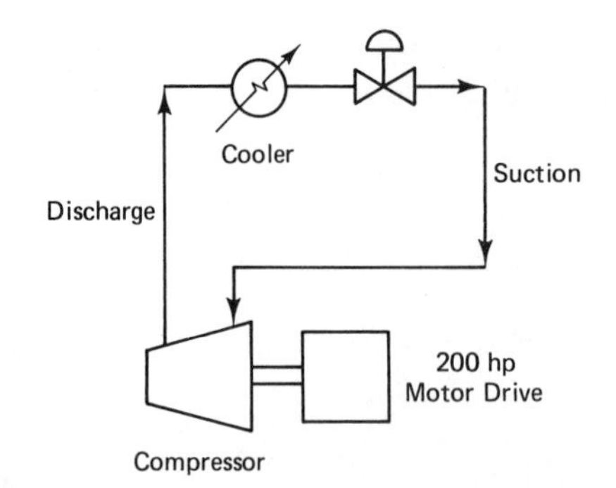

10–5. Determine the standard-state heat of reaction at 25°C for the following reactions:

(a) $C_6H_{12} \longrightarrow C_6H_6 + 3H_2$

(b) $2FeS_2 + \frac{11}{2} O_2 \longrightarrow Fe_2O_3 + 4SO_2$

(c) $NaOH + HCl \longrightarrow NaCl + H_2O$ (L)

(d) $CaC_2 + H_2O$ (L) $\longrightarrow CaO + C_2H_2$

(e) H_2O (g) $+ CO \longrightarrow CO_2 + H_2$

(f) $H_2S + O_2 \longrightarrow SO_2 + H_2$

10–6. Develop a mean molal heat capacity plot for benzene in the gas state using a base temperature of 100°C and covering temperatures up to 1,000 K.

10–7. The condenser at the top of a distillation column is condensing 80 weight percent ethanol-water saturated vapor at a pressure P = 1 atm. A reflux ratio of 1.5 is required by the column, and the distillate product will be a vapor. Determine

(a) The composition of the vapor product.

(b) The heat removed per 100 lb_m of vapor product.

(c) The cooling water required per 100 lb_m of vapor product if the cooling water can be heated from 80°F to a temperature 30°F cooler than the condensing vapors.

Energy Balance Calculations 11

In chapter 10 we showed how energy balances could be used on some simple systems. In this chapter we will study the application of energy balances to somewhat more complex and realistic processes. We can use one energy balance, one total mass balance, and (NC − 1) component balances to describe any system. Don't get confused and try to apply an energy equation for each component; there is only one energy balance that can be applied to a given system.

11.1 HEAT EXCHANGER CALCULATIONS

We begin with an examination of heat exchangers. We have already discussed reboilers, condensers, feed-preheaters, and the like, all of which must be analyzed using energy balances. Some of these heat exchangers involve simply cooling or heating liquids or gases with no changes in phase, i.e., no vaporization or condensation. Others involve phase changes.

The most common type of heat exchanger is a tube-in-shell. In this device, one stream flows through the inside of several tubes in parallel on the tube side of the heat exchanger, while the other stream flows over the outside of the tubes on the shell side of the heat exchanger. Baffles are used on the shell side to make the fluid flow back and forth across the tubes at the desired velocity. Multiple tube passes are used to achieve the desired velocity on the tube side. In a single-pass heat exchanger, the material on the tube side flows in parallel through all the tubes in one direction and then exits. In a two-pass heat exchanger, the stream flows through half of the tubes in one direction and then flows back through the other half of the tubes in the other direction.

11.1.1 Sensible Heat Transfer

Most heat exchangers operate with the hot and cold streams flowing in opposite directions. This countercurrent flow is more efficient in the sense that more heat can be transferred per unit of heat transfer area. Figure 11–1 sketches typical temperature profiles down the length of a heat exchanger (tube-in-tube or tube-in-shell).

The temperature T_H of the hot stream decreases as it flows through the exchanger from right to left in the figure. The temperature T_C of the cold stream increases as it flows in the other direction from left to right. At each and every point in the heat exchanger, the temperature of the hot stream must be greater than the temperature of the cold stream ($T_H > T_C$). However, in a countercurrent heat exchanger the temperature of the cold stream *leaving* the exchanger can be higher than the temperature of the hot stream *leaving*. This cannot occur in a cocurrent heat exchanger, as shown in Figure 11–2 because, in cocurrent flow, the hot and cold streams enter at the same end of the exchanger. Thus, the temperature of the hot stream must always be greater than the temperature of the cold stream.

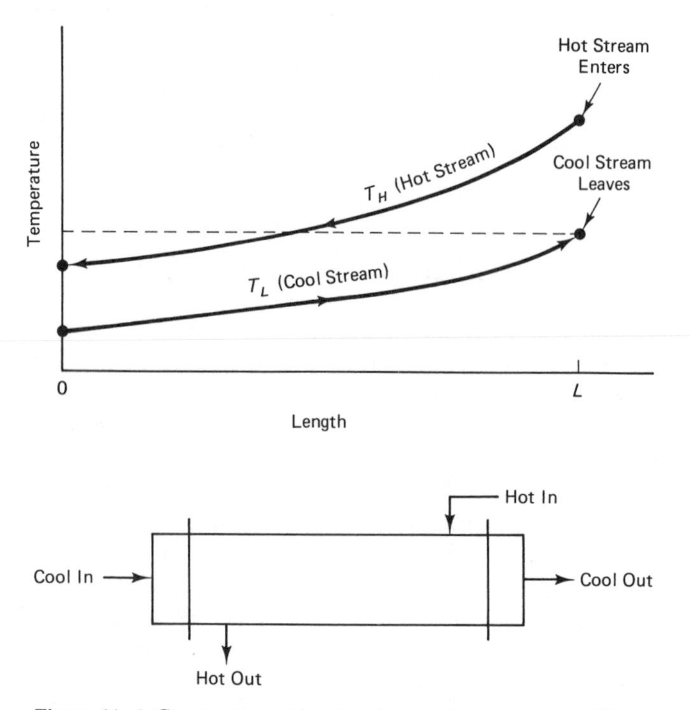

Figure 11–1 Countercurrent heat exchanger temperature profiles.

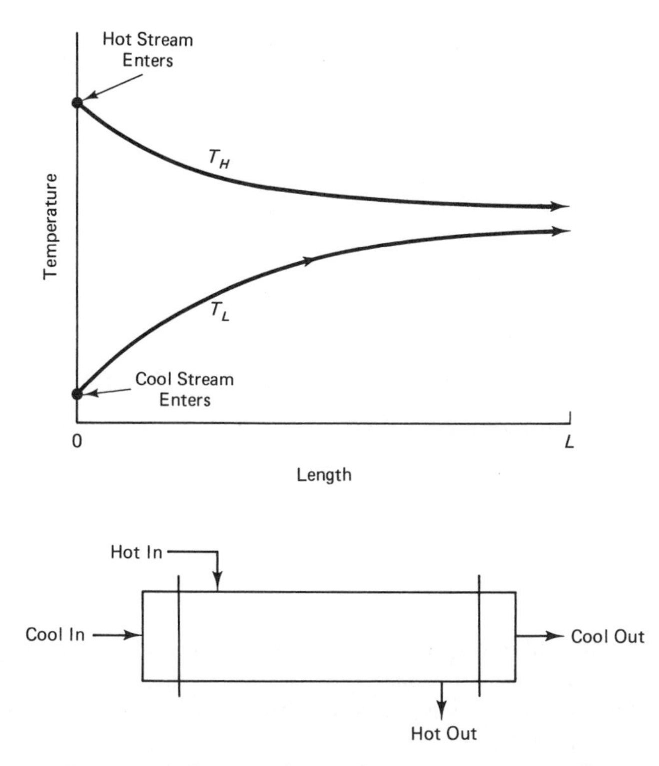

Figure 11–2 Cocurrent heat exchanger temperature profiles.

If neither the hot nor the cool stream is undergoing a phase change, only sensible heat is being exchanged. The amount of heat $-Q$ that is transferred *out of* the hot stream is then given by

$$-Q = -\Delta H_{\text{hot}} \tag{11–1}$$

$$= F_H \int_{T_{H\text{out}}}^{T_{H\text{in}}} (C_p)_{\text{hot}} dT$$

where

F_H = flow rate of the hot stream (kg/hr or lb_m/hr)

$(C_p)_{\text{hot}}$ = heat capacity of the hot stream (kcal/kg °C or Btu/lb_m °F)

$T_{H_{\text{in}}}$ = temperature of the hot stream entering (°C or °F)

$T_{H_{\text{out}}}$ = temperature of the hot stream leaving (°C or °F)

If C_p is constant, equation 11–1 simplifies to

$$-Q = F_H C_p (T_{H_{\text{in}}} - T_{H_{\text{out}}}) \tag{11–2}$$

On the other side of the heat exchanger, the cold stream is being heated. The amount of heat that is transferred *into* the cool stream, assuming sensible heat changes only, is

$$Q = F_C \int_{T_{Cin}}^{T_{Cout}} (C_p)_{cold}\, dT \qquad (11\text{–}3)$$

where

F_C = flow rate of cold stream (kg/hr or lb_m/hr)

$(C_p)_{cold}$ = heat capacity of the cold stream (kcal/kg °C or Btu/lb_m °F)

$T_{C_{in}}$ = temperature of cold stream entering (°C or °F)

$T_{C_{out}}$ = temperature of cold stream leaving (°C or °F)

If C_p is constant, equation 11–3 simplifies to

$$Q = F_C C_p (T_{C_{out}} - T_{C_{in}}) \qquad (11\text{–}4)$$

Of course, the sum of the two Q's must be zero if there are no heat losses from the heat exchanger: the heat that is lost by the hot stream must be transferred into the cold stream. Note that the energy balances are the same for both cocurrent and countercurrent flow arrangements.

Example 11.1.

1,000 lb-mole/hr of CH_4 is being cooled from 2,000°F to 400°F in a countercurrent heat exchanger in which water is the cooling medium. The water is heated from 60°F to 120°F. How much cooling water is required?

Solution:

Mean heat capacities for CH_4 (Figure 10–1) can be used to calculate the heat that must be removed from the gas in cooling it from 2,000°F to 400°F. The two-step procedure discussed in chapter 10 is used: first cool the gas from 2,000°F to 77°F, and then heat it from 77°F to 400°F. For methane, we have

$$C_p^{mean} \text{ for 2,000°F to 77°F} = 15 \text{ Btu/lb-mole °F}$$

$$C_p^{mean} \text{ for 77°F to 400°F} = 9.2 \text{ Btu/lb-mole °F}$$

Thus,

$$Q = (1{,}000 \text{ lb-mole/hr}) [(15)(2{,}000 - 77) - (9.2)(400 - 77)] \text{ Btu/lb-mole}$$

$$= 25.87 \times 10^6 \text{ Btu/hr}$$

The heat capacity of water is about 1 Btu/lb_m °F over the temperature range from

60 to 120°F. So we have

$$Q = F_{\text{water}}(C_p)_{\text{water}}(T_{\text{out}} - T_{\text{in}})$$

$$25.87 \times 10^6 \text{ Btu/hr} = F_{\text{water}}(1 \text{ Btu/lb}_m \text{ °F})(120 - 60\text{°F})$$

$$F_{\text{water}} = 431{,}000 \text{ lb}_m\text{/hr}$$

$$F_{\text{water}} = (431{,}000 \text{ lb}_m\text{/hr})(\text{gallon}/8.33 \text{ lb}_m)(\text{hr}/60 \text{ min})$$

$$F_{\text{water}} = 862 \text{ gpm of cooling water}$$

11.1.2 Condensers and Vaporizers

When vapor entering the condenser is saturated vapor at its dewpoint, and when liquid leaving the condenser is saturated liquid at its bubblepoint, only latent heat of condensation must be removed. However, in some condensers the vapor enters with some superheating and the liquid leaves with some subcooling. In these cases the effects of both latent and sensible heat must be considered. Figure 11–3 illustrates this situation.

Vaporizers (reboilers, evaporators, calandrias, etc.) convert liquid into vapor. Usually the vapor leaving the vaporizer is saturated vapor at its dewpoint that is in phase equilibrium with the liquid that is in the vaporizer. The liquid fed into the vaporizer can be subcooled, and the vapor leaving can be superheated.

Example 11.2.

A distillation column operates adiabatically (without heat losses from the column itself) at atmospheric pressure. A feed of 50 weight percent CH_3OH and 50 weight

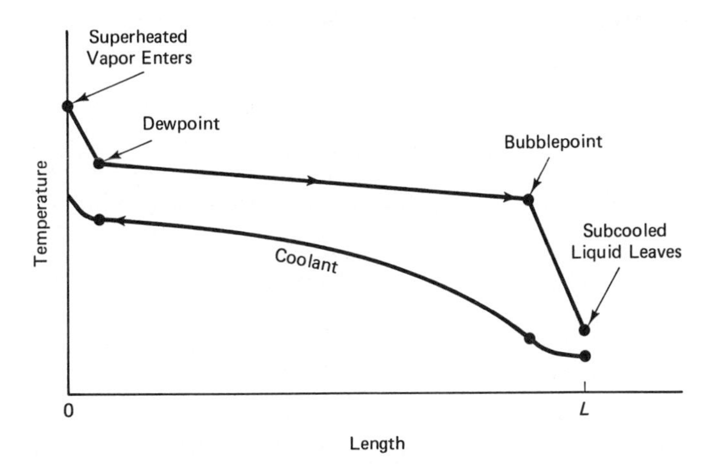

Figure 11–3 Condenser temperature profiles.

percent water enters at 85°C. Saturated liquid products of 90 weight percent CH_3OH and 5 weight percent CH_3OH are withdrawn. The total condenser produces saturated liquid distillate product and reflux. The reflux ratio is 3. If the condenser is cooled by water increasing in temperature from 60°F to 140°F, how much water is required for a feed flow rate of 2,000 lb_m/hr? If the reboiler produces saturated vapor and is a partial reboiler, how much saturated steam at 300°F ($\Delta H_v = 910$ Btu/lb_m) is required?

Solution:

First, we perform total mass and component balances in order to calculate the flow rates of the products.

$$\text{Total mass balance: } F = B + D = 2{,}000 \text{ lb}_m/\text{hr}$$

$$\text{Methanol component balance: } Fz = Bx_B + Dx_D \qquad \text{lb}_m \text{ CH}_3\text{OH/hr}$$

$$(2{,}000)(0.5) = (0.05)(B) + (2{,}000 - B)(0.90)$$

$$B = 941.2 \text{ lb}_m/\text{hr}$$

$$D = 1{,}058.8 \text{ lb}_m/\text{hr}$$

Since the reflux ratio is 3, the reflux flow rate is

$$(3)(1{,}058.8) = 3{,}176.4 \text{ lb}_m/\text{hr}$$

The total amount of material that the condenser must liquify is the sum of the distillate and the reflux:

$$1{,}058.8 + 3{,}176.4 = 4{,}235.2 \text{ lb}_m/\text{hr of 90 weight percent CH}_3\text{OH vapor}$$

From the *Hxy* diagram given in chapter 10 (Figure 10–9), the enthalpies of saturated vapor and saturated liquid at 90 weight percent methanol are

$$H = (323 \text{ cal/g})(\text{Btu}/252 \text{ cal})(454 \text{ g/lb}_m) = 581 \text{ Btu/lb}_m$$

$$h = (40 \text{ cal/g})(\text{Btu}/252 \text{ cal})(454 \text{ g/lb}_m) = 72 \text{ Btu/lb}_m$$

The heat removed in the condenser is

$$Q = (581 - 72 \text{ Btu/lb}_m)(4{,}235.2 \text{ lb}_m/\text{hr})$$

$$= 2.16 \times 10^6 \text{ Btu/hr}$$

The flow rate F_{water} of the cooling water can be calculated from

$$Q = F_{water}(C_p)_{water}(T_{W_{out}} - T_{W_{in}})$$

$$2.16 \times 10^6 \text{ Btu/hr} = F_{water}(1 \text{ Btu/hr °F})(140 - 60\text{°F})$$

$$F_{water} = 27{,}000 \text{ lb}_m/\text{hr}$$

$$(27{,}000 \text{ lb}_m/\text{hr})(\text{gallon}/8.33 \text{ lb}_m)(\text{hr}/60 \text{ min}) = 54 \text{ gpm}$$

In order to get the reboiler energy input, we make an energy balance around the column and reboiler. The system is sketched in Figure 11–4. Enthalpies are read from Figure 10–9.

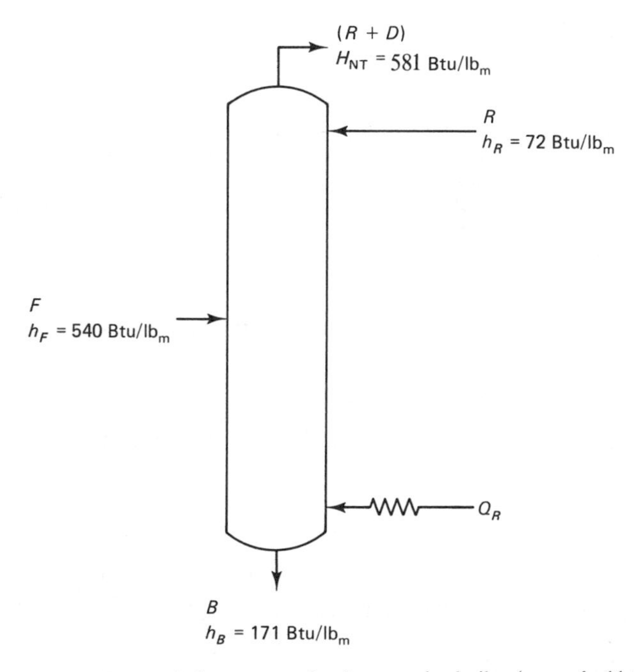

Figure 11–4 Energy balance around column and reboiler (example 11.2).

Feed enthalpy h_F at 50 weight percent methanol and 85°C
(note that this is in the two-phase region)
$$= 300 \text{ cal/g} = 540 \text{ Btu/lb}_m$$

Bottoms enthalpy h_B at 5 weight percent methanol
(on the saturated liquid line with a temperature of 95°C)
$$= 95 \text{ cal/g} = 171 \text{ Btu/lb}_m$$

Reflux enthalpy h_R at 90 weight percent methanol
(on the saturated liquid line with a temperature of about 67°C)
$$= 40 \text{ cal/g} = 72 \text{ Btu/lb}_m$$

Overhead vapor enthalpy H_{NT} at 90 weight percent methanol
(on the saturated vapor line with a temperature of about 71°C)
$$= 323 \text{ cal/g} = 581 \text{ Btu/lb}_m$$

$$\Delta H = Q - W$$

$$(V_{NT}H_{NT} + Bh_B) - (Fh_F + Rh_R) = Q_R - 0 \quad (11\text{–}5)$$

$$(4{,}235.2 \text{ lb}_m/\text{hr})(581 \text{ Btu/lb}_m) + (941.2 \text{ lb}_m/\text{hr})(171 \text{ Btu/lb}_m)$$
$$- (2{,}000 \text{ lb}_m/\text{hr})(540 \text{ Btu/lb}_m) - (3{,}176.4 \text{ lb}_m/\text{hr})(72 \text{ Btu/lb}_m) = Q_R$$

$$Q_R = 1.313 \times 10^6 \text{ Btu/hr}$$

$$\text{Steam flow rate} = (1.313 \times 10^6 \text{ Btu/hr})(\text{lb}_m/910 \text{ Btu}) = 1{,}443 \text{ lb}_m/\text{hr}$$

Example 11.3.

An evaporator brings in brine at 7 weight percent NaCl and 60°F and concentrates it to 35 weight percent NaCl solution at 140°F by evaporating pure water. How much saturated steam at 180°F is needed to produce 20,000 lb_m/hr of concentrated salt solution? Assume that the heat capacities of the solution and of pure water are 1 Btu/lb_m°F and that the heat of vaporization of water is 1,000 Btu/lb_m.

Solution:

Total mass and NaCl component balances give the flow rates of the feed and water vapor.

$$\text{Total mass: } 20{,}000 + V = F$$

$$\text{NaCl component: } (20{,}000)(0.35) + 0 = F\,(0.07)$$

$$F = 100{,}000 \text{ lb}_m/\text{hr}$$

$$V = 80{,}000 \text{ lb}_m/\text{hr}$$

To do an energy balance, we need enthalpy data for salt solutions and for water vapor. We use the simple approximate enthalpy equations

$$h_{\text{solution}} = C_p(T - T_0) = (1)(T - 60) \text{ Btu/lb}_m$$

$$H_v = C_p(T - T_0) + \Delta H = [(1)(T - 60) + 1{,}000] \text{ Btu/lb}_m$$

where a reference temperature of 60°F has been chosen. We then have

$$\text{Enthalpy of the feed} = h_F = (1)(T_F - 60)$$
$$= (1)(60 - 60) = 0 \text{ Btu/lb}_m$$

$$\text{Enthalpy of the product solution} = h_p = (1)(T_P - 60)$$
$$= (1)(140 - 60) = 80 \text{ Btu/lb}_m$$

$$\text{Enthalpy of the water vapor} = H_v = (1)(T_v - 60) + 1{,}000$$
$$= (1)(140 - 60) + 1{,}000 = 1{,}080 \text{ Btu/lb}_m$$

$$\Delta H = Q - W$$

$$(VH_v + Ph_p) - Fh_F = Q \tag{11-6}$$

$$(80{,}000 \text{ lb}_m/\text{hr})(1{,}080 \text{ Btu/lb}_m) + (20{,}000 \text{ lb}_m/\text{hr})(80 \text{ Btu/lb}_m)$$
$$- (100{,}000 \text{ lb}_m/\text{hr})(0 \text{ Btu/lb}_m) = Q$$

$$Q = 88 \times 10^6 \text{ Btu/hr}$$

$$\text{Steam required} = (88 \times 10^6 \text{ Btu/hr})(\text{lb}_m/1{,}000 \text{ Btu}) = 88{,}000 \text{ lb}_m/\text{hr}$$

11.2 FLASH CALCULATIONS INCLUDING ENERGY EQUATIONS

In chapter 7 we examined isothermal flash calculations, in which the temperature and the pressure of the system studied were both specified. Material balances and phase equilibrium relationships were used to calculate the compositions and flow rates of the vapor and liquid streams leaving the system. Now that we know how to do energy balances, we are ready to look at problems in which the temperature is *not* given. One more equation—an energy balance around the system—is needed to define the system.

11.2.1 Types of Calculations

There are several important types of nonisothermal flash calculations that are often encountered in chemical engineering.

Adiabatic Flash The adiabatic flash calculation is the most common type of flash calculation encountered. In adiabatic flashes, no heat is transferred ($Q = 0$). The classical adiabatic flash calculation involves the calculation of the compositions and flow rates of the vapor and liquid streams *and* the calculation of the temperature of the flash from the specified feed and system pressure conditions. We can characterize the situation in terms of what is given and what is unknown:

Given

Feed conditions: Flow rate F (mole/time)
Composition z_j (mole fraction)
Enthalpy h_F (energy/mole)
Pressure P of the system

Unknowns

Vapor and liquid flow rates V and L (mole/time)
Compositions y_j and x_j of the vapor and liquid phases, respectively (mole fraction)
Temperature T of the system

Note that the enthalpies of the various streams are known if their temperatures and compositions are known. We thus let h be the enthalpy of the liquid stream and H be the enthalpy of the vapor stream.

There are 2(NC) + 1 unknown variables (NC − 1 compositions for each phase, 2 flow rates, and 1 temperature). Therefore, we need 2(NC) + 1 equations to specify the system:

$$\text{Total mass balance (1 equation): } F = V + L \tag{11–7}$$

$$\text{Component balances (NC} - 1 \text{ equations): } z_j F = y_j V + x_j L \tag{11–8}$$

$$\text{VLE equations (NC equations): } y_j P = x_j \gamma_j P_j^S \tag{11–9}$$

$$\text{Energy equation (1 equation): } \Delta H = Q - W \tag{11–10}$$

Applying equation 11–10 to the adiabatic flash system gives

$$(VH + Lh) - Fh_F = 0 \tag{11–11}$$

Fixed Rate of Heat Transfer In some situations, a fixed amount of heat is added to the system. The equations describing the system are then identical to those for an adiabatic flash, except that the energy balance includes a heat transfer term Q:

$$(VH + Lh) - Fh_F = Q \tag{11–12}$$

Nonadiabatic Flash with Variable Heat Transfer Rate Suppose the heat transfer rate depends on the flash temperature T. A typical situation is one in which the rate of heat transfer is proportional to the temperature difference between the flash temperature T and a given temperature T_S. In that case,

$$Q = UA(T_S - T) \tag{11–13}$$

where

U = heat transfer coefficient (Btu/ft^2 °F hr or W/m^{2}K)

A = heat transfer area (ft^2 or m^2)

T_S = fixed temperature of the heating or cooling medium (°F or K)

Q = heat transfer rate (Btu/hr or Watts)

11.2.2 Iterative Solution for Multicomponent Systems

In each of the preceding cases, a fairly large set of nonlinear algebraic equations must be solved simultaneously. There are a number of methods of solution, but probably the most straightforward is to use the isothermal flash method of chapter 7 and add an outside convergence loop on temperature to it. In this technique, a flash temperature is guessed, and an isothermal flash calculation is performed at this temperature, giving the compositions and flow rates of liquid and vapor. Enthalpies and heat transfer rates are

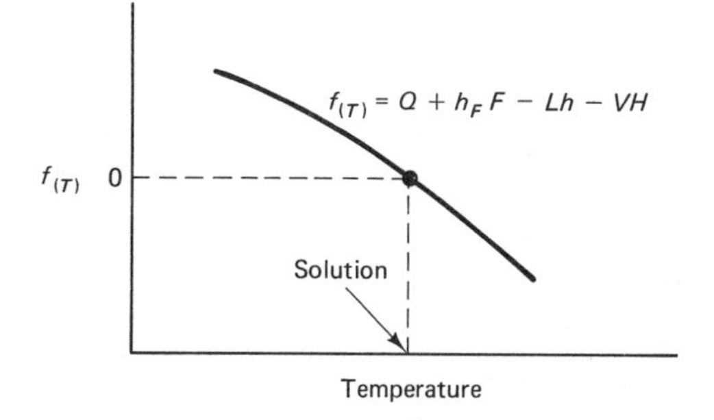

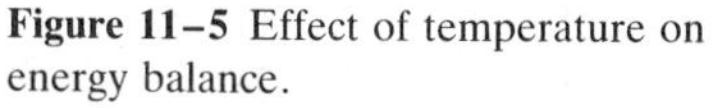
Figure 11–5 Effect of temperature on energy balance.

calculated, and then the energy balance is checked. If it does not balance, a new temperature is guessed.

Note that the higher the temperature, the more flashing that occurs (because of a bigger V/F ratio). The energy balance can be written as a function of temperature thus:

$$f(T) = Q + h_F F - Lh - VH \tag{11–14}$$

Figure 11–5 sketches the shape of this function. We are trying to find the value of the temperature that makes the function equal zero, i.e., we want the root of the function.

A fairly general computer program for performing adiabatic or nonadiabatic flash calculations is given at the end of the chapter. Example 11.4 is done by hand.

Example 11.4.

A liquid hydrocarbon stream in a petroleum refinery flows at a rate of 1,500 lb-mole/hr. Its composition is 40 mole percent C_3H_8, 30 mole percent nC_4H_{10}, and 30 mole percent nC_6H_{14}. Upstream of a flash valve it is all liquid phase at a pressure of 600 psia and a temperature of 200°F. As the liquid flows through the valve, the pressure drops to 200 psia. What are the temperature, flow rates, and compositions of the liquid and vapor phases downstream of the valve if the process is adiabatic? Use simple enthalpy relationships (constant heat capacities and heats of vaporization) and the DePriester charts given in chapter 6 for the hydrocarbon K values.

Solution:

Data

propane $C_V = 22$ Btu/lb-mole °F (vapor)

$C_L = 31$ Btu/lb-mole °F (liquid)

$\Delta H_v = 7{,}380$ Btu/lb-mole

n-butane $C_V = 29$ Btu/lb-mole °F (vapor)

$$C_L = 41 \text{ Btu/lb-mole °F (liquid)}$$

$$\Delta H_v = 8{,}640 \text{ Btu/lb-mole}$$

$$\text{n-hexane } C_V = 43 \text{ Btu/lb-mole °F (vapor)}$$

$$C_L = 51.6 \text{ Btu/lb-mole °F (liquid)}$$

$$\Delta H_v = 9{,}720 \text{ Btu/lb-mole}$$

For the enthalpy calculations, we assume that liquid at 0°F has zero enthalpy ($h = 0$ at 0°F). The enthalpy of the inlet stream at 200°F is

$$Fh_F = (1{,}500 \text{ lb-mole/hr})[(0.4)(31) + (0.3)(41) + (0.3)(51.6) \text{ Btu/lb-mole °F}][200 - 0\text{°F}]$$

$$= 12.05 \times 10^6 \text{ Btu/hr}$$

By an energy balance,

$$Fh_F = Lh_L + VH_V = 12{,}050{,}000 \text{ Btu/hr} \tag{11–15}$$

Now we guess a temperature and do an isothermal flash calculation. (See chapter 7.) At this assumed temperature, we calculate the ratio V/F iteratively, and then the enthalpies of the L and V streams. If equation 11–15 is not satisfied, a new temperature is guessed.

1. First Temperature Guess: $T = 150$°F

   We guess a temperature somewhat lower than the feed temperature because we know the flashing will decrease the temperature. The K values at 150°F are:

	K_{150}	z_j	$K_j z_j$
C_3	1.6	0.4	0.64
nC_4	0.6	0.3	0.18
nC_6	0.11	0.3	0.03
			0.85

Checking whether we are above the bubblepoint temperature, we calculate the summation of the K values times the feed compositions:

$$\sum_{j=1}^{NC} K_j z_j = 0.85$$

Therefore, 150°F is below the bubblepoint temperature. Since nothing flashes at this temperature, $V = 0$. Now we check the energy balance:

$$h_L @ 150\text{°F} = [(0.4)(31) + (0.3)(41) + (0.3)(51.6)][150 - 0]$$

$$= 6{,}027 \text{ Btu/lb-mole}$$

$$Lh_L = (1{,}500 \text{ lb-mole/hr})(6{,}027 \text{ Btu/lb-mole})$$

$$= 9{,}040{,}000 \text{ Btu/hr}$$

But $Fh_F = 12{,}050{,}000$ Btu/hr. Therefore, too little vapor has flashed, and our temperature guess of 150°F is too low.

2. Second Temperature Guess: $T = 180°F$

	K_{180}	z_j	$K_j z_j$	z_j/K_j
C_3	2.0	0.4	0.800	0.200
nC_4	0.8	0.3	0.240	0.375
nC_6	0.16	0.3	0.048	1.875
			1.088	2.450

The summation of the $K_j z_j$ is greater than unity, so we are above the bubblepoint temperature. Also, the summation of the z_j/K_j is above unity, so we are below the dewpoint temperature. Therefore, some flashing will occur.

Now we must iteratively solve for the V/F ratio at the temperature guessed. A ratio of 0.3 is guessed, and the values of x_j are calculated from

$$x_j = \frac{z_j}{1 + (V/F)(K_j - 1)} \tag{11–16}$$

			Guess V/F = 0.3	*Guess V/F = 0.2*	
	z_j	K_{180}	x_j	x_j	y_j
C_3	0.4	2.0	0.308	0.333	0.666
nC_4	0.3	0.8	0.319	0.313	0.250
nC_6	0.3	0.16	0.401	0.361	0.058
			1.028	1.007	0.974

The first guess of $V/F = 0.3$ is too large. So we take a second guess of $V/F = 0.2$, which is pretty close. Accordingly, we assume that $V/F \approx 0.18$. The y_j values are calculated from $y_j = K_j x_j$. Then, normalizing the x and y values so that they add up to unity gives

	x_j	y_j
C_3	0.331	0.683
nC_4	0.311	0.257
nC_6	0.358	0.060
	1.000	1.000

$$V = (0.18)(1{,}500) = 270 \text{ lb-mole/hr of vapor}$$

$$L = 1{,}500 - 270 = 1{,}230 \text{ lb-mole/hr of liquid}$$

$$H_V = (0.683)[7{,}380 + (22)(180)] + (0.257)[8{,}640 + (29)(180)] + (0.060)[9{,}720 + (43)(180)] = 12{,}366 \text{ Btu/lb-mole}$$

$$h_L = [(0.331)(31) + (0.311)(41) + (0.358)(51.6)][180]$$

$$= 7{,}467 \text{ Btu/lb-mole}$$

$$H_V V + h_L L = (12{,}366)(270) + (7{,}467)(1{,}230) = 12.5 \times 10^6 \text{ Btu/hr}$$

Since $h_F F$ = 12,054,000 Btu/hr, our energy balance is not quite satisfied. So we need to guess a slightly lower temperature and repeat the steps. However, for hand calculations, the result is close enough.

11.2.3. Computer Program for Nonisothermal Flash

A FORTRAN program for nonisothermal flash calculations for multicomponent systems is given in Table 11–1, and the results of several test cases are given in Table 11–2. The flash pressure P, the temperature T_F of

TABLE 11–1

FORTRAN PROGRAM FOR NONISOTHERMAL FLASH CALCULATIONS FOR MULTICOMPONENT SYSTEMS

```
C   BINARY TEST CASE: PROPYLENE OXIDE/TERTIARY BUTYL ALCOHOL
C       IDEAL VLE ASSUMED
        PROGRAM TESTFL
        REAL L,MW
        DIMENSION VPC(2,6),Z(6),X(6),Y(6),T1(6),T2(6),P1(6),P2(6),
      + CL(6),CV(6),DH(6),MW(6),PS(6)
        DATA NC,F,Z(1),Z(2),TF,P,Q/2,3500.,.25,.75,250.,40.,0./
        DATA T1(1),T2(1),P1(1),P2(1)/64.,94.,7.737,14.7/
        DATA T1(2),T2(2),P1(2),P2(2)/154.4,215.6,7.737,29.4/
        DATA CL(1),CV(1),DH(1),MW(1)/.56,.35,172.,58.1/
        DATA CL(2),CV(2),DH(2),MW(2)/.8,.43,202.,74.1/
        CALL VAPC(NC,VPC,T1,T2,P1,P2)
        P=35.
        DO 100 NP= 1,2
        TF = 250.
        DO 50 NT=1,2
        TFLASH=208.
        IF(P.LE.0.) STOP
        CALL FLASH(NC,VPC,Z,F,TF,P,Q,PS,TFLASH,X,Y,L,V,CL,CV,DH,MW)
        WRITE(6,1)
      1 FORMAT(/,1X,'     Z1       Z2       F         TF        P          Q')
        WRITE(6,2) (Z(J),J=1,NC),F,TF,P,Q
      2 FORMAT(3X,2F7.4,3F8.1,F10.1)
        WRITE(6,3)
      3 FORMAT(/,8X,'X1      X2      Y1      Y2      L       V       T')
        WRITE(6,4) (X(J),J=1,NC),(Y(J),J=1,NC),L,V,TFLASH
      4 FORMAT(3X,4F7.4,3F7.1)
     50 TF=TF+50.
    100 P=P+10.
        END
C
         SUBROUTINE FLASH(NC,VPC,Z,F,TF,P,Q,PS,T,X,Y,L,V,CL,CV,DH,MW)
         DIMENSION VPC(2,NC),Z(NC),X(NC),Y(NC),PS(NC),CL(NC),CV(NC),
       + DH(NC),MW(NC)
         REAL L,MW,NETQ
```

TABLE 11–1 CONT.

```
C CHECK TO SEE IF YOU ARE IN THE TWO PHASE REGION BY CALCULATING
C     THE AMOUNTS OF HEAT REQUIRED TO COMPLETELY VAPORIZE (QDEW)
C     AND TO COMPLETELY CONDENSE (QBUB) THE FEED.
      TDEW=TF
      TBUB=TF
      CALL BUBPT(NC,VPC,Z,P,TBUB,Y,PS)
      CALL DEWPT(NC,VPC,Z,P,TDEW,X,PS)
      WRITE(6,15)TBUB,TDEW
   15 FORMAT(1X,'TBUB = ',F8.2,'     TDEW =  ',F8.2)
      CALL ENTH(NC,Z,Z,TF,HF,DUM,CL,CV,DH,MW)
      CALL ENTH(NC,Z,Y,TBUB,HLBUB,HVBUB,CL,CV,DH,MW)
      CALL ENTH(NC,X,Z,TDEW,HLDEW,HVDEW,CL,CV,DH,MW)
      QBUB=F*(HLBUB-HF)
      QDEW=F*(HVDEW-HF)
      IF(Q.LT.QBUB)THEN
      V=0.
      L=F
      SUM=0.
      DO 300 J=1,NC
      X(J)=Z(J)
      Y(J)=0.
  300 SUM=SUM+Z(J)*MW(J)*CL(J)
C SUBCOOLED LIQUID
      T=200.+(Q/F+HF)/SUM
      RETURN
      ENDIF
      IF(Q.GT.QDEW)THEN
      V=F
      L=0.
      SUM1=0.
      SUM2=0.
      DO 301 J=1,NC
      X(J)=0.
      Y(J)=Z(J)
      SUM1=SUM1+Z(J)*MW(J)*DH(J)
  301 SUM2=SUM2+Z(J)*MW(J)*CV(J)
C SUPERHEATED VAPOR
      T=(Q/F+HF-SUM1)/SUM2+200.
      RETURN
      ENDIF
      IBUG=40
      FLAGTM=1.
      FLAGTP=1.
      DT=5.
      LOOPT=0
      VF=0.18
C BEGINNING OF TEMPERATURE ITERATION LOOP
  100 DVF=.05
      CALL PSAT(NC,VPC,T,PS)
      LOOPT=LOOPT+1
      IF(LOOPT.GT.IBUG)WRITE(6,77)T,DT,NETQ,VF
      IF(LOOPT.GT.50)THEN
      WRITE(6,3)
    3 FORMAT(1X,'TLOOP')
      STOP
      ENDIF
      FLAGVP=1.
      FLAGVM=1.
      LOOP=0
      IF(T.GT.TDEW)GO TO 60
C BEGINNING OF V/F ITERATION LOOP
  200 XSUM=0.
      LOOP=LOOP+1
      IF(LOOP.GT.50) THEN
      WRITE(6,4)
```

TABLE 11-1 CONT.

```
    4 FORMAT(1X,'FLASH LOOP')
      STOP
      ENDIF
      DO 5 J=1,NC
    5 XSUM=XSUM+Z(J)/(1.+VF*(PS(J)/P-1.))
      IF(LOOP.GT.IBUG) WRITE(6,77)T,DT,NETQ,VF,XSUM
   77 FORMAT(1X,6F15.5)
      IF(ABS(XSUM-1.).LT. .00001) GO TO 40
      IF(XSUM-1.)20,20,30
   20 IF(FLAGVP.LT.0.)DVF=DVF/2.
      VF=VF+DVF
      IF(VF.GT.1.) GO TO 60
      FLAGVM=-1.
      GO TO 200
   30 IF(FLAGVM.LT.0.)DVF=DVF/2.
      VF=VF-DVF
      FLAGVP=-1.
      GO TO 200
C CONVERGED V/F LOOP
   40 DO 41 J=1,NC
      X(J)=Z(J)/(1.+VF*(PS(J)/P-1.))
   41 Y(J)=X(J)*PS(J)/P
      CALL ENTH(NC,X,Y,T,HL,HV,CL,CV,DH,MW)
      V=F*VF
      L=F-V
      NETQ=F*HF+Q-V*HV-L*HL
      IF(ABS(NETQ).LT.1000.) GO TO 1000
      IF(DT.LT. .00001) GO TO 1000
      IF(NETQ)60,60,50
   50 IF(FLAGTP.LT.0.)DT=DT/2.
      T=T+DT
      FLAGTM=-1.
      GO TO 100
   60 IF(FLAGTM.LT.0.) DT=DT/2.
      T=T-DT
      FLAGTP=-1.
      GO TO 100
 1000 RETURN
      END
C
      SUBROUTINE VAPC(NC,VPC,T1,T2,P1,P2)
      DIMENSION VPC(2,NC),T1(NC),T2(NC),P1(NC),P2(NC)
      WRITE(6,1)
    1 FORMAT(//,1X,'VAPOR PRESSURE CONSTANTS')
      DO 10 J=1,NC
      VPC(2,J)=(T2(J)+460.)*(T1(J)+460.)*ALOG(P2(J)/P1(J))
     + /(T1(J)-T2(J))
      VPC(1,J)=ALOG(P1(J)) - VPC(2,J)/(T1(J)+460.)
   10 WRITE(6,2) J,VPC(1,J),VPC(2,J)
    2 FORMAT(5X,I2,3X,2F15.4)
      RETURN
      END
C
      SUBROUTINE PSAT(NC,VPC,T,PS)
      DIMENSION VPC(2,NC),PS(NC)
      DO 10 J=1,NC
   10 PS(J)=EXP(VPC(1,J)+VPC(2,J)/(T+460.))
      RETURN
      END
C
      SUBROUTINE ENTH(NC,X,Y,T,HL,HV,CL,CV,DH,MW)
      DIMENSION X(NC),Y(NC),CL(NC),CV(NC),DH(NC),MW(NC)
      REAL MW
      HL=0.
      HV=0.
```

TABLE 11-1 CON'T.

```
      DO 10 J=1,NC
      HL=HL+CL(J)*X(J)*MW(J)*(T-200.)
   10 HV=HV+(CV(J)*(T-200.)+DH(J))*Y(J)*MW(J)
      RETURN
      END
C
      SUBROUTINE DEWPT(NC,VPC,Y,P,T,X,PS)
      DIMENSION VPC(2,NC),X(NC),Y(NC),PS(NC)
      DT=10.
      F=0.
      DO 10 J=1,NC
      PS(J)=EXP(VPC(1,J)+VPC(2,J)/(T+460.))
      F=F+Y(J)/PS(J)
   10 CONTINUE
      FOLD=1./F-P
   20 T=T+DT
      LOOP=LOOP+1
      IF(LOOP.GT.20)THEN
      WRITE(6,11)
   11 FORMAT(1X,'DEWPT LOOP')
      STOP
      ENDIF
      FNEW=0.
      DO 15 J=1,NC
      PS(J)=EXP(VPC(1,J)+VPC(2,J)/(T+460.))
   15 FNEW=FNEW+Y(J)/PS(J)
      IF(LOOP.GT.15) WRITE(6,1)T,DT,FNEW
    1 FORMAT(1X,3F10.4)
      FNEW=1./FNEW   -   P
      IF(ABS(FNEW).LT.P/10000.)THEN
      DO 17 J=1,NC
   17 X(J)=Y(J)*P/PS(J)
      RETURN
      ENDIF
      DT=-DT*FNEW/(FNEW-FOLD)
      FOLD=FNEW
      GO TO 20
      END
C
      SUBROUTINE BUBPT(NC,VPC,X,P,T,Y,PS)
      DIMENSION VPC(2,NC),X(NC),Y(NC),PS(NC)
      LOOP=0
   20 DF=0.
      F=0.
      DO 10 J=1,NC
      PS(J)=EXP(VPC(1,J)+VPC(2,J)/(T+460.))
      F=F+X(J)*PS(J)
      DF=DF-X(J)*PS(J)*VPC(2,J)/(T+460.)**2
   10 CONTINUE
      LOOP=LOOP+1
      IF(LOOP.GT.20)THEN
      WRITE(6,11)
   11 FORMAT(1X,'BUBPT LOOP')
      STOP
      ENDIF
      F=F-P
      IF(LOOP.GT.15)WRITE(6,1)T,F,DF,LOOP
    1 FORMAT(1X,3F10.5,I3)
      IF(ABS(F/P).LT. .000001)GO TO 100
      T=T-F/DF
      GO TO 20
  100 DO 101 J=1,NC
  101 Y(J)=X(J)*PS(J)/P
      RETURN
      END
```

TABLE 11–2
RESULTS OF TEST CASES FOR PROGRAM IN TABLE 11–1

```
VAPOR PRESSURE CONSTANTS
     1            13.8985      -6210.7230
     2            16.7832      -9054.4985
TBUB =    194.72    TDEW =     214.13

     Z1       Z2       F        TF        P          Q
   .2500    .7500   3500.0    250.0     35.0         .0

       X1       X2       Y1       Y2      L       V       T
   .1886    .8114   .4890   .5110 2784.6   715.4   201.4
TBUB =    194.72    TDEW =     214.13

     Z1       Z2       F        TF        P          Q
   .2500    .7500   3500.0    300.0     35.0         .0

       X1       X2       Y1       Y2      L       V       T
   .1486    .8514   .4109   .5891 2147.1 1352.9   206.0
TBUB =    209.58    TDEW =     227.34

     Z1       Z2       F        TF        P          Q
   .2500    .7500   3500.0    250.0     45.0         .0

       X1       X2       Y1       Y2      L       V       T
   .2054    .7946   .4951   .5049 2960.9   539.1   214.2
TBUB =    209.58    TDEW =     227.35

     Z1       Z2       F        TF        P          Q
   .2500    .7500   3500.0    300.0     45.0         .0

       X1       X2       Y1       Y2      L       V       T
   .1629    .8371   .4177   .5823 2303.3 1196.7   218.8
```

the liquid feed, and the heat transfer rate Q are specified, along with the feed flow rate and composition. The specific numerical example uses propylene oxide and tertiary butyl alcohol, where ideal VLE is assumed. Enthalpies are calculated in the subroutine ENTH from heat capacity and heat of vaporization data. Both bubblepoint and dewpoint subroutines are used to determine whether flashing will occur with the specified heat transfer rate, flash pressure, and feed temperature. Subroutine VAPC calculates the Antoine vapor pressure constants. The main program calls the subroutine FLASH, which will also be used in chapter 13 to do nonequimolal overflow distillation calculations. Figure 11–6 gives a flowchart of the program.

Table 11–2 shows that (1) as the temperature of the liquid feed is raised, more flashing occurs and the flash temperature increases; and (2) as the flash pressure is increased, less flashing occurs and the flash temperature increases. Note that the flash temperature lies between the bubblepoint and the dewpoint temperatures.

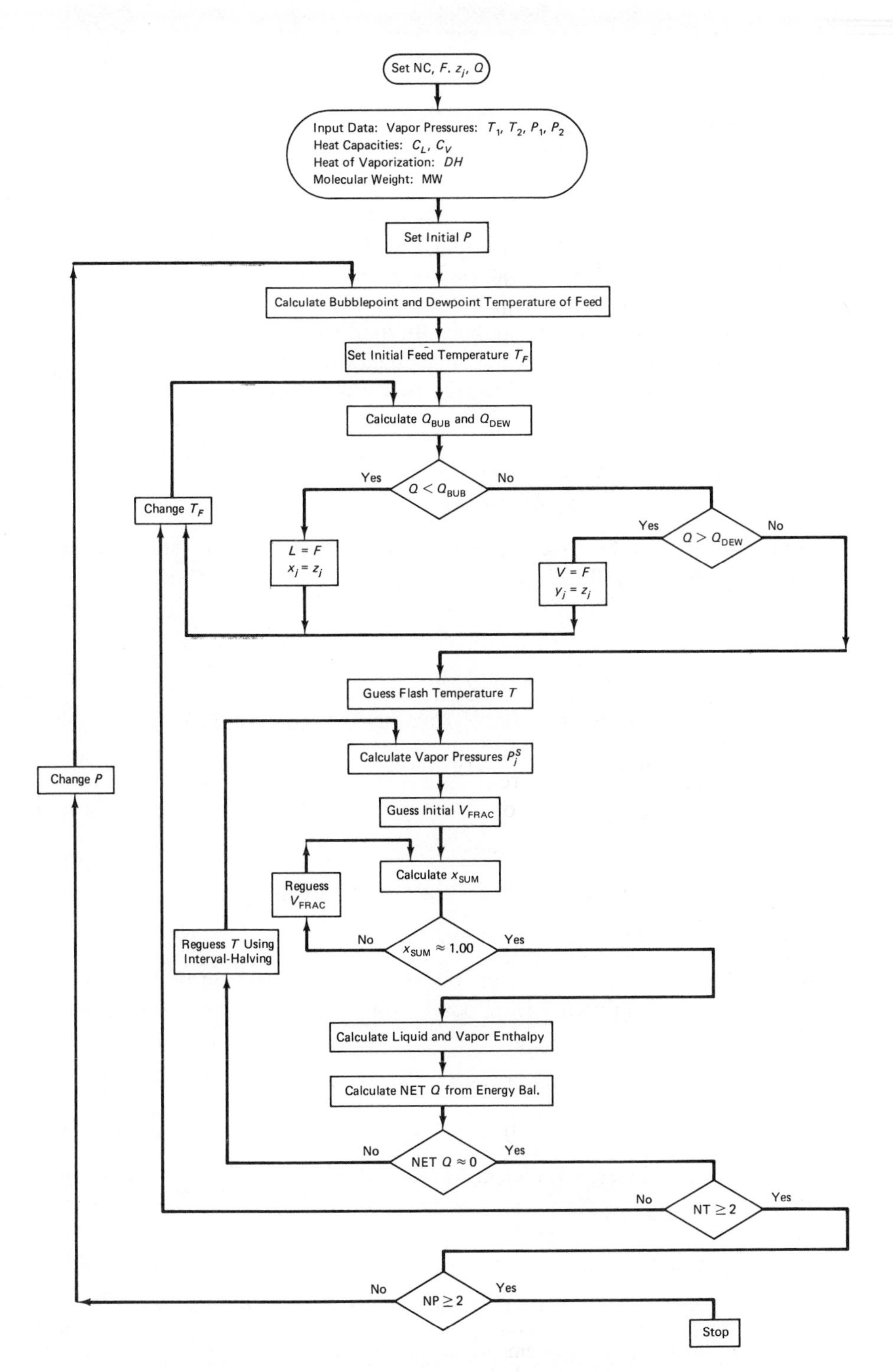

Figure 11–6 Computer flowchart for flash calculation program.

11.3 GRAPHICAL SOLUTION FOR BINARY SYSTEMS

For binary systems, it is easy to do nonisothermal flash calculations on an *Hxy* diagram. The methods are similar to those used in chapter 9 to solve ternary liquid-liquid extraction problems. In the ternary system there were two component-balance equations and one total mass balance that had to be solved simultaneously; in binary nonisothermal flash calculations, there are one component-balance, one total mass balance, and one energy-balance equation. So the same graphical mixing rule method can be used.

11.3.1 Basic Graphical Mixing Techniques on a Binary *Hxy* Diagram

Suppose we have two streams, one liquid and one vapor. Then we can characterize them as follows:

Liquid

Flow rate = L (mole/time)

Composition = x (mole fraction light component)

Enthalpy = h (energy/mole)

Vapor

Flow rate = V (mole/time)

Composition = y (mole fraction light component)

Enthalpy = H (energy/mole)

If we mix the two streams together, the resulting mixture can be located on an *Hxy* diagram by applying total mass, component, and energy balances. We have

$$\text{Total Mass Balance: } L + V = S \tag{11-17}$$

$$\text{Component Balance: } xL + yV = Sz_s \tag{11-18}$$

$$\text{Energy Balance: } hL + HV = Sh_s \tag{11-19}$$

where

S = mole/time of mixture

z_s = composition of mixture (mole fraction light component)

h_s = enthalpy of mixture (energy/mole)

Equation 11–19 assumes adiabatic mixing (no heat added or removed from the system). Combining equations 11–17 and 11–18 gives

$$xL + yV = (L + V)z_s$$

$$(x - z_s)L = (z_s - y)V \tag{11–20}$$

$$\frac{L}{V} = \frac{y - z_s}{z_s - x}$$

Combining equations 11–17 and 11–19 gives

$$hL + HV = (L + V)h_s$$

$$(h_s - h)L = (H - h_s)V \tag{11–21}$$

$$\frac{L}{V} = \frac{H - h_s}{h_s - h}$$

Equating equations 11–20 and 11–21 we have

$$\frac{y - z_s}{z_s - x} = \frac{H - h_s}{h_s - h} = \frac{L}{V} \tag{11–22}$$

Rearranging yields

$$\frac{y - z_s}{H - h_s} = \frac{z_s - x}{h_s - h} \tag{11–23}$$

The point S is sketched on an Hxy diagram in Figure 11–7. S will have coordinates (z_s,h_s). The point L will have coordinates (x,h), and the point V will have coordinates (y,H). The tangents of the two angles α and β are

$$\tan \alpha = \frac{H - h_s}{y - z_s}$$

$$\tan \beta = \frac{h_s - h}{z_s - x}$$

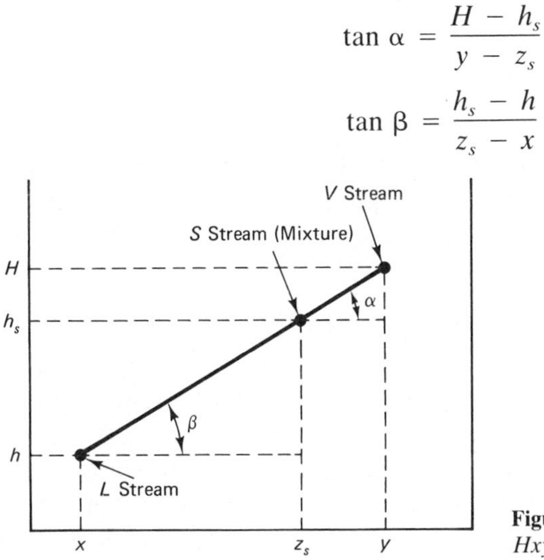

Figure 11–7 Graphical mixing rule on *Hxy* diagram.

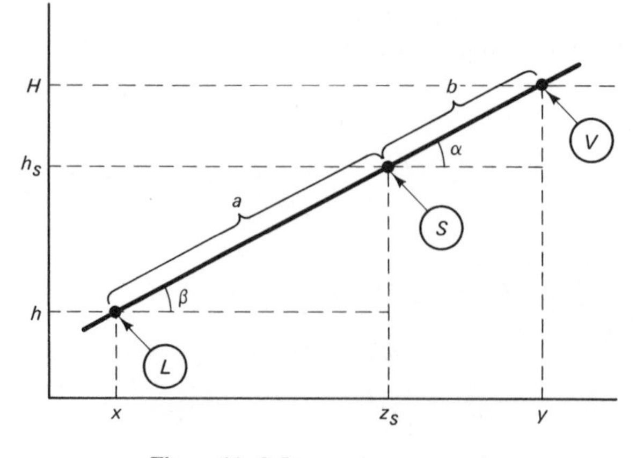

Figure 11–8 Inverse lever-arm rule.

Equation 11–23 shows that tan α = tan β. Therefore, α = β, and the mix point S lies on a straight line between the points L and V. This is exactly the same result we obtained in chapter 9 for ternary liquid-liquid extraction systems.

The *inverse lever-arm rule* can be derived for the *Hxy* diagram. In Figure 11–8, the two triangles shown are similar because the angles α and β are equal. Therefore, the ratio *a*/*b* of the hypotenuses is equal to the ratio of the horizontal sides. From equation 11–22, we have

$$\frac{V}{L} = \frac{z_s - x}{y - z_s} = \text{ratio of horizontal sides}$$

So

$$\frac{a}{b} = \frac{V}{L} \qquad (11\text{–}24)$$

Equation 11–24 is the inverse lever-arm rule. It says, for example, that if we mix 5 moles of liquid with 10 moles of vapor, the mix point S will lie two-thirds of the way from the liquid point to the vapor point, i.e., closer to the vapor point.

11.3.2 Adiabatic Flash on an *Hxy* Diagram

Let us now apply graphical mixing to an adiabatic flash problem. Figure 11–9(a) gives an *Hxy* diagram for a binary feed stream at some high pressure P_F. Let us suppose that the feed is saturated liquid at this pressure and that its composition is z. The enthalpy h_F of the feed can then be read off the diagram on the saturated liquid line at composition z. In a similar manner, Figure 11–9(b) sketches the *Hxy* diagram for the same binary system, but at

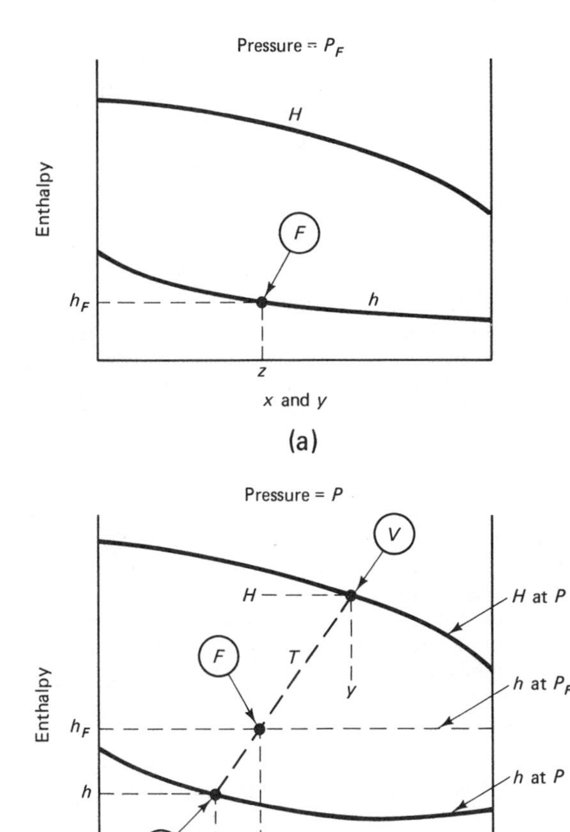

Figure 11–9 *Hxy* diagrams at (a) high and (b) low pressure.

a lower pressure P. The liquid feed is flashed adiabatically into a flash drum operating at pressure P. We wish to determine how much of the feed flashes, what are the compositions of the vapor and liquid streams, and what is the flash temperature.

We begin the procedure by getting the enthalpy of the feed from Figure 11–9(a) at the higher pressure. Then we plot the point F with coordinates (z, h_F) on the *Hxy* diagram at the flash pressure. Next, we find the VLE tie-line that passes through F. The graphical mixing rule says that this straight line must intersect the saturated liquid and saturated vapor lines at the compositions of the two streams leaving the system. Note that for the adiabatic flash calculation performed in subsection 11.2.1, we mixed two streams together to get one total mix stream. Here we are doing the reverse: coming in with one stream and producing two. But the mathematics and graphical procedure are the same. Thus, the compositions x of the liquid and y of the vapor and the flash temperature can be directly and easily read off the *Hxy* diagram, and then the flow rates can be calculated from component and total mass balances.

Example 11.5.

10,000 lb_m/hr of saturated liquid at 1 atmosphere is fed into a flash drum operating at 0.1 atmosphere with no heat addition or removal. The feed stream is a binary mixture of 29 weight percent ethanol and 71 weight percent water. What is the temperature in the flash drum, what are the vapor and liquid flow rates leaving the flash drum, and what are the compositions of the vapor and liquid phases?

Solution:

Figure 11–10 gives the *Hxy* diagrams for ethanol and water at two different pressures: 0.1 and 1 atmosphere. The point F is located on the saturated liquid line at 1 atmosphere. Note that the temperature of the feed (before the flash) can be read off the VLE tie-line to be 185°F.

The VLE tie-line through F is drawn to the saturated liquid and saturated vapor lines at 0.1 atmosphere. The temperature of the flash drum is the temperature on this tie-line at 0.1 atmosphere: 97°F. The compositions are $x = 22$ weight percent and $y = 68$ weight percent ethanol.

The flow rates of liquid and vapor are calculated from total and component balances as follows.

$$\text{Total mass: } F = L + V = 10{,}000 \text{ lb}_m/\text{hr}$$

$$\text{Component (ethanol): } zF = xL + yV$$

$$(0.29)(10{,}000) = (0.22)(10{,}000 - V) + (0.68)(V)$$

$$V = 1{,}522 \text{ lb}_m/\text{hr}$$

$$L = 8{,}478 \text{ lb}_m/\text{hr}$$

Note that about 15 percent of the liquid feed flashed and that, as expected, the temperature dropped from 185°F to 97°F during the flash because of the conversion of sensible to latent heat.

11.3.3 Nonadiabatic Flash on an *Hxy* Diagram

If heat is transferred into the flash drum at a specified rate Q, the energy balance (equation 11–19) will include another term:

$$hL + HV = h_F F + Q \tag{11–25}$$

The right side of this equation can be written

$$h_F F + Q = \left(h_F + \frac{Q}{F}\right)F \tag{11–26}$$

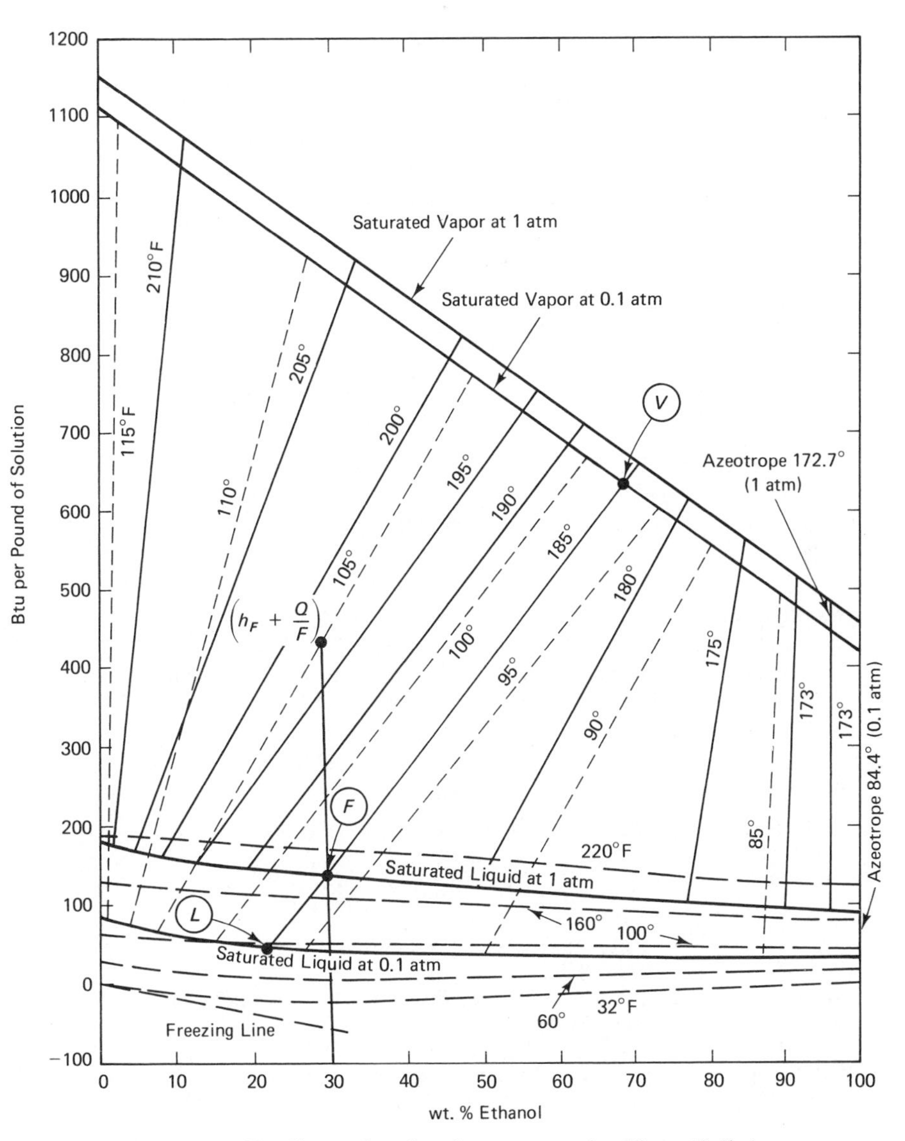

Figure 11–10 *Hxy* diagram for ethanol-water system (see Figure 10–7) showing the location of $(h_F + Q/F)$.

where the quantity $(h_F + Q/F)$ can be considered to be a pseudo feed enthal| h_Δ. We then locate a Δ point that has this enthalpy and a composition z th is the same as the feed. A straight VLE tie-line through this point mu intersect the h curve at x and the H curve at y. This is illustrated in Figu 11–11, where the point $F(z,h_F)$ is raised vertically by a distance Q/F (Bt lb_m).

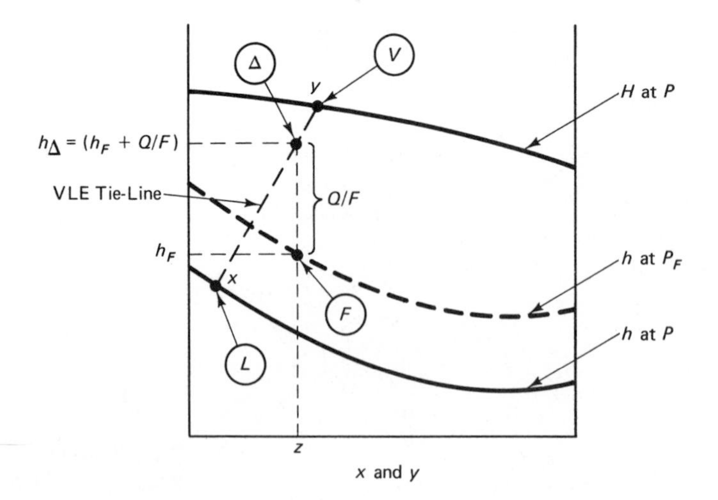

Figure 11–11 Graphical flash calculation.

Example 11.6.

Suppose the flash drum system described in example 11.5 has an internal steam coil which adds 3.1×10^6 Btu/hr of heat. What are the flow rates, compositions, and temperatures of the vapor and liquid streams?

Solution:

The tank is no longer adiabatic, so the pseudo enthalpy point h_Δ must be calculated:

$$h_\Delta = h_F + \frac{Q}{F} = 130 + \frac{(3.1 \times 10^6 \text{ Btu/hr})}{(10{,}000 \text{ lb}_m/\text{hr})}$$

$$= 440 \text{ Btu/lb}_m$$

We thus locate the point Δ at an enthalpy of 440 Btu/lb$_m$ and a composition of 29 weight percent ethanol (the feed composition) on Figure 11–10. The VLE tie-line for 0.1 atmosphere pressure that goes through this point has a temperature of 105°F. This is the temperature of the flash tank. The x and y values at the two ends of this tie-line are $x = 7$ weight percent ethanol and $y = 48$ weight percent ethanol.

Overall and ethanol component balances are solved to get the flow rates:

$$(0.29)(10{,}000) = (0.07)(10{,}000 - V) + (0.48)(V)$$

$$V = 5{,}366 \text{ lb}_m/\text{hr}$$

$$L = 4{,}634 \text{ lb}_m/\text{hr}$$

Note that the addition of heat to the flash tank has resulted in a higher temperature (105°F) than when the process was run adiabatically (97°F). Also, more vapor is formed (5,366 lb$_m$/hr versus 1,522 lb$_m$/hr). This is exactly what we would expect.

11.4 ENERGY BALANCES FOR REACTORS

In chapter 10, sensible heat effects and heats of reaction were discussed. These two concepts can be combined in order to calculate heats of reactions at temperatures other than the standard 25°C. They also can be used to calculate the temperatures that occur in adiabatic and other types of reactors.

An adiabatic reactor example illustrates some of these concepts. Energy balances are a very important part of any analysis of a chemical reactor, particularly when the heat of reaction is large, because reaction rates and equilibrium constants are often very sensitive to temperature changes.

Example 11.7.

A fuel oil of nominal composition $C_{20}H_{40}$ is burned in 50 percent excess air. The process is adiabatic and goes to completion. Fuel and air enter at 77°F. What is the temperature of the flue gases?

Solution:

The balanced reaction is

$$C_{20}H_{40} + 30\ O_2 \longrightarrow 20\ CO_2 + 20\ H_2O_{(g)} \tag{11-27}$$

Note that the water is specified as being in the gaseous phase, not in the liquid phase.

We choose as a basis 1 g-mole of fuel oil. The amount of oxygen fed in is 1.5 times that required. Thus, we have

$$(1.5)(30) = 45 \text{ g-mole of } O_2 \text{ fed in}$$

The amount of nitrogen in the air that comes with the oxygen is

$$(45 \text{ g-mole } O_2)\left(\frac{79 \text{ mole } N_2}{21 \text{ mole } O_2}\right) = 169.3 \text{ g-mole } N_2 \text{ fed in}$$

The oxygen leaving will be the amount fed in minus the amount that reacts:

$$45 - 30 = 15 \text{ g-mole } O_2 \text{ leaving}$$

The hot gases leaving the burner will contain the following materials:

$$N_2 = 169.3 \text{ g-mole}$$

$$O_2 = 15.0 \text{ g-mole}$$

$$CO_2 = 20.0 \text{ g-mole}$$

$$H_2O = 20.0 \text{ g-mole}$$

$$\text{Total} = 224.3 \text{ g-mole of product gases}$$

We will use the same two-step approach used in chapter 10 to calculate the unknown temperature of the hot gases leaving the burner. First, the reactants are converted into products at a constant temperature of 77°F, and the heat of reaction at this temperature is calculated. Then we determine to what temperature the products will be heated by this much heat. Since the burner is adiabatic, Q is zero, and all of the heat of reaction is converted into sensible heat of the products. Figure 11–12 sketches the two steps. We have

$$\Delta H_1 + \Delta H_2 = \Delta H = 0 \tag{11–28}$$

where

$\Delta H_1 = \Delta H_R$ at 77°F

$\Delta H_2 =$ sensible heat of heating the products from 77°F to T

This two-step approach, which does not follow the reaction path, can be used because enthalpy is a state function. The heat of reaction at 77°F can be calculated from the standard heats of formation or from the heat of combustion given in Table 10–4. We will use the latter (ΔH_c). The values for C_{20} are obtained by adding 157,440 for each carbon atom above C_6 to the C_6 value. We get

$$-\Delta H_c = 1{,}002{,}570 + (157{,}440)(14) = 3{,}206{,}700 \text{ cal/g-mole}$$

Since the basis of heat of combustion is liquid water, we must subtract the latent heat of vaporization of water (18,000 Btu/lb-mole), obtaining

$$\begin{aligned}\Delta H_1 &= [-(3{,}206{,}700 \text{ cal/g-mole})(454 \text{ g/lb}_m)(\text{Btu/252 cal}) \\ &\quad + (18{,}000 \text{ Btu/lb-mole})](20 \text{ lb-mole } H_2O) \\ &= -5{,}410{,}000 \text{ Btu/lb-mole fuel oil}\end{aligned}$$

Since mean heat capacities are functions of the final temperature T, an iterative solution is required. We guess a temperature, use Figure 10–2 to read off the mean heat capacities, and calculate ΔH_2. If ΔH_2 does not equal ΔH_1, a new temperature is guessed.

1. Guess $T = 2{,}500$°F

	C_p^{mean} (*Btu/lb-mole °F*)
CO_2	12.4
H_2O	9.7
O_2	8.3
N_2	7.7

$$\begin{aligned}\Delta H_2 &= [(20)(12.4) + (20)(9.7) + (15)(8.3) + (169.3)(7.7)][2{,}500 - 77] \\ &= 4{,}530{,}000 \text{ Btu/lb-mole}\end{aligned}$$

Since ΔH_2 is less than ΔH_1, we must make a higher temperature guess.

2. Guess $T = 3{,}000$°F

The mean heat capacities at 3,000°F are found in Figure 10–2:

$$CO_2 = 12.75, H_2O = 10.1, O_2 = 8.3, \text{ and } N_2 = 7.9 \text{ Btu/lb-mole °F}.$$

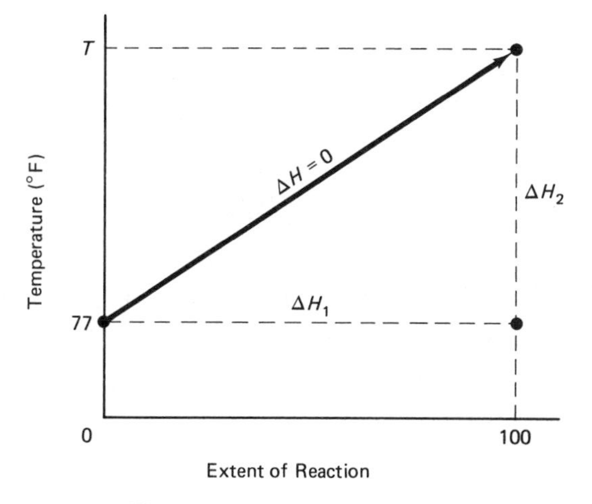

Figure 11–12 Alternative paths.

We thus have

$$\Delta H_2 = [(12.75)(20) + (20)(10.1) + (15)(8.3) + (169.3)(7.9)][3{,}000 - 77]$$
$$= 5{,}600{,}000 \text{ Btu/lb-mole}$$

Interpolating between the two guesses, we get a gas temperature of about 2,900°F.

If the temperature of the reactants had been not 77°F, but some temperature T_{in}, a third step in the ΔH calculation would have been needed. That is, a ΔH_3 obtained by cooling the reactants from T_{in} to 77°F would be added to ΔH_1 and ΔH_2. If the reactor were not adiabatic, the heat transfer rate would have to be included. If the gas did work (as it would in a combustion turbine), the work term would have to be included in the energy balance ($\Delta H = Q - W$).

The application of an energy balance to a liquid-phase reactor with a specified heat of reaction and known heat capacities is given in the next example.

Example 11.8.

A 500-gallon continuous stirred-tank reactor (CSTR) is fed 2,780 lb_m/hr of feed at 70°F with a concentration of 0.5 lb-mole of reactant A/ft^3. Some of the reactant A is consumed in the reactor, producing product B. The concentration of reactant in the stream leaving the reactor is 0.245 lb-mole of A/ft^3 and the temperature of this stream is 140°F. The heat capacity of the reactant and product are both 0.75 Btu/lb_m °F, and their densities are both 50 lb_m/ft^3. The reaction is exothermic, giving off 30,000 Btu/lb-mole of A reacted. How much heat must be removed from the reactor? If cooling water at 70°F is fed into a cooling jacket surrounding the reactor and leaves the jacket at 118.3°F, how much cooling water must be used?

Solution:

First, we calculate the amount of A that has reacted. Then we do an energy balance on the reaction mass in the reactor to determine the heat transfer rate to the cooling jacket. We obtain

$$(2{,}780\ \text{lb}_m/\text{hr})(\text{ft}^3/50\ \text{lb}_m)(0.5 - 0.245\ \text{lb-mole of } A/\text{ft}^3)$$

$$= 14.18\ \text{lb-mole of } A \text{ reacted/hr}$$

$$(14.18\ \text{lb-mole/hr})(30{,}000\ \text{Btu/lb-mole})$$

$$= 0.425 \times 10^6\ \text{Btu/hr heat generated from the reaction}$$

$$\Delta H_R = -0.425 \times 10^6\ \text{Btu/hr}$$

ΔH_R is negative because the chemical energy of the product stream leaving the reactor is lower than the chemical energy of the feed stream—i.e., the reaction is exothermic.

An energy balance yields

$$\Delta H = Q - W$$

$$\Delta H_R + \Delta H_{\text{sensible}} = Q_{\text{jacket}}$$

$$(-0.425 \times 10^6) + (2{,}780\ \text{lb}_m/\text{hr})(0.75\ \text{Btu/lb}_m\ {}^\circ\text{F})(140 - 70^\circ\text{F})$$

$$= (-0.425 + 0.146) \times 10^6\ \text{Btu/hr} = Q_{\text{jacket}}$$

$$Q_{\text{jacket}} = -0.278 \times 10^6\ \text{Btu/hr}$$

All of the heat transferred to the cooling jacket goes into the cooling water to heat it from 70°F up to 118.3°F. Thus,

$$F_w\ (1\ \text{Btu/lb}_m\ {}^\circ\text{F})(118.3 - 70^\circ\text{F}) = 0.278 \times 10^6\ \text{Btu/hr}$$

$$F_w = 5{,}755\ \text{lb}_m/\text{hr}$$

$$= (5{,}755\ \text{lb}_m/\text{hr})(\text{gallon}/8.33\ \text{lb}_m)(\text{hr}/60\ \text{min})$$

$$= 11.5\ \text{gpm of cooling water}$$

PROBLEMS

11–1. A detergent spray drier produces 5,000 lb_m/hr of solid, dried beads containing 5 weight percent moisture by spraying a slurry containing 40 weight percent solids into a cocurrent stream of hot air. This air enters at 400°F with a moisture content of 1 mole of water/100 moles of dry air. The exit air has a moisture content of 10 moles of water/100 moles of dry air.

(a) Calculate the flow rate of inlet moist air in standard cubic feet per minute (SCFM).

(b) Assuming that the dried beads, the exit air, and the inlet slurry are all at the same temperature, calculate this temperature.

Latent heat of vaporization of water = 1,000 Btu/lb_m
Heat capacity of liquid water = 1 Btu/lb_m °F
Heat capacity of water vapor = 0.5 Btu/lb_m °F

Heat capacity of dry air = 7 Btu/lb-mole °F
Molecular weight of dry air = 29 lb_m/lb-mole

11–2. The distillation column of example 11.2 has been revamped such that its distillate product is removed from the reflux drum as a saturated vapor in equilibrium with the liquid reflux in the reflux drum.

(a) Calculate the revised condenser heat removal rate and the cooling water flow rate.

(b) Calculate the revised reboiler heat addition rate.

11–3. 70,000 lb_m/hr of hot oil (c_p = 0.5 Btu/lb_m °F) at 250°F is fed into a cooler. 170.5 gpm of 80°F cooling water is flowing through the cooler in countercurrent flow. The heat transferred from the hot oil into the cooling water can be calculated from the equation

$$Q = UA(\Delta T)_{LM}$$

where

Q = heat transfer rate (Btu/hr)

U = overall heat transfer coefficient

= 120 Btu/hr ft^2 °F

A = heat transfer area = 879 ft^2

$(\Delta T)_{LM}$ = log mean temperature difference

$$= \frac{\Delta T_1 - \Delta T_2}{\ln\left(\dfrac{\Delta T_1}{\Delta T_2}\right)}$$

ΔT_1 = Oil inlet temperature − Water exit temperature

ΔT_2 = Oil exit temperature − Water inlet temperature

Calculate the exit oil and water temperature.

11–4. Methanol synthesis is described in chapter 3. In the flow sheet on p. 54, 8,500 lb-mole/hr of synthesis gas (6,500 lb-mole/hr of H_2, 1,500 of CO, and 500 of CO_2) and 25,000 lb-mole/hr of H_2 recycle are fed to a catalytic reactor at 500°F. All of the CO and CO_2 is converted to methanol in an adiabatic reaction. Determine the reactor outlet temperature.

11–5. Liquid bottoms product from a benzene column (35 lb-mole/hr) at 200°C and composition 1.67 mole percent benzene, 56.66 mole percent toluene, and 41.67 mole percent o-xylene is flashed adiabatically into a toluene column operating at 760 mm Hg. Heat capacities of all components in liquid and vapor phases are 72 Btu/lb-mole °F. Heats of vaporization of all components are 8,000 Btu/lb-mole. VLE data are ideal.

(a) Calculate how much material is flashed.

(b) Calculate the temperature of the feed as it enters the toluene column.

11–6. A methanol-water liquid mixture containing 35 weight percent methanol at 500 psia and 50°C is heated in an exchanger in which steam is condensed. For liquid flow of 1,000 lb_m/hr, 700 lb_m/hr of saturated steam at 175°C are condensed. The liquid mixture is then expanded through a valve to 1 atm pressure. Determine the composition and quantities of liquid and vapor leaving a phase separator just downstream from the valve if the heat of vaporization of saturated steam at 175°C is 795 Btu/lb_m.

11–7. A gas mixture containing 30 mole percent CH_4, 50 mole percent C_2H_4, and 20 mole percent C_2H_6 is cooled to its bubblepoint at 700 psia. The liquid is then expanded to 500 psia in an adiabatic valve. What fraction of the liquid vaporizes, and what is the final fluid temperature? The latent heat of methane is 3,500 Btu/lb-mole, of ethylene is 5,800 Btu/lb-mole, and of ethane is 6,300 Btu/lb-mole.

11–8. 90,000 lb_m/hr of feed at 50°C is fed into a drum equipped with a heating coil. The feed is a binary mixture of 30 weight percent methanol and 70 weight percent water. Low-pressure steam (20 psig) with a saturation temperature of 320°F is condensed inside the heating coil to heat the methanol-water stream. The heat transfer area of the coil is 1,019 ft^2. The overall heat transfer coefficient U is 500 Btu/hr ft^2 °F, and the heat transfer rate Q is given by the equation

$$Q = UA(T_s - T)$$

where

T_s = condensing temperature of steam in the coil

T = temperature of the methanol-water mixture in the drum

The drum operates at atmospheric pressure.

(a) What is the temperature T?

(b) What are the vapor and liquid rates leaving the drum, and what are the compositions of the vapor and liquid?

11–9. In problem 4–26 the reforming of cyclohexane to generate benzene is mentioned as occurring at 900°F.

(a) Is the reaction exothermic or endothermic at 900°F?

(b) If the temperature of the feed gas to the reformer reactor is 900°F and the product gas leaving the reactor is at 900°F, how much heat must be added or removed from the reformer reactor? Express the answer in Btu/hr.

Note: If the mean molar heat capacities of cyclohexane and benzene are needed, use the data in Figure 10–1 together with the following approximations:

$$(C_p)^{mean}_{benzene} = 3\ (C_p)^{mean}_{C_2H_4}$$

$$(C_p)^{mean}_{cyclohexane} = 3.6\ (C_p)^{mean}_{C_2H_4}$$

11–10. A distillation column, operating at 760 mm Hg, is fed 10,000 lb_m/hr of a 40 weight percent solution of methanol in water at 30°C. Distillate product from a total condenser is 98 weight percent methanol, and bottoms product is 2 weight percent methanol. A reflux ratio of 3 is used. The column has a kettle reboiler, reflux is saturated liquid, and heat losses from the column are negligible.

(a) Calculate the heat removed in the condenser.
(b) Calculate the heat added in the reboiler.
(c) What is the vapor boilup rate in the reboiler in lb_m/hr?
(d) How much saturated 50 psig steam must be introduced into the tubes of the reboiler? A steam trap prevents steam vapor from leaving, i.e., all the steam is condensed. Tube side pressure is 40 psig, and steam table data are as follows:

Pressure (psig)	*Enthalpy sat. liquid (Btu/lb_m)*	*Enthalpy sat. vapor (Btu/lb_m)*
50	267.5	1,179.1
40	256.3	1,175.8

11–11. A liquid solution of 40 weight percent methanol and 60 weight percent water is partially vaporized in a furnace. The inlet stream to the furnace is 22,000 lb_m/hr at 50°C, and the exit stream from the furnace is at atmospheric pressure. The fuel burned in the furnace is 1,000 lb_m/hr of liquid toluene. 50 percent excess air is used to produce gaseous H_2O and CO_2. Stack gas leaves the furnace at 800°F. Both inlet air and toluene fuel are fed to the furnace at 25°C.

What is the exit temperature of the methanol-water stream, and what fraction of it has vaporized?

11–12. According to the American Red Cross *Standard First Aid and Personal Safety* manual, 2nd Edition, p. 70 (1979), an average adult breathes 18 times per minute. Approximately one pint ($\frac{1}{8}$ gallon) of air containing 0.04 mole percent CO_2 is inhaled per breath. The gas exhaled contains 4.0 mole percent CO_2.

Calculate the daily Calorie consumption of an average adult. [Note that the "Calorie" (with a capital "C") used in nutrition science is a "kilocalorie" (with a lower case "c") in chemical engineering]. Assume that the body converts carbon to CO_2.

11–13. How much liquid ammonia-water solution must be fed into a flash drum in order to remove 8×10^6 Btu/hr of heat? The feed stream is 50 weight percent NH_3 and is liquid at 90°C. The flash drum operates at 100°C and 10 kg/cm^2 pressure.

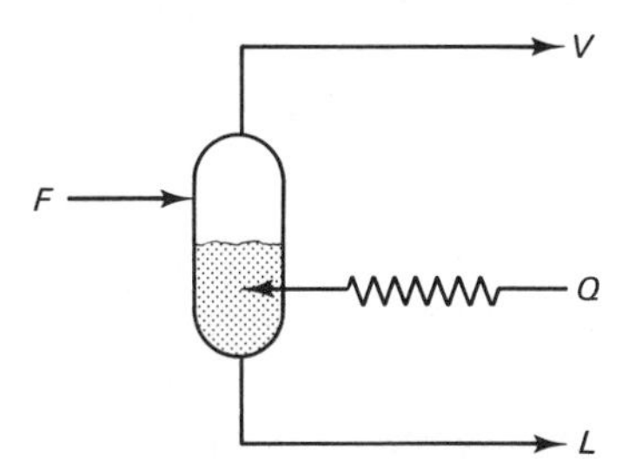

What are the compositions and flow rates of the liquid and vapor streams leaving the flash drum? An *Hxy* diagram for ammonia and water is given below.

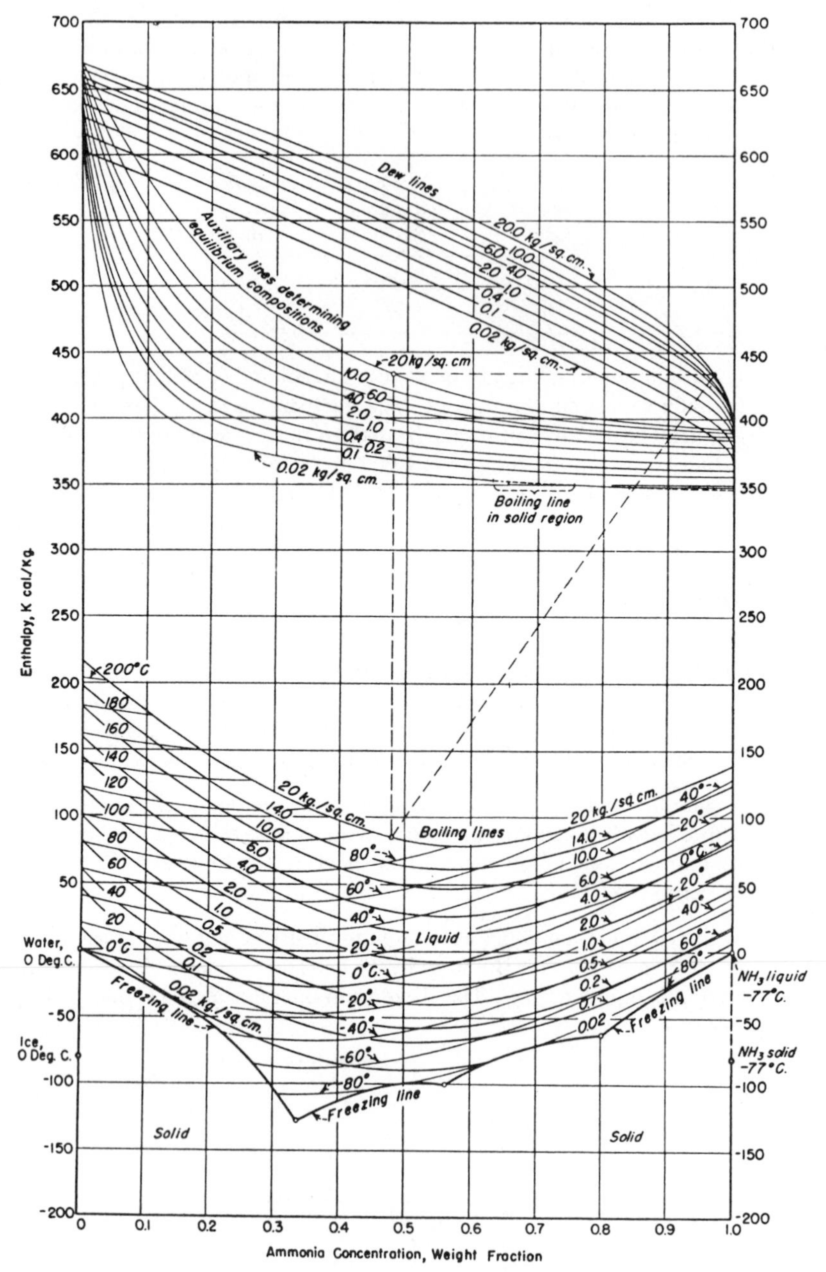

Enthalpy-concentration diagram for aqueous ammonia. Reference states: Enthalpies of liquid water at 0°C and liquid ammonia at −77°C are zero. *From Bosnjakovic,* Technische Thermodynamik, *T. Steinkopff, Leipzig, 1935. Copyright © 1935 John Wiley & Sons, Inc. Reprinted by permission.*

11–14. A mixture of 25 mole percent ethane, 35 mole percent propane, and 40 mole percent isobutane at its bubblepoint at 60 psia is used to absorb heat from another process stream after passing through an adiabatic valve and expanding to 1 atm pressure.

(a) What is the maximum amount of heat absorbable if only saturated vapor mixture exits the heat exchanger?

(b) If half the heat in part (a) is absorbed, what are the compositions and flow rates of the exiting vapor and liquid streams?

(c) What are the temperatures of the exiting stream in (a) and (b)?

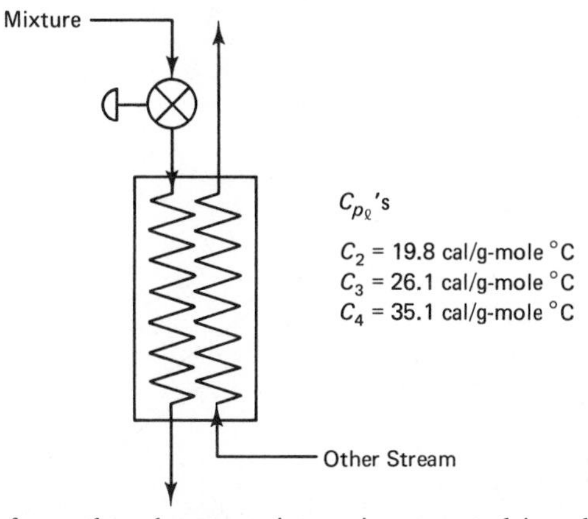

11–15. 50,000 lb_m/hr of an ethanol-water mixture is separated in a heat-integrated distillation column system sketched as follows:

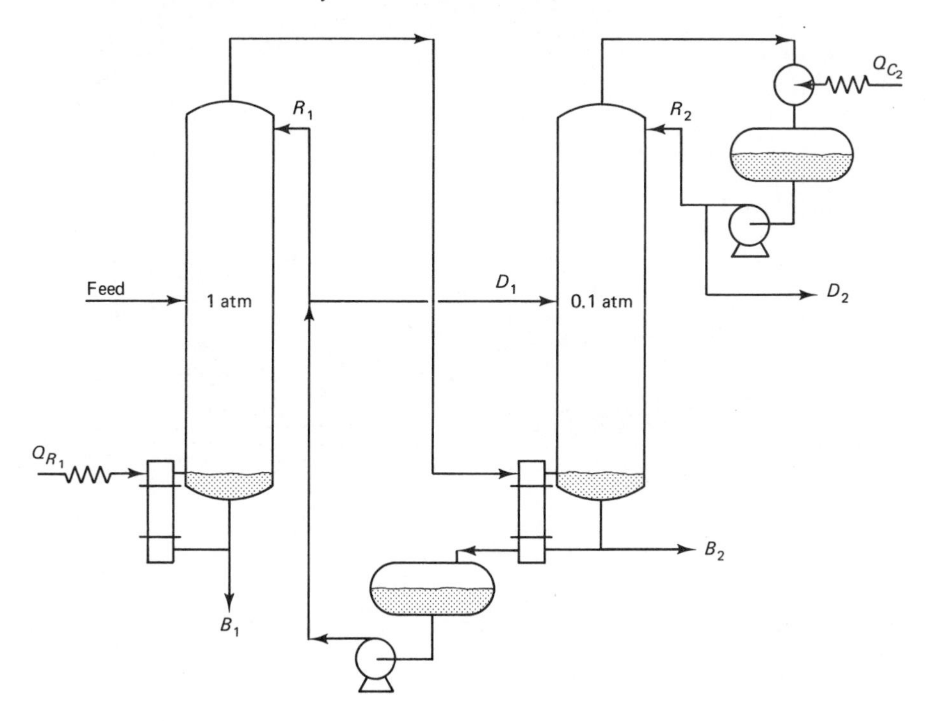

The feed enters the high-pressure column (1 atm) at 195°F. Its composition is 20 weight percent ethanol. Bottoms products from both columns are 1 weight percent ethanol. An *Hxy* diagram for ethanol-water at 0.1 and 1.0 atm is given below.

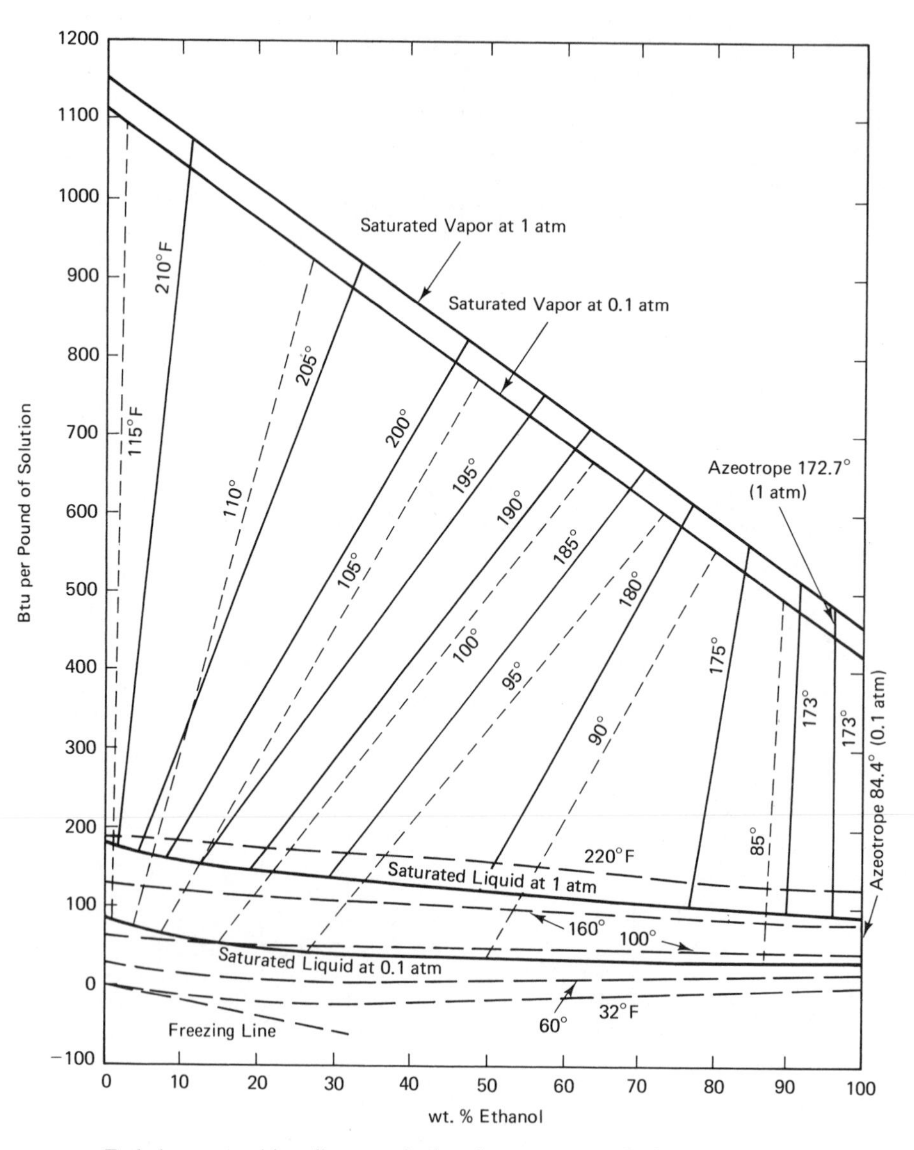

Enthalpy-composition diagram of ethanol-water system. Reference states: pure liquid alcohol and pure liquid water, each at 32°F saturated.

The distillate product from the high-pressure column is 90 weight percent ethanol, which is fed into the low-pressure (0.1 atm) column. Distillate product from the low-pressure column is 99 weight percent ethanol. Assuming saturated liquid products and a reflux ratio (R_2/D_2) in the low-pressure column of 2, determine

(a) The flow rates B_1, D_1, B_2, and D_2.
(b) The temperatures in the base and reflux drums of both columns.
(c) The heat transfer rate Q_{c2} in the low-pressure column condenser.
(d) The heat transfer rate Q_{R1} in the high-pressure column reboiler.

11–16. Hot gases leaving a CO_2 recovery unit at a rate of 2,000 cu ft/min at 350°F and 14.7 psia are passed through a heat recovery train. The gases contain 86 mole percent CO_2 and 14 mole percent H_2O. They leave the unit at 77°F and essentially the same pressure. Since there has been some condensation, the gases now contain only 3.15 mole percent H_2O. How much heat must be removed per minute if the heat of vaporization of steam at 77°F is 1,054 Btu/lb_m?

11–17. A carbon gasification reactor system is shown below.

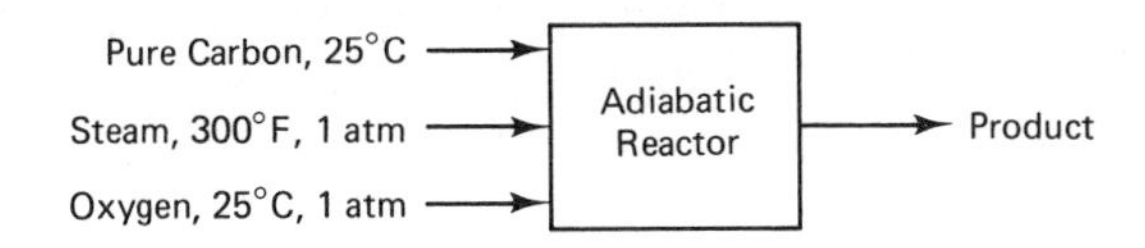

The system operates at atmospheric pressure. Feed rates are

$$\text{pure carbon} = 4.2 \text{ T mole/hr}$$

$$\text{steam} = 8.0 \text{ T mole/hr}$$

$$\text{oxygen} = 1.5 \text{ T mole/hr}$$

where a T mole is a ton mole, or 1,000 kg-mole. The product stream leaving the reactor contains only CO, CO_2, H_2, and H_2O. The molar ratio of CO to CO_2 in the product stream is 1. Assume constant heat capacities as follows:

	C_p *(cal/g-mole °C)*
H_2O (g)	8.8
O_2	7.8
CO	7.6
CO_2	11.4
H_2	7.0

(a) Calculate the composition of the product stream leaving the reactor.
(b) Calculate the temperature of this stream.

(c) If an oxygen feed preheat is added to the system as in the sketch below, what will the temperature of the product stream leaving the heat exchanger be?

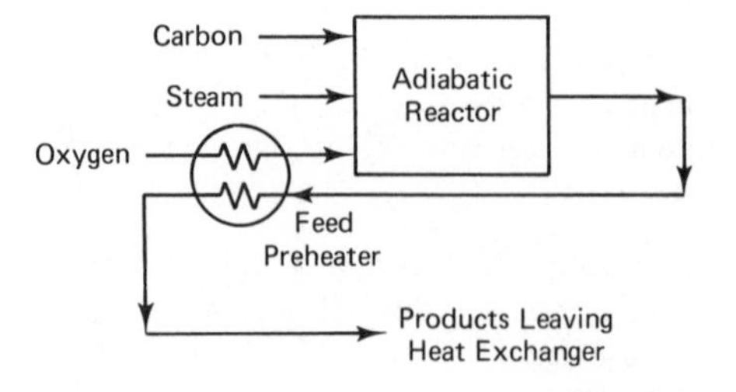

The heat exchanger has an overall heat transfer coefficient of 50 Btu/hr °F ft² and a heat transfer area of 10,000 ft². It is 30 ft long and 8 ft in diameter, with 3,000 1-inch-diameter tubes made of 316 stainless steel.

11–18. In the Texaco partial oxidation process, octane feed stock $C_8H_{18}(g)$ with C_p = 42 cal/g-mole °C and ΔH_f (heat of formation at 25°C) = −49,820 cal/g-mole, is reacted with pure O_2 via the reactions

$$C_8H_{18} + 4O_2 \rightarrow 8\ CO + 9H_2 \qquad \text{(A)}$$

and

$$C_8H_{18} + 8O_2 \rightarrow 8\ CO_2 + 9H_2 \qquad \text{(B)}$$

Octane feed enters the reactor continuously at 1,000 °F, and O_2 feed enters at 77 °F. The composition of the product gas stream is 52.94 mole percent H_2, 42.35 mole percent CO, and 4.71 mole percent CO_2.

(a) On the basis of 100 moles of C_8H_{18} fed in, how many moles of O_2 are fed in?

(b) What fraction of the C_8H_{18} reacts according to equation A?

(c) What is the overall heat of reaction for the two reactions at 25°C?

(d) At what temperature does the product gas leave the adiabatic reactor?

11–19. Brookhaven National Laboratory has recently proposed a process for the formation of methanol by the reaction

$$CO + 2H_2 \longrightarrow CH_3OH(g)$$

using a catalyst in the form of a liquid solution. The catalyst solution is pumped through an external heat exchanger to remove the heat of reaction. The operation is diagrammed in the figure on page 421.

Feed gas enters at 25°C and has a composition of 40 mole percent N_2, 22 mole percent CO and 38 mole percent H_2. Product gas leaves at 100°C.

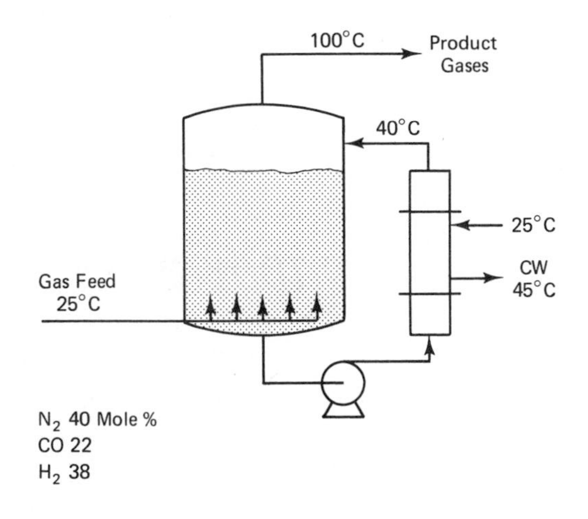

(a) If 95 percent of the H_2 fed to the reactor is converted to CH_3OH, what is the steady-state composition of the product gas stream?

(b) How much energy must be removed from the heat exchanger per 1,000 lb-mole of feed gas? The heat capacity of $CH_3OH(g)$ can be taken as 11.67 Btu/lb-mole °F for this calculation.

(c) If the pump supplies negligible energy to the catalyst solution, at what temperature does the catalyst solution enter the heat exchanger?

(d) The external heat exchanger takes cooling water at 25°C and discharges it at 45°C. The catalyst solution, with C_p = 0.8 Btu/lb_m °F, exits the heat exchanger at 40°C. What is the flow of catalyst solution and of cooling water through the heat exchanger per 1,000 lb-mole of feed gas? What is the heat exchanger surface area required if the overall coefficient of heat transfer is 60 Btu/hr ft^2 °F? (See problem 11–3.)

Evaporation 12

12.1 INTRODUCTION

Evaporation is a process in which a volatile component is removed from a mixture of nonvolatile and volatile components by vaporization. A classical example is the concentration of a salt (NaCl) solution by boiling out the water. The most important characteristic of evaporation is that the vapor from the evaporator (H_2O) contains essentially none of the nonvolatile component (NaCl). This means that a component balance on the nonvolatile component involves only the liquid streams.

In a sense, evaporation is just a special case of nonadiabatic flashing in which the vapor-liquid equilibrium is particularly simple, i.e, the vapor composition is 100 percent volatile component for any liquid composition. Water is by far the most commonly removed volatile component, but other light components are also removed by evaporation. One example is the concentration of a mixture of solid terephthalic acid and liquid acetic acid.

Evaporation is fairly common throughout the chemical process industries in the production of a variety of products: sugar, orange juice, salt, paper, and steel, to name just a few. Waste liquors from papermaking and from metal pickling are concentrated by evaporation. In some cases the solute is concentrated to the point where it begins to crystallize and is remove as a slurry. The evaporator is then sometimes called a *crystallizer*.

Evaporators are extremely energy intensive, consuming large amounts of energy in many processes. Water has a high heat capacity and a large heat of vaporization, and large amounts of water must often be removed. In some cases, especially the food and pharmaceutical industries, the products are temperature sensitive. Means must then be found to evaporate at modest temperature levels, often requiring vacuum operation and large process vessels.

Typically, evaporators are heated by condensing steam on a tube surface, thus boiling the solution being concentrated on the opposite surface of the tube. Low-pressure steam is commonly used. In some cases fuel is burned around tubes in a furnace to produce the required energy. In other cases the fuel is burned while it is in direct contact with the liquid being evaporated. Solar energy is used in several evaporation processes.

12.2 EVAPORATION EQUIPMENT

Evaporation equipment ranges from open shallow pans or ponds (using solar energy) to multiple-effect cascades of specially designed heat transfer units. Schematic diagrams of several commonly used evaporators are shown in Figure 12–1. The horizontal- and vertical-tube evaporators are the oldest designs and are still widely used. Horizontal-tube evaporators are relatively easy to clean, so they are used for scaling applications. Short-tube vertical evaporators are used in sugar-cane concentration.

Long-tube vertical evaporators, with and without forced circulation, have the advantage of developing high liquid velocities in the tubes. This gives high heat transfer coefficients. The forced-circulation type has the further advantages of providing the highest heat transfer rate, a known circulation rate, and reduced exposure to fouling or thermal decomposition. However, it also has higher capital and operating costs due to the circulating pump required.

The long-tube vertical evaporator without forced circulation is the most common type of evaporator, accomplishing over half the total evaporation done. Its biggest use is for the evaporation of black liquor in the pulp and paper industry. In general, it is the cheapest evaporator per unit of evaporating capacity. Usually the unit is operated with the liquid level below the top of the tubes, the entire evaporation being done in a single pass through the tubes. The concentration in the tubes varies from about the feed concentration to the product concentration. This concentration effect, plus the large hydrostatic head, results in a large temperature gradient along the length of the tube. If the feed enters cold at the bottom of the tube, there is a section of rapidly rising temperature; but thereafter the temperature drops to a minimum at the top of the tube.

A falling-film evaporator eliminates the hydrostatic head effect and thus operates at a generally lower temperature than does the long-tube vertical

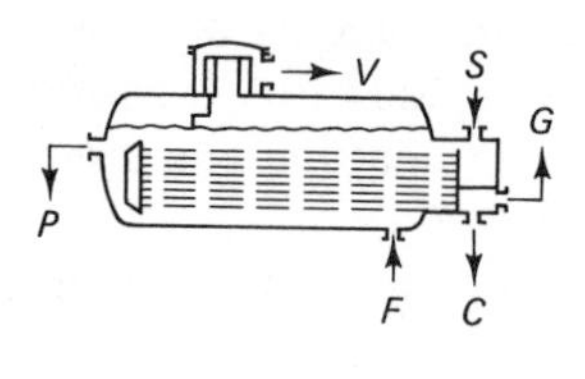

(a)

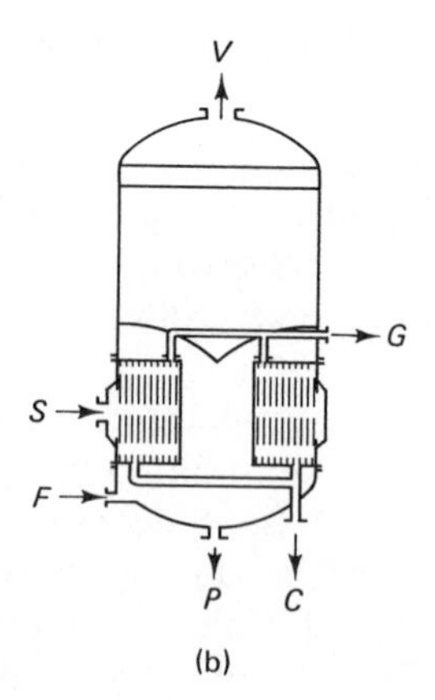

(b)

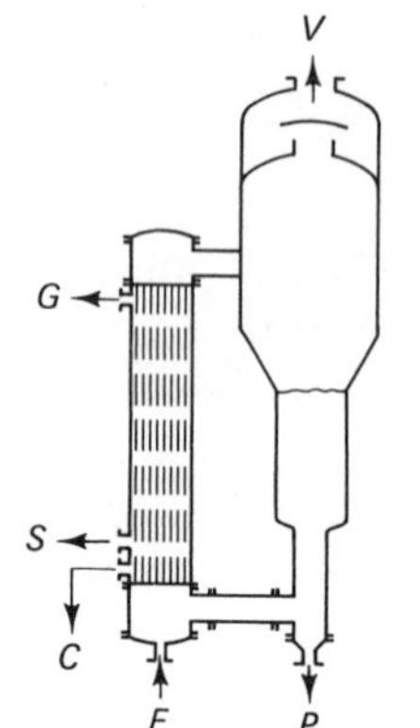

(c)

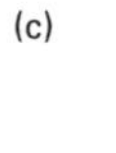

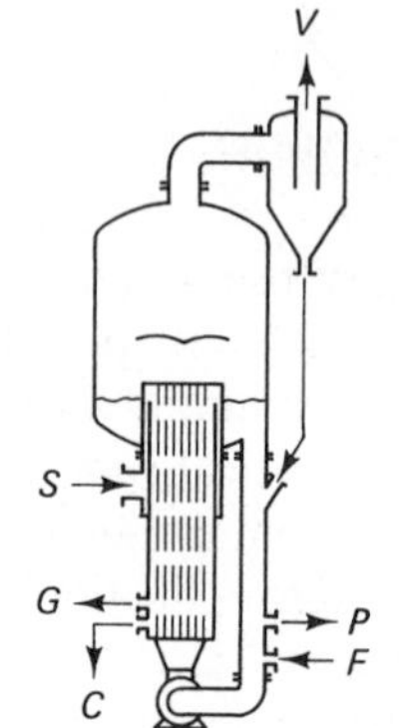

(d)

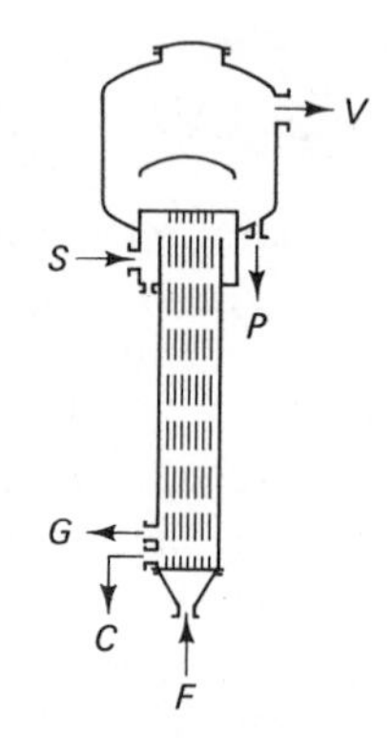

(e)

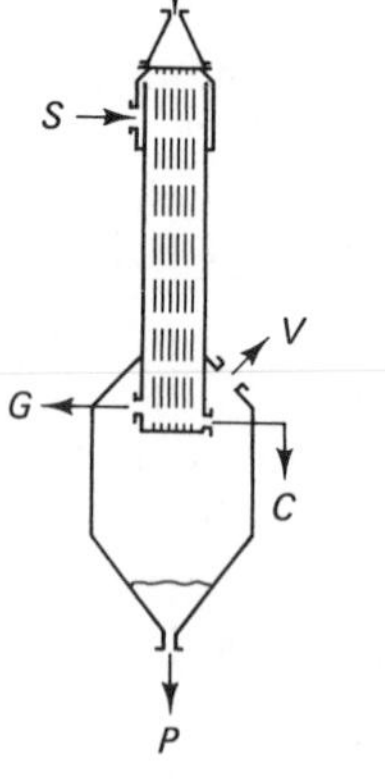

(f)

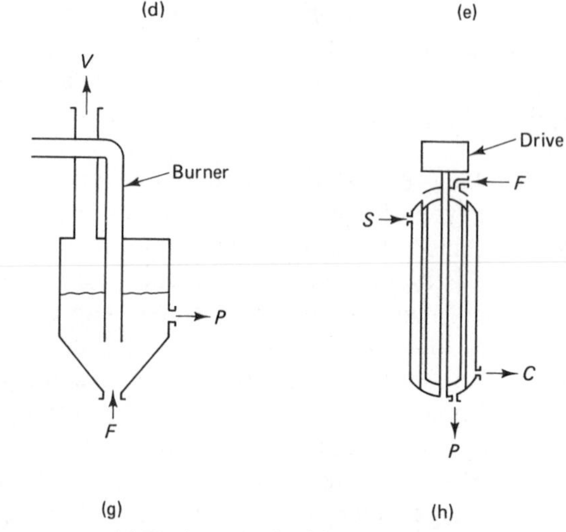

(g)

(h)

Figure 12–1 Types of evaporators. (a) Horizontal tube (b) Short-tube vertical (c) Long-tube vertical (d) Long-tube, forced circulation (e) Long-tube, once through (f) Falling film (g) Submerged combustion (h) Scraped film. Legend: *F*–feed *P*–product *V*–vapor *S*–steam *C*–condensate *G*–gaseous vent

evaporator. The holdup of liquid in the evaporator is also much smaller. For these reasons, falling-film evaporators are used for heat-sensitive materials, such as orange juice.

Materials that become viscous or semisolid during the evaporation process must be mechanically moved during evaporation. Scraped-film evaporators are used in these applications, but they are expensive to build and operate.

The submerged combustion unit avoids the need for a solid heat transfer surface by using the bubble surface of the combustion gases. This unit is especially useful for evaporating highly corrosive or severely scaling materials because there are no solid heat transfer surfaces to corrode or foul.

12.3 HEAT TRANSFER

Whatever the geometry of the evaporator, evaporation involves the transfer of heat from a hot source to an evaporating liquid. The hot source is usually a condensing vapor (e.g., steam) which is often separated from the cooler boiling solution by a solid barrier (the tube wall). Overall, the heat transfer rate can be expressed by the equation

$$Q = UA\,\Delta T \tag{12-1}$$

where

Q = rate of heat transfer (Btu/hr or J/sec)

U = overall heat transfer coefficient (Btu/hr ft^2 °F or W/m^2 °C)

ΔT = temperature difference (driving force) between the hot condensing vapor and the evaporating liquid (°F or °C)

A = surface area through which heat flows (ft^2 or m^2)

A detailed analysis of the heat transfer that goes on in an evaporator is fairly complex and is beyond the scope of this book. The calculation of U involves heat transfer resistances occurring on both the condensing vapor side and the boiling liquid side of the tube wall. We will simply assume that a reasonable value of U is known. U varies inversely with the viscosity of the liquid, which usually increases as the temperature decreases or as the concentration of solute increases.

The calculation of ΔT requires a knowledge of the condensing vapor temperature. Usually the vapor is steam, and there is little subcooling of the condensate. Thus, the condensing steam temperature can be taken directly from the steam tables at the saturation conditions for the condensing pressure.

The temperature of the boiling liquid depends upon the pressure at the point of boiling and, in some systems, on the concentration of solute. This dependence on solute concentration is called *boiling point rise* (BPR). In many evaporators, BPR is negligible and can be ignored. Inclusion of BPR complicates the calculations somewhat, so we will defer a discussion of it until later in the chapter. If BPR is zero, the temperature of the boiling liquid in the evaporator is simply the boiling temperature of the volatile component (water) at the pressure in the evaporator.

In this text we will keep things as simple as possible and assume that the ΔT temperature driving force is simply the condensing steam temperature T_s minus the temperature T in the liquid. Also, we will ignore hydrostatic effects. Under these conditions, the heat transfer rate is given by

$$Q = UA(T_s - T) \tag{12-2}$$

12.4 EVAPORATOR CALCULATIONS WITH NO BPR

The analysis of an evaporator requires the simultaneous solution of a total mass balance, a solute component balance, and an energy balance, together with the equations relating temperature and pressure for the volatile component (vapor pressure relationships).

Figure 12–2 sketches a single-stage evaporator. The feed to the unit is liquid at temperature T_F, flow rate F (mass/time), and nonvolatile component (solute) concentration z (weight fraction solute). Steam is fed into the heating device in the evaporator at a rate of F_S. The pressure of the condensing steam is P_S, which corresponds to a saturation temperature T_S. Vapor is boiled off at a rate V (mass/time). The vapor is pure volatile component, so there is no solute in it ($y = 0$ if y is the concentration of solute in the vapor). Liquid,

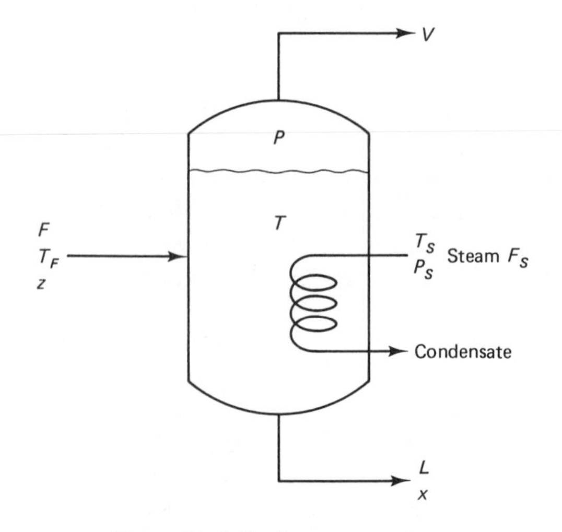

Figure 12–2 Single-stage evaporator.

with a higher concentration of nonvolatile component x (weight fraction solute), is removed at a flow rate L.

The equations describing such a system are

$$\text{Total mass balance: } F = V + L \qquad (12\text{–}3)$$

$$\text{Solute component balance: } zF = Lx \qquad (12\text{–}4)$$

$$\text{Energy balance: } VH + Lh - Fh_F = Q \qquad (12\text{–}5)$$

where

H = enthalpy of the pure vapor product

h = enthalpy of the liquid product

h_F = enthalpy of the liquid feed

The enthalpies can be calculated from the temperatures and compositions of the streams. If the vapor is pure water, its enthalpy can be looked up in the steam tables (Appendix A–2) for saturated vapor (if there is no BPR) at the appropriate pressure. If the solution is liquid water plus a nonvolatile component whose heat capacity is approximately that of water, the solution enthalpies can also be found using the steam tables for saturated liquid water at the appropriate temperature.

Example 12.1.

A falling-film evaporator is used to concentrate orange juice from 12 weight percent solids to 48 weight percent solids with no boiling point rise. Liquid enthalpies can be assumed to be those of pure water. The evaporator operates at 40°C, with feed entering at 30°C. Saturated steam is used at 60°C. The overall heat transfer coefficient is 350 Btu/hr ft^2 °F. Calculate the heat transfer area and the steam flow rate required to produce 3,000 lb$_m$/hr of orange juice concentrate.

Solution:

A solids balance gives the flow rate of the feed required to make 3,000 lb$_m$/hr of liquid product:

$$(0.12)(F) = (0.48)(3{,}000)$$

$$F = 12{,}000 \text{ lb}_m/\text{hr}$$

$$V = 12{,}000 - 3{,}000 = 9{,}000 \text{ lb}_m/\text{hr}$$

Enthalpies from the steam tables are

Feed: h_F at 30°C (86°F) = 54.00 Btu/lb$_m$

Vapor: H at 40°C (104°F) = 1,106.9 Btu/lb$_m$

(note that the pressure in the evaporator is 1.0695 psia, the saturation pressure of water at 104°F)

Liquid: h at 40°C (104°F) = 71.96 Btu/lb$_m$

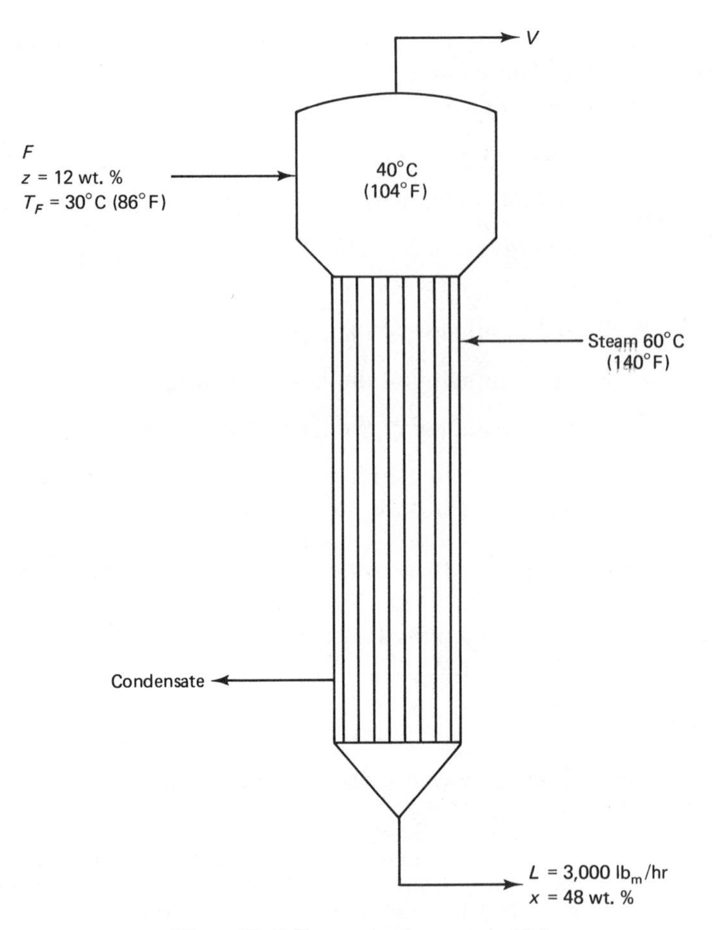

Figure 12–3 Evaporator in example 12.1

The energy equation is used to calculate the heat input to the evaporator:

$$\Delta H = Q - W$$

$$(VH + Lh) - Fh_F = Q$$

$$(9{,}000)(1{,}106.9) + (3{,}000)(71.96) - (12{,}000)(54.00) = Q$$

$$Q = 9.53 \times 10^6 \text{ Btu/hr}$$

The heat transfer area is calculated from $Q = UA(T_S - T)$, where $T_S = 60°\text{C} = 140°\text{F}$ is the temperature of the steam:

$$A = \frac{(9.53 \times 10^6 \text{ Btu/hr})}{(350 \text{ Btu/hr ft}^2 \text{ °F})(140 - 104\text{°F})}$$

$$= 726 \text{ ft}^2$$

The amount of steam consumed can be calculated from an energy balance on the steam side of the evaporator. We have

$$\Delta H = Q - W$$

$$F_S h_c - F_S H_S = Q$$

where

F_S = steam flow rate (lb_m/hr)

H_S = enthalpy of saturated steam vapor entering at 140°F

= 1,122.00 Btu/lb_m

h_c = enthalpy of saturated liquid condensate at 140°F

= 107.89 Btu/lb_m

$Q = -9.53 \times 10^6$ Btu/hr

(note that Q is negative because heat is transferred out of the steam system)

Thus,

$$F_S(107.89 - 1{,}122.00 \text{ Btu/lb}_m) = -9.53 \times 10^6 \text{ Btu/hr}$$

$$F_S = 9{,}397 \text{ lb}_m\text{/hr}$$

12.5 EVAPORATION CALCULATIONS WITH BPR

Most solutions that are concentrated in evaporators are mixtures of water and a nonvolatile salt. The boiling temperature of the solution sometimes depends on the salt concentration. The difference between the temperature T of the boiling solution and the temperature T_W of boiling pure water at the same pressure is the BPR. Thus,

$$T = T_W + \text{BPR} \tag{12-6}$$

BPR is a function of the solute concentration. A graph called a *Dühring chart* is commonly used to determine BPR. Figure 12–4 gives a Dühring chart for the NaOH-water system.

Suppose we want to find the BPR of a 50 weight percent solution of NaOH in an evaporator operating at a pressure of 0.5069 psia. We use the steam tables to find the boiling point of pure water at 0.5069 psia, which turns out to be 80°F. A vertical line at 80°F on the abscissa in Figure 12–4 then intersects the 50 weight percent NaOH line at 150°F, so this is the temperature of the boiling 50 weight percent solution at 0.5069 psia. Consequently, the BPR at this concentration is 150 − 80 = 70°F.

The temperature of a boiling 20 weight percent solution at the same pressure is about 90°F. The BPR at this concentration is 90 − 80 = 10°F.

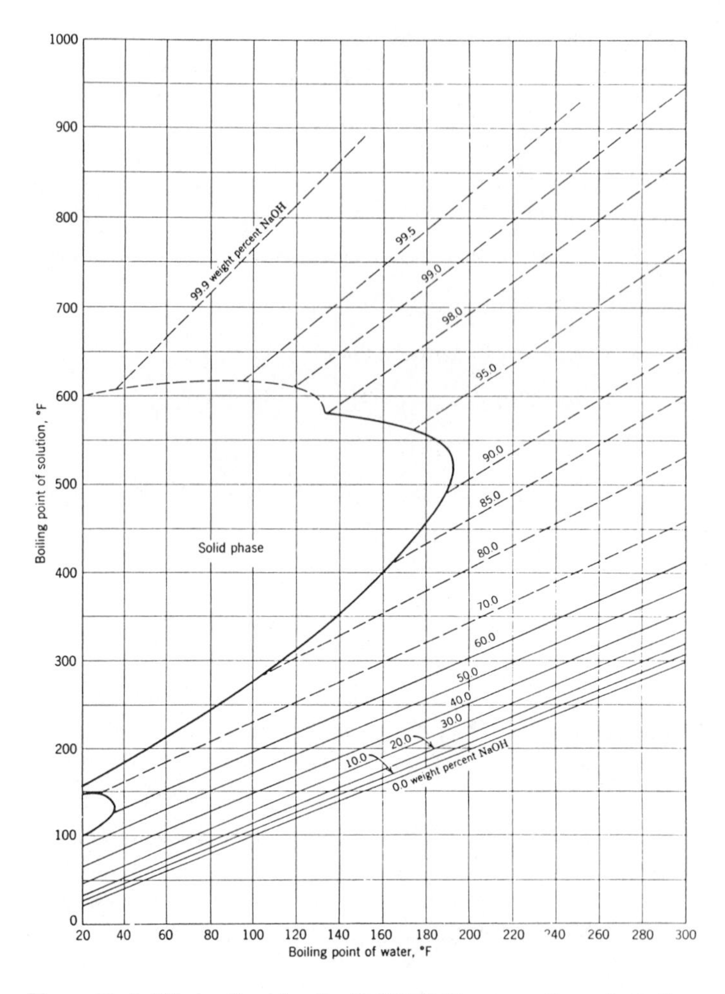

Figure 12–4 Dühring lines for the NaOH-H_2O system. *From A. S. Foust et al.,* Principles of Unit Operations, Second Edition. *Copyright © 1980 John Wiley & Sons, Inc. Reprinted by permission.*

Clearly, then, BPR is a fairly strong function of concentration in the NaOH system.

Each chemical system has its own Dühring chart. As an alternative to multiple Dühring charts, the nomograph in Figure 12–5 can be used to determine BPR for a number of common solutes in water solutions. The vertical line on the right side of the figure has two scales: solution temperature and boiling point rise. The vertical line on the left is a reference line, while the diagonal line is a concentration line. Figure 12–5(a) is for high-solids concentrations; Figure 12–5(b) is for low-solids concentrations.

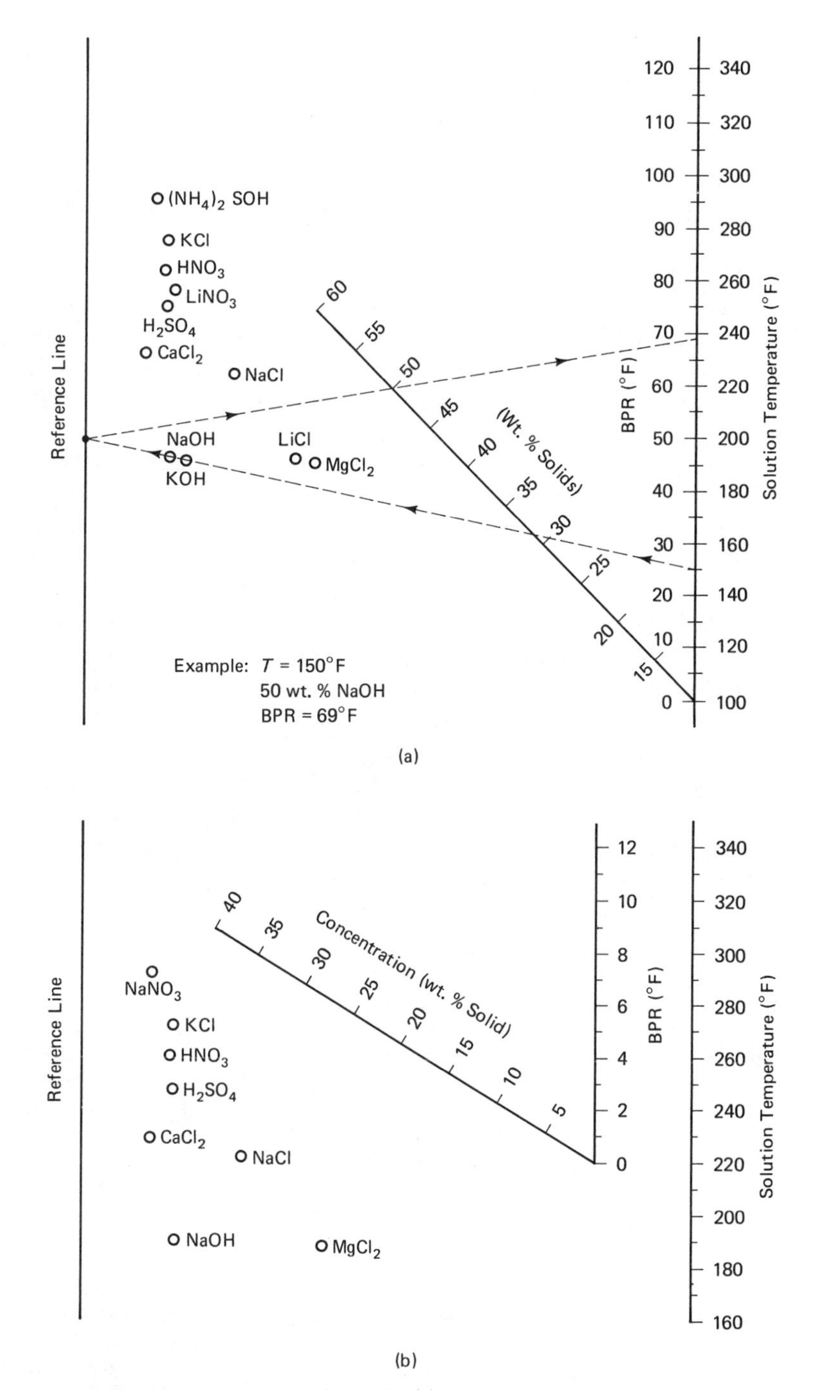

Figure 12–5 Nomograph for BPR. (a) High-concentration scale (b) Low-concentration scale.

To illustrate the use of such a nomograph, let us consider the 50 weight percent NaOH example just discussed. A line is drawn from the solution temperature (150°F on the vertical right-hand scale) through the point for the solute (NaOH) to the left-hand vertical reference line. From this intersection point, a line is then drawn through the concentration line (50 weight percent) back to the right-hand vertical line, and the BPR (70°F) is read off the BPR scale. This value matches the Dühring chart results.

Note that we have to know the solution temperature in order to use Figure 12–5. Typically, however, this is unknown. What is known is the pressure of the evaporator and the corresponding boiling temperature of pure water. Hence, a trial-and-error procedure must be used. A BPR is guessed and added to the pure water boiling temperature to get the solution temperature. Then Figure 12–5 is used to get a new value of BPR.

The advantage of Figure 12–5 is that it gives data for a number of solutes. Its disadvantage is that an iterative solution is needed. One way around this problem is to use the figure to generate a Dühring chart for the system you are interested in. The Dühring charts can be calculated directly. A solution temperature and a concentration are specified, and the BPR is determined from the figure. The pure water boiling point is then calculated using the formula $T_W = T - \text{BPR}$. This gives one point on the Dühring chart. Picking other solution temperatures and calculating pure water boiling points for the same concentration generates a constant-concentration line on the chart. The procedure is repeated for other concentrations.

One of the most interesting aspects of BPR is that the water boiling off the solution is superheated. That is, its temperature is above the saturation temperature of water at the pressure in the evaporator. Thus, the superheated steam tables must be used to find the vapor enthalpy.

Example 12.2.

A $MgCl_2$ solution is evaporated in a short-tube vertical evaporator. The feed concentration is 10 weight percent solids, and the feed temperature is 80°F. The concentrated liquid product from the evaporator is 35 weight percent solids. The pressure in the evaporator is 5 psia. Saturated steam at 30 psig is used as the energy source. The $MgCl_2$ solution has a heat capacity that is 0.95 Btu/lb_m °F at 10 weight percent and 0.85 at 35 weight percent. The overall heat transfer coefficient is 400 Btu/hr ft^2 °F. What is the heat transfer area required to produce 2,500 lb_m/hr of concentrated product?

Solution:

The pressure in the evaporator is 5 psia. This corresponds to a boiling point of pure water of 162°F. However, the solution in the evaporator will boil at a higher temperature than this because of boiling point rise. Figure 12–5 is used to determine this temperature at a $MgCl_2$ concentration of 35 weight percent. The BPR is found by trial and error to be 32°F, giving a solution temperature of 194°F. This is also the temperature of the superheated water vapor leaving the evaporator.

A total mass balance and a $MgCl_2$ component balance are used to calculate the flow rates. We have

Total

$$F = V + L = V + 2{,}500\ \text{lb}_\text{m}/\text{hr}$$

Component

$$zF = Lx$$
$$(0.10)(F) = (2{,}500)(0.35)$$
$$F = 8{,}750\ \text{lb}_\text{m}/\text{hr}$$
$$V = 6{,}259\ \text{lb}_\text{m}/\text{hr}$$

Enthalpies of the three process streams must now be determined. The enthalpy of the water vapor can be determined from the superheated steam tables by interpolating between the enthalpies at 180°F and 200°F for 5 psia to get the enthalpy at 194°F. We have, At 5 psia,

$$H \text{ at } 180°\text{F} = 1{,}139.4\ \text{Btu/lb}_\text{m}$$
$$H \text{ at } 200°\text{F} = 1{,}148.8\ \text{Btu/lb}_\text{m}$$
$$H \text{ at } 194°\text{F} = 1{,}139.4 + (194 - 180)\left(\frac{1{,}148.8 - 1{,}139.4}{20}\right)$$
$$= 1{,}145.98\ \text{Btu/lb}_\text{m}$$

The enthalpies of the solutions can be calculated from their heat capacities. The same reference state as used in the steam tables must be employed. (The enthalpy of liquid water at 32°F is zero.) We obtain

$$h_{35\%} = (0.85)(T - 32)$$
$$= (0.85)(194 - 32) = 137.7\ \text{Btu/lb}_\text{m}$$
$$h_{10\%} = (0.95)(T_F - 32)$$
$$= (0.95)(80 - 32) = 45.6\ \text{Btu/lb}_\text{m}$$

Now an energy balance can be written around the evaporator to calculate the heat transfer rate Q:

$$\Delta H = Q - W$$
$$(VH + Lh_{35\%}) - F\,h_{10\%} = Q$$
$$Q = (6{,}250\ \text{lb}_\text{m}/\text{hr})(1{,}145.98\ \text{Btu/lb}_\text{m}) + (2{,}500\ \text{lb}_\text{m}/\text{hr})(137.7\ \text{Btu/lb}_\text{m})$$
$$- (8{,}750\ \text{lb}_\text{m}/\text{hr})(45.6\ \text{Btu/lb}_\text{m}) = 7.108 \times 10^6\ \text{Btu/hr}$$

The temperature of the saturated condensing steam at 30 psig (44.7 psia) is found to be 274°F from the saturated steam tables. Since the solution is at 194°F, the temperature differential driving force ΔT is 274 − 194 = 80°F. The area for heat transfer can now be calculated as

$$A = \frac{Q}{U\Delta T} = \frac{7.108 \times 10^6\ \text{Btu/hr}}{(400\ \text{Btu/hr ft}^2\ °\text{F})(80°\text{F})} = 222\ \text{ft}^2$$

12.6 MULTIPLE-EFFECT EVAPORATORS

Since evaporators are major consumers of energy, engineers have developed many alternative evaporator configurations that use energy more efficiently. One of the most important of these is the multiple-effect evaporator. In multiple-effect evaporator systems, the vapor that is boiled from one evaporator is used as the heat source in another evaporator. The pressures in the evaporators have to be adjusted so that the temperature differentials are big enough for the required heat transfer rates.

Figure 12–6 illustrates two different types of multiple-effect evaporator systems. In the *forward-feed* system, the feed is introduced into the first stage, where steam is used to remove some of the water. The vapor from stage 1 is then used in stage 2 as the heat source. The liquid from stage 1 is fed into stage 2 as its feed stream. Clearly, the temperature in stage 2 must be lower

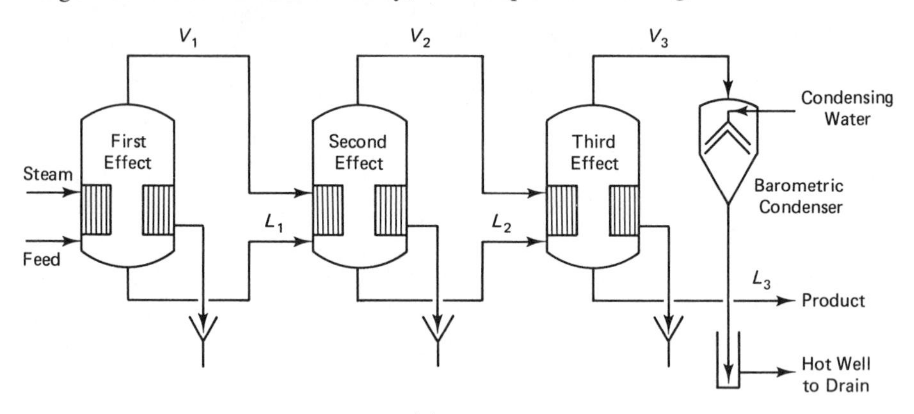

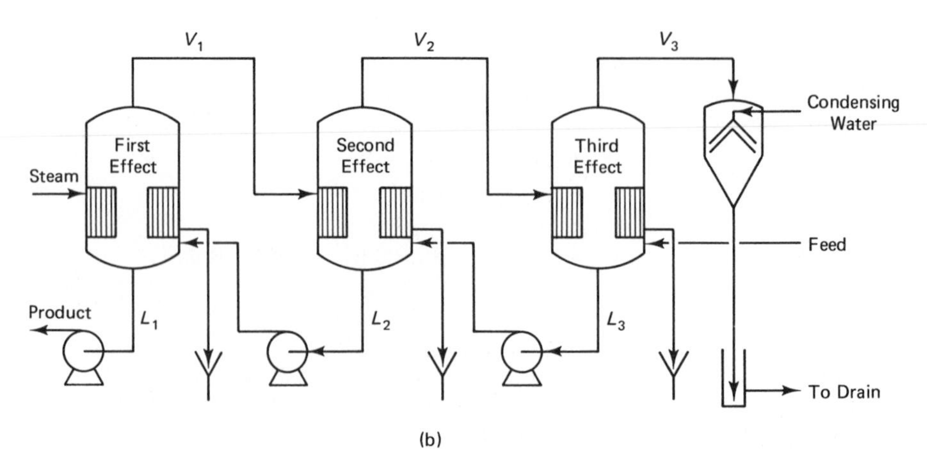

Figure 12–6 Multiple-effect evaporators. (a) Three-effect evaporator system, forward feed (b) Three-effect evaporator system, backward flow.

than the temperature of the condensing water vapor coming from stage 1. Thus, the pressure in stage 1 must be higher than the pressure in stage 2. In addition, the solids concentration in stage 2 is higher than it is in stage 1, so there will be more boiling point rise. The effects of BPR on the solution temperatures must accordingly be included in the determination of the pressure levels that are required to achieve the necessary temperature differential driving forces.

Stage 3 is similarly fed liquid from stage 2 and heated by the vapor from stage 2. The final product leaving stage 3 has the highest solids concentration, but is at the lowest temperature and pressure.

Figure 12–6 also shows a *backward-feed* system. In such a system, the energy and the process stream flows run countercurrent to each other. The feed enters the last stage, and the concentrated product leaves from the first stage. The highest temperature occurs where the solids concentration is the highest. Often this is desirable because it tends to keep heat transfer coefficients high despite the increase in viscosity due to higher solids concentrations.

The choice of the best configuration depends upon the effect of solution temperature on heat transfer coefficients, scaling, and corrosion, and on the temperature sensitivity of the materials being handled. Forward flow is used if temperatures have to be kept low as concentrations increase to prevent thermal degradation. Reverse feed is used if viscosity effects on heat transfer coefficients are strong.

Cross-flow configurations are also used in some evaporators. The feed is split into several streams that are fed into each stage. The liquid from each stage is then removed as the final product. The solids concentrations in each stage in this configuration are the same, but the temperatures are different from stage to stage.

The equations describing each individual stage are essentially the same as equations 12–3 through 12–5. Total mass, solid component, and energy balances must be used together with the physical property data relating BPR to concentration and enthalpies to concentration, temperature, and pressure.

The basic equations are the same for any configuration. The feed stream to any stage, however, will be different depending on the flow configuration: forward, reverse, or cross-flow. The solution of the system of simultaneous equations is usually an iterative one for two reasons: first, most of the physical property data are given in tabular or graphical form and are nonlinear; and second, the parameters that are known are often at opposite ends of the cascade of evaporators, preventing a sequential approach to the problem, i.e., starting at one end and working your way from stage to stage.

There are both design and rating problems in evaporation, just as there are in distillation. The classical design problem is to determine the conditions in each stage and the heat transfer areas required. Typically, the areas in all vessels are kept the same so that the vessels are physically similar.

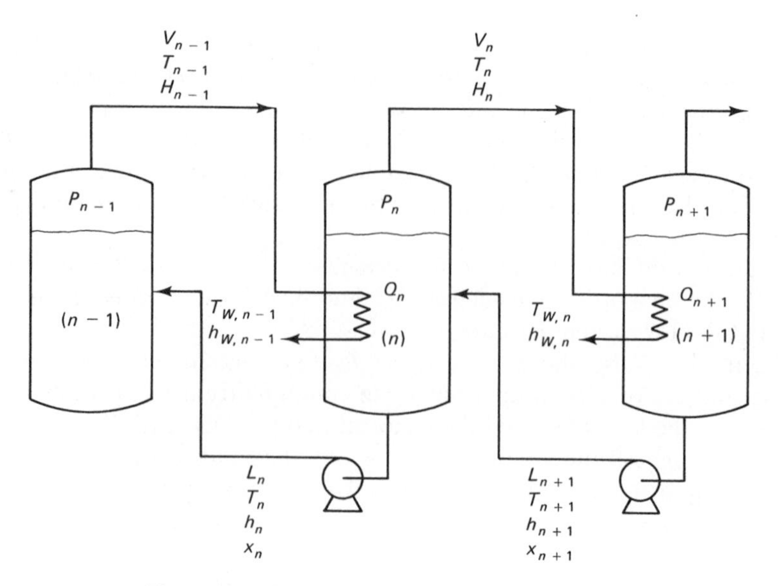

Figure 12–7 Stage n in a multiple-effect evaporator.

Figure 12–7 gives a sketch of the nth stage in a reverse-feed, multiple-effect evaporator system. The equations describing this stage are:

$$\text{Total mass balance: } L_{n+1} = V_n + L_n \tag{12–7}$$

$$\text{Solids component balance: } L_{n+1}x_{n+1} = L_n x_n \tag{12–8}$$

Energy balance (on evaporator contents):

$$\Delta H = Q - W$$

$$(V_n H_n + L_n h_n) - (L_{n+1}h_{n+1}) = Q_n \tag{12–9}$$

$$\text{Heat transfer: } Q_n = U_n A_n (T_{W,n-1} - T_n) \tag{12–10}$$

Energy balance (on hot side of heat exchanger):

$$Q_n = V_{n-1}(H_{n-1} - h_{W,n-1}) \tag{12–11}$$

where

L_n = liquid solution flow rate from stage n (mass/time)

V_n = pure water vapor flow rate from stage n (mass/time)

x_n = solute concentration in stage n (weight fraction solute)

T_n = temperature in stage n

H_n = enthalpy of the pure water vapor (superheated if BPR is significant) at T_n and the pressure P_n of stage n (energy/mass)

h_n = enthalpy of the liquid solution in stage n (energy/mass)

Q_n = heat transfer rate in stage n (energy/time)

$T_{W,n-1}$ = temperature of the pure liquid water condensate leaving the hot side of the heat exchanger in stage n

$h_{W,n-1}$ = enthalpy of the pure liquid water condensate leaving the hot side of the heat exchanger in stage n at temperature $T_{W,n-1}$ (energy/mass)

A_n = heat transfer area in stage n (area)

U_n = overall heat transfer coefficient in stage n (energy/time-area-temperature differential)

Note that the vapor from the upstream evaporator enters the hot side of the heat exchanger as a superheated vapor at T_{n-1} and pressure P_{n-1}. It condenses at the same pressure P_{n-1}, but at a lower temperature $T_{W,n-1}$, the saturation temperature of pure water at P_{n-1}.

Normally each evaporator is the same size, so the areas are equal, i.e., $A_1 = A_2 = A_3 = \ldots = A_N$. The heat transfer coefficients usually vary from stage to stage as temperatures and concentrations change.

Equations like 12–7 through 12–11 can be written for each stage in the evaporator. Then they, together with equations giving BPR and enthalpy data, must be solved simultaneously. The procedure is illustrated in example 12.3 for a two-stage, backward-feed evaporator. For a more general treatment, see the paper by R. N. Lambert, D. D. Joye, and F. W. Koko. *Ind. Eng. Chem. Research* 1987, 26, pp. 100, 104.

A two-stage, backward-feed evaporator system is shown in Figure 12–8.

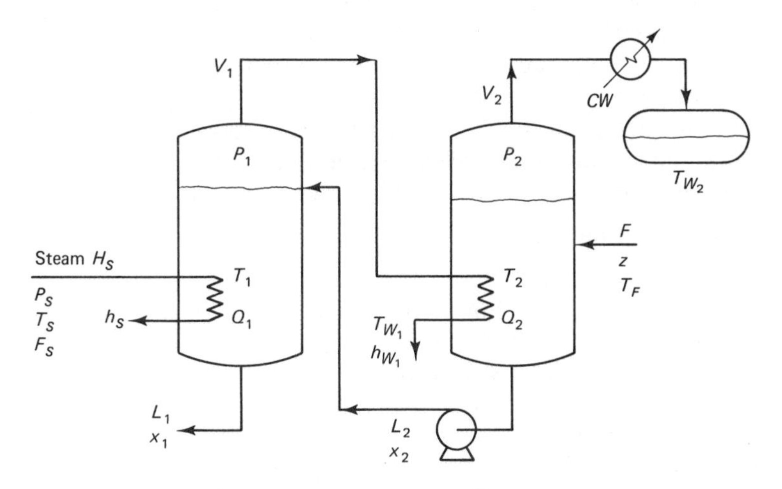

Figure 12–8 Two-stage, reverse-flow evaporator.

The equations describing this system are as follows:

Stage 1

$$L_2 = V_1 + L_1 \tag{12–12}$$

$$L_2 x_2 = L_1 x_1 \tag{12–13}$$

$$L_1 h_1 + V_1 H_1 - L_2 h_2 = Q_1 \tag{12–14}$$

$$Q_1 = U_1 A_1 (T_s - T_1) \tag{12–15}$$

$$Q_1 = F_s (H_s - h_s) \tag{12–16}$$

Stage 2

$$F = V_2 + L_2 \tag{12–17}$$

$$Fz = L_2 x_2 \tag{12–18}$$

$$L_2 h_2 + V_2 H_2 - F h_F = Q_2 \tag{12–19}$$

$$Q_2 = U_2 A_2 (T_{W_1} - T_2) \tag{12–20}$$

$$Q_2 = V_1 (H_1 - h_{W_1}) \tag{12–21}$$

In addition to the mass, component, energy, and heat transfer equations given earlier, physical property data (enthalpies of vapor and liquid streams as functions of temperature, composition, and pressure) and phase equilibrium data (BPR data as a function of concentration and temperature) are required.

Suppose now that we are given the following parameters in the design of a two-effect reverse-flow evaporator system:

- Feed flow rate F, concentration z, temperature T_F
- Steam pressure P_s
- Condensing temperature T_{W_2} in the last-stage condenser (typically water cooled)
- Concentration x_1 of the final product leaving stage 1
- Overall heat transfer coefficients U_n in both stages
- Areas in each stage, equal but unknown ($A_1 = A_2 = A$)

Our job is to determine the following:

- Area A in each evaporator
- Temperatures T_1 and T_2 in each stage
- Concentration x_2 in stage 2
- Pressures P_1 and P_2 in each stage
- Liquid and vapor flow rates L_1, L_2, V_1, and V_2
- Heat transfer rates Q_1 and Q_2
- Steam flow rate F_s

Note that we are given the steam pressure at one end of the cascade and a condensing temperature at the other end. Thus, an iterative solution is usually required.

Equation 12–20 can be solved for area to yield

$$A = \frac{Q_2}{U_2(T_{W_1} - T_2)} \tag{12–22}$$

The right-hand side is substituted into equation 12–15, together with the relationship $T_{W_1} = T_1 - \text{BPR}_1$, to obtain

$$(U_1 Q_2/U_2) \frac{(T_s - T_1)}{(T_1 - \text{BPR}_1 - T_2)} = Q_1 \tag{12–23}$$

Solving for T_1 then gives

$$T_1 = \frac{T_s + C\,(\text{BPR}_1 + T_2)}{1 + C} \tag{12–24}$$

where

$$C = \frac{U_2 Q_1}{U_1 Q_2} \tag{12–25}$$

Now we can outline the solution procedure:

1. Calculate L_1 from an overall solute balance:

$$L_1 = \frac{Fz}{x_1} \tag{12–26}$$

2. Using steam tables, determine P_2 from T_{W_2}
3. Calculate the total vapor flow rate $(V_1 + V_2)$ from an overall mass balance:

$$V_1 + V_2 = F - L_1 \tag{12–27}$$

4. *Guess* a value of V_1 (vapor flow rate); a reasonable initial guess might be

$$V_1 = \frac{(V_1 + V_2)}{2} \tag{12–28}$$

5. Calculate V_2 from an overall mass balance:

$$V_2 = F - L_1 - V_1 \tag{12–29}$$

6. Calculate

$$L_2 = F - V_2 \tag{12–30}$$

7. Calculate

$$x_2 = \frac{Fz}{L_2} \tag{12–31}$$

8. Using a Dühring chart for the solute, determine T_2 in stage 2 from T_{W_2} and x_2. (T_{W_2} is the boiling point temperature of pure water at P_2.)
9. Calculate the boiling point rise in stage 2:

$$\text{BPR}_2 = T_2 - T_{W_2} \tag{12–32}$$

10. Look up the enthalpies H_2, h_2, and h_F from concentrations (x_2 and z), temperatures (T_2 and T_F), and pressure (P_2).
11. Calculate Q_2 from the energy balance on stage 2 (equation 12–19):

$$Q_2 = V_2H_2 + L_2h_2 - Fh_F$$

12. Calculate T_1 by iteratively solving equation 12–24 as follows:
    (a) Guess a value for T_1^{guess}.
    (b) Using a solution temperature of T_1^{guess} and concentration x_1, determine T_{W_1} from a Dühring chart, and then calculate $\text{BPR}_1 = T_1^{\text{guess}} - T_{W_1}$.
    (c) Look up enthalpy h_1 at temperature T_1^{guess} and concentration x_1.
    (d) Look up P_1 and h_{W_1} at temperature T_{W_1} in the saturated steam tables.
    (e) Look up H_1 at pressure P_1 and temperature T_1^{guess} in the superheated steam tables.
    (f) Calculate Q_1 from equation 12–14.
    (g) Calculate the parameter C from equation 12–25.
    (h) Calculate T_1^{calc} from equation 12–24:

$$T_1^{\text{calc}} = \frac{T_s + C(\text{BPR}_1 + T_2)}{1 + C}$$

    (i) If the guessed and the calculated values of T_1 are not close enough, go back to (a).
13. Calculate heat transfer areas from equations 12–15 and 12–20:

$$A_1 = \frac{Q_1}{U_1(T_s - T_1)}$$

$$A_2 = \frac{Q_2}{U_2(T_{W_1} - T_2)}$$

14. If the areas are not equal, reguess V_1 using the following equation, and go back to step 5:

$$V_1 = V_1^{\text{old}} \left(\frac{A_2}{A_1}\right) \tag{12–33}$$

This procedure may look a little involved, but each of the iterations converges quite quickly (only one or two iterations are usually necessary) since the enthalpies and BPRs are not strong functions of temperature. The following example illustrates the method.

Example 12.3.

Caustic is concentrated in a two-effect reverse-flow evaporator to a concentration of 50 weight percent NaOH. Feed flow rate is 10,000 lb_m/hr, feed concentration is 10 weight percent NaOH, and feed temperature is 100°F. The temperature in the condenser on stage 2 is 110°F. Heat transfer coefficients are 300 and 400 Btu/hr ft^2 °F in the first and second stages, respectively. Saturated steam is used in stage 1 at 25 psia, and the heat transfer areas in the two stages are the same. Calculate what these areas are and all the process conditions throughout the system. Figure 12–4 is a Dühring chart for NaOH-water solutions; Figure 12–9 is an enthalpy diagram for NaOH solutions.

Solution:

We will follow the procedure just outlined. From the data given, we know that $x_1 = 0.50$ and $T_{W_2} = 110$°F. Then:

1. $L_1 = Fz/x_1 = (10{,}000)(0.10)/0.50 = 2{,}000$ lb_m/hr.
2. The steam tables give $P_2 = 1.274$ psia at $T_{W_2} = 110$°F.
3. The total vapor flow rate is $V_1 + V_2 = F - L_1 = 10{,}000 - 2{,}000 = 8{,}000$ lb_m/hr.
4. Guess $V_1 = \frac{1}{2}(V_1 + V_2) = \frac{1}{2}(8{,}000) = 4{,}000$ lb_m/hr.
5. $V_2 = 8{,}000 - 4{,}000 = 4{,}000$ lb_m/hr.
6. $L_2 = F - V_2 = 10{,}000 - 4{,}000 = 6{,}000$ lb_m/hr.
7. $x_2 = Fz/L_2 = (10{,}000)(0.10)/6{,}000 = 0.167 = 16.7$ weight percent NaOH.
8. The Dühring chart in Figure 12–4 shows that the temperature T_2 of a 16.7 weight percent NaOH solution is 120°F when it is under a pressure P_2 = 1.274 psia where pure water boils at a temperature T_{W_2} of 110°F. Therefore, $T_2 = 120$°F.
9. $BPR_2 = 120 - 110 = 10$°F.
10. Enthalpies from Figure 12–9 are:

Feed enthalpy h_F at 100°F and 10 weight percent = 60 Btu/lb_m

Liquid enthalpy h_2 at 120°F and 16.7 weight percent = 75 Btu/lb_m

Enthalpy from the superheated steam tables is:

Vapor enthalpy H_2 at 120°F and 1.2748 psia = 1,114 Btu/lb_m

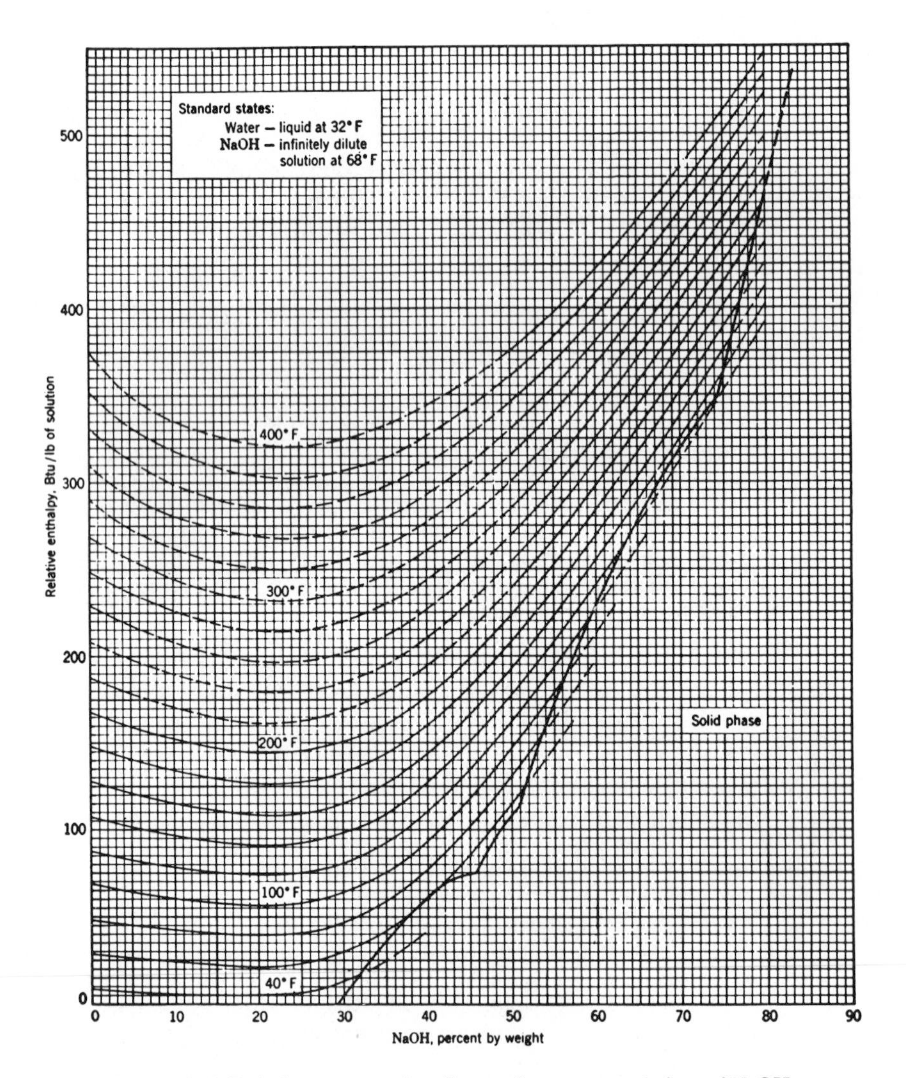

Figure 12–9 Enthalpy-concentration diagram for aqueous solutions of NaOH under a total pressure of one atmosphere. The reference state for water is taken as liquid water at 32°F under its own vapor pressure. This reference state is identical with the one used in most steam tables. For sodium hydroxide, the reference state is that of an infinitely dilute solution at 68°F. *From W. L. McCabe*, Trans. Am. Inst. Chem. Engrs., *31, 129 (1935) by permission.*

11. $Q_2 = V_2H_2 + L_2h_2 - Fh_F = (4{,}000)(1{,}114) + (6{,}000)(75) - (10{,}000)(60) = 4.306 \times 10^6$ Btu/hr.
12. (a) Guess $T_1 = 220°F$.
    (b) The Dühring chart (Figure 12–4) shows that a 50 weight percent NaOH solution boiling at a temperature of 220°F is at the same pressure as pure water boiling at 140°F. Hence, $T_{W_1} = 140°F$ and $BPR_1 = 220 - 140 = 80°F$.
    (c) From Figure 12–9, h_1 at 220°F and 50 weight percent = 240 Btu/lb$_m$.
    (d) From the saturated steam tables at $T_{W_1} = 140°F$, $P_1 = 2.8886$ psia and $h_{W_1} = 107.89$ Btu/lb$_m$.
    (e) From the superheated steam tables at $P_1 = 2.8886$ psia and $T_1 = 220°F$, $H_1 = 1{,}158.9$ Btu/lb$_m$.
    (f) $Q_1 = V_1H_1 + L_1h_1 - L_2h_2 = (4{,}000)(1{,}158.9) + (2{,}000)(240) - (6{,}000)(75) = 4.666 \times 10^6$ Btu/hr.
    (g) $C = (U_2Q_1)/(U_1Q_2) = (400)(4.666 \times 10^6)/[(300)(4.306 \times 10^6)] = 1.445$.
    (h) $T_1^{calc} = [T_s + C\,(BPR_1 + T_2)]/(1 + C) = \dfrac{[240 + (1.445)(80 + 120)]}{(1 + 1.445)} = 216.3°F$.
    (i) Since this value is pretty close to our guess of 220°F, we will use a T_2 of 216°F.

13. $A_1 = (4.666 \times 10^6)/[(300)(240.07 - 216)] = 656\ ft^2$
    $A_2 = (4.306 \times 10^6)/[(400)(140 - 120)] = 538\ ft^2$.
14. Reguessing V_1 yields $V_1 = V_1^{old}(A_2/A_1) = (4{,}000)(538/656) = 3{,}280$ lb$_m$/hr.

Repeating the procedure for this value of V_1

$$V_2 = 4{,}720\ lb_m/hr$$

$$L_2 = 5{,}280\ lb_m/hr$$

$$x_2 = 0.189$$

$$Q_2 = 5.054 \times 10^6\ Btu/hr$$

$$Q_1 = 3.885 \times 10^6\ Btu/hr$$

$$C = 1.025$$

$$T_1^{calc} = 220°F$$

$$A_1 = 648\ ft^2$$

$$A_2 = 631\ ft^2$$

This result is close enough for hand calculations.

The procedure illustrated must be appropriately modified if there are more than two stages or if forward-feed or cross-feed flow patterns are used. But the basic equations describing the systems are similar.

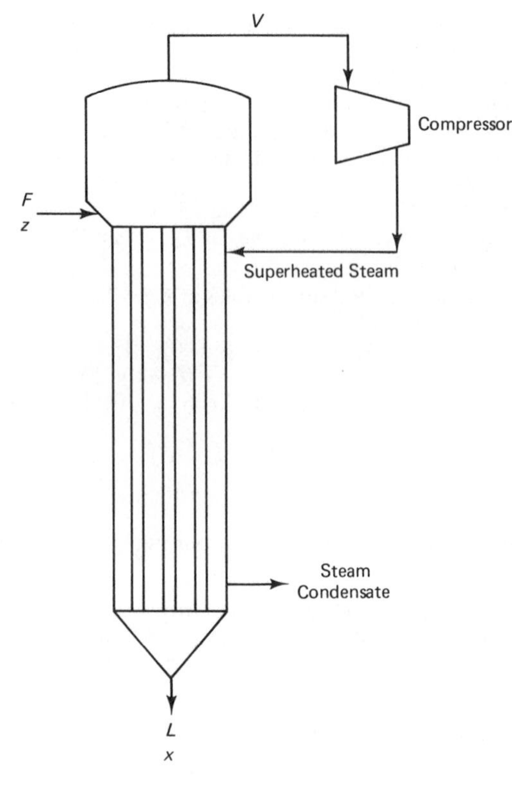

Figure 12–10 Vapor recompression evaporator.

12.7 VAPOR RECOMPRESSION

Just as in distillation columns, vapor recompression is sometimes used in evaporators to reduce energy consumption. Vapor recompression systems are especially desirable when space or capital cost must be minimized, such as in the shipboard generation of potable water from seawater.

Figure 12–10 sketches a vapor recompression evaporator. The vapor from the evaporator is compressed up to a high enough pressure so that it can be condensed by transferring heat into the evaporator. The bigger the BPR, the higher the compressor discharge pressure required.

Example 12.4.

A vapor recompression evaporator is used to generate potable water for use on a cruise ship. 60,000 gal/day of water are required. Seawater is 3.5 weight percent salts (mostly NaCl). The evaporator produces a concentrated liquid brine that is 10 weight percent salt. The seawater is warmed to 164°F by hot product streams.

The evaporator operates at 8 psia, and the discharge pressure of the compressor is 15 psia. The vapor from the compressor is at 300°F. The overall heat transfer coefficient is 600 Btu/hr ft^2 °F. Calculate the heat transfer area and the work performed by the compressor.

Solution:

Total mass and salt component balances give the flow rates of the seawater feed F and concentrated brine solution L. The amount of vapor that must be produced is

$$(60{,}000 \text{ gallon/day})(8.33 \text{ lb}_m/\text{gallon})(\text{day}/24 \text{ hr}) = 20{,}825 \text{ lb}_m/\text{hr}$$

Thus, we have

$$F = L + 20{,}825$$
$$(0.035)(F) = (L)(0.10)$$

Solving simultaneously gives

$$F = 32{,}038 \text{ lb}_m/\text{hr of salt water feed}$$
$$L = 11{,}213 \text{ lb}_m/\text{hr of concentrated brine solution}$$

From the steam tables, the saturation temperature of water at 8 psia is 183°F. The BPR of a 10 weight percent NaCl solution can be found from Figure 12–5 (the low scale) to be about 3°F with a salt solution temperature of 186°F. The enthalpy of the slightly superheated vapor leaving the evaporator at 186°F and 8 psia is calculated by interpolating in the superheated steam tables:

$$H \text{ at } 186°\text{F and 8 psia}$$
$$= 1{,}139.3 + \frac{(186 - 183)(1{,}147.5 - 1{,}139.3)}{200 - 183}$$
$$= 1{,}140.8 \text{ Btu/lb}_m$$

The enthalpy of the superheated vapor leaving the compressor at 300°F and 15 psia is 1,192.8 Btu/lb$_m$. The work put into the steam as it goes through the compressor is, then,

$$W = (20{,}825 \text{ lb}_m/\text{hr})(1{,}192.8 - 1{,}140.8 \text{ Btu/lb}_m)$$
$$= 1.083 \times 10^6 \text{ Btu/hr}$$

So we have

$$(1.083 \times 10^6 \text{ Btu/hr})(\text{hr}/60 \text{ min})[\text{HP}/(42.4 \text{ Btu/min})] = 426 \text{ HP}$$

The steam from the compressor condenses at 15 psia, leaving as saturated liquid with an enthalpy $h = 181.1$ Btu/lb$_m$. The heat removed from the condensing steam is

$$(20{,}825 \text{ lb}_m/\text{hr})(1{,}192.8 - 181.1 \text{ Btu/lb}_m)$$
$$= 21.07 \times 10^6 \text{ Btu/hr}$$

Most of the heat is transferred at the saturation temperature at 15 psia, which is 213°F. The heat transfer area required is

$$A = \frac{Q}{U(T_s - T)}$$

$$= \frac{(21.07 \times 10^6 \text{ Btu/hr})}{(600 \text{ Btu/hr ft}^{2}\text{°F})(213 - 186\text{°F})}$$

$$= 1{,}300 \text{ ft}^2$$

An energy balance on the evaporator itself can be done to see if the energy transferred from the compressed steam is equal to what is required. The equation is

$$Q = VH + Lh - Fh_F$$

For the dilute solutions involved, we will assume that we can use the enthalpy of pure water for the enthalpy of the salt solutions. We thus obtain

$$h \text{ at } 186\text{°F} = 153.94 \text{ Btu/lb}_m$$

$$h_F \text{ at } 164\text{°F} = 131.89 \text{ Btu/lb}_m$$

$$Q = (20{,}825)(1{,}140.8) + (11{,}213)(153.95) - (32{,}038)(131.89)$$

$$= 21.26 \times 10^6 \text{ Btu/hr}$$

This value is quite close to the heat given off by the condensing steam. Note that the two heat transfer rates must be exactly equal in the operating unit. If they momentarily weren't, the work put into the compressor or feed preheating or evaporator conditions would change until an exact balance was achieved.

PROBLEMS

12–1. Compare the energy consumptions of single- and double-effect evaporators required to concentrate 100,000 lb_m/hr of a 13 weight percent sugar-water solution to 60 weight percent sugar at 90°F. Assume negligible heat-of-mixing effects. The temperature in the final condenser is 150°F, and heat capacities of the sugar-water solution can be assumed to be linear functions of the sugar concentration, with the following data given:

x (wt. % sugar)	c_p *(Btu/lb$_m$ °F)*
0	1
60	0.66

The BPR, which can also be assumed to be linearly related to sugar concentration, is 6°F at 60 weight percent sugar.

Use a 20°F temperature difference driving force in the double-effect reboiler/condenser, and assume backward feed.

12–2. A mixture of 20 weight percent H_2SO_4 and 80 weight percent water at 100°F is fed into an evaporator system at a rate of 60,000 lb_m/hr. The evaporator produces 80 weight percent H_2SO_4. The condenser can cool to 120°F.

Assume that H_2SO_4 is nonvolatile and that the activity coefficient of H_2O out of an 80 weight percent H_2SO_4 solution is 0.103. A feed preheater is used to conserve energy. Concentrated solution from the evaporator is run through a countercurrent heat exchanger and cooled to 120°F, giving a 20°F ΔT approach on the cold end of the heat exchanger.

Calculate the temperature of the feed entering the evaporator (i.e., leaving the preheater), the heat transfer rate in the preheater, and the flow rate of 50 psia steam (at 400°F) used in the evaporator. Assume that condensate leaves with 20°F of subcooling and that the steam pressure in the shell side of the evaporator reboiler is 41.85 psia.

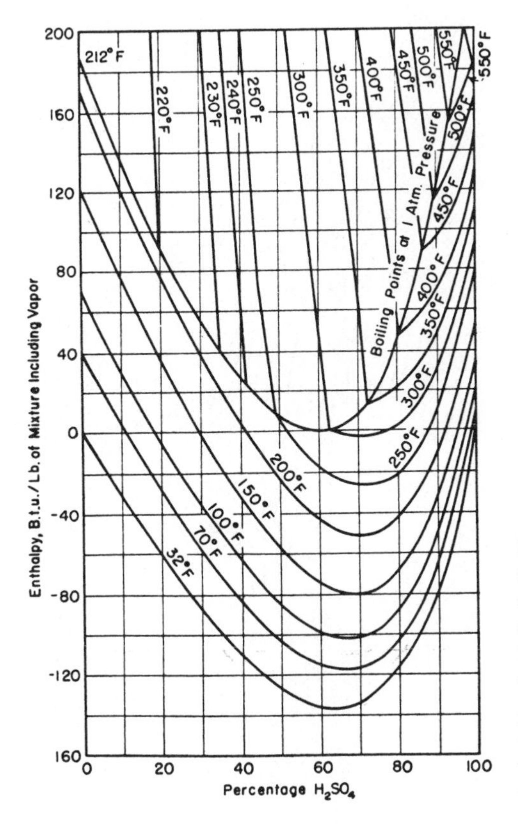

Enthalpy-concentration diagram for aqueous sulfuric acid at 1 atm. Reference states: Enthalpies of pure liquid components at 32°F and vapor pressures are zero. In the two-phase region the upper ends of the tie-lines are assumed to be steam. *From Hougen and Watson,* Chemical Process Principles, Part I. *Copyright © 1943 John Wiley & Sons, Inc. Reprinted by permission.*

12–3. Orange juice is to be concentrated in a single-effect evaporator with vapor recompression. The feed juice is 12 weight percent sugar, the rest of it being water and insignificant amounts of flavor and fiber. The product is to be 40 weight percent sugar. The evaporation temperature is held at 75°F. The vaporized steam is compressed to 5 psia from its original pressure of 0.45 psia, at which value H_V = 1,095.1 Btu/lb_m and h_L = 44.4 Btu/lb_m. After compres-

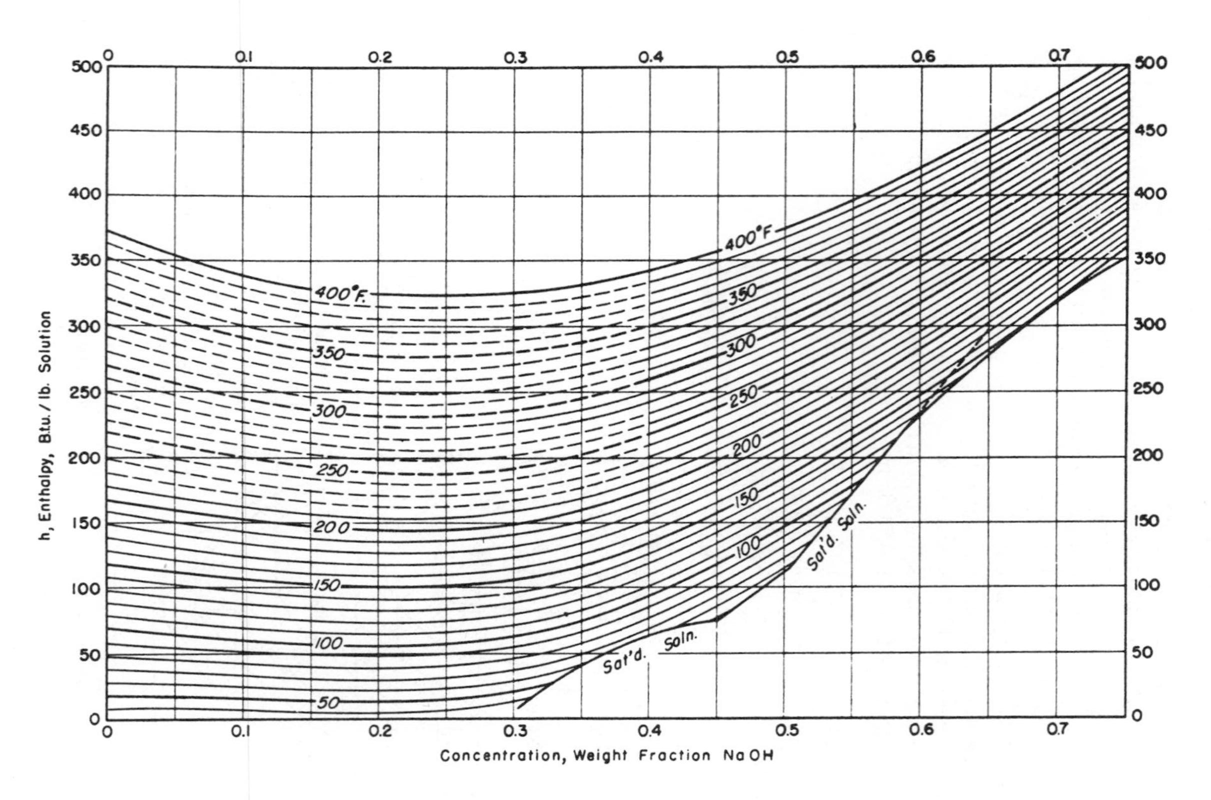

Enthalpy-concentration diagram for aqueous sodium hydroxide at 1 atm (Problem 12–6). Reference states: Enthalpy of liquid water at 32°F and vapor pressure is zero; partial molal enthalpy of infinitely dilute NaOH solution at 64°F and 1 atm is zero. *From W. L. McCabe,* Trans. Am. Inst. Chem. Engrs., *31, 129 (1935) by permission.*

sion, the steam temperature is 600°F. The overall heat transfer coefficient for the evaporator is expected to be 250 Btu/hr ft^2 °F. See problem 12–1 for BPR data.

Determine, on the basis of 1 ton/hr of orange juice concentrate,

(a) The evaporator surface area.

(b) The power required by the compressor.

12–4. A three-effect evaporator is run in *parallel feed* mode. That is, fresh feed which is 90 weight percent water and 10 weight percent nonvolatile solids is fed in at 120°F to each effect. Product streams from all effects are 60 weight percent solids. Assume that there is no BPR and that there is a 40°F temperature driving-force difference in each heat exchanger. Condensate temperature in the final effect is 120°F.

Assuming that heat capacities are constant at 1 Btu/lb$_m$ °F and that the latent heat of vaporization of water is 1,000 Btu/lb$_m$ over the range of temperatures involved, calculate the feed rate to each stage if 10,000 lb$_m$/hr of saturated steam is used in the first stage.

12–5. Seawater consisting of 3.5 weight percent NaCl is evaporated in a two-effect, backward-feed evaporator using 50 psia steam as its energy source. The feed enters the second effect at 80°F, and the product contains 20 weight percent NaCl. It is desired to produce 3 tons/hr of concentrate. The overall heat transfer coefficients are 650 Btu/hr ft^2 °F in the first effect and 700 Btu/hr ft^2 °F in the second effect. Also, a barometric condenser produces a 2.1 psia pressure in the second effect. Calculate the required evaporator heat transfer areas, which are identical in the two effects.

12–6. Spent NaOH is to be concentrated from 10 weight percent NaOH to 50 weight percent NaOH in a double-effect forward-feed long-tube vertical evaporator. Solution enters at 100°F, and the second effect is maintained at 1 psia. Overall heat transfer coefficients are 600 Btu/hr ft^2 °F and 350 Btu/hr ft^2 °F for the first and second effects, respectively. Determine the heat transfer areas required for the two identical effects if 10 tons/hr of dilute NaOH solution is to be concentrated, using 5 psig saturated steam to heat the first evaporator. (See figure on page 448.)

12–7. Skim milk is concentrated from 9 weight percent solids to 20 weight percent solids in a double-effect forward-feed evaporator. Each evaporator has a heat transfer surface of 1,000 ft^2. Milk is fed in at 80°F, and the heat is supplied by 50 psia steam. The condenser at the second effect maintains a pressure of 5 psia. BPR is not significant, and the heat capacity of (liquid) milk is 0.95 at 9 weight percent solids and 0.90 at 20 weight percent solids. Determine the production capacity if the overall heat transfer coefficient is 600 Btu/hr ft^2 °F in the first effect and 450 Btu/hr ft^2 °F in the second. (See figure at top of page 450.)

12–8. Animal glue solution is concentrated in a three-effect forward-feed vertical tube evaporator system. 60,000 gal/day of glue solution at 6.5 weight percent glue is fed in to produce a 55 weight percent concentration product. The

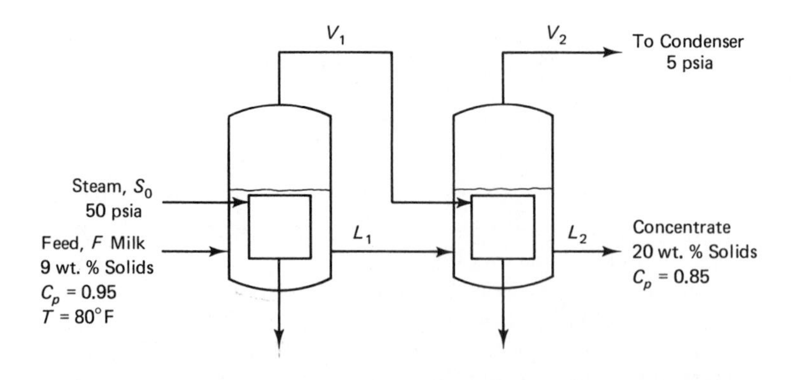

solution increases in viscosity as evaporation proceeds, but has no BPR and exhibits thermal properties that are essentially those of water. The feed enters at 150°F, and the first-effect evaporator is heated with 30 psia saturated steam. The vapor leaves the third effect at 100°F. The effects are of equal heat transfer surface area, but the increasing viscosity produces decreasing overall heat transfer coefficients in successive effects. From experience, U_1 = 500 Btu/hr ft^2 °F, U_2 = 450 Btu/hr ft^2 °F, and U_3 = 350 Btu/hr ft^2 °F. Determine the heat transfer areas and the steam consumption for this system.

12–9. An evaporator operating at 10 psia is fed 1,000 lb$_m$/hr of 10 weight percent NaOH solution at 100°F. The vapors from the evaporation are compressed to 120 psia and leave the compressor at 500°F, at which condition they are returned as hot steam to the evaporator steam tubes. The product from the

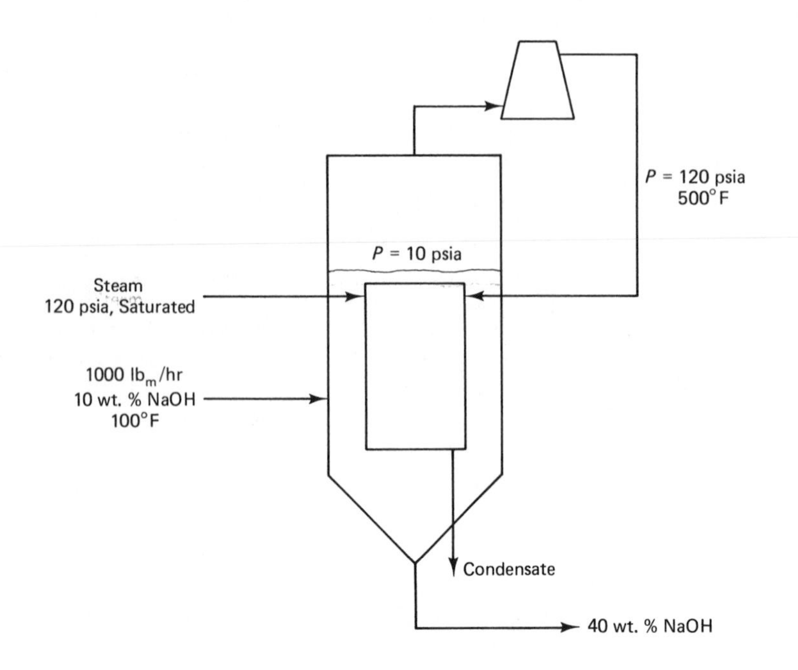

evaporator is 40 weight percent NaOH solution. If necessary, additional steam that is saturated at 120 psia can be fed into the evaporator steam tubes. Determine whether the additional steam is necessary, and, if so, how much will be used. If it is determined by the process operator that no added steam will be used, at what temperature should the compressed steam be returned to the steam tubes in the evaporator?

12–10. A 10 weight percent NaOH aqueous solution is concentrated to 40 weight percent in an evaporator. Water vapor from the evaporator is condensed at 100°F. Feed flow rate is 10,000 lb_m/hr and feed temperature at 100°F. Saturated steam at 5 psig is used to heat the evaporator. Figure 12–4 and Figure 12–9 give Dühring and enthalpy diagrams for NaOH-H_2O.

(a) If the feed is introduced directly into the evaporator, how much steam is required?

(b) Suppose two feed preheaters are used, as sketched in the figure below. The first recovers heat from the concentrated liquid product and the second from the steam condensate. Assume that the temperature differences between hot and cold streams in the cold ends of the heat exchangers are 10°F.

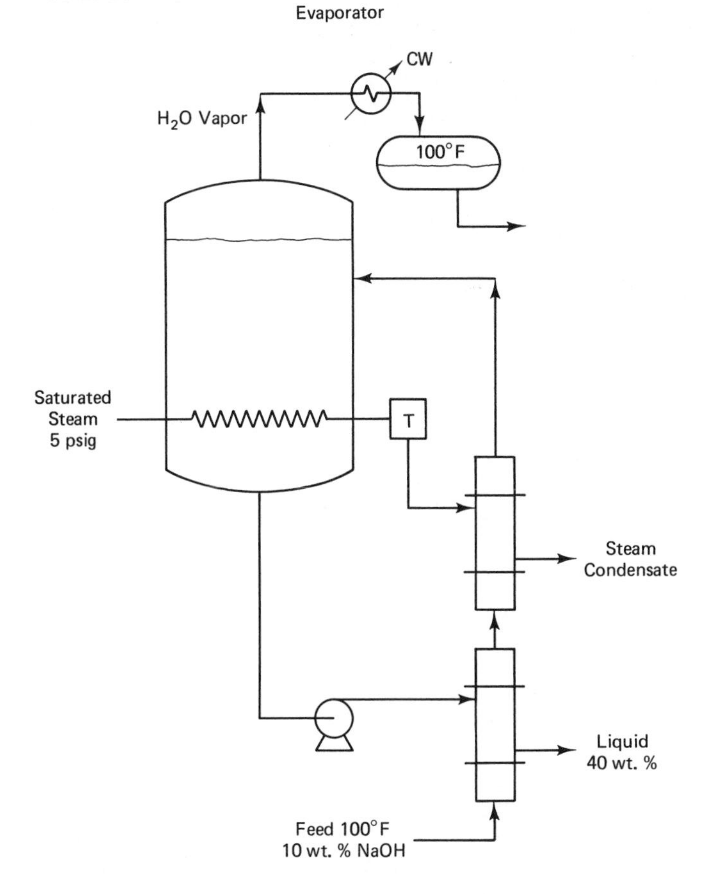

Calculate the heat transfer rate in both heat exchangers and the amount of steam required. The steam condensate leaves the evaporator (and enters the feed preheater) as saturated liquid at 5 psig.

12–11. A feed stream with a flow rate of 60,000 lb_m/hr and a temperature of 70°F, containing 20 weight percent solids, is fed into a single-stage evaporator operating at 5 psia. Saturated steam at 20 psia is used. The overall heat transfer coefficient is 500 Btu/hr °F ft^2 and the heat transfer area is 1,000 ft^2.

The heat capacities of the feed and the liquid product are 0.8 and 0.9, respectively. Boiling point rise is negligible.

Calculate the concentration of solid in the liquid stream leaving the evaporator and the flow rates of the liquid and vapor streams.

12–12. A 20 weight percent NaOH solution at 80°F is fed into a single-stage evaporator with heat transfer area of 600 ft^2 and overall heat transfer coefficient of 400 Btu/hr °F ft^2. Saturated steam at 10 psig is used. The evaporator runs at 2 psia. The concentration of the liquid product from the evaporator is 50 weight percent NaOH.

(a) What are the BPR and evaporator temperatures?
(b) What is the feed flow rate to the evaporator?
(c) How much steam is used?

Binary Nonequimolal Overflow Distillation 13

13.1 INTRODUCTION

In chapter 8 we discussed distillation under the assumption of equimolal overflow, i.e., vapor and liquid molar flow rates in any section of the distillation column are the same on all trays in that section. In this chapter we will not make that assumption. Instead, energy balances on all trays will be used to calculate the liquid and vapor flow rates on all trays. This *nonequimolal overflow* (NEMO) model of the column is more rigorous and should be used in those columns where there are

1. Large temperature differences from the top of the column to the bottom (where sensible heat effects are important).
2. Significant differences in the molar heats of vaporization of the components.
3. Large heats of mixing.

The graphical calculation procedure that includes energy balances is called the *Ponchon method*. The graphical solution uses an *Hxy* diagram on which several Δ points are drawn, using the same graphical mixing rule that was used in nonadiabatic flash calculations and in liquid-liquid extraction calculations.

It might be useful to summarize at this point both the equimolal overflow McCabe-Thiele method (chapter 8) and the nonequimolal overflow Ponchon method so that you can compare them directly. Both methods use VLE data and feed conditions to calculate a minimum reflux ratio. Then component balances (together with energy balances in the Ponchon method) are used with VLE relationships to step up the column.

13.1.1 Review of McCabe-Thiele Method

1. Limitations
   (a) Limited to binary systems only.
   (b) Requires equimolal overflow (gives straight operating lines).
2. Design Method
   (a) *Find the minimum reflux ratio.* Usually, this is obtained from the intersection of the feed thermal condition q line with the VLE equilibrium curve.
   (b) *Set the actual reflux ratio.* Typically, this is some preset multiple of the minimum reflux ratio, for example 1.1 times the minimum.
   (c) *Draw lines representing component balances.* These are the straight operating lines for each section of the column.
   (d) *Step off stages.* Use alternately the operating line and the VLE curve. Use the stripping operating line from x_B up to where it intersects the q line. Then use the rectifying operating line.

13.1.2 Summary of Ponchon Method

1. Limitations: Limited to binary systems.
2. Design Procedure (proof given in section 13.2)
   (a) *Find the minimum reflux ratio.* Usually, this is done by extending the VLE tie-line on an *Hxy* diagram that goes through the feed point (z, h_F) to a vertical line drawn at the distillate composition x_D. As shown in Figure 13–1, the coordinates of the point of intersection are (x_D, h_Δ^{min}). The minimum reflux ratio is calculated from

$$\text{RR}_{min} = \frac{h_\Delta^{min} - H_{NT}}{H_{NT} - h_D} \tag{13-1}$$

where

$$H_{NT} = \text{enthalpy of vapor leaving the top tray}$$

$$h_D = \text{enthalpy of liquid distillate product}$$

   (b) *Set the actual reflux ratio.* Typically, this is 1.1 times the minimum. Then calculate the vertical coordinate h_Δ of a Δ point:

$$h_\Delta = H_{NT} + \text{RR}\,(H_{NT} - h_D) \tag{13-2}$$

   (c) *Locate the Δ and Δ_B points using component and energy balances.* Draw a straight line from the Δ point through the feed point to

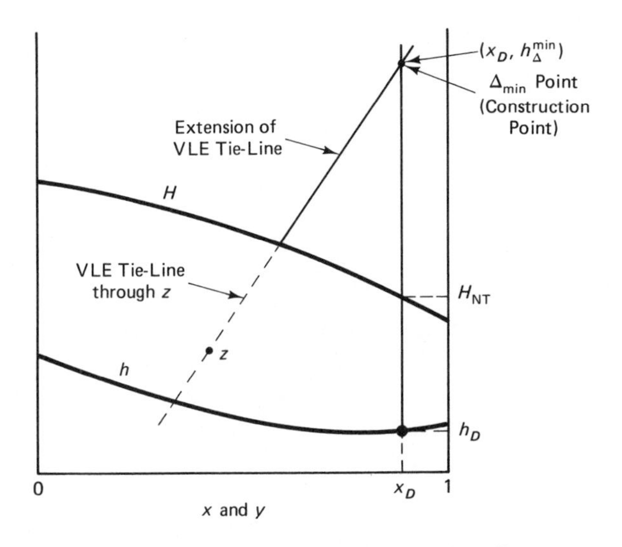

Figure 13–1 Minimum reflux ratio on Ponchon diagram.

a vertical line though the bottoms composition x_B. (See Figure 13–2.) The point of intersection is called the Δ_B point.

(d) *Step off stages.* Beginning at the bottom of the column, use the Δ_B point to construct operating lines in the stripping section and the VLE tie-lines for phase equilibrium. When a VLE tie-line is used that crosses the line that joins the Δ and Δ_B points, this is the optimum feed tray. Now use the Δ point to construct operating lines in the rectifying section.

13.2 PONCHON DIAGRAMS FOR SIMPLE COLUMNS

13.2.1 Rectifying Section

The standard simple column has one feed, a total condenser, and a partial reboiler, and makes two products (distillate and bottoms). (See Figure 13–3.) Consider the top part of the column. We cut the column above the nth tray, as shown in Figure 13–4, and do mass, component, and energy balances around this system.

We assume that we know Q_c, D, x_D, h_D, L_{n+1}, x_{n+1}, and h_{n+1}. We wish to calculate V_n, y_n, and H_n. The problem is mathematically similar to the mixing problem discussed in chapter 11. The only difference is that one of the streams is flowing out instead of in. This can be taken care of simply by using a negative sign, as illustrated in Figure 13–5.

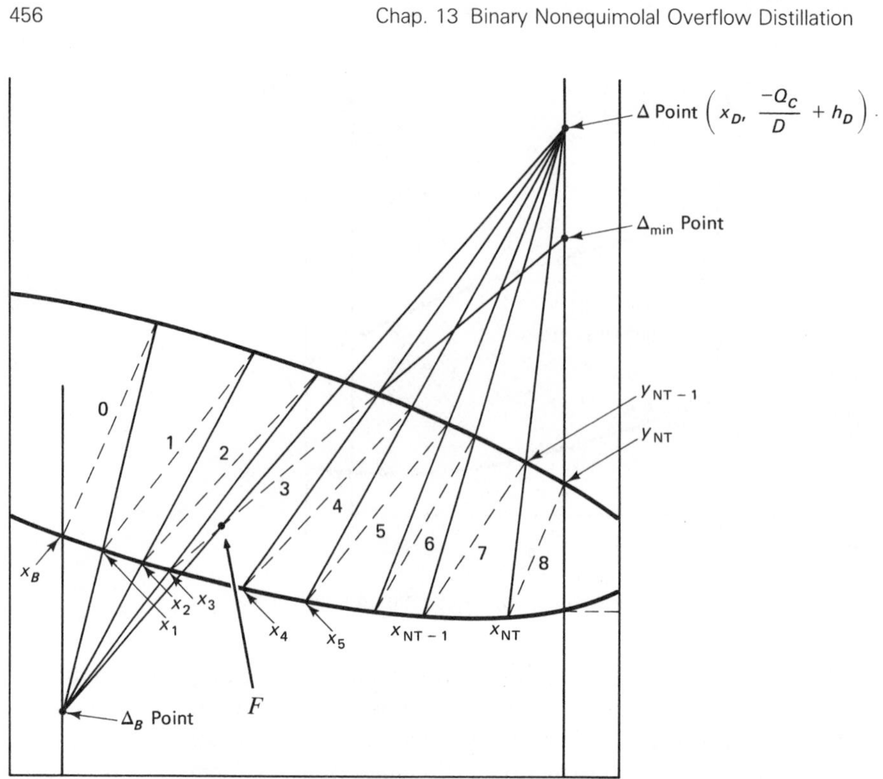

Figure 13–2 Stepping off stages on Ponchon diagram.

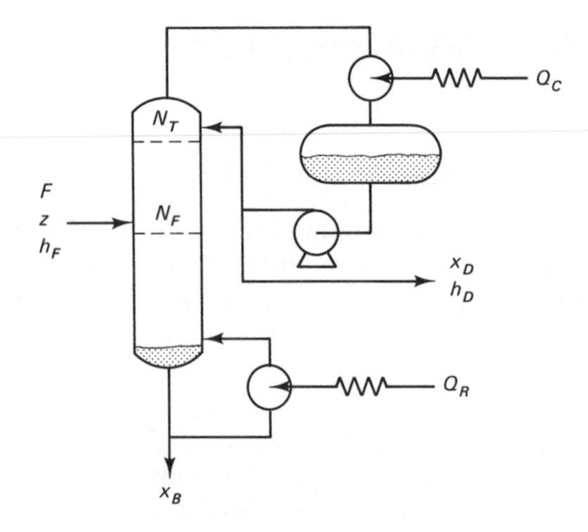

Figure 13–3 Distillation column nomenclature.

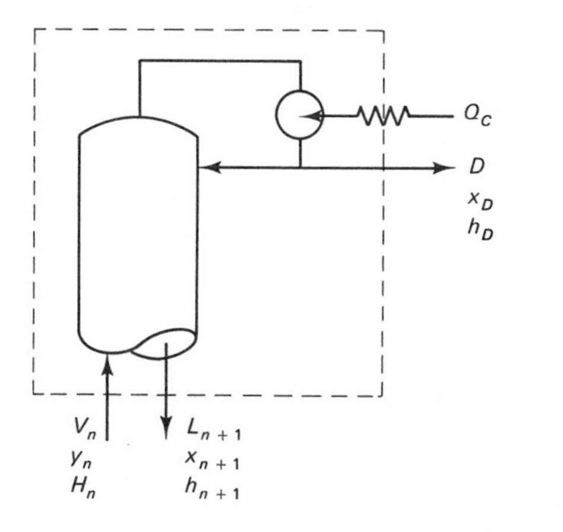

Figure 13–4 Top section of the distillation column.

Instead of using S for summation, it is convenient to use the term Δ for difference. Δ, a pseudo flow rate, is the difference between the vapor and liquid streams passing each other in the rectifying section of the column. It is defined by

$$\Delta = V_n - L_{n+1} \tag{13–3}$$

A total mass balance, a light component balance, and an energy balance around the system under consideration give

$$\text{Total: } V_n - L_{n+1} = D \tag{13–4}$$

$$\text{Component: } V_n y_n - L_{n+1} x_{n+1} = D x_D \tag{13–5}$$

$$\text{Energy: } V_n H_n - L_{n+1} h_{n+1} = -Q_c + D h_D = D\left(h_D - \frac{Q_c}{D}\right) \tag{13–6}$$

The right-hand side of equation 13–6 is written in the form shown because we want to define a pseudo enthalpy h_Δ and a pseudo composition x_Δ.

Comparing equations 13–3 and 13–4, we see that $\Delta = D$. Therefore, we define h_Δ and x_Δ such that they give us the three equations

$$V_n - L_{n+1} = \Delta \tag{13–7}$$

$$V_n y_n - L_{n+1} x_{n+1} = \Delta x_\Delta \tag{13–8}$$

$$V_n H_n - L_{n+1} h_{n+1} = \Delta h_\Delta \tag{13–9}$$

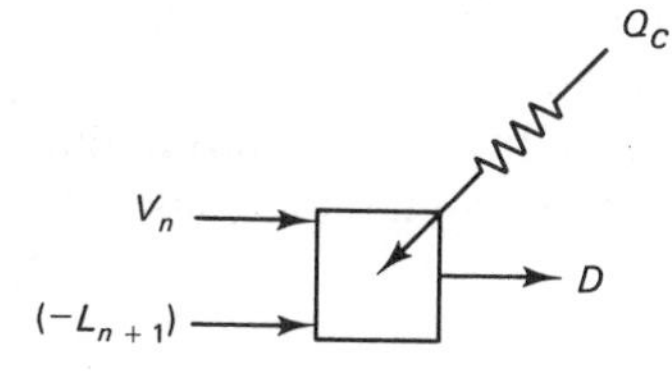

Figure 13–5 System equivalent to the top of a distillation column.

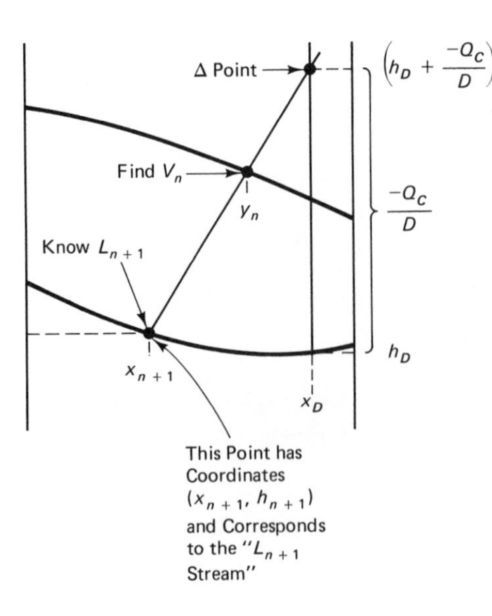

Figure 13–6 Use of Δ point.

The coordinates of the Δ point are

$$x_\Delta = x_D \tag{13–10}$$

$$h_\Delta = h_D - \frac{Q_c}{D} \tag{13–11}$$

The graphical mixing rule tells us that if we mix the V_n and L_{n+1} streams together (with L_{n+1} taken to be negative) to get the Δ stream, the coordinates of this Δ stream will lie on a straight line joining the points V_n and L_{n+1}. Thus, the Δ point can be used to do component and energy balances. The method is similar to the use of operating lines in the McCabe-Thiele procedure to do component balances.

Figure 13–6 illustrates the use of a Δ point. If we know the point L_{n+1} and the Δ point on an *Hxy* diagram, we can simply draw a straight line joining the two. The intersection of this line with the saturated vapor line gives us the point V_n. Thus, y_n and H_n can be read off the graph, and then the flow rate V_n can be calculated by means of a component balance.

The straight line that we drew can be considered an operating line. It passes through the coordinates (x_{n+1}, h_{n+1}) and (y_n, H_n). The Δ point permits us to relate the compositions and enthalpies of passing streams in the column. In effect, we are solving simultaneously mass, component, and energy balances graphically.

13.2.2 Condenser

Balances around the condenser and reflux drum, sketched in Figure 13–7, give

$$V_{NT} = R + D \tag{13–12}$$

$$V_{NT}(h_D - H_{NT}) = Q_c \tag{13–13}$$

where

R = reflux flow rate

D = distillate flow rate

Solving for R yields

$$R = \frac{-Q_c - D(H_{NT} - h_D)}{H_{NT} - h_D} \tag{13–14}$$

The reflux ratio RR = R/D can then be expressed as

$$\text{RR} = \frac{(h_D - Q_c/D) - H_{NT}}{H_{NT} - h_D} \tag{13–15}$$

$$= \frac{h_\Delta - H_{NT}}{H_{NT} - h_D} \tag{13–16}$$

Thus, the reflux ratio is the ratio of the distances a and b shown in Figure 13–8. Note that the point h_D lies on the saturated liquid line if there is no subcooling. If the distillate and the reflux are subcooled, the point will lie below the saturated liquid line.

13.2.3 Stripping Section

Similar balances in the stripping section lead to the Δ_B point which is used to relate the compositions of passing streams in that section. The coordinates of the Δ_B point are

$$\text{composition } x_{\Delta B} = x_B \tag{13–17}$$

$$\text{enthalpy } h_{\Delta B} = h_B - \frac{Q_R}{B} \tag{13–18}$$

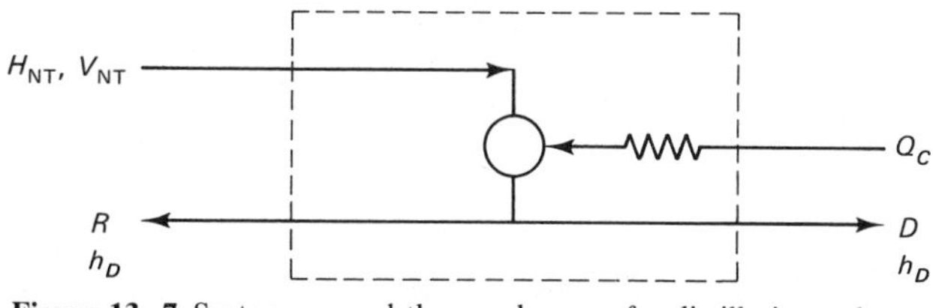

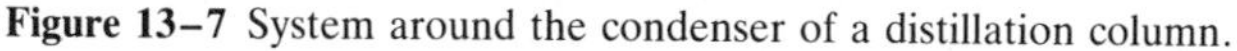

Figure 13–7 System around the condenser of a distillation column.

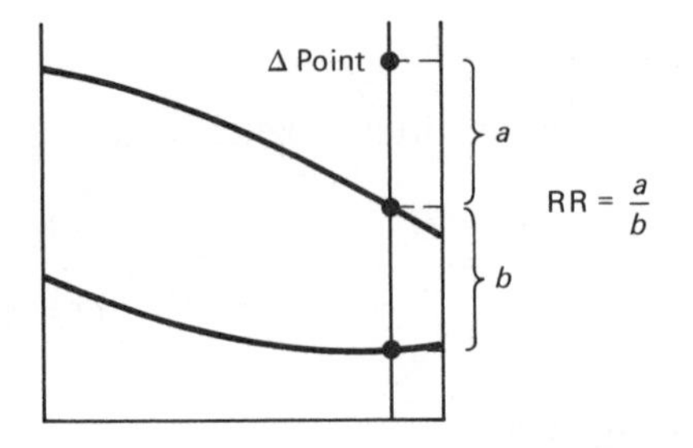

Figure 13–8 Reflux ratio from Δ point.

where

$$Q_R = \text{heat transfer rate in the reboiler}$$

$$B = \text{bottoms product flow rate}$$

$$h_B = \text{enthalpy of liquid bottoms product}$$

13.2.4 Whole Column

Mass, component, and energy balances around the entire system as sketched in Figure 13–9(a) are

$$\text{Total: } F = B + D \tag{13–19}$$

$$\text{Component: } zF = Bx_B + Dx_D \tag{13–20}$$

$$\text{Energy: } h_F F = B\left(h_B - \frac{Q_R}{B}\right) + D\left(h_D - \frac{Q_c}{D}\right) \tag{13–21}$$

These three equations tell us that mixing a stream B with enthalpy $(h_B - Q_R/B)$ and composition x_B together with a stream D with enthalpy $(h_D - Q_c/D)$ and composition x_D will give a stream F which has an enthalpy h_F and composition z. Thus, a straight line from the Δ point to the Δ_B point must go through the feed point, as shown in Figure 13–9(b). The point F can lie in the two-phase region, in the subcooled-liquid region, or in the superheated-vapor region. This overall balance line replaces the q line used in the McCabe-Thiele method.

Example 13.1.

A methanol-water separation is achieved in a distillation column operating at atmospheric pressure. Feed flow rate is 3,000 kg/hr, and feed composition is 60 weight percent methanol. The feed enters the column at a pressure of 1 atmosphere and

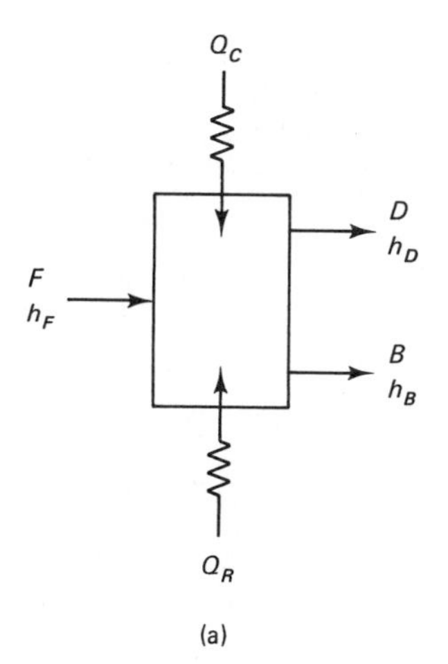

(a)

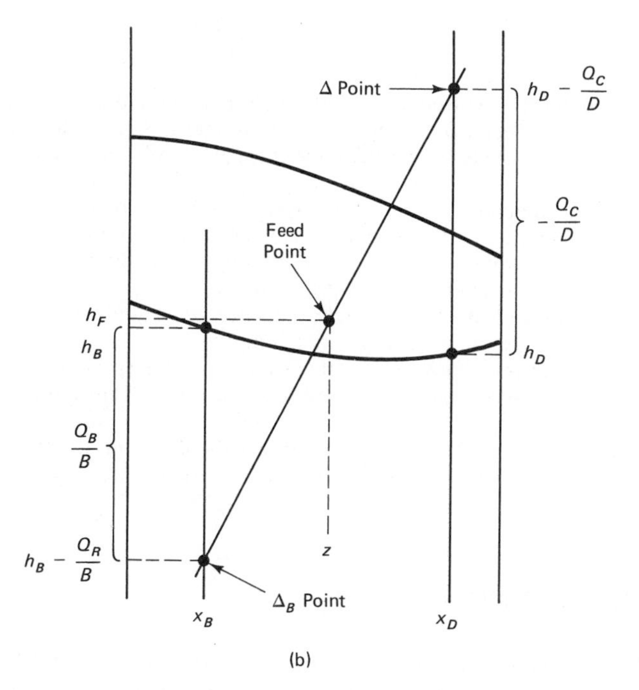

(b)

Figure 13–9 Overall column. (a) System (b) *Hxy* diagram.

a temperature of 75°C. Determine the number of trays required and the optimum feed tray to make 95 weight percent methanol distillate and 5 weight percent methanol bottoms when a reflux ratio of 0.75 is used. Assume theoretical trays, a total condenser, a partial reboiler, and saturated reflux.

Solution:

Total mass and component balances give the product flow rates.

Total

$$F = D + B$$

Component

$$zF = x_D D + x_B B$$

$$(0.60)(3{,}000) = (0.95)(D) + (0.05)(B)$$

$$D = 1{,}833 \text{ kg/hr}$$

$$B = 1{,}167 \text{ kg/hr}$$

Since the reflux ratio is given, the Δ point can be located on the *Hxy* diagram for methanol-water at atmospheric pressure (see Figure 13–10) by using equation 13–2:

$$h_\Delta = H_{\text{NT}} + \text{RR}(H_{\text{NT}} - h_D)$$

$$= 310 + 0.75\,(310 - 50) = 505 \text{ cal/g}$$

The feed point is plotted at $z = 60$ weight percent and a temperature of 75°C. Note that the feed is partially vaporized. A straight line from the Δ point through the feed point intersects the vertical line through x_B at -500 cal/g. This is the Δ_B point.

Stages are stepped off by starting at $x_B = 5$ weight percent on the saturated liquid line in Figure 13–10. The vapor leaving the base of the column has a composition $y_B = 31$ weight percent. An operating line is drawn from y_B to the Δ_B point (the dashed straight line), and the composition $x_1 = 19$ weight percent is read off the point of intersection with the saturated liquid curve. Then the VLE tie-line is used to get $y_1 = 55$ weight percent.

This stepping procedure is continued for five steps (partial reboiler plus four trays). Between trays 4 and 5 the feed line is crossed. The optimum feed tray is tray 4.

To get x_5 from y_4, the Δ point is used, not the Δ_B point. Stepping is continued until x_D is reached. Eight trays plus the partial reboiler are required.

13.3 LIMITING CONDITIONS

In the McCabe-Thiele method, the minimum reflux ratio and the minimum number of stages can be easily visualized and graphically calculated on an *xy* diagram. These limiting conditions represent useful information about the constraints on the separation. In the Ponchon method, the same limiting

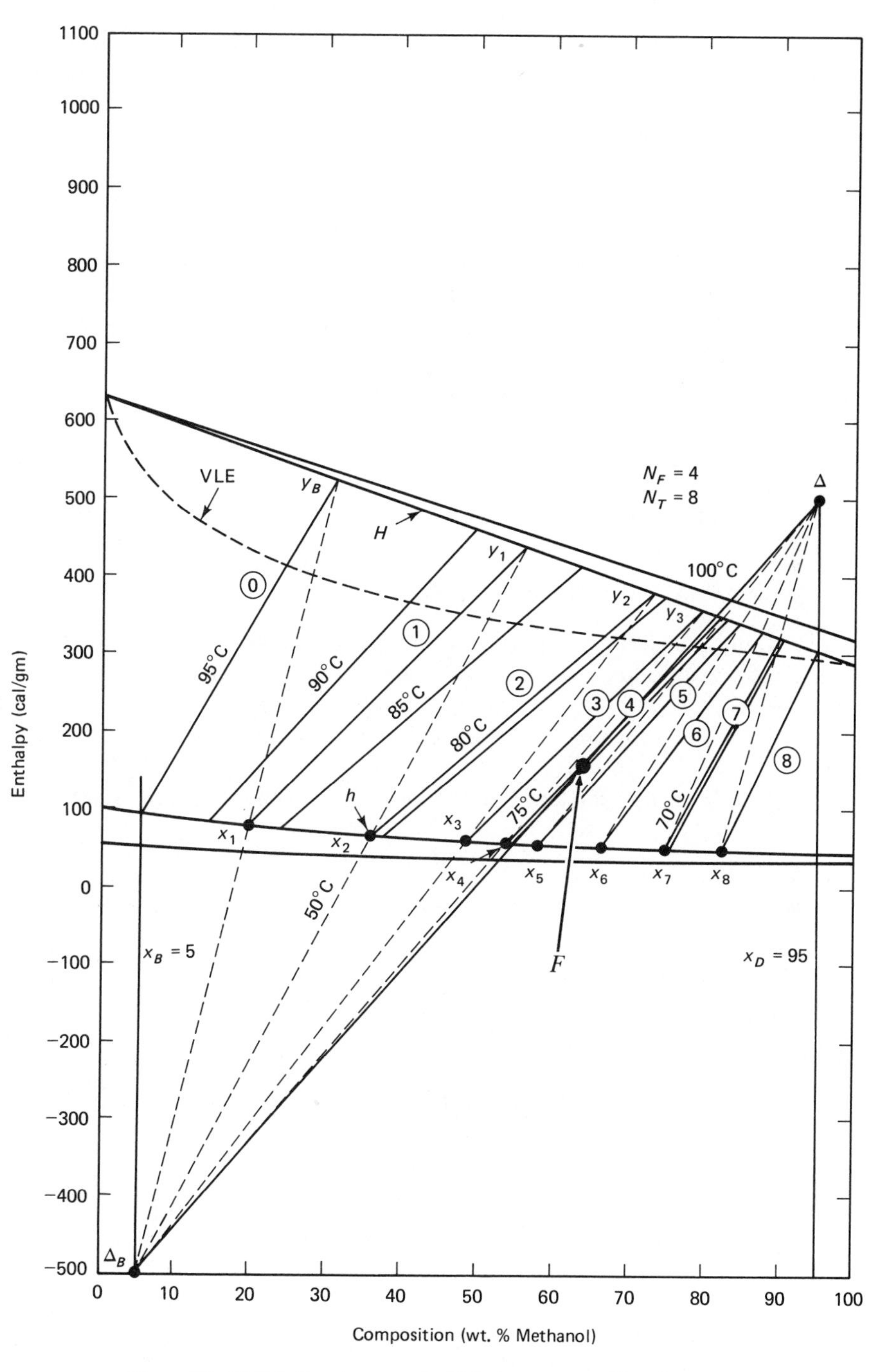

Figure 13–10 Solution to example 13.1 shown on an *Hxy* diagram.

conditions can be determined with essentially the same ease, and they yield the same useful information.

13.3.1 Minimum Reflux Ratio

If we extend the VLE tie-lines to intersect with a vertical line through x_D, the highest intersection point establishes the minimum Δ point. As sketched in Figure 13–11(a), the highest point usually occurs for the tie-line through the feed point. This is equivalent to establishing the minimum reflux ratio in the McCabe-Thiele method by finding the rectifying operating line that intersects the feed q line on the VLE curve. However, there are some systems in which the highest point of intersection on the vertical line through x_D may occur on the extension of a tie-line that is not the one that goes through the feed point.

Once the minimum Δ point is established, the minimum reflux ratio can be calculated from equation 13–1.

13.3.2 Minimum Number of Trays

The minimum number of trays occurs at an infinite reflux ratio. If the reflux ratio is infinite, the condenser heat removal must be infinite (or D must be zero). Therefore, the Δ point must be located at infinity. The operating lines on a Ponchon diagram are vertical lines at an infinite reflux ratio (total reflux).

Figure 13–11(b) sketches the stepping-off of the minimum number of

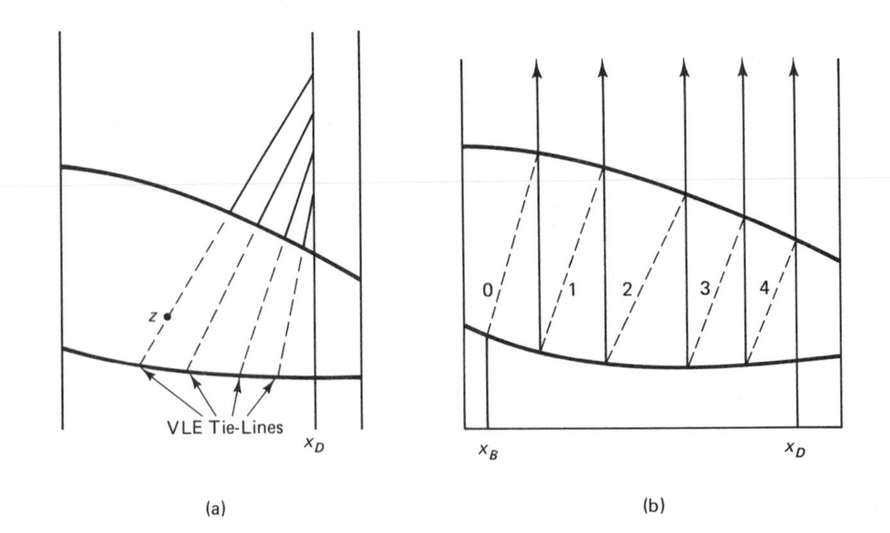

Figure 13–11 Limiting conditions. (a) Minimum reflux ratio (b) Total reflux.

stages. Vertical operating lines and the VLE tie-lines are used. For the system shown, four trays plus a reboiler are required.

13.4 COMPLEX COLUMNS

So far we have applied the Ponchon method to just the "plain vanilla" distillation column with a single feed and liquid distillate and bottoms products. However, there are no inherent limitations to the method that would restrict us to just these simple columns.

Complex columns may have multiple feeds, multiple products, and multiple heat input or heat removal locations. The Ponchon method becomes a little more complex, but the procedure is essentially the same. Total mass, component, and energy balances are written around each section of the column, giving a different Δ point for each section. Several important cases will be considered.

13.4.1 Partial Condenser

Many columns produce a distillate product as a vapor from the reflux drum, as illustrated in Figure 13–12. The flow rate of this vapor product is D, its composition is y_D, and its enthalpy is H_D. The D point on the Hxy diagram is located on the saturated vapor line.

The Δ point is located at $(y_D, H_D - Q_c/D)$. Note that the liquid in the reflux drum is in equilibrium with the vapor product. This liquid has a composition x_R which is located at the other end of the tie-line from y_D. The reflux drum gives an additional stage of separation because of the partial condenser.

Occasionally one sees a distillation column where the product is removed as a vapor directly from the overhead vapor line, before the condenser. In this case the condenser is a total condenser, producing liquid reflux at the same composition as the vapor product. The reflux drum does not provide an additional stage. The Ponchon diagram looks almost the same as the liquid distillate case, except that Q_c is reduced because the distillate product is not condensed.

Other variations on the theme are also possible. For example, some columns take both vapor and liquid products. This obviously changes the Ponchon diagram. It is instructive to sketch the diagram for this kind of system.

13.4.2 Dual Feeds

Many columns have more than a single feed stream. If the compositions of the two feed streams are identical, they can be combined and fed onto a single tray. However, if the compositions are not the same (as is usual), they

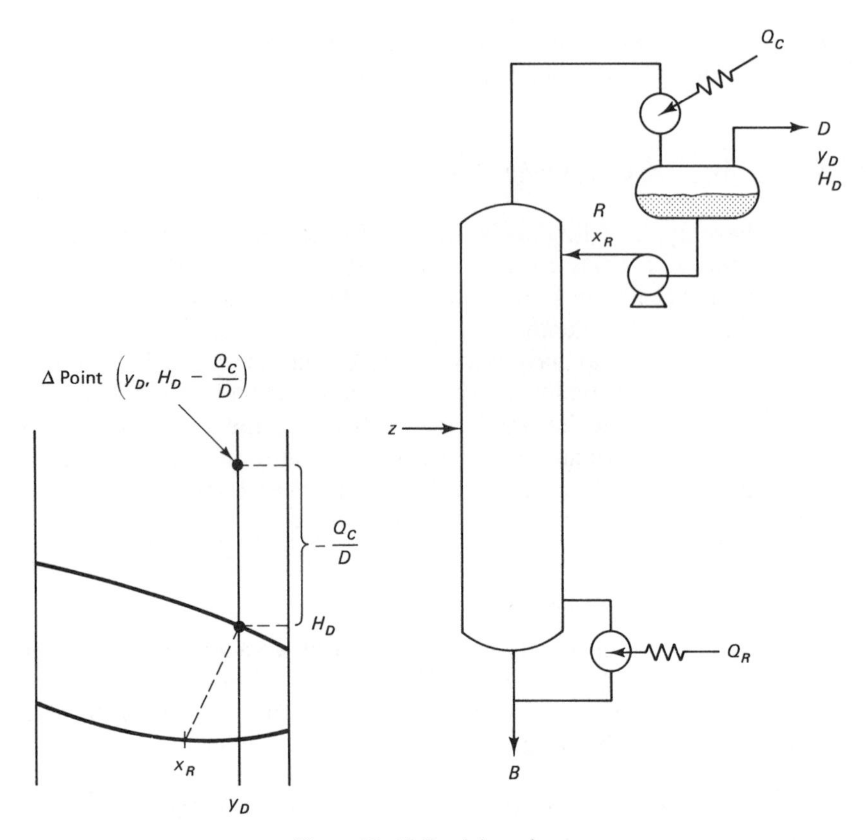

Figure 13–12 Partial condenser.

should be fed onto different feed trays in order to consume the minimum amount of energy.

Figure 13–13 shows a column with feed F_1 introduced on a tray low in the column and feed F_2 introduced on a tray higher in the column. The compositions are z_1 and z_2, where z_2 is greater than z_1.

The column now has three different sections: a rectifying section at the top above the top feed, a middle section between the feed trays, and a stripping section below the lower feed. Therefore, we need three Δ points to analyze it. Let us call these points Δ, Δ_M, and Δ_B. Clearly, the Δ and Δ_B points will be located at the same coordinates as in a conventional column:

$$\Delta \text{ point} = \left(x_D, h_D - \frac{Q_c}{D}\right)$$

$$\Delta_B \text{ point} = \left(x_B, h_B - \frac{Q_R}{B}\right)$$

To find the location of the Δ_M point, we write the balance equations and rearrange them so that they have the form

$$V_n - L_{n+1} = \Delta_M = D - F_2 \tag{13–22}$$

$$V_n y_n - L_{n+1} x_{n+1} = \Delta_M x_{\Delta_M} = D x_D - F_2 z_2 \tag{13–23}$$

$$V_n H_n - L_{n+1} h_{n+1} = \Delta_M h_{\Delta_M} = -Q_c + D h_D - F_2 h_{F_2} \tag{13–24}$$

The pseudo flow rate Δ_M, the pseudo composition x_{Δ_M}, and the pseudo enthalpy h_{Δ_M} define the coordinates of the Δ_M point that is used to step from tray to tray in the middle section of the column. The coordinates of the Δ_M point are

$$x_{\Delta_M} = \frac{D x_D - F_2 z_2}{D - F_2} \tag{13–25}$$

$$h_{\Delta_M} = \frac{-Q_c + D h_D - F_2 h_{F_2}}{D - F_2} \tag{13–26}$$

This procedure is used for any complex column to locate the Δ points for each section. Remember that the coordinates of the Δ points are pseudo compositions, so there is no restriction that they must have values between 0 and 1 only.

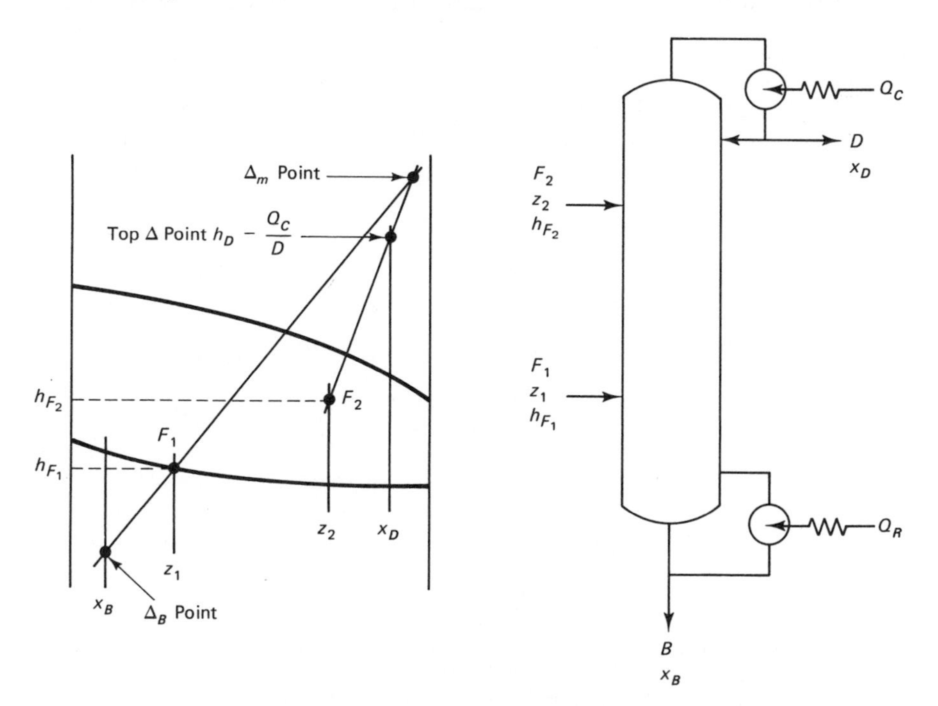

Figure 13–13 Two feeds.

13.4.3 Intermediate Reboilers and Condensers and Sidestreams

The procedure just described is used for each section of the column. The following example should help to show the details of the method.

Example 13.2.

30,000 lb_m/hr of feed at 80°C is fed into an atmospheric distillation column. Feed composition is 45 weight percent methanol and 55 weight percent water. The distillate product from a total condenser is 98 weight percent methanol, and the bottoms product from the base of the column is 4 weight percent methanol. The partial reboiler in the base of the column has a heat duty of 5.65×10^6 Btu/hr. There is another intermediate reboiler up in the stripping section which uses lower pressure steam and has a heat duty of 9.13×10^6 Btu/hr. This intermediate reboiler is located where the liquid composition on the tray is approximately 10 weight percent methanol. Feed is introduced on the optimum feed tray. Without assuming equimolal overflow, calculate the number of trays, the condenser duty, and the thermal condition of the feed.

Solution:

Since equimolal overflow cannot be assumed, we must use a Ponchon *Hxy* diagram for the design of the column. The two reboilers will result in three different column sections and three Δ points: Δ above the feed, Δ_M between the feed and the intermediate reboiler, and Δ_B below the intermediate reboiler. Figure 13–14 shows the general layout of the column. Figure 13–15 gives the *Hxy* diagram for methanol and water at atmospheric pressure.

The Δ_B point will be located at $(x_B, h_B - Q_B/B)$. We first calculate product flow rates D and B from overall total and component balances:

$$F = B + D$$

$$zF = x_B B + x_D D$$

$$30{,}000 = B + D$$

$$(0.45)(30{,}000) = (0.04)(B) + (0.98)(D)$$

$$D = 13{,}085 \text{ lb}_m\text{/hr} \qquad B = 16{,}915 \text{ lb}_m\text{/hr}$$

$$\frac{Q_B}{B} = \frac{(5.65 \times 10^6 \text{ Btu/hr})}{(16{,}915 \text{ lb}_m\text{/hr})} = 334 \text{ Btu/lb}_m$$

h_B at 4 weight percent on the saturated liquid line = 80 cal/g

$$h_B - \frac{Q_B}{B} = 80 - (334 \text{ Btu/lb}_m)(\text{lb}_m/454 \text{ g})(252 \text{ cal/Btu})$$

$$= -106 \text{ cal/g}$$

The Δ_B point is thus located at 4 weight percent and -106 cal/g. It is used to draw the operating lines in the lower section of the column. The operating line connecting Δ_B and y_B intersects the saturated liquid line (see Figure 13–15) at $x_1 = 12$ weight percent. Therefore, we should locate the intermediate reboiler at tray 1.

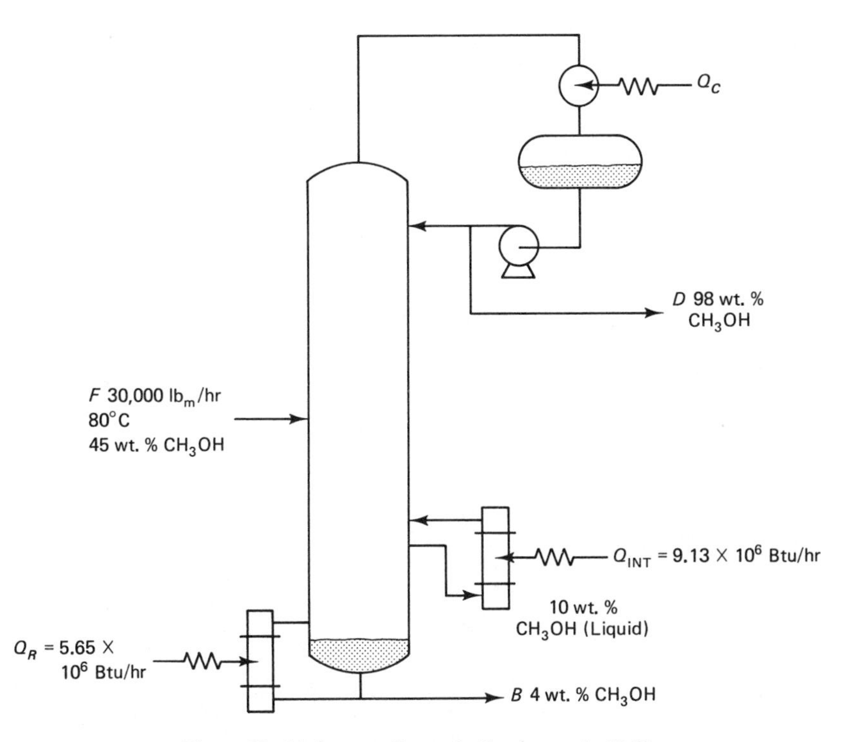

Figure 13–14 Intermediate reboiler (example 13.2).

For the section above the intermediate reboiler, the mass, component, and energy equations are written in standard format to determine the location of the Δ_M point:

$$L_{n+1} - V_n = B = \Delta_M \tag{13–27}$$

$$L_{n+1}x_{n+1} - V_n y_n = Bx_B = \Delta_M x_{\Delta M} \tag{13–28}$$

$$L_{n+1}h_{n+1} - V_n H_n = Bh_B - Q_R - Q_{\text{int}} = \Delta_M h_{\Delta M} \tag{13–29}$$

Thus, the coordinates of the Δ_M point are

$$x_{\Delta M} = x_B = 4 \text{ wt. } \%$$

$$h_{\Delta M} = h_B - \frac{(Q_R + Q_{\text{int}})}{B}$$

$$= 80 \text{ cal/g} - \left[\frac{(5.65 + 9.13)(10^6 \text{ Btu/hr})}{(16{,}915 \text{ lb}_m\text{/hr})}\right] (\text{lb}_m/454 \text{ g})(252 \text{ cal/Btu})$$

$$= -405 \text{ cal/g}$$

We use the Δ_M point to draw operating lines on Figure 13–15 to get x_2 and x_3. The optimum feed tray is tray 3, since we cross the straight line connecting the feed point to the Δ_M point.

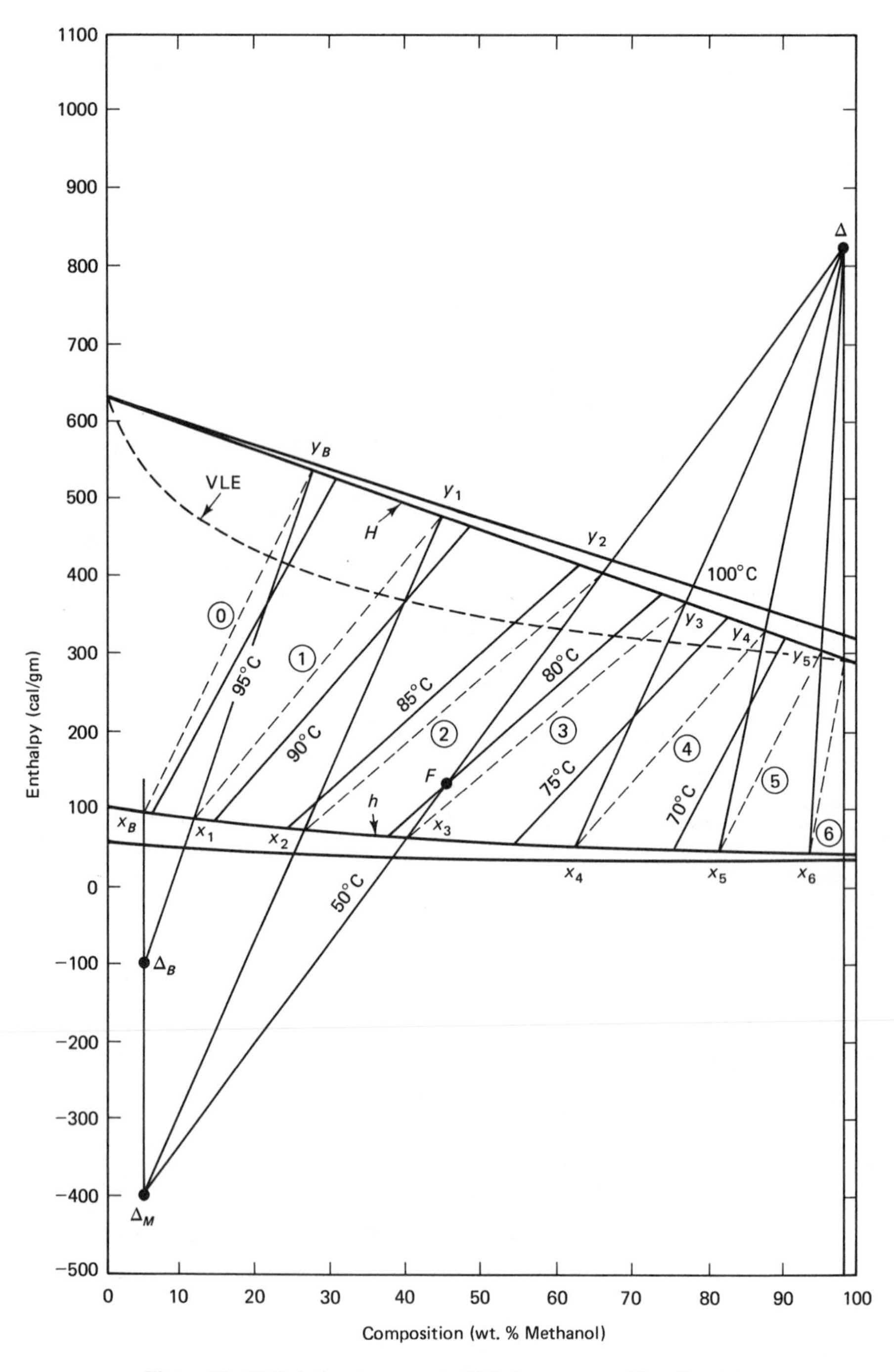

Figure 13–15 Solution to example 13.2 shown on an *Hxy* diagram.

The Δ point for the top section of the column can be calculated by determining Q_c from an overall energy balance. Alternatively, we can get the Δ point graphically by drawing a straight line from the Δ_M point through the feed point at (z,h_F) to a vertical line through x_D.

Stepping is continued up the column using the Δ point. A total of six trays plus the total reboiler are needed.

The Δ point is located at an enthalpy of 830 cal/g. We can solve for Q_c from

$$h_\Delta = h_D - \frac{Q_c}{D}$$

We have

$$830 \text{ cal/g} = 50 \text{ cal/g} - \left[\frac{Q_c \text{ Btu/hr}}{13{,}085 \text{ lb}_\text{m}\text{/hr}}\right]\left(\frac{252 \text{ cal}}{\text{Btu}}\right)\left(\frac{\text{lb}_\text{m}}{454 \text{ g}}\right)$$

$$Q_c = -18.4 \times 10^6 \text{ Btu/hr}$$

The feed is 45 weight percent methanol at 80°C and is in the two-phase region. We can read the compositions of the liquid and vapor phases at 80°C from the ends of the VLE tie-line:

$$x_F = 37 \text{ wt. \%} \qquad y_F = 76 \text{ wt. \%}$$

Let F_L and F_V be the flow rates of the liquid and vapor feed, respectively. Then

$$30{,}000 = F_L + F_V$$

$$(30{,}000)(0.45) = (F_L)(0.37) + (F_V)(0.76)$$

$$F_L = 23{,}840 \text{ lb}_\text{m}\text{/hr}$$

Therefore, the feed thermal condition is

$$\frac{F_L}{F} = \frac{23{,}840}{30{,}000} = 0.79 \quad (79\% \text{ liquid})$$

13.5 NEMO COMPUTER PROGRAM

The equations describing the nonequimolal overflow distillation column, including the energy balances, can be quite conveniently solved on a digital computer. A FORTRAN program that solves the rating problem for a binary NEMO column is given in Table 13–1. The equations and their solution are described below, assuming that the pressures on all trays are known.

First, let us look at the base of the column. Figure 13–16 sketches the system. Suppose we know B, x_B, Q_R, and P_B (the pressure in the column base). Then a bubblepoint calculation will give us y_B and T_B, and enthalpy data will give us h_B and H_B.

The unknown variables are V_B, L_1, and x_1. The variables T_1 and h_1 can be calculated from x_1 once it is known. Since there are three unknowns, three

TABLE 13–1
NEMO BINARY DISTILLATION RATING PROGRAM

```
C  NON-EQUIMOLAL OVERFLOW BINARY DISTILLATION RATING PROGRAM
C     GIVEN: FEED CONDITIONS
C            PRODUCT SPECIFICATIONS:  XD AND XB
C            FEED TRAY NUMBER AND TOTAL NUMBER OF TRAYS
C            PRESSURE IN CONDENSER AND TRAY PRESSURE DROP
C     CALCULATE:HEAT INPUT REQUIRED TO MAKE THE SEPARATION
C  EXAMPLE IS PROPYLENE OXIDE/TERTIARY BUTYL ALCOHOL SYSTEM
C   IDEAL VLE
C
      REAL L,MW,LF
      DIMENSION VPC(2,6),Z(6),XF(6),YF(6),T1(6),T2(6),P1(6),P2(6),
     + CL(6),CV(6),DH(6),MW(6),PS(6)
      DIMENSION X(100),Y(100),T(100),P(100),L(100),V(100),
     + XX(2),YY(2),HL(100),HV(100)
      DATA NC,F,Z(1),Z(2),TF,QF/2,3500.,.25,.75,250., 0./
      DATA NT,NF,XB,XD,QR,PC,TB,TNT/33,25,.00001,.9999,
     + 30000000.,35.,240.,150./
      DATA T1(1),T2(1),P1(1),P2(1)/64.,94.,7.737,14.7/
      DATA T1(2),T2(2),P1(2),P2(2)/154.4,215.6,7.737,29.4/
      DATA CL(1),CV(1),DH(1),MW(1)/.56,.35,172.,58.1/
      DATA CL(2),CV(2),DH(2),MW(2)/.8,.43,202.,74.1/
      CALL VAPC(NC,VPC,T1,T2,P1,P2)
      TFLASH=TF-40.
C  SET PRESSURES ON ALL TRAYS
      P(NT)=PC
      DO 16 N=NT-1,1,-1
   16 P(N)=P(N+1)+.1
      PB=P(1)+.1
C  DO FLASH CALCULATION ON FEED
      CALL FLASH(NC,VPC,Z,F,TF,P(NF),QF,PS,TFLASH,XF,YF,LF,VF,CL
     + ,CV,DH,MW)
      WRITE(6,1)
    1 FORMAT(/,1X,'     Z1      Z2      F       TF      P(NF)       QF')
      WRITE(6,2)  (Z(J),J=1,NC),F,TF,P(NF),QF
    2 FORMAT(3X,2F7.4,3F8.1,F10.1)
      WRITE(6,3)
    3 FORMAT(/,5X,'XF1    XF2    YF1    YF2    LF     VF     TFLASH')
      WRITE(6,4)  (XF(J),J=1,NC),(YF(J),J=1,NC),LF,VF,TFLASH
    4 FORMAT(3X,4F7.4,3F7.1)
      CALL ENTH(NC,Z,YF,TF,HF,DUM,CL,CV,DH,MW)
C INITIAL GUESSES OF TRAY TEMPERATURES
      DT=(TB-TNT)/NT
      T(1)=TB+DT
      DO 17 N=2,NT
   17 T(N)=T(N-1)+DT
      D=F*(Z(1)-XB)/(XD-XB)
      B=F-D
      DQR=QR/10.
      LOOP=0
      FLAGP=1.
      X(1)=XB
      XX(1)=XB
      XX(2)=1.-XB
      CALL BUBPT(NC,VPC,XX,PB,TB,YY,PS)
      CALL ENTH(NC,XX,YY,TB,HLB,HVB,CL,CV,DH,MW)
      YB=YY(1)
C  BEGINNING OF ITERATIVE QR LOOP
  100 LOOP=LOOP+1
      IF(LOOP.GT.40)WRITE(6,77)LOOP,QR
   77 FORMAT(1X,I3,3X,F10.0)
      IF(LOOP.GT.50)THEN
      WRITE(6,10)
   10 FORMAT(1X,'QLOOP')
```

TABLE 13-1 CONT.

```
      STOP
      ENDIF
C  BEGINNING OF ITERATIVE X(1) LOOP, USING SUCCESSIVE SUBSTITUTION
   15 XX(1)=X(1)
      XX(2)=1.-X(1)
      CALL BUBPT(NC,VPC,XX,P(1),T(1),YY,PS)
      CALL ENTH(NC,XX,YY,T(1),HL(1),HV(1),CL,CV,DH,MW)
      Y(1)=YY(1)
      VB=(QR+B*(HL(1)-HLB))/(HVB-HL(1))
      L(1)=VB+B
      XCALC= (VB*YB+B*XB)/L(1)
      IF(ABS(X(1)-XCALC).LT. .0000001) GO TO 20
      X(1)=XCALC
      GO TO 15
C  CONVERGED X(1) LOOP
C
C  STEPPING UP COLUMN, TRAY BY TRAY
C
   20 DO 30 N=1,NT-1
      X(N+1)=X(N)
   25 CALL BUBPT(NC,VPC,XX,P(N+1),T(N+1),YY,PS)
      CALL ENTH(NC,XX,YY,T(N+1),HL(N+1),HV(N+1),CL,CV,DH,MW)
      Y(N+1)=YY(1)
      V(N)=(QR+B*(HL(N+1)-HLB))/(HV(N)-HL(N+1))
      IF(N.GE.NF) V(N)=V(N)+F*(HF-HL(N+1))/(HV(N)-HL(N+1))
      L(N+1)=V(N)+B
      IF(N.GE.NF) L(N+1)=L(N+1)-F
      XCALC=(V(N)*Y(N)+B*XB)/L(N+1)
      IF(N.GE.NF) XCALC=XCALC-F*Z(1)/L(N+1)
      IF(ABS(X(N+1)-XCALC).LT. .0000001) GO TO 28
      X(N+1)=XCALC
      XX(1)=X(N+1)
      XX(2)=1.-X(N+1)
      GO TO25
   28 IF(X(N+1).GT.XD) GO TO 70
C  CHECK TO MAKE SURE THAT COMPOSITIONS ALWAYS INCREASE AS
C        WE GO UP THE COLUMN
      IF(X(N+1).LT.X(N)) GO TO 60
   30 CONTINUE
      IF(LOOP.GT.40)WRITE(6,78)Y(NT)
   78 FORMAT(11X,F8.6)
      IF(ABS(Y(NT)-XD).LT. .000001) THEN
      N=0
      TC=T(NT)
      CALL BUBPT(NC,VPC,YY,PC,TC,XX,PS)
      CALL ENTH(NC,YY,YY,TC,HLC,DUM,CL,CV,DH,MW)
      V(NT)=(QR+B*(HLC-HLB)+F*(HF-HLC))/(HV(NT)-HLC)
      R=V(NT)-D
      RR=R/D
      WRITE(6,31)
   31 FORMAT(/,4X,'N      T         X         Y         L         V    QR'
      WRITE(6,32)N,TB,XB,YB,B,VB,QR/1000000.
   32 FORMAT(3X,I3,3X,F7.2,2F9.6,2F8.2,F9.4)
      DO 33 N=1,NT
      IF(N.EQ.NF)WRITE(6,35)N,T(N),X(N),Y(N),L(N),V(N)
   35 FORMAT(1X,'NF',I3,3X,F7.2,2F9.6,2F8.2)
   33 IF(N.NE.NF)WRITE(6,32)N,T(N),X(N),Y(N),L(N),V(N)
      WRITE(6,*)'                          XD        D       R      RR'
      WRITE(6,34)XD,D,R,RR
   34 FORMAT(16X,F9.6,3F8.2)
      STOP
      ENDIF
      IF(Y(NT)-XD)60,60,70
```

TABLE 13-1 CONT.

```
   60 IF(FLAGP.LT.0.)DQR=DQR/2.
      QR=QR+DQR
      FLAGM=-1.
      GO TO 100
   70 IF(FLAGM.LT.0.)DQR=DQR/2.
      QR=QR-DQR
      FLAGP=-1.
      GO TO 100
      END

      SUBROUTINE FLASH(NC,VPC,Z,F,TF,P,Q,PS,T,X,Y,L,V,CL,CV,DH,MW)
      DIMENSION VPC(2,NC),Z(NC),X(NC),Y(NC),PS(NC),CL(NC),CV(NC),
     + DH(NC),MW(NC)
      REAL L,MW,NETQ
      TDEW=TF
      TBUB=TF
      CALL BUBPT(NC,VPC,Z,P,TBUB,Y,PS)
      CALL DEWPT(NC,VPC,Z,P,TDEW,X,PS)
      WRITE(6,15)TBUB,TDEW
   15 FORMAT(1X,'TBUB = ',F8.2,'     TDEW =   ',F8.2)
      CALL ENTH(NC,Z,Z,TF,HF,DUM,CL,CV,DH,MW)
      CALL ENTH(NC,Z,Y,TBUB,HLBUB,HVBUB,CL,CV,DH,MW)
      CALL ENTH(NC,X,Z,TDEW,HLDEW,HVDEW,CL,CV,DH,MW)
      QBUB=F*(HLBUB-HF)
      QDEW=F*(HVDEW-HF)
C CHECK TO SEE IF FEED IS SUBCOOLED
      IF(Q.LT.QBUB)THEN
      V=0.
      L=F
      SUM=0.
      DO 300 J=1,NC
      X(J)=Z(J)
      Y(J)=0.
  300 SUM=SUM+Z(J)*MW(J)*CL(J)
      T=200.+(Q/F+HF)/SUM
      RETURN
      ENDIF
C CHECK TO SEE IF FEED IS SUPERHEATED VAPOR
      IF(Q.GT.QDEW)THEN
      V=F
      L=0.
      SUM1=0.
      SUM2=0.
      DO 301 J=1,NC
      X(J)=0.
      Y(J)=Z(J)
      SUM1=SUM1+Z(J)*MW(J)*DH(J)
  301 SUM2=SUM2+Z(J)*MW(J)*CV(J)
      T=(Q/F+HF-SUM1)/SUM2+200.
      RETURN
      ENDIF
      IBUG=40
      FLAGTM=1.
      FLAGTP=1.
      DT=5.
      LOOPT=0
      VF=0.18
C OUTSIDE CONVERGENCE LOOP FOR TEMPERATURE
  100 DVF=.05
      CALL PSAT(NC,VPC,T,PS)
      LOOPT=LOOPT+1
      IF(LOOPT.GT.IBUG)WRITE(6,77)T,DT,NETQ,VF
      IF(LOOPT.GT.50)THEN
```

TABLE 13-1 CONT.

```
      WRITE(6,3)
    3 FORMAT(1X,'TLOOP')
      STOP
      ENDIF
      FLAGVP=1.
      FLAGVM=1.
      LOOP=0
      IF(T.GT.TDEW)GO TO 60
C INSIDE CONVERGENCE LOOP FOR V/F RATIO (ISOTHERMAL FLASH)
  200 XSUM=0.
      LOOP=LOOP+1
      IF(LOOP.GT.50) THEN
      WRITE(6,4)
    4 FORMAT(1X,'FLASH LOOP')
      STOP
      ENDIF
      DO 5 J=1,NC
    5 XSUM=XSUM+Z(J)/(1.+VF*(PS(J)/P-1.))
      IF(LOOP.GT.IBUG) WRITE(6,77)T,DT,NETQ,VF,XSUM
   77 FORMAT(1X,6F15.5)
      IF(ABS(XSUM-1.).LT. .00001) GO TO 40
      IF(XSUM-1.)20,20,30
   20 IF(FLAGVP.LT.0.)DVF=DVF/2.
      VF=VF+DVF
      IF(VF.GT.1.) GO TO 60
      FLAGVM=-1.
      GO TO 200
   30 IF(FLAGVM.LT.0.)DVF=DVF/2.
      VF=VF-DVF
      FLAGVP=-1.
      IF(VF.LT.0.) GO TO 20
      GO TO 200
   40 DO 41 J=1,NC
      X(J)=Z(J)/(1.+VF*(PS(J)/P-1.))
   41 Y(J)=X(J)*PS(J)/P
      CALL ENTH(NC,X,Y,T,HL,HV,CL,CV,DH,MW)
      V=F*VF
      L=F-V
      NETQ=F*HF+Q-V*HV-L*HL
      IF(ABS(NETQ).LT.1000.) GO TO 1000
      IF(DT.LT. .00001) GO TO 1000
      IF(NETQ)60,60,50
   50 IF(FLAGTP.LT.0.)DT=DT/2.
      T=T+DT
      FLAGTM=-1.
      GO TO 100
   60 IF(FLAGTM.LT.0.) DT=DT/2.
      T=T-DT
      FLAGTP=-1.
      GO TO 100
 1000 RETURN
      END

      SUBROUTINE VAPC(NC,VPC,T1,T2,P1,P2)
      DIMENSION VPC(2,NC),T1(NC),T2(NC),P1(NC),P2(NC)
      WRITE(6,1)
    1 FORMAT(//,1X,'VAPOR PRESSURE CONSTANTS')
      DO 10 J=1,NC
      VPC(2,J)=(T2(J)+460.)*(T1(J)+460.)*ALOG(P2(J)/P1(J))
     + /(T1(J)-T2(J))
      VPC(1,J)=ALOG(P1(J)) - VPC(2,J)/(T1(J)+460.)
   10 WRITE(6,2) J,VPC(1,J),VPC(2,J)
    2 FORMAT(5X,I2,3X,2F15.4)
```

TABLE 13–1 CONT.

```
      RETURN
      END

      SUBROUTINE PSAT(NC,VPC,T,PS)
      DIMENSION VPC(2,NC),PS(NC)
      DO 10 J=1,NC
   10 PS(J)=EXP(VPC(1,J)+VPC(2,J)/(T+460.))
      RETURN
      END

      SUBROUTINE ENTH(NC,X,Y,T,HL,HV,CL,CV,DH,MW)
      DIMENSION X(NC),Y(NC),CL(NC),CV(NC),DH(NC),MW(NC)
      REAL MW
      HL=0.
      HV=0.
      DO 10 J=1,NC
      HL=HL+CL(J)*X(J)*MW(J)*(T-200.)
   10 HV=HV+(CV(J)*(T-200.)+DH(J))*Y(J)*MW(J)
      RETURN
      END

      SUBROUTINE DEWPT(NC,VPC,Y,P,T,X,PS)
      DIMENSION VPC(2,NC),X(NC),Y(NC),PS(NC)
      DT=10.
      F=0.
      DO 10 J=1,NC
      PS(J)=EXP(VPC(1,J)+VPC(2,J)/(T+460.))
      F=F+Y(J)/PS(J)
   10 CONTINUE
      FOLD=1./F-P
   20 T=T+DT
      LOOP=LOOP+1
      IF(LOOP.GT.20)THEN
      WRITE(6,11)
   11 FORMAT(1X,'DEWPT LOOP')
      STOP
      ENDIF
      FNEW=0.
      DO 15 J=1,NC
      PS(J)=EXP(VPC(1,J)+VPC(2,J)/(T+460.))
   15 FNEW=FNEW+Y(J)/PS(J)
      IF(LOOP.GT.15) WRITE(6,1)T,DT,FNEW
    1 FORMAT(1X,3F10.4)
      FNEW=1./FNEW  -  P
      IF(ABS(FNEW).LT.P/10000.)THEN
      DO 17 J=1,NC
   17 X(J)=Y(J)*P/PS(J)
      RETURN
      ENDIF
      DT=-DT*FNEW/(FNEW-FOLD)
      FOLD=FNEW
      GO TO 20
      END

      SUBROUTINE BUBPT(NC,VPC,X,P,T,Y,PS)
      DIMENSION VPC(2,NC),X(NC),Y(NC),PS(NC)
      LOOP=0
   20 DF=0.
      F=0.
      DO 10 J=1,NC
      PS(J)=EXP(VPC(1,J)+VPC(2,J)/(T+460.))
      F=F+X(J)*PS(J)
      DF=DF-X(J)*PS(J)*VPC(2,J)/(T+460.)**2
```

TABLE 13-1 CONT.

```
 10 CONTINUE
    LOOP=LOOP+1
    IF(LOOP.GT.20)THEN
    WRITE(6,11)
 11 FORMAT(1X,'BUBPT LOOP')
    STOP
    ENDIF
    F=F-P
    IF(LOOP.GT.15)WRITE(6,1)T,F,DF,LOOP
  1 FORMAT(1X,3F10.5,I3)
    IF(ABS(F/P).LT. .0000001)GO TO 100
    T=T-F/DF
    GO TO 20
100 DO 101 J=1,NC
101 Y(J)=X(J)*PS(J)/P
    RETURN
    END
```

equations are necessary. As usual, these are the total mass, component, and energy balances, given by

$$L_1 = V_B + B \tag{13–30}$$

$$x_1 L_1 = y_B V_B + x_B B \tag{13–31}$$

and

$$Q_R + h_1 L_1 = H_B V_B + h_B B \tag{13–32}$$

One convenient way to solve these simultaneous equations is to combine equations 13–30 and 13–32 to solve for V_B. We obtain

$$V_B = \frac{Q_R - B(h_B - h_1)}{(H_B - h_1)} \tag{13–33}$$

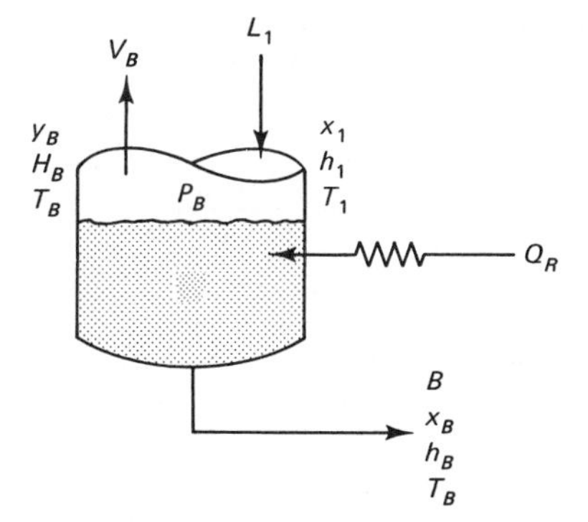

Figure 13–16 Base of distillation column.

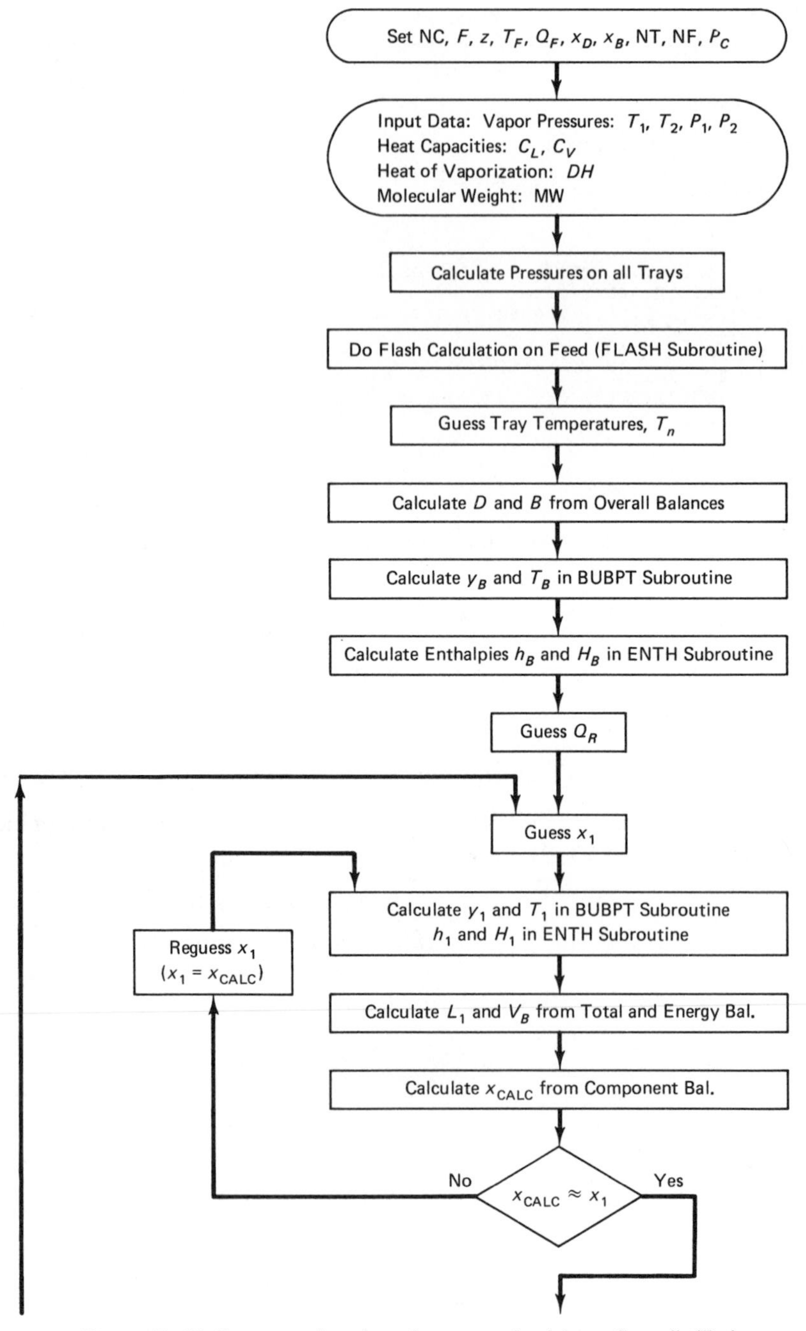

Figure 13–17 Computer flowchart for nonequimolal-overflow distillation rating program.

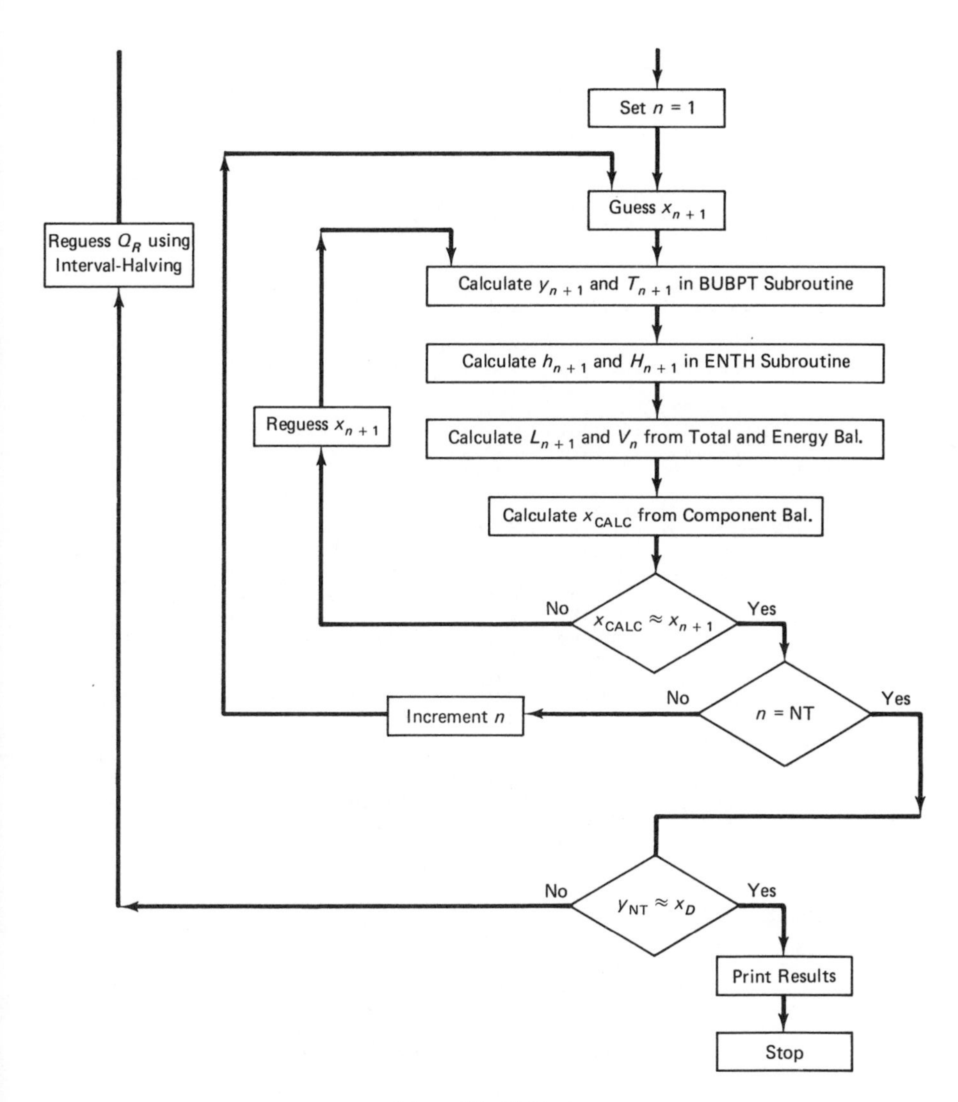

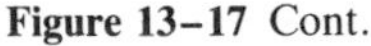
Figure 13–17 Cont.

An iterative procedure is then used to solve for x_1:

1. Guess a value of x_1; we know it must be greater than x_B.
2. Calculate T_1 and y_1 using a bubblepoint calculation for the liquid of composition x_1 at pressure P_1.
3. Use enthalpy data to get h_1 and H_1.
4. Calculate V_B from equation 13–33.
5. Calculate L_1 from equation 13–30.

TABLE 13–2
RESULTS FROM PROGRAM OF TABLE 13–1

```
VAPOR PRESSURE CONSTANTS
     1              13.8985      -6210.7230
     2              16.7832      -9054.4985
TBUB =    196.04     TDEW =      215.30

  Z1      Z2        F        TF      P(NF)          QF
 .2500   .7500   3500.0    250.0     35.8           .0

 XF1     XF2      YF1     YF2     LF      VF      TFLASH
 .1901   .8099   .4895   .5104 2799.6   700.4   202.5

   N       T         X          Y        L       V       QR
   0     229.20   .000010   .000035 2624.94 2030.04  28.7988
   1     229.06   .000021   .000072 4654.98 2027.98
   2     228.92   .000037   .000128 4652.92 2025.91
   3     228.78   .000061   .000213 4650.85 2023.83
   4     228.64   .000098   .000341 4648.77 2021.74
   5     228.49   .000154   .000536 4646.68 2019.62
   6     228.34   .000239   .000830 4644.56 2017.47
   7     228.19   .000366   .001274 4642.41 2015.28
   8     228.02   .000559   .001947 4640.22 2013.02
   9     227.84   .000851   .002962 4637.96 2010.66
  10     227.65   .001290   .004493 4635.60 2008.15
  11     227.42   .001953   .006799 4633.09 2005.44
  12     227.15   .002950   .010262 4630.37 2002.40
  13     226.81   .004446   .015439 4627.34 1998.92
  14     226.38   .006680   .023126 4623.86 1994.79
  15     225.81   .009992   .034423 4619.73 1989.79
  16     225.04   .014848   .050775 4614.73 1983.67
  17     223.99   .021861   .073922 4608.61 1976.24
  18     222.57   .031756   .105670 4601.18 1967.52
  19     220.70   .045277   .147350 4592.46 1957.92
  20     218.33   .062957   .198997 4582.86 1948.36
  21     215.48   .084784   .258528 4573.30 1940.15
  22     212.27   .109880   .321584 4565.09 1934.47
  23     208.96   .136448   .382555 4559.41 1931.73
  24     205.82   .162184   .436441 4556.67 1931.40
NF 25    203.08   .185010   .480350 4556.34 2634.54
  26     198.86   .221957   .544214 1759.48 2651.95
  27     188.66   .319803   .678041 1776.89 2735.96
  28     170.49   .526692   .849757 1860.90 2944.85
  29     152.67   .786280   .955198 2069.79 3153.40
  30     144.08   .938029   .989441 2278.33 3245.86
  31     141.51   .985580   .997689 2370.80 3271.17
  32     140.78   .996882   .999508 2396.11 3276.83
  33     140.50   .999365   .999900 2401.77 3279.13
                    XD          D       R       RR
                  .999900   875.06 2404.07     2.75
```

6. Calculate x_1 from equation 13–31:

$$x_1 = \frac{y_B V_B + x_B B}{L_1} \tag{13–34}$$

7. If the value of x_1 calculated from equation 13–34 is not close enough to the value of x_1 guessed in step 1, reguess x_1 and go to step 2.

Since the physical properties are not strong functions of x_1, the direct substitution of the calculated value of x_1 back into step 2 works very well.

After the value of x_1 is gotten by iteration, the equations describing a system where we cut the column above tray 1 are written and solved in a similar iterative manner for the three new unknowns x_2, L_2, and V_1. This procedure is then repeated up the entire column.

The program of Table 13–1 is a NEMO rating program for a binary system that does these tray-to-tray mass, component, and energy balance calculations. The specific example is propylene oxide–tertiary butyl alcohol in a 33-tray column, fed in on tray 25. The product purities are specified, and the program solves iteratively for the reboiler heat input Q_R required to make the separation.

The adiabatic flash calculation on the feed is done using the Subroutine FLASH discussed in chapter 11. Subroutine BUBPT is used to calculate y and T from the given x and P. Figure 13–17 gives a flowchart of the program.

Results for a 25 weight percent propylene oxide feed are given in Table 13–2.

PROBLEMS

13–1. Calculate the minimum reflux ratio and the number of theoretical trays required to achieve the separation specified below with a reflux ratio of 1, (a) assuming equimolal overflow and (b) *not* assuming equimolal overflow.

Feed: 54 weight percent methanol in a binary methanol-water system.
Saturated liquid at column pressure of 760 mm Hg.
Distillate: 98 weight percent methanol from a total condenser.
Reflux is saturated liquid.
Bottom: 3 weight percent methanol, with a partial reboiler.
Physical Property Data:
Enthalpy: See figure on page 482.
VLE:

mole % methanol	
x	y
2	13.4
4	23.0
6	30.4
8	36.5
10	41.8
15	51.7
20	57.9
30	66.5
40	72.9
50	77.9
60	82.5
70	87
80	91.5
90	95.8

13–2. Using the enthalpy-composition diagram given in the figure below for the methanol-water system, calculate the number of stages required and the feed plate location to produce 98 weight percent methanol liquid product and 5 weight percent methanol bottoms product from a feed of 45 weight percent methanol entering the column at 50°C. The column has a total condenser and a partial reboiler, and operates at 1.5 times minimum reflux. What is the

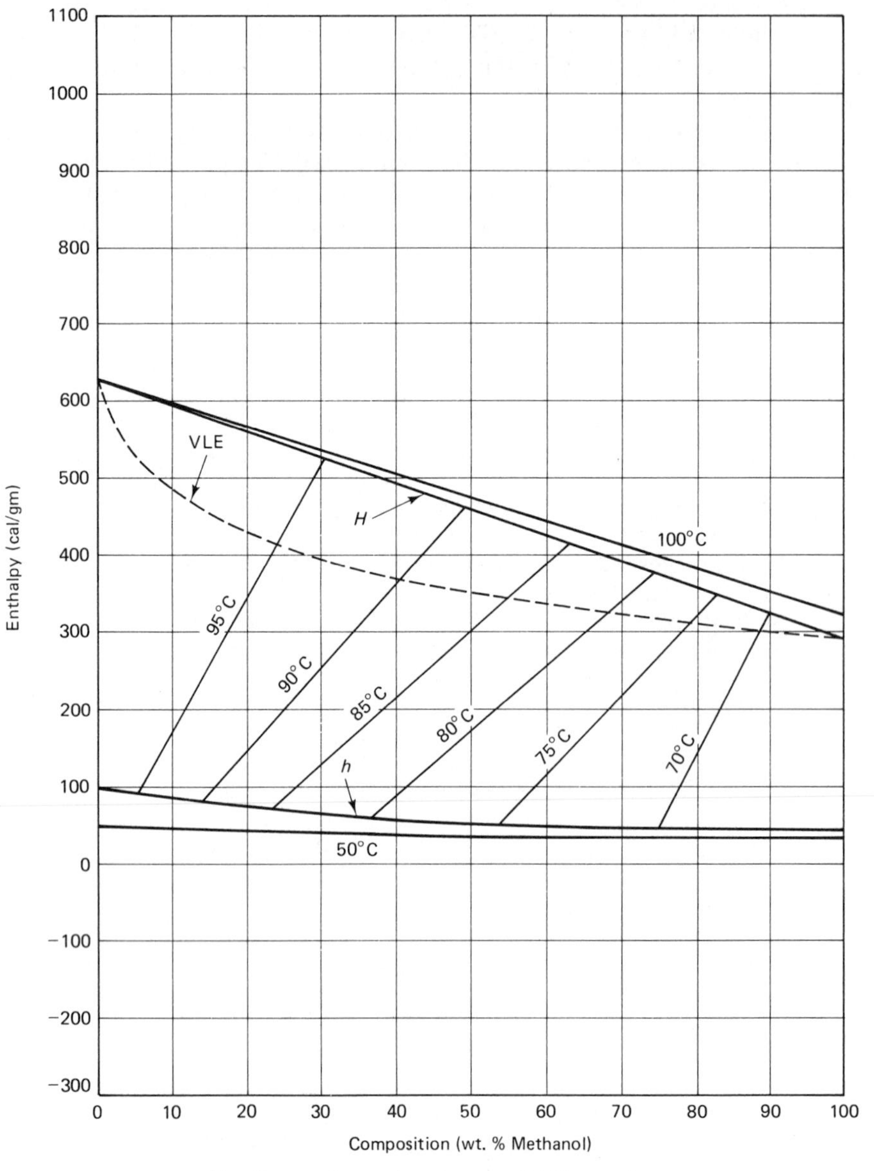

Hxy diagram for methanol-water @ 760 mm Hg.

reboiler heat duty in Btu/hr if the feed rate is 5,000 lb_m/hr? What is the condenser duty?

13–3. The column of the preceding problem was designed 20 years ago and has been operating ever since. Energy costs have risen, and it is now desired to reduce operating costs. One suggestion is to heat the feed to 200°F using waste heat, thus, perhaps, operating at a lower reflux ratio and reducing the reboiler duty. Can this be done? Will we have to change the column feed point? What will be the new reboiler heat duty?

13–4. Calculate the number of theoretical trays required to produce a 98 weight percent methanol vapor distillate product from a partial condenser and a 12 weight percent methanol liquid product that is withdrawn from tray 2 of a distillation column. Equimolal overflow cannot be assumed in this binary methanol-water system operating at atmospheric pressure. Feed is 45 weight percent methanol and is 22 percent vaporized. Feed rate is 100 lb_m/min. Heat added to the system in the total reboiler at the base of the column is 31,615 Btu/minute. What is the composition of the liquid in the reboiler?

13–5. Vapor from the top of a distillation column passes through two condensers in series. The first is a partial condenser producing D_1 lb_m/hr of saturated liquid with a composition x_1 weight fraction of the light component. The uncondensed vapor is in equilibrium with the liquid leaving the condenser. The second condenser is a total condenser producing D_2 lb_m/hr of saturated liquid with composition x_2. A binary system is being separated.

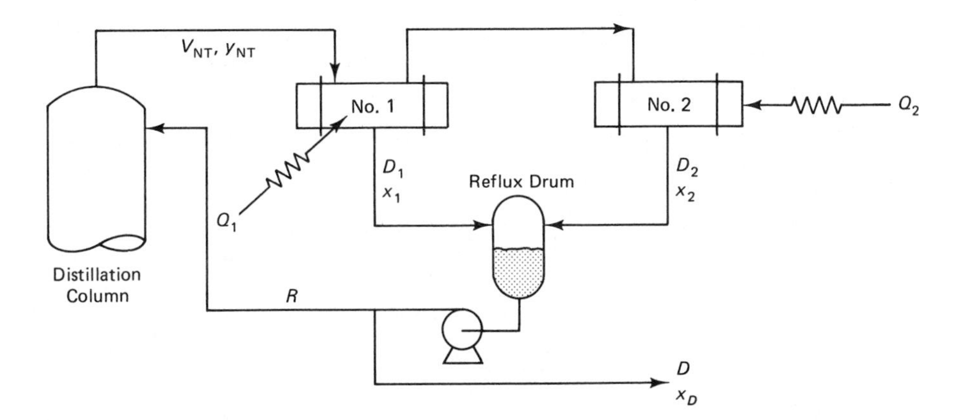

The liquid streams from both condensers are mixed together in the reflux drum. Locate these various streams on an *Hxy* diagram (sketch your own), and show how you would calculate the heat removal rates Q_1 and Q_2, in both condensers, given V_{NT}, y_{NT}, and D_1.

13–6. The temperature in the top of a distillation column is 40°F. In order to be able to use cooling water at 80°F to condense this vapor, a gas compressor is used to boost the pressure up to the point where the bubblepoint temperature of the overhead vapor is 110°F. The work of compression is W Btu/hr.

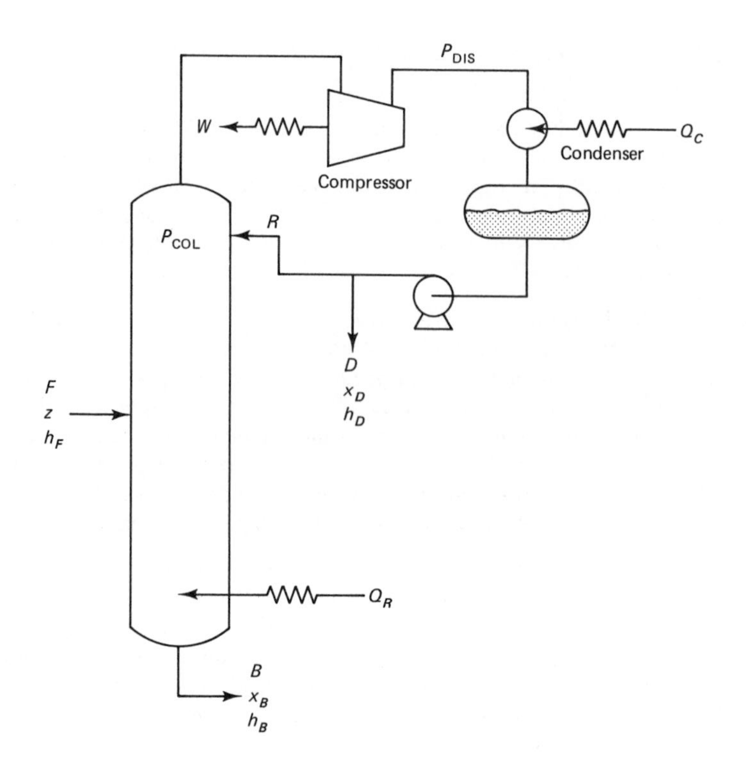

Show, on an *Hxy* diagram, how you would locate the "Δ point" for constructing operating lines in the rectifying section of this nonequimolal overflow column.

13–7. Design two alternative distillation columns, one with a feed preheater and one without. Each column has a feed rate of 60,000 lb_m/hr of 60 weight percent methanol and 40 weight percent water at 50°C. Both columns produce saturated liquid distillate and bottoms products of 95 weight percent methanol and 15 weight percent methanol, respectively. In the system with the feed preheater, the feed stream is preheated to 85°C before being introduced into the column. Each column operates at atmospheric pressure.

Without assuming equimolal overflow,

(a) Calculate the minimum reflux ratios for each column.

(b) Set the actual reflux ratio at 1.2 times the minimum, and calculate the condenser and reboiler heat duties.

(c) Calculate the heat input rate to the feed preheater.

13–8. A methanol-water mixture containing 40 weight percent CH_3OH is fed to a laboratory distillation column. This liquid, originally at 50°C, is pumped at a rate of 2 gpm through a heater where its temperature is increased to 85°C. In this condition, it enters on the third theoretical tray of the six-theoretical tray column. The column has a total condenser and a partial reboiler. Saturated liquid reflux to the column is also pumped and is fixed at a rate of 2 gpm.

Both the column and the preheater operate at 1 atm. Distillate flow rate is 1 gpm.

(a) How much heat is absorbed by the feed in the heater before it enters the column?

(b) What are the compositions and flow rates of bottoms and distillate products?

(c) How much heat is supplied by the partial reboiler at the bottom of the column?

Use the following additional data:

Specific heats of methanol-water solutions		*Densities of methanol-water solutions*	
mole % CH_3OH	*C_p, cal/g°C*	*wt. % CH_3OH*	*ρ, g/cc*
5.88	0.995	0	0.998
12.3	0.98	10	0.982
27.3	0.92	20	0.969
45.8	0.83	30	0.954
69.6	0.726	40	0.935
100	0.617	50	0.914
		60	0.891
		70	0.868
		80	0.843
		90	0.818
		100	0.789

For water, 1 gal weighs 8.33 lb_m.

13–9. An absorption refrigeration system using an NH_3-H_2O system is shown on page 486. An *Hxy* diagram is given on page 487. The purpose of the cycle is to absorb 8×10^6 Btu/hr of energy as heat into the refrigerator at a temperature of about 0°C. For the conditions shown, calculate

(a) The amount of 97.5 weight percent NH_3 going to the refrigerator.

(b) The concentration and quantity of strong liquor fed to the column.

(c) The minimum reflux ratio. Assume that the strong liquor enters the column as a saturated liquid.

(d) The number of theoretical stages in the column and the number of such stages below the feed point if the reflux ratio is 1.5 times the minimum.

(e) The heat transfer rate Q_R in the reboiler.

13–10. A stripping column is fed on the top tray with 30,000 lb_m/hr of a binary mixture of 50 weight percent methanol and 50 weight percent water at 50°C. The column operates at atmospheric pressure, producing saturated-liquid bottoms product that is 10 weight percent methanol and saturated-vapor overhead product that is 75 weight percent methanol. The stripping column has a partial

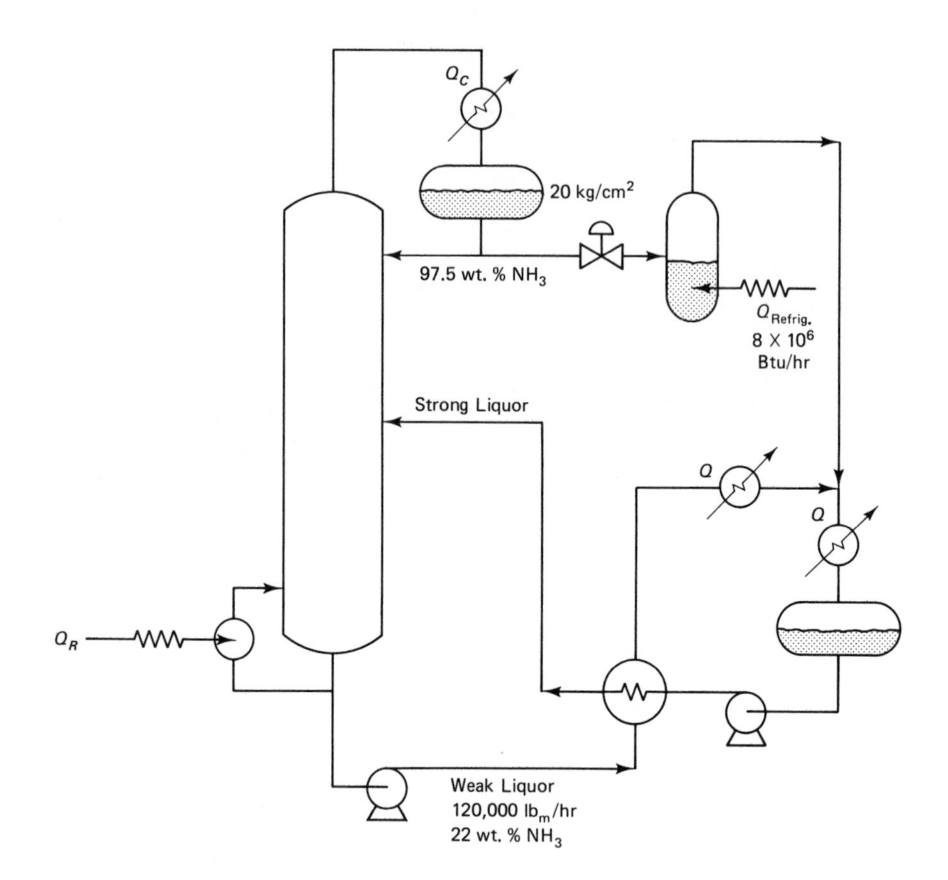

reboiler and 100 percent efficient trays. If equimolal overflow *cannot* be assumed, calculate the number of trays and the reboiler heat input.

13–11. A saturated vapor stream containing 50 weight percent ethanol is fed into the bottom of a rectifying column. The total condenser at the top returns some reflux liquid to the column and produces a distillate product of 85 weight percent ethanol. The column pressure is 1 atm. An *Hxy* diagram is given on page 489.

(a) What is the minimum reflux ratio possible? At this reflux ratio, what would be the composition of the liquid bottoms product from the column?

(b) If we use 120 percent of the minimum reflux, what is the bottoms product composition and how many ideal stages are there in the column?

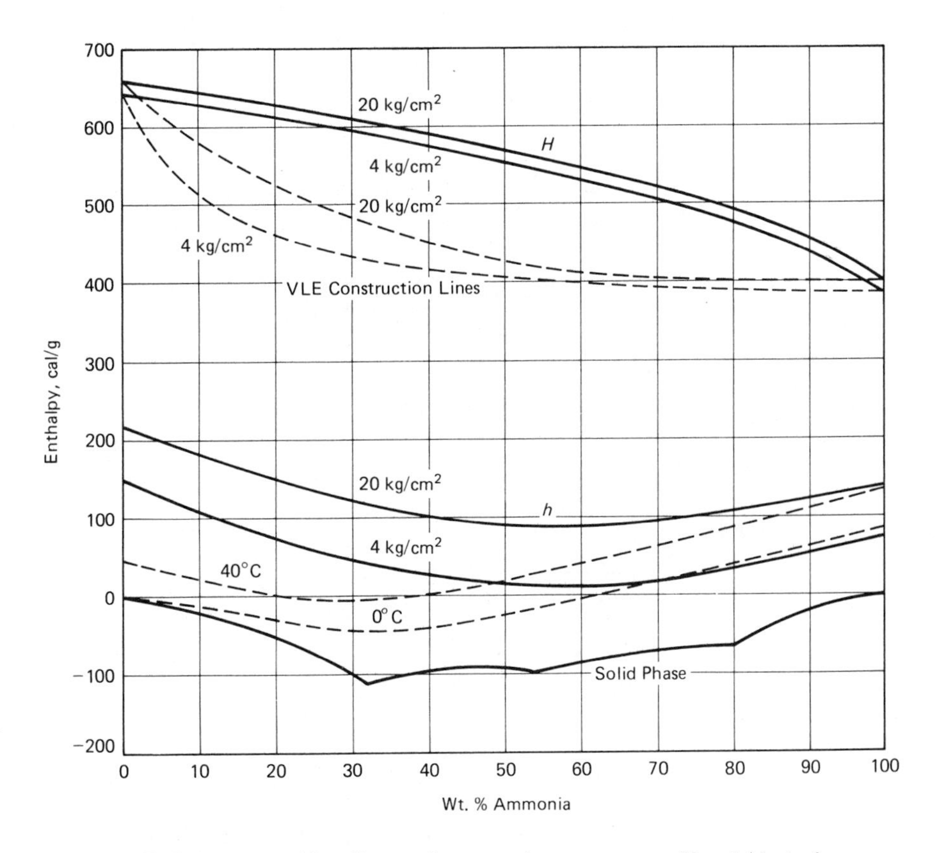

Enthalpy-composition diagram for ammonia-water system, 20 and 4 kg/cm^2 pressures.

13–12. Ethanol and water are distilled in a column that makes a distillate product of 90 weight percent C_2H_5OH and a vapor sidestream product of 70 weight percent C_2H_5OH. See the figure on page 488. The column operates at atmospheric pressure. The feed of 10,000 lb_m/hr of 45 weight percent C_2H_5OH enters at 185°F.

The mass flow rates of the distillate and sidestream products are equal. The bottoms product is 5 weight percent C_2H_5OH. The reflux ratio is 2. Assume a total condenser, partial reboiler, saturated liquid reflux, and theoretical trays. Equimolal overflow cannot be assumed. An *Hxy* diagram is given on page 489.

(a) Calculate the reboiler duty and the condenser duty in Btu/hr.

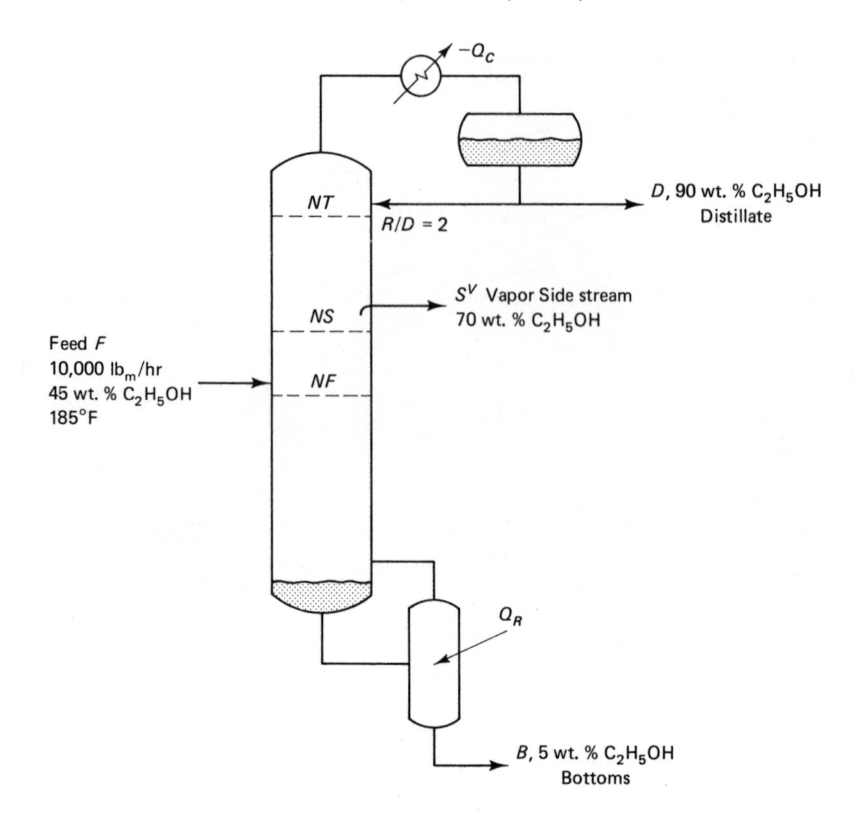

(b) Locate the Δ points necessary to calculate the stages throughout the column.

(c) Calculate the number of trays in the column.

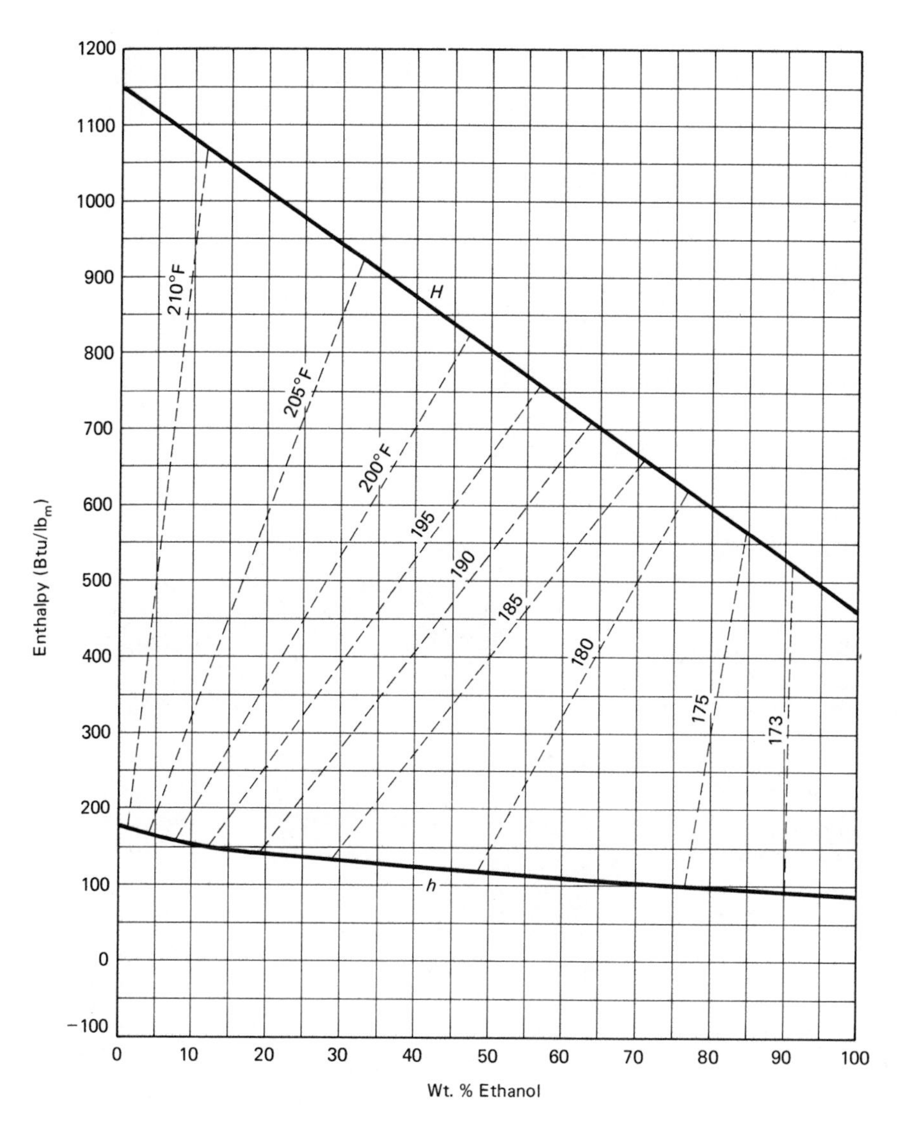
1200
1100
1000
900
800
700
600
500
400
300
200
100
0
−100
0
10
20
30
40
50
60
70
80
90
100
Enthalpy (Btu/lbm)
Wt. % Ethanol
H
h
210°F
205°F
200°F
195
190
185
180
175
173

APPENDIX

Units and Conversion Factors A.1

TABLE A.1–1**

CONVERSION FACTORS TO SI FOR SELECTED QUANTITIES

To convert from	To	Multiply by	
barrel (for petroleum, 42 gal)	meter³ (m^3)	1.5898729	E − 01
British thermal unit (Btu, International Table)	joule (J)	1.0550559	E + 03
Btu/lb_m °F (heat capacity)	joule/kilogram-kelvin (J/kg K)	4.1868000*	E + 03
Btu/hour	watt (W)	2.9307107	E − 01
Btu/second	watt (W)	1.0550559	E + 03
Btu/ft² hr °F (heat transfer coefficient)	joule/meter²-second-kelvin (J/m^2 s K)	5.6782633	E + 00
Btu/ft² hour (heat flux)	joule/meter²-second (J/m^2 s)	3.1545907	E + 00
Btu/ft hr °F (thermal conductivity)	joule/meter-second-kelvin (J/m s K)	1.7307347	E + 00
calorie (International Table)	joule (J)	4.1868000*	E + 00
cal/g °C	joule/kilogram-kelvin (J/kg K)	4.1868000*	E + 03
centimeter	meter (m)	1.0000000*	E − 02
centimeter of mercury (0°C)	pascal (Pa)	1.3332237	E + 03
centimeter of water (4°C)	pascal (Pa)	9.80638	E + 01
centipoise	pascal-second (Pa s)	1.0000000*	E − 03
centistoke	meter²/second (m^2/s)	1.0000000*	E − 06
degree Fahrenheit (°F)	kelvin (K)	$t_K = (t_F + 459.67)/1.8$	
degree Rankine (°R)	kelvin (K)	$t_K = t_R/1.8$	
dyne	newton (N)	1.0000000*	E − 05
erg	joule (J)	1.0000000*	E − 07
farad (International of 1948)	farad (F)	9.99505	E − 01
fluid ounce (U.S.)	meter³ (m^3)	2.9573530	E − 05
foot	meter (m)	3.0480000*	E − 01
foot (U.S. Survey)	meter (m)	3.0480061	E − 01
foot of water (39.2°F)	pascal (Pa)	2.98898	E + 03
foot²	meter² (m^2)	9.2903040*	E − 02
foot/second²	meter/second² (m/s^2)	3.0480000*	E − 01
foot²/hour	meter²/second (m^2/s)	2.5806400*	E − 05
foot-pound-force	joule (J)	1.3558179	E + 00
foot²/second	meter²/second (m^2/s)	9.2903040*	E − 02
foot³	meter³ (m^3)	2.8316847	E − 02
gallon (U.S. liquid)	meter³ (m^3)	3.7854118	E − 03

To convert from	to	Multiply by	
gram	kilogram (kg)	1.0000000*	E − 03
horsepower (550 ft-lb_F/s)	watt (W)	7.4569987	E + 02
inch	meter (m)	2.5400000*	E − 02
inch of mercury (60°F)	pascal (Pa)	3.37685	E + 03
inch of water (60°F)	pascal (Pa)	2.48843	E + 02
$inch^2$	$meter^2$ (m^2)	6.4516000*	E − 04
$inch^3$	$meter^3$ (m^3)	1.6387064*	E − 05
kilocalorie	joule (J)	4.1868000*	E + 03
kilogram-force (kgf)	newton (N)	9.8066500*	E + 00
micron	meter (m)	1.0000000*	E − 06
mil	meter (m)	2.5400000*	E − 05
mile (U.S. Statute)	meter (m)	1.6093440*	E + 03
mile/hour	meter/second (m/s)	4.4704000*	E − 01
millimeter of mercury (0°C)	pascal (Pa)	1.3332237	E + 02
ohm (International of 1948)	ohm (Ω)	1.000495	E + 00
ounce-mass (avoirdupois)	kilogram (kg)	2.8349523	E − 02
ounce (U.S. fluid)	$meter^3$ (m^3)	2.9573530	E − 05
pint (U.S. liquid)	$meter^3$ (m^3)	4.7317647	E − 04
poise (absolute viscosity)	pascal-second (Pa s)	1.0000000*	E − 01
poundal	newton (N)	1.3825495	E − 01
pound-force (lb_F avoirdupois)	newton (N)	4.4482216	E + 00
pound-force-second/ft^2	pascal-second (Pa s)	4.7880258	E + 01
pound-mass (lb_m avoirdupois)	kilogram (kg)	4.5359237*	E − 01
pound-mass/$foot^3$	kilogram/$meter^3$ (kg/m^3)	1.6018463	E + 01
pound-mass/foot-second	pascal-second (Pa s)	1.4881639	E + 00
psi	pascal (Pa)	6.8947573	E + 03
quart (U.S. liquid)	$meter^3$ (m^3)	9.4635295	E − 04
slug	kilogram (kg)	1.4593903	E + 01
stoke (kinematic viscosity)	$meter^2$/second (m^2/s)	1.0000000*	E − 04
ton (long, 2240 lb_m)	kilogram (kg)	1.0160469	E + 03
ton (short, 2000 lb_m)	kilogram (kg)	9.0718474*	E + 02
torr (mm Hg, 0°C)	pascal (Pa)	1.3332237	E + 02
volt (International of 1948)	volt (absolute) (V)	1.000330	E + 00
watt (International of 1948)	watt (W)	1.000165	E + 00
watt-hour	joule (J)	3.6000000*	E + 03

*An asterisk after the seventh decimal place indicates the conversion factor is exact and all subsequent digits are zero.
**Adapted from *Chemical Engineering Progress*, August 1977, p. 137.

TABLE A.1–2*
FUNDAMENTAL CONSTANTS AND CONVERSION FACTORS

Gas law constant R

Numerical value	*Units*
1.9872	cal/g-mole K
1.9872	Btu/lb-mole °R
82.057	cm^3 atm/g-mole K
8314.34	J/kg-mole K
82.057×10^{-3}	m^3 atm/kg-mole K
8314.34	kg m^2/s^2 kg-mole K
10.731	ft^3 $lb_f/in.^2$ lb-mole °R
0.7302	ft^3 atm/lb-mole °R
1545.3	ft lb_f/lb-mole °R
8314.34	m^3 Pa/kg-mole K

Volume and Density

1 g-mole ideal gas at 0°C, 760 mm Hg = 22.4140 liters = 22,414 cm^3
1 lb-mole ideal gas at 0°C, 760 mm Hg = 359.05 ft^3
1 kg-mole ideal gas at 0°C, 760 mm Hg = 22.414 m^3
Density of dry air at 0°C, 760 mm Hg = 1.2929 g/liter
= 0.080711 lb_m/ft^3
Molecular weight of air = 28.97 lb_m/lb-mole
1 g/cm^3 = 62.43 lb_m/ft^3 = 1000 kg/m^3
1 g/cm^3 = 8.345 lb_m/U.S. gal
1 lb_m/ft^3 = 16.0185 kg/m^3

Length

1 in. = 2.540 cm
100 cm = 1 m (meter)
1 micron = 10^{-6} m = 10^{-4} cm = 10^{-3} mm = 1 μm (micrometer)
1 Å (angstrom) = 10^{-10} m = 10^{-4} μm
1 mile = 5280 ft
1 m = 3.2808 ft = 39.37 in.

Mass

1 lb_m = 453.59 g = 0.45359 kg
1 lb_m = 16 oz = 7000 grains
1 kg = 1000 g = 2.2046 lb_m
1 ton (short) = 2000 lb_m
1 ton (long) = 2240 lb_m
1 ton (metric) = 1000 kg

*Adapted from *Transport Processes: Momentum, Heat and Mass*, C. J. Geankoplis, Allyn and Bacon, Inc., Boston, 1983.

TABLE A.1-2 CONT.

Standard Acceleration of Gravity

g = 9.80665 m/s^2
g = 980.665 cm/s^2
g = 32.174 ft/s^2
g_c (gravitational conversion factor) = 32.1740 lb$_m$ ft/lb$_f$ s^2
= 980.665 g$_m$ cm/g$_f$ s^2

Volume

1 liter = 1000 cm^3
1 in.^{3} = 16.387 cm^3
1 ft^3 = 28.317 liters
1 ft^3 = 0.028317 m^3
1 ft^3 = 7.481 U.S. gal
1 m^3 = 264.17 U.S. gal
1 m^3 = 1000 liters
1 U.S. gal = 4 qt
1 U.S. gal = 3.7854 liters
1 U.S. gal = 3785.4 cm^3
1 British gal = 1.20094 U.S. gal

Force

1 g cm/s^2 (dyn) = 10^{-5} kg m/s^2 N (newton)
1 g cm/s^2 = 7.2330 × 10^{-5} lb$_m$ ft/s^2 (poundal)
1 kg m/s^2 = 1 N (newton)
1 lb$_f$ = 4.4482 N
1 g cm/s^2 = 2.2481 × 10^{-6} lb$_f$

Pressure

1 bar = 1 × 10^5 Pa (pascal) = 1 × 10^5 N/m^2
1 psia = 1 lb$_f$/in.^{2}
1 psia = 2.0360 in. Hg at 0°C
1 psia = 2.311 ft H_2O at 70°F
1 psia = 51.715 mm Hg at 0°C (ρ_{Hg} = 13.5955 g/cm^3)
1 atm = 14.696 psia = 1.01325 × 10^5 N/m^2 = 1.01325 bars
1 atm = 760 mm Hg at 0°C = 1.01325 × 10^5 Pa
1 atm = 29.921 in. Hg at 0°C
1 atm = 33.90 ft H_2O at 4°C
1 psia = 6.89476 × 10^4 g/cm s^2
1 psia = 6.89476 × 10^4 dyn/cm^2
1 dyn/cm^2 = 2.0886 × 10^{-3} lb$_f$/ft^2
1 psia = 6.89476 × 10^3 N/m^2
1 lb$_f$/ft^2 = 4.7880 × 10^2 dyn/cm^2 = 47.880 N/m^2
1 mm Hg (0°C) = 1.333224 × 10^2 N/m^2 = 0.1333224 kPa

Power

1 hp = 0.74570 kW
1 hp = 550 ft lb$_f$/s
1 hp = 0.7068 Btu/s
1 watt (W) = 14.34 cal/min
1 Btu/hr = 0.29307 W (watt)
1 J/s (joule/s) = 1 W
1 kW = 1.34 hp

TABLE A.1-2 CONT.

Heat, Energy, Work

1 J = 1 N m = 1 kg m^2/s^2
1 kg m^2/s^2 J (joule) = 10^7 g cm^2/s^2 (erg)
1 Btu = 1,055.06 J = 1.05506 kJ
1 Btu = 252.16 cal (thermochemical)
1 kcal (thermochemical) = 1000 cal = 4.1840 kJ
1 cal (thermochemical) = 4.1840 J
1 cal (IT) = 4.1868 J
1 Btu = 251.996 cal (IT)
1 Btu = 778.17 ft lb$_f$
1 hp hr = 0.7457 kW hr
1 hp hr = 2544.5 Btu
1 ft lb$_f$ = 1.35582 J
1 ft lb$_f$/lb$_m$ = 2.9890 J/kg

Thermal Conductivity

1 Btu/hr ft °F = 4.1365 × 10^{-3} cal/s cm °C
1 Btu/hr ft °F = 1.73073 W/m K

Heat-Transfer Coefficient

1 Btu/hr ft^2 °F = 1.3571 × 10^{-4} cal/s cm^2 °C
1 Btu/hr ft^2 °F = 5.6783 × 10^{-4} W/cm^2 °C
1 Btu/hr ft^2 °F = 5.6783 W/m^2 K
1 kcal/hr m^2 °F = 0.2048 Btu/hr ft^2 °F

Viscosity

1 cp = 10^{-2} g/cm s (poise)
1 cp = 2.4191 lb$_m$/ft hr
1 cp = 6.7197 × 10^{-4} lb$_m$/ft s
1 cp = 10^{-3} Pa s = 10^{-3} kg/m s = 10^{-3} N s/m^2
1 cp = 2.0886 × 10^{-5} lb$_f$ s/ft^2
1 Pa s = 1 N s/m^2 = 1 kg/m s = 1000 cp

Diffusivity

1 cm^2/s = 3.875 ft^2/hr
1 cm^2/s = 10^{-4} m^2/s
1 m^2/hr = 10.764 ft^2/hr
1 m^2/s = 3.875 × 10^4 ft^2/hr
1 centistoke = 10^{-2} cm^2/s
1 Pa s = 1 N s/m^2 = 1 kg/m s = 1000 cp

Mass Flux and Molar Flux

1 g/s cm^2 = 7.3734 × 10^3 lb$_m$/ft^2
1 g-mole/s cm^2 = 7.3734 × 10^3 lb-mole/hr ft^2
1 g-mole/s cm^2 = 10 kg-mole/s m^2 = 1 × 10^4 g-mole/s m^2
1 lb-mole/ft^2 = 1.3562 × 10^{-3} kg-mole/s m^2

TABLE A.1-2 CONT.

Heat Flux and Heat Flow

1 Btu/hr ft^2 = 3.1546 W/m^2
1 Btu/hr = 0.29307 W

Heat Capacity and Enthalpy

1 Btu/lb_m °F = 4.1868 kJ/kg K
1 Btu/lb_m °F = 1.000 cal/g °C
1 Btu/lb_m = 2326.0 J/kg
1 ft lb_f/lb_m = 2.9890 J/kg
1 cal (IT)/g °C = 4.1868 kJ/kg K

Mass-Transfer Coefficient

1 k_c cm/s = 10^{-2} m/s
1 k_c ft/hr = 8.4668×10^{-5} m/s
1 k_x g-mole/s cm^2 mol frac = 10-kg mole/s m^2 mole frac
1 k_x g-mole/s m^2 mole frac = 1×10^4 g-mole/s m^2 mole frac
1 k_x lb-mole/hr ft^2 mole frac = 1.3562×10^{-3} kg-mole/s m^2 mole frac
1 $k_x a$ lb-mole/hr ft^3 mole frac = 4.449×10^{-3} kg-mole/s m^3 mole frac
1 k_G kg-mole/s m^2 atm = 0.98692×10^{-5} kg-mole/s m^2 Pa
1 $k_G a$ kg-mole/s m^3atm = 0.98692×10^{-5} kg-mole/s m^3 Pa

Steam Tables

A.2

TABLE A.2–1
SATURATED STEAM TABLES

Temp. fahr.	Abs. pressure Lb_f/sq. in.	Enthalpy Btu/lb_m Sat. liquid	Evap.	Sat. vapor	Temp. fahr.	Abs. pressure Lb_f/sq. in.	Enthalpy Btu/lb_m Sat. liquid	Evap.	Sat. vapor
T	P				T	P			
32°	0.08854	0.00	1075.8	1075.8	55°	0.2141	23.07	1062.7	1085.8
33	0.09223	1.01	1075.2	1076.2	56	0.2220	24.06	1062.2	1086.3
34	0.09603	2.02	1074.7	1076.7	57	0.2302	25.06	1061.6	1086.7
35°	0.09995	3.02	1074.1	1077.1	58	0.2386	26.06	1061.0	1087.1
36	0.10401	4.03	1073.6	1077.6	59	0.2473	27.06	1060.5	1087.6
37	0.10821	5.04	1073.0	1078.0	60°	0.2563	28.06	1059.9	1088.0
38	0.11256	6.04	1072.4	1078.4	61	0.2655	29.06	1059.3	1088.4
39	0.11705	7.04	1071.9	1078.9	62	0.2751	30.05	1058.8	1088.9
40°	0.12170	8.05	1071.3	1079.3	63	0.2850	31.05	1058.2	1089.3
41	0.12652	9.05	1070.7	1079.7	64	0.2951	32.05	1057.6	1089.7
42	0.13150	10.05	1070.1	1080.2	65°	0.3056	33.05	1057.1	1090.2
43	0.13665	11.06	1069.5	1080.6	66	0.3164	34.05	1056.5	1090.6
44	0.14199	12.06	1068.9	1081.0	67	0.3276	35.05	1056.0	1091.0
45°	0.14752	13.06	1068.4	1081.5	68	0.3390	36.04	1055.5	1091.5
46	0.15323	14.06	1067.8	1081.9	69	0.3509	37.04	1054.9	1091.9
47	0.15914	15.07	1067.3	1082.4	70°	0.3631	38.04	1054.3	1092.3
48	0.16525	16.07	1066.7	1082.8	71	0.3756	39.04	1053.8	1092.8
49	0.17157	17.07	1066.1	1083.2	72	0.3886	40.04	1053.2	1093.2
50°	0.17811	18.07	1065.6	1083.7	73	0.4019	41.03	1052.6	1093.6
51	0.18486	19.07	1065.0	1084.1	74	0.4156	42.03	1052.1	1094.1
52	0.19182	20.07	1064.4	1084.5	75°	0.4298	43.03	1051.5	1094.5
53	0.19900	21.07	1063.9	1085.0	76	0.4443	44.03	1050.9	1094.9
54	0.20642	22.07	1063.3	1085.4	77	0.4593	45.02	1050.4	1095.4

TABLE A.2–1 CONT.

Temp. fahr.	Abs. pressure $\frac{Lb_f}{sq.\ in.}$	Enthalpy Btu/lb_m Sat. liquid	Evap.	Sat. vapor
T	P			
78	0.4747	46.02	1049.8	1095.8
79	0.4906	47.02	1049.2	1096.2
80°	0.5069	48.02	1048.6	1096.6
81	0.5237	49.02	1048.1	1097.1
82	0.5410	50.01	1047.5	1097.5
83	0.5588	51.01	1046.9	1097.9
84	0.5771	52.01	1046.4	1098.4
85°	0.5959	53.00	1045.8	1098.8
86	0.6152	54.00	1045.2	1099.2
87	0.6351	55.00	1044.7	1099.7
88	0.6556	56.00	1044.1	1100.1
89	0.6766	56.99	1043.5	1100.5
90°	0.6982	57.99	1042.9	1100.9
91	0.7204	58.99	1042.4	1101.4
92	0.7432	59.99	1041.8	1101.8
93	0.7666	60.98	1041.2	1102.2
94	0.7906	61.98	1040.7	1102.6
95°	0.8153	62.98	1040.1	1103.1
96	0.8407	63.98	1039.5	1103.5
97	0.8668	64.97	1038.9	1103.9
98	0.8935	65.97	1038.4	1104.4
99	0.9210	66.97	1037.8	1104.8
100°	0.9492	67.97	1037.2	1105.2
101	0.9781	68.96	1036.6	1105.6
102	1.0078	69.96	1036.1	1106.1
103	1.0382	70.96	1035.5	1106.5
104	1.0695	71.96	1034.9	1106.9
105°	1.1016	72.95	1034.3	1107.3
106	1.1345	73.95	1033.8	1107.8
107	1.1683	74.95	1033.3	1108.2
108	1.2029	75.95	1032.7	1108.6
109	1.2384	76.94	1032.1	1109.0
110°	1.2748	77.94	1031.6	1109.5
111	1.3121	78.94	1031.0	1109.9
112	1.3504	79.94	1030.4	1110.3
113	1.3896	80.94	1029.8	1110.7
114	1.4298	81.93	1029.2	1111.1
115°	1.4709	82.93	1028.7	1111.6
116	1.5130	83.93	1028.1	1112.0
117	1.5563	84.93	1027.5	1112.4
118	1.6006	85.92	1026.9	1112.8
119	1.6459	86.92	1026.3	1113.2
120°	1.6924	87.92	1025.8	1113.7
121	1.7400	88.92	1025.2	1114.1
122	1.7888	89.92	1024.6	1114.5
123	1.8387	90.91	1024.0	1114.9
124	1.8897	91.91	1023.4	1115.3
125°	1.9420	92.91	1022.9	1115.8
126	1.9955	93.91	1022.3	1116.2
127	2.0503	94.91	1021.7	1116.6
128	2.1064	95.91	1021.1	1117.0
129	2.1638	96.90	1020.5	1117.4
130°	2.2225	97.90	1020.0	1117.9
131	2.2826	98.90	1019.4	1118.3
132	2.3440	99.90	1018.8	1118.7
133	2.4069	100.90	1018.2	1119.1
134	2.4712	101.90	1017.6	1119.5
135°	2.5370	102.90	1017.0	1119.9
136	2.6042	103.90	1016.4	1120.3
137	2.6729	104.89	1015.9	1120.8
138	2.7432	105.89	1015.3	1121.2
139	2.8151	106.89	1014.7	1121.6
140°	2.8886	107.89	1014.1	1122.0
141	2.9637	108.89	1013.5	1122.4
142	3.0404	109.89	1012.9	1122.8
143	3.1188	110.89	1012.3	1123.2
144	3.1990	111.89	1011.7	1123.6
145°	3.281	112.89	1011.2	1124.1
146	3.365	113.89	1010.6	1124.5
147	3.450	114.89	1010.0	1124.9
148	3.537	115.89	1009.4	1125.3
149	3.627	116.89	1008.8	1125.7
150°	3.718	117.89	1008.2	1126.1
151	3.811	118.89	1007.6	1126.5
152	3.906	119.89	1007.0	1126.9
153	4.003	120.89	1006.4	1127.3
154	4.102	121.89	1005.8	1127.7
155°	4.203	122.89	1005.2	1128.1
156	4.306	123.89	1004.7	1128.6
157	4.411	124.89	1004.1	1129.0
158	4.519	125.89	1003.5	1129.4
159	4.629	126.89	1002.9	1129.8
160°	4.741	127.89	1002.3	1130.2
161	4.855	128.89	1001.7	1130.6
162	4.971	129.89	1001.1	1131.0
163	5.090	130.89	1000.5	1131.4
164	5.212	131.89	999.9	1131.8
165°	5.335	132.89	999.3	1132.2
166	5.461	133.89	998.7	1132.6
167	5.590	134.89	998.1	1133.0
168	5.721	135.90	997.5	1133.4
169	5.855	136.90	996.9	1133.8
170°	5.992	137.90	996.3	1134.2
171	6.131	138.90	995.7	1134.6
172	6.273	139.90	995.1	1135.0
173	6.417	140.90	994.5	1135.4
174	6.565	141.90	993.9	1135.8
175°	6.715	142.91	993.3	1136.2
176	6.868	143.91	992.7	1136.6
177	7.024	144.91	992.1	1137.0
178	7.183	145.91	991.5	1137.4
179	7.345	146.92	990.8	1137.7
180°	7.510	147.92	990.2	1138.1
181	7.678	148.92	989.6	1138.5

Table A.2–1 Cont.

Temp. fahr.	Abs. pressure Lb$_f$/sq. in.	Enthalpy Btu/lb$_m$		
		Sat. liquid	Evap.	Sat. vapor
T	P			
182	7.850	149.92	989.0	1138.9
183	8.024	150.93	988.4	1139.3
184	8.202	151.93	987.8	1139.7
185°	8.383	152.93	987.2	1140.1
186	8.567	153.94	986.6	1140.5
187	8.755	154.94	986.0	1140.9
188	8.946	155.94	985.4	1141.3
189	9.141	156.95	984.8	1141.7
190°	9.339	157.95	984.1	1142.0
191	9.541	158.95	983.4	1142.4
192	9.746	159.96	982.8	1142.8
193	9.955	160.96	982.2	1143.2
194	10.168	161.97	981.6	1143.6
195°	10.385	162.97	981.0	1144.0
196	10.605	163.97	980.4	1144.4
197	10.830	164.98	979.7	1144.7
198	11.058	165.98	979.1	1145.1
199	11.290	166.99	978.5	1145.5
200°	11.526	167.99	977.9	1145.9
202	12.011	170.00	976.6	1146.6
204	12.512	172.02	975.4	1147.4
206	13.031	174.03	974.2	1148.2
208	13.568	176.04	972.9	1148.9
210°	14.123	178.05	971.6	1149.7
212	14.696	180.07	970.3	1150.4
214	15.289	182.08	969.0	1151.1
216	15.901	184.10	967.8	1151.9
218	16.533	186.11	966.5	1152.6
220°	17.186	188.13	965.2	1153.4
222	17.861	190.15	963.9	1154.1
224	18.557	192.17	962.6	1154.8
226	19.275	194.18	961.3	1155.5
228	20.016	196.20	960.1	1156.3
230°	20.780	198.23	958.8	1157.0
232	21.567	200.25	957.4	1157.7
234	22.379	202.27	956.1	1158.4
236	23.217	204.29	954.8	1159.1
238	24.080	206.32	953.5	1159.8
240°	24.969	208.34	952.2	1160.5
242	25.884	210.37	950.8	1161.2
244	26.827	212.39	949.5	1161.9
246	27.798	214.42	948.2	1162.6
248	28.797	216.45	946.8	1163.3
250°	29.825	218.48	945.5	1164.0
252	30.884	220.51	944.2	1164.7
254	31.973	222.54	942.8	1165.3
256	33.093	224.58	941.4	1166.0
258	34.245	226.61	940.1	1166.7
260°	35.429	228.64	938.7	1167.3
262	36.646	230.68	937.3	1168.0
264	37.897	232.72	936.0	1168.7
266	39.182	234.76	934.5	1169.3
268	40.502	236.80	933.2	1170.0
270°	41.858	238.84	931.8	1170.6
272	43.252	240.88	930.3	1171.2
274	44.682	242.92	929.0	1171.9
276	46.150	244.96	927.5	1172.5
278	47.657	247.01	926.1	1173.1
280°	49.203	249.06	924.7	1173.8
282	50.790	251.10	923.3	1174.4
284	52.418	253.15	921.8	1175.0
286	54.088	255.20	920.4	1175.6
288	55.800	257.26	918.9	1176.2
290°	57.556	259.31	917.5	1176.8
292	59.356	261.36	916.0	1177.4
294	61.201	263.42	914.6	1178.0
296	63.091	265.48	913.1	1178.6
298	65.028	267.53	911.6	1179.1
300°	67.013	269.59	910.1	1179.7
302	69.046	271.66	908.6	1180.3
304	71.127	273.72	907.2	1180.9
306	73.259	275.78	905.6	1181.4
308	75.442	277.85	904.1	1182.0
310°	77.68	279.92	902.6	1182.5
312	79.96	281.99	901.0	1183.1
314	82.30	284.06	899.5	1183.6
316	84.70	286.13	898.0	1184.1
318	87.15	288.20	896.5	1184.7
320°	89.66	290.28	894.9	1185.2
322	92.22	292.36	893.3	1185.7
324	94.84	294.43	891.8	1186.2
326	97.52	296.52	890.2	1186.7
328	100.26	298.60	888.6	1187.2
330°	103.06	300.68	887.0	1187.7
332	105.92	302.77	885.4	1188.2
334	108.85	304.86	883.8	1188.7
336	111.84	306.95	882.2	1189.2
338	114.89	309.04	880.6	1189.6
340°	118.01	311.13	879.0	1190.1
342	121.20	313.23	877.4	1190.6
344	124.45	315.33	875.7	1191.0
346	127.77	317.43	874.1	1191.5
348	131.17	319.53	872.4	1191.9
350°	134.63	321.63	870.7	1192.3
352	138.16	323.74	869.1	1192.8
354	141.77	325.85	867.3	1193.2
356	145.45	327.96	865.6	1193.6
358	149.21	330.07	863.9	1194.0
360°	153.04	332.18	862.2	1194.4
362	156.95	334.30	860.5	1194.8
364	160.93	336.42	858.8	1195.2
366	165.00	338.54	857.1	1195.6
368	169.15	340.66	855.3	1196.0
370°	173.37	342.79	853.5	1196.3
372	177.68	344.91	851.8	1196.7
374	182.07	347.04	850.0	1197.0

TABLE A.2–1 CONT.

Temp. fahr.	Abs. pressure $\frac{Lb_f}{sq.\ in.}$	Enthalpy Btu/lb_m Sat. liquid	Evap.	Sat. vapor	Temp. fahr.	Abs. pressure $\frac{Lb_f}{sq.\ in.}$	Enthalpy Btu/lb_m Sat. liquid	Evap.	Sat. vapor
T	P				T	P			
376	186.55	349.18	848.2	1197.4	398	241.73	372.80	827.9	1200.7
378	191.12	351.31	846.4	1197.7	400°	247.31	374.97	826.0	1201.0
380°	195.77	353.45	844.6	1198.1	405	261.71	380.39	821.2	1201.6
382	200.50	355.59	842.8	1198.4	410	276.75	385.83	816.3	1202.1
384	205.33	357.73	841.0	1198.7	415	292.45	391.29	811.3	1202.6
386	210.25	359.88	839.1	1199.0	420	308.83	396.77	806.3	1203.1
388	215.26	362.02	837.3	1199.3	425°	325.92	402.27	801.2	1203.5
390°	220.37	364.17	835.4	1199.6	430	343.72	407.79	796.0	1203.8
392	225.56	366.33	833.6	1199.9	435	362.27	413.34	790.8	1204.1
394	230.85	368.48	831.7	1200.2	440	381.59	418.90	785.4	1204.3
396	236.24	370.64	829.9	1200.5	445	401.68	424.49	780.0	1204.5

TABLE A.2–2
SUPERHEATED VAPOR ENTHALPY (Btu/lb_m)

Abs. press. lb_f/sq. in. (sat. temp.)	Sat. liquid	Sat. vapor	Temperature—degrees fahrenheit 120°	140°	160°	180°	200°	220°	240°	260°	280°	300°	320°	340°	360°	380°
1 (101.74)	69.7	1106.0	1114.3	1123.3	1132.4	1141.4	1150.4	1159.5	1168.5	1177.6	1186.7	1195.8	1204.9	1214.1	1223.3	1232.5
2 (126.08)	94.0	1116.2		1122.6	1131.8	1140.9	1150.0	1159.1	1168.2	1177.3	1186.5	1195.6	1204.8	1213.9	1223.1	1232.3
3 (141.48)	109.4	1122.6			1131.2	1140.4	1149.6	1158.8	1167.9	1177.1	1186.2	1195.4	1204.6	1213.8	1223.0	1232.2
4 (152.97)	120.9	1127.3			1130.6	1139.9	1149.2	1158.5	1167.6	1176.8	1186.0	1195.2	1204.4	1213.6	1222.8	1232.1
5 (162.24)	130.1	1131.1				1139.4	1148.8	1158.1	1167.3	1176.5	1185.7	1195.0	1204.2	1213.4	1222.7	1231.9
6 (170.06)	138.0	1134.2				1138.9	1148.3	1157.7	1167.0	1176.3	1185.5	1194.7	1204.0	1213.2	1222.5	1231.8
7 (176.85)	144.8	1136.9				1138.4	1147.9	1157.4	1166.7	1176.0	1185.3	1194.5	1203.8	1213.1	1222.3	1231.6
8 (182.86)	150.8	1139.3					1147.5	1157.0	1166.3	1175.7	1185.0	1194.3	1203.6	1212.9	1222.2	1231.5
9 (188.28)	156.2	1141.4					1147.0	1156.6	1166.0	1175.4	1184.8	1194.1	1203.4	1212.7	1222.0	1231.4
10 (193.21)	161.2	1143.3					1146.6	1156.2	1165.7	1175.1	1184.5	1193.9	1203.2	1212.5	1221.9	1231.2
11 (197.75)	165.7	1145.0					1146.1	1155.8	1165.4	1174.9	1184.3	1193.6	1203.0	1212.4	1221.7	1231.1
12 (201.96)	170.0	1146.6						1155.4	1165.0	1174.6	1184.0	1193.4	1202.8	1212.2	1221.6	1230.9
13 (205.88)	173.9	1148.1						1155.1	1164.7	1174.3	1183.8	1193.2	1202.6	1212.0	1221.4	1230.8
14 (209.56)	177.6	1149.5						1154.6	1164.4	1174.0	1183.5	1193.0	1202.4	1211.8	1221.2	1230.6
14.696 (212.00)	180.1	1150.4						1154.4	1164.2	1173.8	1183.3	1192.8	1202.3	1211.7	1221.1	1230.5
15 (213.03)	181.1	1150.8						1154.3	1164.1	1173.7	1183.2	1192.8	1202.2	1211.7	1221.1	1230.5
16 (216.32)	184.4	1152.0						1153.8	1163.7	1173.4	1183.0	1192.5	1202.0	1211.5	1220.9	1230.3

TABLE A.2–2 CONT.
SUPERHEATED VAPOR ENTHALPY (Btu/lb_m)

Abs. press. lb_f/sq. in (sat. temp.)	*Temperature—degrees fahrenheit* 400°	420°	440°	460°	480°	500°	600°	700°	800°	900°	1000°	1100°	1200°	1300°	1400°	1600°
1 (101.74)	1241.7	1251.0	1260.3	1269.6	1278.9	1288.3	1335.7	1383.8	1432.8	1482.7	1533.5	1585.2	1637.7	1691.2	1745.7	1857.5
2 (126.08)	1241.6	1250.9	1260.2	1269.5	1278.8	1288.2	1335.6	1383.8	1432.8	1482.7	1533.5	1585.1	1637.7	1691.2	1745.7	1857.4
3 (141.48)	1241.5	1250.7	1260.0	1269.4	1278.7	1288.1	1335.5	1383.7	1432.8	1482.7	1533.4	1585.1	1637.7	1691.2	1745.7	1857.4
4 (152.97)	1241.3	1250.6	1259.9	1269.3	1278.7	1288.1	1335.5	1383.7	1432.7	1482.6	1533.4	1585.1	1637.7	1691.2	1745.7	1857.4
5 (162.24)	1241.2	1250.5	1259.8	1269.2	1278.6	1288.0	1335.4	1383.6	1432.7	1482.6	1533.4	1585.1	1637.7	1691.2	1745.7	1857.4
6 (170.06)	1241.1	1250.4	1259.7	1269.1	1278.5	1287.9	1335.3	1383.6	1432.6	1482.6	1533.4	1585.0	1637.6	1691.1	1745.6	1857.4
7 (176.85)	1240.9	1250.3	1259.6	1269.0	1278.4	1287.8	1335.3	1383.5	1432.6	1482.5	1533.3	1585.0	1637.6	1691.1	1745.6	1857.4
8 (182.86)	1240.8	1250.1	1259.5	1268.9	1278.3	1287.7	1335.2	1383.5	1432.6	1482.5	1533.3	1585.0	1637.6	1691.1	1745.6	1857.4
9 (188.28)	1240.7	1250.0	1259.4	1268.8	1278.2	1287.6	1335.2	1383.4	1432.5	1482.5	1533.3	1585.0	1637.6	1691.1	1745.6	1857.4
10 (193.21)	1240.6	1249.9	1259.3	1268.7	1278.1	1287.5	1335.1	1383.4	1432.5	1482.4	1533.2	1585.0	1637.6	1691.1	1745.6	1857.3
11 (197.75)	1240.4	1249.8	1259.2	1268.6	1278.0	1287.4	1335.0	1383.3	1432.4	1482.4	1533.2	1584.9	1637.6	1691.1	1745.6	1857.3
12 (201.96)	1240.3	1249.7	1259.0	1268.5	1277.9	1287.3	1335.0	1383.3	1432.4	1482.4	1533.2	1584.9	1637.5	1691.0	1745.5	1857.3
13 (205.88)	1240.2	1249.5	1258.9	1268.4	1277.8	1287.3	1334.9	1383.2	1432.4	1482.3	1533.2	1584.9	1637.5	1691.0	1745.5	1857.3
14 (209.56)	1240.0	1249.4	1258.8	1268.3	1277.7	1287.2	1334.8	1383.2	1432.3	1482.3	1533.1	1584.9	1637.5	1691.0	1745.5	1857.3
14.696 (212.00)	1239.9	1249.3	1258.8	1268.2	1277.6	1287.1	1334.8	1383.2	1432.3	1482.3	1533.1	1584.8	1637.5	1691.0	1745.5	1857.3
15 (213.03)	1239.9	1249.3	1258.7	1268.2	1277.6	1287.1	1334.8	1383.1	1432.3	1482.3	1533.1	1584.8	1637.5	1691.0	1745.5	1857.3
16 (216.32)	1239.8	1249.2	1258.6	1268.0	1277.5	1287.0	1334.7	1383.1	1432.3	1482.2	1533.1	1584.8	1637.4	1691.0	1745.5	1857.3

Table A.2–2 Cont.
Superheated Vapor Enthalpy (Btu/lb_m)

Abs. press. lb_f/sq. in. (sat. temp.)	Sat. liquid	Sat. vapor	Temperature—degrees fahrenheit 220°	230°	240°	250°	260°	270°	280°	290°	300°	320°	340°	360°	380°	400°
17 (219.44)	187.6	1153.1	1153.4	1158.4	1163.4	1168.3	1173.1	1177.9	1182.7	1187.5	1192.3	1201.8	1211.3	1220.7	1230.2	1239.6
18 (222.41)	190.6	1154.2		1158.0	1163.0	1167.9	1172.8	1177.7	1182.5	1187.3	1192.1	1201.6	1211.1	1220.6	1230.0	1239.5
19 (225.24)	193.4	1155.3		1157.7	1162.7	1167.6	1172.5	1177.4	1182.2	1187.1	1191.9	1201.4	1210.9	1220.4	1229.9	1239.4
20 (227.96)	196.2	1156.3		1157.3	1162.3	1167.3	1172.2	1177.1	1182.0	1186.8	1191.6	1201.2	1210.8	1220.3	1229.7	1239.2
21 (230.57)	198.8	1157.2			1162.0	1167.0	1171.9	1176.8	1181.7	1186.6	1191.4	1201.0	1210.6	1220.1	1229.6	1239.1
22 (233.07)	201.3	1158.1			1161.6	1166.6	1171.6	1176.5	1181.4	1186.3	1191.2	1200.8	1210.4	1219.9	1229.5	1239.0
23 (235.49)	203.8	1159.0			1161.2	1166.3	1171.3	1176.3	1181.2	1186.1	1190.9	1200.6	1210.2	1219.8	1229.3	1238.8
24 (237.82)	206.1	1159.8			1160.9	1166.0	1171.0	1176.0	1180.9	1185.8	1190.7	1200.4	1210.0	1219.6	1229.2	1238.7
25 (240.07)	208.4	1160.6				1165.6	1170.7	1175.7	1180.6	1185.6	1190.5	1200.2	1209.8	1219.4	1229.0	1238.5
26 (242.25)	210.6	1161.3				1165.3	1170.4	1175.4	1180.4	1185.3	1190.2	1200.0	1209.7	1219.3	1228.9	1238.4
27 (244.36)	212.8	1162.0				1165.0	1170.1	1175.1	1180.1	1185.1	1190.0	1199.8	1209.5	1219.1	1228.7	1238.3
28 (246.41)	214.8	1162.7				1164.6	1169.7	1174.8	1179.8	1184.8	1189.8	1199.6	1209.3	1218.9	1228.6	1238.1
29 (248.40)	216.9	1163.4				1164.3	1169.4	1174.5	1179.6	1184.6	1189.5	1199.3	1209.1	1218.8	1228.4	1238.0
30 (250.33)	218.8	1164.1					1169.1	1174.2	1179.3	1184.3	1189.3	1199.1	1208.9	1218.6	1228.3	1237.9
31 (252.22)	220.7	1164.7					1168.8	1173.9	1179.0	1184.0	1189.0	1198.9	1208.7	1218.4	1228.1	1237.7
32 (254.05)	222.6	1165.4					1168.5	1173.6	1178.7	1183.8	1188.8	1198.7	1208.5	1218.3	1227.9	1237.6
33 (255.84)	224.4	1166.0					1168.1	1173.3	1178.4	1183.5	1188.6	1198.5	1208.3	1218.1	1227.8	1237.4
34 (257.58)	226.2	1166.5					1167.8	1173.0	1178.2	1183.3	1188.3	1198.3	1208.2	1217.9	1227.6	1237.3

Table A.2–2 Cont.
Superheated Vapor Enthalpy (Btu/lb_m)

Abs. press. lb_f/sq. in (sat. temp.)	Temperature—degrees fahrenheit															
	420°	440°	460°	480°	500°	550°	600°	650°	700°	800°	900°	1000°	1100°	1200°	1400°	1600°
17 (219.44)	1249.1	1258.5	1267.9	1277.4	1286.9	1310.7	1334.6	1358.7	1383.0	1432.2	1482.2	1533.1	1584.8	1637.4	1745.5	1857.3
18 (222.41)	1248.9	1258.4	1267.8	1277.3	1286.8	1310.6	1334.6	1358.7	1383.0	1432.2	1482.2	1533.0	1584.8	1637.4	1745.5	1857.2
19 (225.24)	1248.8	1258.3	1267.7	1277.2	1286.7	1310.5	1334.5	1358.6	1382.9	1432.1	1482.1	1533.0	1584.7	1637.4	1745.4	1857.2
20 (227.96)	1248.7	1258.2	1267.6	1277.1	1286.6	1310.5	1334.4	1358.6	1382.9	1432.1	1482.1	1533.0	1584.7	1637.4	1745.4	1857.2
21 (230.57)	1248.6	1258.1	1267.5	1277.0	1286.5	1310.4	1334.4	1358.5	1382.8	1432.0	1482.1	1532.9	1584.7	1637.3	1745.4	1857.2
22 (233.07)	1248.4	1257.9	1267.4	1276.9	1286.4	1310.3	1334.3	1358.5	1382.8	1432.0	1482.0	1532.9	1584.7	1637.3	1745.4	1857.2
23 (235.49)	1248.3	1257.8	1267.3	1276.8	1286.4	1310.2	1334.2	1358.4	1382.7	1432.0	1482.0	1532.9	1584.6	1637.3	1745.4	1857.2
24 (237.82)	1248.2	1257.7	1267.2	1276.7	1286.3	1310.2	1334.2	1358.4	1382.7	1431.9	1482.0	1532.9	1584.6	1637.3	1745.4	1857.2
25 (240.07)	1248.1	1257.6	1267.1	1276.6	1286.2	1310.1	1334.1	1358.3	1382.6	1431.9	1481.9	1532.8	1584.6	1637.3	1745.3	1857.2
26 (242.25)	1248.0	1257.5	1267.0	1276.5	1286.1	1310.0	1334.0	1358.2	1382.6	1431.8	1481.9	1532.8	1584.6	1637.2	1745.3	1857.1
27 (244.36)	1247.8	1257.4	1266.9	1276.4	1286.0	1309.9	1334.0	1358.2	1382.5	1431.8	1481.9	1532.8	1584.6	1637.2	1745.3	1857.1
28 (246.41)	1247.7	1257.2	1266.8	1276.3	1285.9	1309.9	1333.9	1358.1	1382.5	1431.8	1481.8	1532.8	1584.5	1637.2	1745.3	1857.1
29 (248.40)	1247.6	1257.1	1266.7	1276.2	1285.8	1309.8	1333.9	1358.1	1382.4	1431.7	1481.8	1532.7	1584.5	1637.2	1745.3	1857.1
30 (250.33)	1247.5	1257.0	1266.6	1276.2	1285.7	1309.7	1333.8	1358.0	1382.4	1431.7	1481.8	1532.7	1584.5	1637.2	1745.3	1857.1
31 (252.22)	1247.3	1256.9	1266.5	1276.1	1285.6	1309.6	1333.7	1358.0	1382.3	1431.6	1481.7	1532.7	1584.5	1637.1	1745.2	1857.1
32 (254.05)	1247.2	1256.8	1266.4	1276.0	1285.5	1309.6	1333.7	1357.9	1382.3	1431.6	1481.7	1532.6	1584.4	1637.1	1745.2	1857.1
33 (255.84)	1247.1	1256.7	1266.3	1275.9	1285.4	1309.5	1333.6	1357.8	1382.2	1431.6	1481.7	1532.6	1584.4	1637.1	1745.2	1857.1
34 (257.78)	1246.9	1256.6	1266.2	1275.8	1285.4	1309.4	1333.5	1357.8	1382.2	1431.5	1481.6	1532.6	1584.4	1637.1	1745.2	1857.0

TABLE A.2–2 CONT.
SUPERHEATED VAPOR ENTHALPY (Btu/lb_m)

Abs. press. lb_f/sq. in. (sat. temp.)	Sat. liquid	Sat. vapor	Temperature—degrees fahrenheit 260°	270°	280°	290°	300°	320°	340°	360°	380°	400°	420°	440°	460°	480°
35 (259.28)	227.9	1167.1	1167.5	1172.7	1177.9	1183.0	1188.1	1198.1	1208.0	1217.8	1227.5	1237.2	1246.8	1256.4	1266.1	1275.7
36 (260.95)	229.6	1167.6		1172.4	1177.6	1182.7	1187.8	1197.9	1207.8	1217.6	1227.3	1237.0	1246.7	1256.3	1266.0	1275.6
37 (262.57)	231.3	1168.2		1172.1	1177.3	1182.5	1187.6	1197.6	1207.6	1217.4	1227.2	1236.9	1246.6	1256.2	1265.8	1275.5
38 (264.16)	232.9	1168.7		1171.8	1177.0	1182.2	1187.3	1197.4	1207.4	1217.3	1227.0	1236.7	1246.4	1256.1	1265.7	1275.4
39 (265.72)	234.5	1169.2		1171.5	1176.7	1181.9	1187.1	1197.2	1207.2	1217.1	1226.9	1236.6	1246.3	1256.0	1265.6	1275.3
40 (267.25)	236.0	1169.7		1171.2	1176.5	1181.7	1186.8	1197.0	1207.0	1216.9	1226.7	1236.5	1246.2	1255.9	1265.5	1275.2
41 (268.74)	237.6	1170.2		1170.9	1176.2	1181.4	1186.6	1196.8	1206.8	1216.7	1226.6	1236.3	1246.1	1255.8	1265.4	1275.1
42 (270.21)	239.0	1170.7			1175.9	1181.1	1186.3	1196.6	1206.6	1216.6	1226.4	1236.2	1245.9	1255.6	1265.3	1275.0
43 (271.64)	240.5	1171.1			1175.6	1180.9	1186.1	1196.3	1206.4	1216.4	1226.3	1236.1	1245.8	1255.5	1265.2	1274.9
44 (273.05)	242.0	1171.6			1175.3	1180.6	1185.8	1196.1	1206.2	1216.2	1226.1	1235.9	1245.7	1255.4	1265.1	1274.8
45 (274.44)	243.4	1172.0			1175.0	1180.3	1185.6	1195.9	1206.0	1216.0	1225.9	1235.8	1245.6	1255.3	1265.0	1274.7
46 (275.80)	244.8	1172.4			1174.7	1180.0	1185.3	1195.7	1205.8	1215.9	1225.8	1235.6	1245.4	1255.2	1264.9	1274.6
47 (277.13)	246.1	1172.9			1174.4	1179.8	1185.1	1195.5	1205.6	1215.7	1225.6	1235.5	1245.3	1255.1	1264.8	1274.5
48 (278.45)	247.5	1173.3			1174.1	1179.5	1184.8	1195.2	1205.4	1215.5	1225.5	1235.4	1245.2	1254.9	1264.7	1274.4
49 (279.74)	248.8	1173.7			1173.8	1179.2	1184.6	1195.0	1205.2	1215.3	1225.3	1235.2	1245.0	1254.8	1264.6	1274.3
50 (281.01)	250.1	1174.1				1178.9	1184.3	1194.8	1205.0	1215.2	1225.2	1235.1	1244.9	1254.7	1264.5	1274.2
51 (282.26)	251.4	1174.4				1178.6	1184.0	1194.6	1204.9	1215.0	1225.0	1234.9	1244.8	1254.6	1264.4	1274.1

Table A.2–2 Cont.
Superheated Vapor Enthalpy (Btu/lb_m)

Abs. press. lb_f/sq. in (sat. temp.)	*Temperature—degrees fahrenheit* 500°	520°	540°	560°	580°	600°	650°	700°	750°	800°	900°	1000°	1100°	1200°	1400°	1600°
35 (259.28)	1285.3	1294.9	1304.5	1314.1	1323.8	1333.5	1357.7	1382.1	1406.7	1431.5	1481.6	1532.6	1584.4	1637.1	1745.2	1857.0
36 (260.95)	1285.2	1294.8	1304.4	1314.1	1323.7	1333.4	1357.7	1382.1	1406.7	1431.5	1481.6	1532.5	1584.3	1637.0	1745.2	1857.0
37 (262.57)	1285.1	1294.7	1304.3	1314.0	1323.6	1333.3	1357.6	1382.0	1406.6	1431.4	1481.5	1532.5	1584.3	1637.0	1745.2	1857.0
38 (264.16)	1285.0	1294.6	1304.3	1313.9	1323.6	1333.3	1357.6	1382.0	1406.6	1431.4	1481.5	1532.5	1584.3	1637.0	1745.1	1857.0
39 (265.72)	1284.9	1294.5	1304.2	1313.8	1323.5	1333.2	1357.5	1381.9	1406.6	1431.3	1481.5	1532.4	1584.3	1637.0	1745.1	1857.0
40 (267.25)	1284.8	1294.5	1304.1	1313.8	1323.4	1333.1	1357.4	1381.9	1406.5	1431.3	1481.4	1532.4	1584.3	1637.0	1745.1	1857.0
41 (268.74)	1284.7	1294.4	1304.0	1313.7	1323.4	1333.1	1357.4	1381.8	1406.5	1431.3	1481.4	1532.4	1584.2	1636.9	1745.1	1857.0
42 (270.21)	1284.6	1294.3	1303.9	1313.6	1323.3	1333.0	1357.3	1381.8	1406.4	1431.2	1481.4	1532.4	1584.2	1636.9	1745.1	1856.9
43 (271.64)	1284.5	1294.2	1303.9	1313.5	1323.2	1332.9	1357.3	1381.7	1406.4	1431.2	1481.3	1532.3	1584.2	1636.9	1745.1	1856.9
44 (273.05)	1284.4	1294.1	1303.8	1313.5	1323.2	1332.9	1357.2	1381.7	1406.3	1431.1	1481.3	1532.3	1584.2	1636.9	1745.0	1856.9
45 (274.44)	1284.4	1294.0	1303.7	1313.4	1323.1	1332.8	1357.1	1381.6	1406.3	1431.1	1481.3	1532.3	1584.1	1636.9	1745.0	1856.9
46 (275.80)	1284.3	1293.9	1303.6	1313.3	1323.0	1332.7	1357.1	1381.6	1406.2	1431.0	1481.3	1532.3	1584.1	1636.8	1745.0	1856.9
47 (277.13)	1284.2	1293.9	1303.5	1313.2	1323.0	1332.7	1357.0	1381.5	1406.2	1431.0	1481.2	1532.2	1584.1	1636.8	1745.0	1856.9
48 (278.45)	1284.1	1293.8	1303.5	1313.2	1322.9	1332.6	1357.0	1381.5	1406.1	1431.0	1481.2	1532.2	1584.1	1636.8	1745.0	1856.9
49 (279.74)	1284.0	1293.7	1303.4	1313.1	1322.8	1332.5	1356.9	1381.4	1406.1	1430.9	1481.2	1532.2	1584.0	1636.8	1745.0	1856.9
50 (281.01)	1283.9	1293.6	1303.3	1313.0	1322.7	1332.5	1356.9	1381.4	1406.1	1430.9	1481.1	1532.1	1584.0	1636.8	1745.0	1856.8
51 (282.26)	1283.8	1293.5	1303.2	1312.9	1322.7	1332.4	1356.8	1381.3	1406.0	1430.9	1481.1	1532.1	1584.0	1636.7	1744.9	1856.8

TABLE A.2–2 CONT.
SUPERHEATED VAPOR ENTHALPY (Btu/lb_m)

Abs. press. lb_f/sq. in. (sat. temp.)	*Sat. liquid*	*Sat. vapor*	*290°*	*300°*	*320°*	*340°*	*360°*	*380°*	*400°*	*420°*	*440°*	*460°*	*480°*	*500°*	*520°*	*540°*
				Temperature—degrees fahrenheit												
52 (283.49)	252.6	1174.8	1178.4	1183.8	1194.3	1204.7	1214.8	1224.8	1234.8	1244.6	1254.5	1264.2	1274.0	1283.7	1293.4	1303.1
53 (284.70)	253.9	1175.2	1178.1	1183.5	1194.1	1204.5	1214.6	1224.7	1234.6	1244.5	1254.3	1264.1	1273.9	1283.6	1293.3	1303.1
54 (285.90)	255.1	1175.6	1177.8	1183.2	1193.9	1204.3	1214.4	1224.5	1234.5	1244.4	1254.2	1264.0	1273.8	1283.5	1293.3	1303.0
55 (287.07)	256.3	1175.9	1177.5	1183.0	1193.6	1204.0	1214.3	1224.4	1234.3	1244.3	1254.1	1263.9	1273.7	1283.4	1293.2	1302.9
56 (288.23)	257.5	1176.3	1177.2	1182.7	1193.4	1203.8	1214.1	1224.2	1234.2	1244.1	1254.0	1263.8	1273.6	1283.3	1293.1	1302.8
57 (289.37)	258.7	1176.6	1176.9	1182.5	1193.2	1203.6	1213.9	1224.0	1234.1	1244.0	1253.9	1263.7	1273.5	1283.2	1293.0	1302.7
58 (290.50)	259.8	1176.9		1182.2	1192.9	1203.4	1213.7	1223.9	1233.9	1243.9	1253.8	1263.6	1273.4	1283.1	1292.9	1302.7
59 (291.61)	261.0	1177.3		1181.9	1192.7	1203.2	1213.5	1223.7	1233.8	1243.7	1253.6	1263.5	1273.3	1283.1	1292.8	1302.6
60 (292.71)	262.1	1177.6		1181.6	1192.5	1203.0	1213.4	1223.6	1233.6	1243.6	1253.5	1263.4	1273.2	1283.0	1292.7	1302.5
61 (293.79)	263.2	1177.9		1181.4	1192.2	1202.8	1213.2	1223.4	1233.5	1243.5	1253.4	1263.3	1273.1	1282.9	1292.7	1302.4
62 (294.85)	264.3	1178.2		1181.1	1192.0	1202.6	1213.0	1223.2	1233.3	1243.3	1253.3	1263.1	1273.0	1282.8	1292.6	1302.3
63 (295.90)	265.4	1178.5		1180.8	1191.8	1202.4	1212.8	1223.1	1233.2	1243.2	1253.2	1263.0	1272.9	1282.7	1292.5	1302.3
64 (296.94)	266.4	1178.8		1180.5	1191.5	1202.2	1212.6	1222.9	1233.0	1243.1	1253.0	1262.9	1272.8	1282.6	1292.4	1302.2
65 (297.97)	267.5	1179.1		1180.3	1191.3	1202.0	1212.5	1222.7	1232.9	1242.9	1252.9	1262.8	1272.7	1282.5	1292.3	1302.1
66 (298.99)	268.5	1179.4		1180.0	1191.1	1201.8	1212.3	1222.6	1232.7	1242.8	1252.8	1262.7	1272.6	1282.4	1292.2	1302.0
67 (299.99)	269.6	1179.7		1179.7	1190.8	1201.6	1212.1	1222.4	1232.6	1242.7	1252.7	1262.6	1272.5	1282.3	1292.1	1301.9
68 (300.98)	270.6	1180.0			1190.6	1201.4	1211.9	1222.2	1232.4	1242.5	1252.5	1262.5	1272.4	1282.2	1292.0	1301.9
69 (301.96)	271.6	1180.3			1190.3	1201.2	1211.7	1222.1	1232.3	1242.4	1252.4	1262.4	1272.3	1282.1	1291.9	1301.8

TABLE A.2–2 CONT.
SUPERHEATED VAPOR ENTHALPY (Btu/lb_m)

Abs. press. lb_f/sq. in (sat. temp.)	*Temperature—degrees fahrenheit*															
	560°	*580°*	*600°*	*650°*	*700°*	*750°*	*800°*	*850°*	*900°*	*950°*	*1000°*	*1050°*	*1100°*	*1200°*	*1400°*	*1600°*
52 (283.49)	1312.9	1322.6	1332.3	1356.7	1381.3	1406.0	1430.8	1455.8	1481.1	1506.5	1532.1	1557.9	1584.0	1636.7	1744.9	1856.8
53 (284.70)	1312.8	1322.5	1332.3	1356.7	1381.2	1405.9	1430.8	1455.8	1481.0	1506.4	1532.1	1557.9	1583.9	1636.7	1744.9	1856.8
54 (285.90)	1312.7	1322.5	1332.2	1356.6	1381.2	1405.9	1430.7	1455.8	1481.0	1506.4	1532.0	1557.9	1583.9	1636.7	1744.9	1856.8
55 (287.07)	1312.6	1322.4	1332.1	1356.6	1381.1	1405.8	1430.7	1455.7	1481.0	1506.4	1532.0	1557.9	1583.9	1636.7	1744.9	1856.8
56 (288.23)	1312.6	1322.3	1332.1	1356.5	1381.1	1405.8	1430.7	1455.7	1480.9	1506.3	1532.0	1557.8	1583.9	1636.6	1744.9	1856.8
57 (289.37)	1312.5	1322.2	1332.0	1356.5	1381.0	1405.7	1430.6	1455.7	1480.9	1506.3	1532.0	1557.8	1583.9	1636.6	1744.8	1856.8
58 (290.50)	1312.4	1322.2	1331.9	1356.4	1381.0	1405.7	1430.6	1455.6	1480.9	1506.3	1531.9	1557.8	1583.8	1636.6	1744.8	1856.7
59 (291.61)	1312.3	1322.1	1331.9	1356.3	1380.9	1405.7	1430.5	1455.6	1480.8	1506.3	1531.9	1557.7	1583.8	1636.6	1744.8	1856.7
60 (292.71)	1312.3	1322.0	1331.8	1356.3	1380.9	1405.6	1430.5	1455.6	1480.8	1506.2	1531.9	1557.7	1583.8	1636.6	1744.8	1856.7
61 (293.79)	1312.2	1322.0	1331.7	1356.2	1380.8	1405.6	1430.4	1455.5	1480.8	1506.2	1531.8	1557.7	1583.8	1636.5	1744.8	1856.7
62 (294.85)	1312.1	1321.9	1331.7	1356.2	1380.8	1405.5	1430.4	1455.5	1480.7	1506.2	1531.8	1557.7	1583.7	1636.5	1744.8	1856.7
63 (295.90)	1312.0	1321.8	1331.6	1356.1	1380.7	1405.5	1430.4	1455.4	1480.7	1506.1	1531.8	1557.6	1583.7	1636.5	1744.8	1856.7
64 (296.94)	1312.0	1321.8	1331.5	1356.1	1380.7	1405.4	1430.3	1455.4	1480.7	1506.1	1531.8	1557.6	1583.7	1636.5	1744.7	1856.7
65 (297.97)	1311.9	1321.7	1331.5	1356.0	1380.6	1405.4	1430.3	1455.4	1480.6	1506.1	1531.7	1557.6	1583.7	1636.4	1744.7	1856.7
66 (298.99)	1311.8	1321.6	1331.4	1355.9	1380.6	1405.3	1430.2	1455.3	1480.6	1506.1	1531.7	1557.6	1583.6	1636.4	1744.7	1856.6
67 (299.99)	1311.7	1321.5	1331.3	1355.9	1380.5	1405.3	1430.2	1455.3	1480.6	1506.0	1531.7	1557.5	1583.6	1636.4	1744.7	1856.6
68 (300.98)	1311.7	1321.5	1331.3	1355.8	1380.5	1405.2	1430.2	1455.3	1480.5	1506.0	1531.6	1557.5	1583.6	1636.4	1744.7	1856.6
69 (301.96)	1311.6	1321.4	1331.2	1355.8	1380.4	1405.2	1430.1	1455.2	1480.5	1506.0	1531.6	1557.5	1583.6	1636.4	1744.7	1856.6

Table A.2–2 Cont.

Superheated Vapor Enthalpy (Btu/lb_m)

Abs. press. lb_f/sq. in. (sat. temp.)	Sat. liquid	Sat. vapor	310°	320°	330°	340°	350°	360°	370°	380°	390°	400°	420°	440°	460°	480°
			Temperature—degrees fahrenheit													
70 (302.92)	272.6	1180.6	1184.5	1190.1	1195.6	1201.0	1206.3	1211.5	1216.7	1221.9	1227.0	1232.1	1242.3	1252.3	1262.2	1272.2
71 (303.88)	273.6	1180.8	1184.3	1189.9	1195.3	1200.7	1206.1	1211.4	1216.6	1221.8	1226.9	1232.0	1242.1	1252.2	1262.1	1272.1
72 (304.83)	274.6	1181.1	1184.0	1189.6	1195.1	1200.5	1205.9	1211.2	1216.4	1221.6	1226.7	1231.9	1242.0	1252.0	1262.0	1271.9
73 (305.76)	275.5	1181.3	1183.8	1189.4	1194.9	1200.3	1205.7	1211.0	1216.2	1221.4	1226.6	1231.7	1241.9	1251.9	1261.9	1271.8
74 (306.68)	276.5	1181.6	1183.5	1189.1	1194.7	1200.1	1205.5	1210.8	1216.0	1221.3	1226.4	1231.6	1241.7	1251.8	1261.8	1271.7
75 (307.60)	277.4	1181.9	1183.2	1188.9	1194.4	1199.9	1205.3	1210.6	1215.9	1221.1	1226.3	1231.4	1241.6	1251.7	1261.7	1271.6
76 (308.50)	278.4	1182.1	1183.0	1188.6	1194.2	1199.7	1205.1	1210.4	1215.7	1220.9	1226.1	1231.3	1241.5	1251.6	1261.6	1271.5
77 (309.40)	279.3	1182.4	1182.7	1188.4	1194.0	1199.5	1204.9	1210.2	1215.5	1220.8	1225.9	1231.1	1241.3	1251.4	1261.5	1271.4
78 (310.29)	280.2	1182.6		1188.1	1193.7	1199.2	1204.7	1210.0	1215.3	1220.6	1225.8	1231.0	1241.2	1251.3	1261.3	1271.3
79 (311.16)	281.1	1182.8		1187.9	1193.5	1199.0	1204.5	1209.8	1215.2	1220.4	1225.6	1230.8	1241.1	1251.2	1261.2	1271.2
80 (312.03)	282.0	1183.1		1187.6	1193.3	1198.8	1204.3	1209.7	1215.0	1220.3	1225.5	1230.7	1240.9	1251.1	1261.1	1271.1
81 (312.89)	282.9	1183.3		1187.4	1193.0	1198.6	1204.1	1209.5	1214.8	1220.1	1225.3	1230.5	1240.8	1250.9	1261.0	1271.0
82 (313.74)	283.8	1183.5		1187.1	1192.8	1198.4	1203.9	1209.3	1214.6	1219.9	1225.2	1230.3	1240.6	1250.8	1260.9	1270.9
83 (314.59)	284.7	1183.8		1186.9	1192.6	1198.2	1203.7	1209.1	1214.4	1219.7	1225.0	1230.2	1240.5	1250.7	1260.8	1270.8
84 (315.42)	285.5	1184.0		1186.6	1192.3	1197.9	1203.5	1208.9	1214.3	1219.6	1224.8	1230.0	1240.4	1250.6	1260.7	1270.7
85 (316.25)	286.4	1184.2		1186.4	1192.1	1197.7	1203.2	1208.7	1214.1	1219.4	1224.7	1229.9	1240.2	1250.4	1260.6	1270.6
86 (317.07)	287.2	1184.4		1186.1	1191.9	1197.5	1203.0	1208.5	1213.9	1219.2	1224.5	1229.7	1240.1	1250.3	1260.4	1270.5

Table A.2–2 Cont.
Superheated Vapor Enthalpy (Btu/lb_m)

Abs. press. $lb_f/sq.$ in (sat. temp.)	Temperature—degrees fahrenheit															
	500°	520°	540°	560°	580°	600°	650°	700°	750°	800°	850°	900°	1000°	1200°	1400°	1600°
70 (302.92)	1282.0	1291.9	1301.7	1311.5	1321.3	1331.1	1355.7	1380.4	1405.2	1430.1	1455.2	1480.5	1531.6	1636.3	1744.6	1856.6
71 (303.88)	1281.9	1291.8	1301.6	1311.4	1321.2	1331.1	1355.7	1380.3	1405.1	1430.0	1455.2	1480.4	1531.6	1636.3	1744.6	1856.6
72 (304.83)	1281.8	1291.7	1301.5	1311.4	1321.2	1331.0	1355.6	1380.3	1405.1	1430.0	1455.1	1480.4	1531.5	1636.3	1744.6	1856.6
73 (305.76)	1281.7	1291.6	1301.4	1311.3	1321.1	1330.9	1355.5	1380.2	1405.0	1430.0	1455.1	1480.4	1531.5	1636.3	1744.6	1856.6
74 (306.68)	1281.6	1291.5	1301.4	1311.2	1321.0	1330.9	1355.5	1380.2	1405.0	1429.9	1455.0	1480.3	1531.5	1636.3	1744.6	1856.6
75 (307.60)	1281.5	1291.4	1301.3	1311.1	1321.0	1330.8	1355.4	1380.1	1404.9	1429.9	1455.0	1480.3	1531.5	1636.2	1744.6	1856.5
76 (308.50)	1281.4	1291.3	1301.2	1311.0	1320.9	1330.7	1355.4	1380.1	1404.9	1429.8	1455.0	1480.3	1531.4	1636.2	1744.5	1856.5
77 (309.40)	1281.4	1291.2	1301.1	1311.0	1320.8	1330.7	1355.3	1380.0	1404.8	1429.8	1454.9	1480.2	1531.4	1636.2	1744.5	1856.5
78 (310.29)	1281.3	1291.2	1301.0	1310.9	1320.7	1330.6	1355.3	1380.0	1404.8	1429.8	1454.9	1480.2	1531.4	1636.2	1744.5	1856.5
79 (311.16)	1281.2	1291.1	1301.0	1310.8	1320.7	1330.5	1355.2	1379.9	1404.8	1429.7	1454.9	1480.2	1531.3	1636.2	1744.5	1856.5
80 (312.03)	1281.1	1291.0	1300.9	1310.7	1320.6	1330.5	1355.1	1379.9	1404.7	1429.7	1454.8	1480.1	1531.3	1636.2	1744.5	1856.5
81 (312.89)	1281.0	1290.9	1300.8	1310.7	1320.5	1330.4	1355.1	1379.8	1404.7	1429.6	1454.8	1480.1	1531.3	1636.1	1744.5	1856.5
82 (313.74)	1280.9	1290.8	1300.7	1310.6	1320.5	1330.3	1355.0	1379.8	1404.6	1429.6	1454.8	1480.1	1531.3	1636.1	1744.4	1856.5
83 (314.59)	1280.8	1290.7	1300.6	1310.5	1320.4	1330.3	1355.0	1379.7	1404.6	1429.6	1454.7	1480.0	1531.2	1636.1	1744.4	1856.4
84 (315.42)	1280.7	1290.6	1300.5	1310.4	1320.3	1330.2	1354.9	1379.7	1404.5	1429.5	1454.7	1480.0	1531.2	1636.1	1744.4	1856.4
85 (316.25)	1280.6	1290.5	1300.5	1310.4	1320.2	1330.1	1354.8	1379.6	1404.5	1429.5	1454.6	1480.0	1531.2	1636.0	1744.4	1856.4
86 (317.07)	1280.5	1290.4	1300.4	1310.3	1320.2	1330.1	1354.8	1379.6	1404.4	1429.4	1454.6	1479.9	1531.2	1636.0	1744.4	1856.4

TABLE A.2–2 CONT.
SUPERHEATED VAPOR ENTHALPY (Btu/lb_m)

Abs. press. lb_f/sq. in. (sat. temp.)	Sat. liquid	Sat. vapor	Temperature—degrees fahrenheit 320°	330°	340°	350°	360°	370°	380°	390°	400°	420°	440°	460°	480°	500°
87 (317.88)	288.1	1184.6	1185.9	1191.6	1197.3	1202.8	1208.3	1213.7	1219.1	1224.3	1229.6	1240.0	1250.2	1260.3	1270.4	1280.4
88 (318.68)	288.9	1184.8	1185.6	1191.4	1197.1	1202.6	1208.1	1213.5	1218.9	1224.2	1229.4	1239.8	1250.1	1260.2	1270.3	1280.3
89 (319.48)	289.7	1185.1	1185.4	1191.1	1196.8	1202.4	1207.9	1213.4	1218.7	1224.0	1229.3	1239.7	1249.9	1260.1	1270.2	1280.2
90 (320.27)	290.6	1185.3		1190.9	1196.6	1202.2	1207.7	1213.2	1218.6	1223.9	1229.1	1239.5	1249.8	1260.0	1270.1	1280.1
91 (321.06)	291.4	1185.5		1190.7	1196.4	1202.0	1207.5	1213.0	1218.4	1223.7	1229.0	1239.4	1249.7	1259.9	1270.0	1280.0
92 (321.83)	292.2	1185.7		1190.4	1196.2	1201.8	1207.3	1212.8	1218.2	1223.5	1228.8	1239.3	1249.6	1259.8	1269.9	1279.9
93 (322.60)	293.0	1185.9		1190.2	1195.9	1201.6	1207.1	1212.6	1218.0	1223.4	1228.7	1239.1	1249.4	1259.6	1269.8	1279.8
94 (323.36)	293.8	1186.1		1189.9	1195.7	1201.4	1206.9	1212.4	1217.8	1223.2	1228.5	1239.0	1249.3	1259.5	1269.7	1279.7
95 (324.12)	294.6	1186.2		1189.7	1195.5	1201.2	1206.7	1212.2	1217.7	1223.0	1228.3	1238.8	1249.2	1259.4	1269.5	1279.6
96 (324.87)	295.3	1186.4		1189.4	1195.3	1201.0	1206.5	1212.1	1217.5	1222.9	1228.2	1238.7	1249.1	1259.3	1269.4	1279.5
97 (325.61)	296.1	1186.6		1189.2	1195.0	1200.7	1206.3	1211.9	1217.3	1222.7	1228.0	1238.6	1248.9	1259.2	1269.3	1279.4
98 (326.35)	296.9	1186.8		1189.0	1194.8	1200.5	1206.1	1211.7	1217.2	1222.6	1227.9	1238.4	1248.8	1259.1	1269.2	1279.3
99 (327.08)	297.6	1187.0		1188.7	1194.6	1200.3	1205.9	1211.5	1217.0	1222.4	1227.7	1238.3	1248.7	1258.9	1269.1	1279.2
100 (327.81)	298.4	1187.2		1188.5	1194.3	1200.1	1205.7	1211.3	1216.8	1222.2	1227.6	1238.1	1248.6	1258.8	1269.0	1279.1
102 (329.25)	299.9	1187.5		1188.0	1193.9	1199.7	1205.3	1210.9	1216.4	1221.9	1227.3	1237.9	1248.3	1258.6	1268.8	1278.9
104 (330.66)	301.4	1187.9			1193.4	1199.2	1204.9	1210.6	1216.1	1221.6	1226.9	1237.6	1248.0	1258.4	1268.6	1278.7
106 (332.05)	302.8	1188.2			1193.0	1198.8	1204.5	1210.2	1215.7	1221.2	1226.6	1237.3	1247.8	1258.1	1268.4	1278.6
108 (333.42)	304.3	1188.6			1192.5	1198.4	1204.1	1209.8	1215.4	1220.9	1226.3	1237.0	1247.5	1257.9	1268.2	1278.4

Table A.2–2 Cont.
Superheated Vapor Enthalpy (Btu/lb_m)

Abs. press. lb_f/sq. in. (sat. temp)	Temperature—degrees fahrenheit 520°	540°	560°	580°	600°	650°	700°	750°	800°	850°	900°	950°	1000°	1200°	1400°	1600°
87 (317.88)	1290.4	1300.3	1310.2	1320.1	1330.0	1354.7	1379.5	1404.4	1429.4	1454.6	1479.9	1505.4	1531.1	1636.0	1744.4	1856.4
88 (318.68)	1290.3	1300.2	1310.1	1320.0	1329.9	1354.7	1379.5	1404.4	1429.4	1454.5	1479.9	1505.4	1531.1	1636.0	1744.4	1856.4
89 (319.48)	1290.2	1300.1	1310.0	1320.0	1329.9	1354.6	1379.4	1404.3	1429.3	1454.5	1479.8	1505.4	1531.1	1636.0	1744.3	1856.4
90 (320.27)	1290.1	1300.0	1310.0	1319.9	1329.8	1354.6	1379.4	1404.3	1429.3	1454.5	1479.8	1505.3	1531.0	1635.9	1744.3	1856.4
91 (321.06)	1290.0	1300.0	1309.9	1319.8	1329.7	1354.5	1379.3	1404.2	1429.2	1454.4	1479.8	1505.3	1531.0	1635.9	1744.3	1856.3
92 (321.83)	1289.9	1299.9	1309.8	1319.7	1329.7	1354.4	1379.3	1404.2	1429.2	1454.4	1479.7	1505.3	1531.0	1635.9	1744.3	1856.3
93 (322.60)	1289.8	1299.8	1309.7	1319.7	1329.6	1354.4	1379.2	1404.1	1429.2	1454.4	1479.7	1505.2	1531.0	1635.9	1744.3	1856.3
94 (323.36)	1289.7	1299.7	1309.7	1319.6	1329.5	1354.3	1379.2	1404.1	1429.1	1454.3	1479.7	1505.2	1530.9	1635.9	1744.3	1856.3
95 (324.12)	1289.6	1299.6	1309.6	1319.5	1329.4	1354.3	1379.1	1404.0	1429.1	1454.3	1479.6	1505.2	1530.9	1635.8	1744.2	1856.3
96 (324.87)	1289.6	1299.6	1309.5	1319.4	1329.4	1354.2	1379.1	1404.0	1429.0	1454.3	1479.6	1505.2	1530.9	1635.8	1744.2	1856.3
97 (325.61)	1289.5	1299.5	1309.4	1319.4	1329.3	1354.1	1379.0	1403.9	1429.0	1454.2	1479.6	1505.1	1530.9	1635.8	1744.2	1856.3
98 (326.35)	1289.4	1299.4	1309.4	1319.3	1329.2	1354.1	1379.0	1403.9	1429.0	1454.2	1479.5	1505.1	1530.8	1635.8	1744.2	1856.3
99 (327.08)	1289.3	1299.3	1309.3	1319.2	1329.2	1354.0	1378.9	1403.9	1428.9	1454.1	1479.5	1505.1	1530.8	1635.8	1744.2	1856.2
100 (327.81)	1289.2	1299.2	1309.2	1319.2	1329.1	1354.0	1378.9	1403.8	1428.9	1454.1	1479.5	1505.0	1530.8	1635.7	1744.2	1856.2
102 (329.25)	1289.0	1299.1	1309.0	1319.0	1329.0	1353.9	1378.7	1403.7	1428.8	1454.0	1479.4	1505.0	1530.7	1635.7	1744.1	1856.2
104 (330.66)	1288.8	1298.9	1308.9	1318.9	1328.8	1353.7	1378.6	1403.6	1428.7	1454.0	1479.3	1504.9	1530.7	1635.7	1744.1	1856.2
106 (332.05)	1288.7	1298.7	1308.7	1318.7	1328.7	1353.6	1378.5	1403.5	1428.6	1453.9	1479.3	1504.9	1530.6	1635.6	1744.1	1856.2
108 (333.42)	1288.5	1298.5	1308.6	1318.6	1328.6	1353.5	1378.4	1403.4	1428.6	1453.8	1479.2	1504.8	1530.6	1635.6	1744.0	1856.1

Table A.2–2 Cont.
Superheated Vapor Enthalpy (Btu/lb_m)

Abs. press. lb_f/sq. in. (sat. temp.)	Sat. liquid	Sat. vapor	Temperature—degrees fahrenheit													
			340°	350°	360°	370°	380°	390°	400°	420°	440°	460°	480°	500°	520°	540°
110 (334.77)	305.7	1188.9	1192.0	1197.9	1203.7	1209.4	1215.0	1220.5	1226.0	1236.7	1247.3	1257.7	1268.0	1278.2	1288.3	1298.4
112 (336.11)	307.1	1189.2	1191.5	1197.5	1203.3	1209.0	1214.7	1220.2	1225.7	1236.4	1247.0	1257.4	1267.7	1278.0	1288.1	1298.2
114 (337.42)	308.4	1189.5	1191.1	1197.0	1202.9	1208.6	1214.3	1219.9	1225.4	1236.2	1246.8	1257.2	1267.5	1277.8	1287.9	1298.0
116 (338.72)	309.8	1189.8	1190.6	1196.6	1202.5	1208.2	1213.9	1219.5	1225.0	1235.9	1246.5	1257.0	1267.3	1277.6	1287.7	1297.9
118 (339.99)	311.1	1190.1	1190.1	1196.2	1202.1	1207.9	1213.6	1219.2	1224.7	1235.6	1246.2	1256.7	1267.1	1277.4	1287.6	1297.7
120 (341.25)	312.4	1190.4		1195.7	1201.6	1207.5	1213.2	1218.8	1224.4	1235.3	1246.0	1256.5	1266.9	1277.2	1287.4	1297.5
122 (342.50)	313.8	1190.7		1195.2	1201.2	1207.1	1212.8	1218.5	1224.0	1235.0	1245.7	1256.3	1266.7	1277.0	1287.2	1297.4
124 (343.72)	315.0	1190.9		1194.8	1200.8	1206.7	1212.5	1218.1	1223.7	1234.7	1245.4	1256.0	1266.4	1276.8	1287.0	1297.2
126 (344.94)	316.3	1191.2		1194.3	1200.4	1206.3	1212.1	1217.8	1223.4	1234.4	1245.2	1255.8	1266.2	1276.6	1286.8	1297.0
128 (346.13)	317.6	1191.5		1193.9	1199.9	1205.9	1211.7	1217.4	1223.1	1234.1	1244.9	1255.5	1266.0	1276.4	1286.6	1296.8
130 (347.32)	318.8	1191.7		1193.4	1199.5	1205.5	1211.3	1217.1	1222.7	1233.8	1244.6	1255.3	1265.8	1276.2	1286.5	1296.7
132 (348.48)	320.0	1192.0		1192.9	1199.1	1205.1	1211.0	1216.7	1222.4	1233.5	1244.4	1255.1	1265.6	1276.0	1286.3	1296.5
134 (349.64)	321.2	1192.2		1192.5	1198.7	1204.7	1210.6	1216.4	1222.1	1233.2	1244.1	1254.8	1265.4	1275.8	1286.1	1296.3
136 (350.78)	322.4	1192.5			1198.2	1204.3	1210.2	1216.0	1221.7	1232.9	1243.8	1254.6	1265.1	1275.6	1285.9	1296.2
138 (351.91)	323.6	1192.7			1197.8	1203.9	1209.8	1215.7	1221.4	1232.6	1243.6	1254.3	1264.9	1275.4	1285.7	1296.0
140 (353.02)	324.8	1193.0			1197.3	1203.5	1209.4	1215.3	1221.1	1232.3	1243.3	1254.1	1264.7	1275.2	1285.5	1295.8
142 (354.12)	326.0	1193.2			1196.9	1203.0	1209.1	1215.0	1220.7	1232.0	1243.0	1253.8	1264.5	1275.0	1285.3	1295.6

Table A.2–2 Cont.
Superheated Vapor Enthalpy (Btu/lb_m)

Abs. press. lb_f/sq. in (sat. temp.)	Temperature—degrees fahrenheit															
	560°	580°	600°	620°	640°	660°	680°	700°	750°	800°	850°	900°	1000°	1200°	1400°	1600°
110 (334.77)	1308.4	1318.4	1328.4	1338.4	1348.4	1358.4	1368.4	1378.3	1403.4	1428.5	1453.7	1479.2	1530.5	1635.5	1744.0	1856.1
112 (336.11)	1308.3	1318.3	1328.3	1338.3	1348.3	1358.3	1368.2	1378.2	1403.3	1428.4	1453.7	1479.1	1530.4	1635.5	1744.0	1856.1
114 (337.42)	1308.1	1318.1	1328.2	1338.2	1348.1	1358.1	1368.1	1378.1	1403.2	1428.3	1453.6	1479.0	1530.4	1635.4	1744.0	1856.1
116 (338.72)	1307.9	1318.0	1328.0	1338.0	1348.0	1358.0	1368.0	1378.0	1403.1	1428.2	1453.5	1479.0	1530.3	1635.4	1743.9	1856.0
118 (339.99)	1307.8	1317.8	1327.9	1337.9	1347.9	1357.9	1367.9	1377.9	1403.0	1428.2	1453.5	1478.9	1530.3	1635.4	1743.9	1856.0
120 (341.25)	1307.6	1317.7	1327.7	1337.8	1347.8	1357.8	1367.8	1377.8	1402.9	1428.1	1453.4	1478.8	1530.2	1635.3	1743.9	1856.0
122 (342.50)	1307.5	1317.6	1327.6	1337.6	1347.7	1357.7	1367.7	1377.7	1402.8	1428.0	1453.3	1478.8	1530.2	1635.3	1743.8	1856.0
124 (343.72)	1307.3	1317.4	1327.5	1337.5	1347.5	1357.6	1367.6	1377.6	1402.7	1427.9	1453.2	1478.7	1530.1	1635.2	1743.8	1855.9
126 (344.94)	1307.2	1317.3	1327.3	1337.4	1347.4	1357.5	1367.5	1377.5	1402.6	1427.8	1453.2	1478.6	1530.1	1635.2	1743.8	1855.9
128 (346.13)	1307.0	1317.1	1327.2	1337.2	1347.3	1357.3	1367.4	1377.4	1402.5	1427.7	1453.1	1478.6	1530.0	1635.2	1743.7	1855.9
130 (347.32)	1306.8	1317.0	1327.0	1337.1	1347.2	1357.2	1367.3	1377.3	1402.5	1427.7	1453.0	1478.5	1529.9	1635.1	1743.7	1855.9
132 (348.48)	1306.7	1316.8	1326.9	1337.0	1347.1	1357.1	1367.2	1377.2	1402.4	1427.6	1452.9	1478.4	1529.9	1635.1	1743.7	1855.8
134 (349.64)	1306.5	1316.7	1326.8	1336.9	1346.9	1357.0	1367.0	1377.1	1402.3	1427.5	1452.9	1478.4	1529.8	1635.0	1743.6	1855.8
136 (350.78)	1306.4	1316.5	1326.6	1336.7	1346.8	1356.9	1366.9	1377.0	1402.2	1427.4	1452.8	1478.3	1529.8	1635.0	1743.6	1855.8
138 (351.91)	1306.2	1316.4	1326.5	1336.6	1346.7	1356.8	1366.8	1376.9	1402.1	1427.3	1452.7	1478.2	1529.7	1635.0	1743.6	1855.8
140 (353.02)	1306.0	1316.2	1326.4	1336.5	1346.6	1356.6	1366.7	1376.8	1402.0	1427.3	1452.6	1478.2	1529.7	1634.9	1743.5	1855.7
142 (354.12)	1305.9	1316.1	1326.2	1336.3	1346.4	1356.5	1366.6	1376.7	1401.9	1427.2	1452.6	1478.1	1529.6	1634.9	1743.5	1855.7

Table A.2–2 Cont.
Superheated Vapor Enthalpy (Btu/lb$_m$)

Abs. press. lb$_f$/sq. in. (sat. temp.)	Sat. liquid	Sat. vapor	*Temperature—degrees fahrenheit* 360°	370°	380°	390°	400°	420°	440°	460°	480°	500°	520°	540°	560°	580°
144 (355.21)	327.1	1193.4	1196.5	1202.6	1208.7	1214.6	1220.4	1231.7	1242.8	1253.6	1264.2	1274.8	1285.2	1295.5	1305.7	1315.9
146 (356.29)	328.3	1193.6	1196.0	1202.2	1208.3	1214.2	1220.0	1231.4	1242.5	1253.3	1264.0	1274.5	1285.0	1295.3	1305.6	1315.8
148 (357.36)	329.4	1193.9	1195.6	1201.8	1207.9	1213.9	1219.7	1231.1	1242.2	1253.1	1263.8	1274.3	1284.8	1295.1	1305.4	1315.6
150 (358.42)	330.5	1194.1	1195.1	1201.4	1207.5	1213.5	1219.4	1230.8	1242.0	1252.9	1263.6	1274.1	1284.6	1295.0	1305.2	1315.5
152 (359.46)	331.6	1194.3	1194.7	1200.9	1207.1	1213.1	1219.0	1230.5	1241.7	1252.6	1263.4	1273.9	1284.4	1294.8	1305.1	1315.3
154 (360.49)	332.7	1194.5		1200.5	1206.7	1212.8	1218.7	1230.2	1241.4	1252.4	1263.1	1273.7	1284.2	1294.6	1304.9	1315.2
156 (361.52)	333.8	1194.7		1200.1	1206.3	1212.4	1218.3	1229.9	1241.1	1252.1	1262.9	1273.5	1284.0	1294.4	1304.8	1315.0
158 (362.53)	334.9	1194.9		1199.7	1205.9	1212.0	1218.0	1229.6	1240.9	1251.9	1262.7	1273.3	1283.8	1294.3	1304.6	1314.9
160 (363.53)	335.9	1195.1		1199.2	1205.5	1211.6	1217.6	1229.3	1240.6	1251.6	1262.4	1273.1	1283.7	1294.1	1304.4	1314.7
162 (364.53)	337.0	1195.3		1198.8	1205.1	1211.3	1217.3	1229.0	1240.3	1251.4	1262.2	1272.9	1283.5	1293.9	1304.3	1314.6
164 (365.51)	338.0	1195.5		1198.4	1204.7	1210.9	1216.9	1228.6	1240.0	1251.1	1262.0	1272.7	1283.3	1293.7	1304.1	1314.4
166 (366.48)	339.0	1195.7		1197.9	1204.3	1210.5	1216.6	1228.3	1239.7	1250.9	1261.8	1272.5	1283.1	1293.6	1303.9	1314.3
168 (367.45)	340.1	1195.8		1197.5	1203.9	1210.1	1216.2	1228.0	1239.5	1250.6	1261.5	1272.3	1282.9	1293.4	1303.8	1314.1
170 (368.41)	341.1	1196.0		1197.1	1203.5	1209.7	1215.8	1227.7	1239.2	1250.4	1261.3	1272.1	1282.7	1293.2	1303.6	1314.0
172 (369.35)	342.1	1196.2		1196.6	1203.1	1209.4	1215.5	1227.4	1238.9	1250.1	1261.1	1271.9	1282.5	1293.0	1303.5	1313.8
174 (370.29)	343.1	1196.4			1202.7	1209.0	1215.1	1227.1	1238.6	1249.9	1260.8	1271.6	1282.3	1292.9	1303.3	1313.7
176 (371.22)	344.1	1196.5			1202.3	1208.6	1214.8	1226.8	1238.3	1249.6	1260.6	1271.4	1282.1	1292.7	1303.1	1313.5
178 (372.14)	345.1	1196.7			1201.8	1208.2	1214.4	1226.4	1238.1	1249.3	1260.4	1271.2	1281.9	1292.5	1303.0	1313.4

Table A.2-2 Cont.
Superheated Vapor Enthalpy (Btu/lb_m)

Abs. press. lb_f/sq. in (sat. temp.)	Temperature—degrees fahrenheit 600°	620°	640°	660°	680°	700°	750°	800°	850°	900°	950°	1000°	1100°	1200°	1400°	1600°
144 (355.21)	1326.1	1336.2	1346.3	1356.4	1366.5	1376.6	1401.8	1427.1	1452.5	1478.0	1503.7	1529.6	1581.8	1634.8	1743.5	1855.7
146 (356.29)	1325.9	1336.1	1346.2	1356.3	1366.4	1376.5	1401.7	1427.0	1452.4	1478.0	1503.6	1529.5	1581.8	1634.8	1743.5	1855.7
148 (357.36)	1325.8	1335.9	1346.1	1356.2	1366.3	1376.4	1401.6	1426.9	1452.4	1477.9	1503.6	1529.4	1581.7	1634.8	1743.4	1855.6
150 (358.42)	1325.7	1335.8	1346.0	1356.1	1366.2	1376.3	1401.5	1426.9	1452.3	1477.8	1503.5	1529.4	1581.7	1634.7	1743.4	1855.6
152 (359.46)	1325.5	1335.7	1345.8	1356.0	1366.1	1376.2	1401.4	1426.8	1452.2	1477.8	1503.5	1529.3	1581.6	1634.7	1743.4	1855.6
154 (360.49)	1325.4	1335.6	1345.7	1355.8	1365.9	1376.1	1401.4	1426.7	1452.1	1477.7	1503.4	1529.3	1581.6	1634.6	1743.3	1855.6
156 (361.52)	1325.2	1335.4	1345.6	1355.7	1365.8	1376.0	1401.3	1426.6	1452.1	1477.6	1503.3	1529.2	1581.5	1634.6	1743.3	1855.5
158 (362.53)	1325.1	1335.3	1345.5	1355.6	1365.7	1375.9	1401.2	1426.5	1452.0	1477.6	1503.3	1529.2	1581.5	1634.6	1743.3	1855.5
160 (363.53)	1325.0	1335.2	1345.3	1355.5	1365.6	1375.7	1401.1	1426.4	1451.9	1477.5	1503.2	1529.1	1581.4	1634.5	1743.2	1855.5
162 (364.53)	1324.8	1335.0	1345.2	1355.4	1365.5	1375.6	1401.0	1426.4	1451.8	1477.4	1503.2	1529.1	1581.4	1634.5	1743.2	1855.5
164 (365.51)	1324.7	1334.9	1345.1	1355.3	1365.4	1375.5	1400.9	1426.3	1451.8	1477.4	1503.1	1529.0	1581.3	1634.4	1743.2	1855.4
166 (366.48)	1324.5	1334.8	1345.0	1355.1	1365.3	1375.4	1400.8	1426.2	1451.7	1477.3	1503.0	1529.0	1581.3	1634.4	1743.1	1855.4
168 (367.45)	1324.4	1334.6	1344.8	1355.0	1365.2	1375.3	1400.7	1426.1	1451.6	1477.2	1503.0	1528.9	1581.3	1634.4	1743.1	1855.4
170 (368.41)	1324.2	1334.5	1344.7	1354.9	1365.1	1375.2	1400.6	1426.0	1451.5	1477.2	1502.9	1528.8	1581.2	1634.3	1743.1	1855.4
172 (369.35)	1324.1	1334.4	1344.6	1354.8	1365.0	1375.1	1400.5	1426.0	1451.5	1477.1	1502.9	1528.8	1581.2	1634.3	1743.0	1855.3
174 (370.29)	1324.0	1334.2	1344.5	1354.7	1364.8	1375.0	1400.4	1425.9	1451.4	1477.0	1502.8	1528.7	1581.1	1634.2	1743.0	1855.3
176 (371.22)	1323.8	1334.1	1344.3	1354.5	1364.7	1374.9	1400.3	1425.8	1451.3	1477.0	1502.7	1528.7	1581.1	1634.2	1743.0	1855.3
178 (372.14)	1323.7	1334.0	1344.2	1354.4	1364.6	1374.8	1400.2	1425.7	1451.3	1476.9	1502.7	1528.6	1581.0	1634.2	1743.0	1855.3

TABLE A.2-2 CONT.
SUPERHEATED VAPOR ENTHALPY (Btu/lb_m)

Abs. press. lb_f/sq. in. (sat. temp.)	*Sat. liquid*	*Sat. vapor*	*Temperature—degrees fahrenheit* 380°	390°	400°	420°	440°	460°	480°	500°	520°	540°	560°	580°	600°	620°
180 (373.06)	346.0	1196.9	1201.4	1207.8	1214.0	1226.1	1237.8	1249.1	1260.2	1271.0	1281.7	1292.3	1302.8	1313.2	1323.5	1333.8
182 (373.96)	347.0	1197.0	1201.0	1207.4	1213.7	1225.8	1237.5	1248.8	1259.9	1270.8	1281.5	1292.1	1302.6	1313.0	1323.4	1333.7
184 (374.86)	348.0	1197.2	1200.6	1207.0	1213.3	1225.5	1237.2	1248.6	1259.7	1270.6	1281.4	1292.0	1302.5	1312.9	1323.3	1333.6
186 (375.75)	348.9	1197.3	1200.2	1206.6	1212.9	1225.2	1236.9	1248.3	1259.5	1270.4	1281.2	1291.8	1302.3	1312.7	1323.1	1333.4
188 (376.64)	349.9	1197.5	1199.7	1206.2	1212.6	1224.8	1236.6	1248.1	1259.2	1270.2	1281.0	1291.6	1302.1	1312.6	1323.0	1333.3
190 (377.51)	350.8	1197.6	1199.3	1205.8	1212.2	1224.5	1236.3	1247.8	1259.0	1270.0	1280.8	1291.4	1302.0	1312.4	1322.8	1333.2
192 (378.38)	351.7	1197.8	1198.9	1205.4	1211.8	1224.2	1236.0	1247.5	1258.8	1269.7	1280.6	1291.2	1301.8	1312.3	1322.7	1333.0
194 (379.24)	352.6	1197.9	1198.4	1205.0	1211.5	1223.8	1235.8	1247.3	1258.5	1269.5	1280.4	1291.1	1301.6	1312.1	1322.5	1332.9
196 (380.10)	353.6	1198.1		1204.6	1211.1	1223.5	1235.5	1247.0	1258.3	1269.3	1280.2	1290.9	1301.5	1312.0	1322.4	1332.8
198 (380.95)	354.5	1198.2		1204.2	1210.7	1223.2	1235.2	1246.8	1258.1	1269.1	1280.0	1290.7	1301.3	1311.8	1322.3	1332.6
200 (381.79)	355.4	1198.4		1203.8	1210.3	1222.9	1234.9	1246.5	1257.8	1268.9	1279.8	1290.5	1301.1	1311.7	1322.1	1332.5
202 (382.62)	356.2	1198.5		1203.4	1210.0	1222.5	1234.6	1246.2	1257.6	1268.7	1279.6	1290.4	1301.0	1311.5	1322.0	1332.4
204 (383.45)	357.1	1198.6		1203.0	1209.6	1222.2	1234.3	1246.0	1257.3	1268.5	1279.4	1290.2	1300.8	1311.3	1321.8	1332.2
206 (384.27)	358.0	1198.7		1202.6	1209.2	1221.9	1234.0	1245.7	1257.1	1268.3	1279.2	1290.0	1300.6	1311.2	1321.7	1332.1
208 (385.09)	358.9	1198.9		1202.2	1208.8	1221.5	1233.7	1245.4	1256.9	1268.0	1279.0	1289.8	1300.5	1311.0	1321.5	1331.9
210 (385.90)	359.8	1199.0		1201.8	1208.4	1221.2	1233.4	1245.2	1256.6	1267.8	1278.8	1289.6	1300.3	1310.9	1321.4	1331.8
212 (386.70)	360.6	1199.1		1201.4	1208.0	1220.9	1233.1	1244.9	1256.4	1267.6	1278.6	1289.4	1300.1	1310.7	1321.2	1331.7

Table A.2–2 Cont.
Superheated Vapor Enthalpy (Btu/lb_m)

Abs. press. lb_f/sq. in (sat. temp.)	Temperature—degrees fahrenheit														
	640°	660°	680°	700°	750°	800°	850°	900°	950°	1000°	1050°	1100°	1200°	1400°	1600°
180 (373.06)	1344.1	1354.3	1364.5	1374.7	1400.2	1425.6	1451.2	1476.8	1502.6	1528.6	1554.7	1581.0	1634.1	1742.9	1855.2
182 (373.96)	1344.0	1354.2	1364.4	1374.6	1400.1	1425.5	1451.1	1476.8	1502.6	1528.5	1554.6	1580.9	1634.1	1742.9	1855.2
184 (374.86)	1343.8	1354.1	1364.3	1374.5	1400.0	1425.5	1451.0	1476.7	1502.5	1528.5	1554.6	1580.9	1634.0	1742.9	1855.2
186 (375.75)	1343.7	1354.0	1364.2	1374.4	1399.9	1425.4	1451.0	1476.6	1502.4	1528.4	1554.5	1580.8	1634.0	1742.8	1855.2
188 (376.64)	1343.6	1353.8	1364.1	1374.3	1399.8	1425.3	1450.9	1476.6	1502.4	1528.3	1554.5	1580.8	1634.0	1742.8	1855.1
190 (377.51)	1343.5	1353.7	1363.9	1374.2	1399.7	1425.2	1450.8	1476.5	1502.3	1528.3	1554.4	1580.7	1633.9	1742.8	1855.1
192 (378.38)	1343.3	1353.6	1363.8	1374.1	1399.6	1425.1	1450.7	1476.4	1502.3	1528.2	1554.4	1580.7	1633.9	1742.7	1855.1
194 (379.24)	1343.2	1353.5	1363.7	1374.0	1399.5	1425.1	1450.7	1476.4	1502.2	1528.2	1554.3	1580.6	1633.8	1742.7	1855.1
196 (380.10)	1343.1	1353.4	1363.6	1373.9	1399.4	1425.0	1450.6	1476.3	1502.1	1528.1	1554.3	1580.6	1633.8	1742.7	1855.0
198 (380.95)	1342.9	1353.2	1363.5	1373.7	1399.3	1424.9	1450.5	1476.2	1502.1	1528.1	1554.2	1580.5	1633.8	1742.6	1855.0
200 (381.79)	1342.8	1353.1	1363.4	1373.6	1399.2	1424.8	1450.4	1476.2	1502.0	1528.0	1554.2	1580.5	1633.7	1742.6	1855.0
202 (382.62)	1342.7	1353.0	1363.3	1373.5	1399.1	1424.7	1450.4	1476.1	1502.0	1528.0	1554.1	1580.5	1633.7	1742.6	1855.0
204 (383.45)	1342.6	1352.9	1363.2	1373.4	1399.0	1424.6	1450.3	1476.0	1501.9	1527.9	1554.1	1580.4	1633.6	1742.5	1854.9
206 (384.27)	1342.4	1352.8	1363.1	1373.3	1399.0	1424.6	1450.2	1476.0	1501.8	1527.8	1554.0	1580.4	1633.6	1742.5	1854.9
208 (385.09)	1342.3	1352.6	1362.9	1373.2	1398.9	1424.5	1450.2	1475.9	1501.8	1527.8	1554.0	1580.3	1633.6	1742.5	1854.9
210 (385.90)	1342.2	1352.5	1362.8	1373.1	1398.8	1424.4	1450.1	1475.8	1501.7	1527.7	1553.9	1580.3	1633.5	1742.5	1854.9
212 (386.70)	1342.1	1352.4	1362.7	1373.0	1398.7	1424.3	1450.0	1475.8	1501.7	1527.7	1553.9	1580.2	1633.5	1742.4	1854.8

Table A.2–2 Cont.
Superheated Vapor Enthalpy (Btu/lb_m)

Abs. press. lb_f/sq. in. (sat. temp.)	Sat. liquid	Sat. vapor	Temperature—degrees fahrenheit 390°	400°	420°	440°	460°	480°	500°	520°	540°	560°	580°	600°	620°	640°
214 (387.50)	361.5	1199.2	1200.9	1207.6	1220.5	1232.8	1244.6	1256.1	1267.4	1278.4	1289.3	1300.0	1310.6	1321.1	1331.5	1341.9
216 (388.29)	362.3	1199.4	1200.5	1207.2	1220.2	1232.5	1244.4	1255.9	1267.2	1278.2	1289.1	1299.8	1310.4	1321.0	1331.4	1341.8
218 (389.08)	363.2	1199.5	1200.1	1206.9	1219.8	1232.2	1244.1	1255.7	1266.9	1278.0	1288.9	1299.6	1310.3	1320.8	1331.3	1341.7
220 (389.86)	364.0	1199.6	1199.7	1206.5	1219.5	1231.9	1243.8	1255.4	1266.7	1277.8	1288.7	1299.5	1310.1	1320.7	1331.1	1341.6
222 (390.63)	364.8	1199.7		1206.1	1219.1	1231.6	1243.6	1255.2	1266.5	1277.6	1288.5	1299.3	1310.0	1320.5	1331.0	1341.4
224 (391.40)	365.7	1199.8		1205.7	1218.8	1231.3	1243.3	1254.9	1266.3	1277.4	1288.3	1299.1	1309.8	1320.4	1330.9	1341.3
226 (392.17)	366.5	1199.9		1205.3	1218.5	1231.0	1243.0	1254.7	1266.1	1277.2	1288.2	1299.0	1309.6	1320.2	1330.7	1341.2
228 (392.93)	367.3	1200.0		1204.9	1218.1	1230.7	1242.8	1254.5	1265.9	1277.0	1288.0	1298.8	1309.5	1320.1	1330.6	1341.0
230 (393.68)	368.1	1200.1		1204.5	1217.8	1230.4	1242.5	1254.2	1265.6	1276.8	1287.8	1298.6	1309.3	1319.9	1330.5	1340.9
232 (394.43)	368.9	1200.2		1204.1	1217.4	1230.1	1242.2	1254.0	1265.4	1276.6	1287.6	1298.5	1309.2	1319.8	1330.3	1340.8
234 (395.17)	369.7	1200.3		1203.7	1217.1	1229.8	1241.9	1253.7	1265.2	1276.4	1287.4	1298.3	1309.0	1319.6	1330.2	1340.7
236 (395.91)	370.5	1200.4		1203.3	1216.7	1229.5	1241.7	1253.5	1265.0	1276.2	1287.2	1298.1	1308.9	1319.5	1330.0	1340.5
238 (396.64)	371.3	1200.5		1202.9	1216.4	1229.1	1241.4	1253.2	1264.7	1276.0	1287.1	1297.9	1308.7	1319.3	1329.9	1340.4
240 (397.37)	372.1	1200.6		1202.5	1216.0	1228.8	1241.1	1253.0	1264.5	1275.8	1286.9	1297.8	1308.5	1319.2	1329.8	1340.3
242 (398.10)	372.9	1200.7		1202.1	1215.7	1228.5	1240.8	1252.7	1264.3	1275.6	1286.7	1297.6	1308.4	1319.0	1329.6	1340.1
244 (398.82)	373.7	1200.8		1201.6	1215.3	1228.2	1240.6	1252.5	1264.1	1275.4	1286.5	1297.4	1308.2	1318.9	1329.5	1340.0
246 (399.53)	374.5	1200.9		1201.2	1214.9	1227.9	1240.3	1252.2	1263.8	1275.2	1286.3	1297.3	1308.1	1318.8	1329.4	1339.9
248 (400.24)	375.2	1201.0			1214.6	1227.6	1240.0	1252.0	1263.6	1275.0	1286.1	1297.1	1307.9	1318.6	1329.2	1339.8

TABLE A.2–2 CONT.
SUPERHEATED VAPOR ENTHALPY (Btu/lb_m)

Abs. press. lb_f/sq. in (sat. temp.)	Temperature—degrees fahrenheit															
	660°	680°	700°	720°	740°	760°	780°	800°	850°	900°	950°	1000°	1100°	1200°	1400°	1600°
214 (387.50)	1352.3	1362.6	1372.9	1383.2	1393.4	1403.7	1414.0	1424.2	1449.9	1475.7	1501.6	1527.6	1580.2	1633.4	1742.4	1854.8
216 (388.29)	1352.2	1362.5	1372.8	1383.1	1393.3	1403.6	1413.9	1424.1	1449.9	1475.6	1501.5	1527.6	1580.1	1633.4	1742.4	1854.8
218 (389.08)	1352.0	1362.4	1372.7	1383.0	1393.2	1403.5	1413.8	1424.1	1449.8	1475.6	1501.5	1527.5	1580.1	1633.4	1742.3	1854.8
220 (389.86)	1351.9	1362.3	1372.6	1382.9	1393.1	1403.4	1413.7	1424.0	1449.7	1475.5	1501.4	1527.5	1580.0	1633.3	1742.3	1854.7
222 (390.63)	1351.8	1362.2	1372.5	1382.8	1393.0	1403.3	1413.6	1423.9	1449.6	1475.4	1501.4	1527.4	1580.0	1633.3	1742.3	1854.7
224 (391.40)	1351.7	1362.0	1372.4	1382.7	1393.0	1403.2	1413.5	1423.8	1449.6	1475.4	1501.3	1527.3	1579.9	1633.2	1742.2	1854.7
226 (392.17)	1351.6	1361.9	1372.3	1382.6	1392.9	1403.1	1413.4	1423.7	1449.5	1475.3	1501.2	1527.3	1579.9	1633.2	1742.2	1854.7
228 (392.93)	1351.5	1361.8	1372.2	1382.5	1392.8	1403.1	1413.4	1423.7	1449.4	1475.2	1501.2	1527.2	1579.8	1633.1	1742.2	1854.6
230 (393.68)	1351.3	1361.7	1372.0	1382.4	1392.7	1403.0	1413.3	1423.6	1449.3	1475.2	1501.1	1527.2	1579.8	1633.1	1742.1	1854.6
232 (394.43)	1351.2	1361.6	1371.9	1382.3	1392.6	1402.9	1413.2	1423.5	1449.3	1475.1	1501.0	1527.1	1579.7	1633.1	1742.1	1854.6
234 (395.17)	1351.1	1361.5	1371.8	1382.2	1392.5	1402.8	1413.1	1423.4	1449.2	1475.0	1501.0	1527.1	1579.7	1633.0	1742.1	1854.6
236 (395.91)	1351.0	1361.4	1371.7	1382.1	1392.4	1402.7	1413.0	1423.3	1449.1	1475.0	1500.9	1527.0	1579.7	1633.0	1742.0	1854.5
238 (396.64)	1350.9	1361.2	1371.6	1382.0	1392.3	1402.6	1412.9	1423.2	1449.0	1474.9	1500.9	1527.0	1579.6	1632.9	1742.0	1854.5
240 (397.37)	1350.7	1361.1	1371.5	1381.9	1392.2	1402.5	1412.8	1423.2	1449.0	1474.8	1500.8	1526.9	1579.6	1632.9	1742.0	1854.5
242 (398.10)	1350.6	1361.0	1371.4	1381.8	1392.1	1402.4	1412.7	1423.1	1448.9	1474.8	1500.7	1526.8	1579.5	1632.9	1742.0	1854.5
244 (398.82)	1350.5	1360.9	1371.3	1381.7	1392.0	1402.3	1412.7	1423.0	1448.8	1474.7	1500.7	1526.8	1579.5	1632.8	1741.9	1854.4
246 (399.53)	1350.4	1360.8	1371.2	1381.6	1391.9	1402.2	1412.6	1422.9	1448.7	1474.6	1500.6	1526.7	1579.4	1632.8	1741.9	1854.4
248 (400.24)	1350.3	1360.7	1371.1	1381.5	1391.8	1402.2	1412.5	1422.8	1448.7	1474.6	1500.6	1526.7	1579.4	1632.7	1741.9	1854.4

Table A.2–2 Cont.
Superheated Vapor Enthalpy (Btu/lb_m)

Abs. press. lb_f/sq. in. (sat. temp.)	*Sat. liquid*	*Sat. vapor*	*Temperature—degrees fahrenheit* 420°	440°	460°	480°	500°	520°	540°	560°	580°	600°	620°	640°	660°	680°
250 (400.95)	376.0	1201.1	1214.2	1227.3	1239.7	1251.7	1263.4	1274.8	1285.9	1296.9	1307.7	1318.5	1329.1	1339.6	1350.1	1360.6
255 (402.70)	377.9	1201.3	1213.3	1226.5	1239.0	1251.1	1262.8	1274.3	1285.5	1296.5	1307.3	1318.1	1328.7	1339.3	1349.8	1360.3
260 (404.42)	379.8	1201.5	1212.4	1225.7	1238.3	1250.5	1262.3	1273.8	1285.0	1296.0	1306.9	1317.7	1328.4	1339.0	1349.5	1360.0
265 (406.11)	381.6	1201.7	1211.5	1224.9	1237.6	1249.9	1261.7	1273.2	1284.5	1295.6	1306.5	1317.3	1328.0	1338.7	1349.2	1359.7
270 (407.78)	383.4	1201.9	1210.6	1224.1	1236.9	1249.2	1261.1	1272.7	1284.0	1295.2	1306.1	1317.0	1327.7	1338.3	1348.9	1359.4
275 (409.43)	385.2	1202.1	1209.6	1223.3	1236.2	1248.6	1260.5	1272.2	1283.6	1294.7	1305.7	1316.6	1327.3	1338.0	1348.6	1359.1
280 (411.05)	387.0	1202.3	1208.7	1222.4	1235.4	1247.9	1260.0	1271.7	1283.1	1294.3	1305.3	1316.2	1327.0	1337.7	1348.3	1358.9
285 (412.65)	388.7	1202.4	1207.7	1221.6	1234.7	1247.3	1259.4	1271.1	1282.6	1293.9	1304.9	1315.8	1326.6	1337.4	1348.0	1358.6
290 (414.23)	390.5	1202.6	1206.8	1220.8	1234.0	1246.6	1258.8	1270.6	1282.1	1293.4	1304.5	1315.5	1326.3	1337.0	1347.7	1358.3
295 (415.79)	392.2	1202.7	1205.8	1219.9	1233.3	1246.0	1258.2	1270.1	1281.6	1293.0	1304.1	1315.1	1325.9	1336.7	1347.4	1358.0
300 (417.33)	393.8	1202.8	1204.8	1219.1	1232.5	1245.3	1257.6	1269.5	1281.1	1292.5	1303.7	1314.7	1325.6	1336.4	1347.1	1357.7
305 (418.85)	395.4	1203.0	1203.8	1218.2	1231.8	1244.7	1257.0	1269.0	1280.7	1292.1	1303.3	1314.3	1325.2	1336.0	1346.8	1357.4
310 (420.35)	397.1	1203.1		1217.4	1231.0	1244.0	1256.4	1268.5	1280.2	1291.6	1302.9	1313.9	1324.9	1335.7	1346.4	1357.1
315 (421.83)	398.8	1203.2		1216.5	1230.3	1243.3	1255.8	1267.9	1279.7	1291.2	1302.5	1313.6	1324.5	1335.4	1346.1	1356.8
320 (423.29)	400.4	1203.4		1215.6	1229.5	1242.6	1255.2	1267.4	1279.2	1290.7	1302.0	1313.2	1324.2	1335.0	1345.8	1356.5
325 (424.74)	402.0	1203.5		1214.8	1228.7	1241.9	1254.6	1266.8	1278.7	1290.3	1301.6	1312.8	1323.8	1334.7	1345.5	1356.2
330 (426.16)	403.6	1203.6		1213.9	1228.0	1241.3	1254.0	1266.3	1278.2	1289.8	1301.2	1312.4	1323.5	1334.4	1345.2	1356.0
335	405.1	1203.6		1213.0	1227.2	1240.6	1253.4	1265.7	1277.7	1289.4	1300.8	1312.0	1323.1	1334.0	1344.9	1355.7

TABLE A.2–2 CONT.
SUPERHEATED VAPOR ENTHALPY (Btu/lb_m)

Abs. press. lb_f/sq. in (sat. temp.)	Temperature—degrees fahrenheit														
	700°	720°	740°	760°	780°	800°	850°	900°	950°	1000°	1050°	1100°	1200°	1400°	1600°
250 (400.95)	1371.0	1381.4	1391.7	1402.1	1412.4	1422.7	1448.6	1474.5	1500.5	1526.6	1552.9	1579.3	1632.7	1741.8	1854.4
255 (402.70)	1370.7	1381.1	1391.5	1401.8	1412.2	1422.5	1448.4	1474.3	1500.3	1526.5	1552.8	1579.2	1632.6	1741.8	1854.3
260 (404.42)	1370.4	1380.8	1391.2	1401.6	1412.0	1422.3	1448.2	1474.2	1500.2	1526.3	1552.6	1579.1	1632.5	1741.7	1854.2
265 (406.11)	1370.2	1380.6	1391.0	1401.4	1411.7	1422.1	1448.0	1474.0	1500.0	1526.2	1552.5	1579.0	1632.4	1741.6	1854.2
270 (407.78)	1369.9	1380.3	1390.7	1401.1	1411.5	1421.9	1447.8	1473.8	1499.9	1526.1	1552.4	1578.8	1632.3	1741.5	1854.1
275 (409.43)	1369.6	1380.1	1390.5	1400.9	1411.3	1421.7	1447.7	1473.7	1499.7	1525.9	1552.3	1578.7	1632.2	1741.4	1854.0
280 (411.05)	1369.4	1379.8	1390.3	1400.7	1411.1	1421.5	1447.5	1473.5	1499.6	1525.8	1552.1	1578.6	1632.1	1741.4	1854.0
285 (412.65)	1369.1	1379.6	1390.0	1400.4	1410.9	1421.3	1447.3	1473.3	1499.4	1525.6	1552.0	1578.5	1632.0	1741.3	1853.9
290 (414.23)	1368.8	1379.3	1389.8	1400.2	1410.6	1421.1	1447.1	1473.2	1499.3	1525.5	1551.9	1578.4	1631.9	1741.2	1853.9
295 (415.79)	1368.5	1379.1	1389.5	1400.0	1410.4	1420.9	1446.9	1473.0	1499.1	1525.4	1551.7	1578.3	1631.8	1741.1	1853.8
300 (417.33)	1368.3	1378.8	1389.3	1399.8	1410.2	1420.6	1446.7	1472.8	1499.0	1525.2	1551.6	1578.1	1631.7	1741.0	1853.7
305 (418.85)	1368.0	1378.5	1389.0	1399.5	1410.0	1420.4	1446.5	1472.7	1498.8	1525.1	1551.5	1578.0	1631.6	1741.0	1853.7
310 (420.35)	1367.7	1378.3	1388.8	1399.3	1409.8	1420.2	1446.4	1472.5	1498.7	1524.9	1551.4	1577.9	1631.5	1740.9	1853.6
315 (421.83)	1367.4	1378.0	1388.6	1399.1	1409.5	1420.0	1446.2	1472.3	1498.5	1524.8	1551.2	1577.8	1631.4	1740.8	1853.5
320 (423.29)	1367.2	1377.8	1388.3	1398.8	1409.3	1419.8	1446.0	1472.1	1498.4	1524.7	1551.1	1577.7	1631.3	1740.7	1853.5
325 (424.74)	1366.9	1377.5	1388.1	1398.6	1409.1	1419.6	1445.8	1472.0	1498.2	1524.5	1551.0	1577.5	1631.2	1740.7	1853.4
330 (426.16)	1366.6	1377.2	1387.8	1398.4	1408.9	1419.4	1445.6	1471.8	1498.0	1524.4	1550.8	1577.4	1631.1	1740.6	1853.4
335 (427.58)	1366.3	1377.0	1387.6	1398.1	1408.7	1419.2	1445.4	1471.6	1497.9	1524.2	1550.7	1577.3	1631.0	1740.5	1853.3

Table A.2–2 Cont.
Superheated Vapor Enthalpy (Btu/lb$_m$)

Abs. press. lb$_f$/sq. in. (sat. temp.)	Sat. liquid	Sat. vapor	Temperature—degrees fahrenheit 430°	440°	450°	460°	470°	480°	490°	500°	520°	540°	560°	580°	600°	620°
340 (428.97)	406.7	1203.7	1204.5	1212.1	1219.4	1226.4	1233.2	1239.9	1246.4	1252.8	1265.2	1277.2	1288.9	1300.4	1311.6	1322.7
345 (430.35)	408.2	1203.8		1211.2	1218.5	1225.6	1232.5	1239.2	1245.8	1252.1	1264.6	1276.7	1288.4	1299.9	1311.2	1322.4
350 (431.72)	409.7	1203.9		1210.3	1217.7	1224.8	1231.7	1238.5	1245.1	1251.5	1264.1	1276.2	1288.0	1299.5	1310.9	1322.0
355 (433.06)	411.2	1204.0		1209.3	1216.8	1224.0	1231.0	1237.8	1244.4	1250.9	1263.5	1275.7	1287.5	1299.1	1310.5	1321.7
360 (434.40)	412.7	1204.1		1208.4	1215.9	1223.2	1230.2	1237.1	1243.8	1250.3	1263.0	1275.2	1287.1	1298.7	1310.1	1321.3
365 (435.72)	414.1	1204.1		1207.5	1215.1	1222.4	1229.5	1236.4	1243.1	1249.6	1262.4	1274.7	1286.6	1298.2	1309.7	1320.9
370 (437.03)	415.6	1204.2		1206.5	1214.2	1221.6	1228.7	1235.7	1242.4	1249.0	1261.8	1274.1	1286.1	1297.8	1309.3	1320.6
375 (438.32)	417.0	1204.3		1205.6	1213.3	1220.7	1227.9	1234.9	1241.8	1248.4	1261.2	1273.6	1285.6	1297.4	1308.9	1320.2
380 (439.60)	418.5	1204.3		1204.6	1212.4	1219.9	1227.2	1234.2	1241.1	1247.7	1260.7	1273.1	1285.2	1297.0	1308.5	1319.8
385 (440.86)	419.9	1204.3			1211.5	1219.1	1226.4	1233.5	1240.4	1247.1	1260.1	1272.6	1284.7	1296.5	1308.1	1319.5
390 (442.12)	421.3	1204.4			1210.6	1218.2	1225.6	1232.8	1239.7	1246.4	1259.5	1272.1	1284.2	1296.1	1307.7	1319.1
395 (443.36)	422.7	1204.4			1209.7	1217.4	1224.8	1232.0	1239.0	1245.8	1258.9	1271.5	1283.8	1295.7	1307.3	1318.7
400 (444.59)	424.0	1204.5			1208.8	1216.5	1224.0	1231.3	1238.3	1245.1	1258.3	1271.0	1283.3	1295.2	1306.9	1318.3
405 (445.81)	425.4	1204.5			1207.9	1215.7	1223.2	1230.5	1237.6	1244.5	1257.8	1270.5	1282.8	1294.8	1306.5	1318.0
410 (447.01)	426.8	1204.5			1206.9	1214.8	1222.4	1229.8	1236.9	1243.8	1257.2	1270.0	1282.3	1294.3	1306.1	1317.6
415 (448.21)	428.1	1204.6			1206.0	1214.0	1221.6	1229.0	1236.2	1243.2	1256.6	1269.4	1281.8	1293.9	1305.7	1317.2
420 (449.39)	429.4	1204.6			1205.0	1213.1	1220.8	1228.3	1235.5	1242.5	1256.0	1268.9	1281.4	1293.5	1305.3	1316.9
425 (450.57)	430.7	1204.6				1212.2	1220.0	1227.5	1234.8	1241.8	1255.4	1268.3	1280.9	1293.0	1304.9	1316.5

TABLE A.2–2 CONT.
SUPERHEATED VAPOR ENTHALPY (Btu/lb_m)

Abs. press. lb_f/sq. in (sat. temp.)	*Temperature—degrees fahrenheit*															
	640°	*660°*	*680°*	*700°*	*720°*	*740°*	*760°*	*780°*	*800°*	*850°*	*900°*	*950°*	*1000°*	*1200°*	*1400°*	*1600°*
340 (428.97)	1333.7	1344.6	1355.4	1366.1	1376.7	1387.3	1397.9	1408.4	1419.0	1445.2	1471.5	1497.7	1524.1	1630.9	1740.4	1853.2
345 (430.35)	1333.4	1344.3	1355.1	1365.8	1376.5	1387.1	1397.7	1408.2	1418.8	1445.0	1471.3	1497.6	1524.0	1630.8	1740.3	1853.2
350 (431.72)	1333.0	1343.9	1354.8	1365.5	1376.2	1386.8	1397.4	1408.0	1418.5	1444.8	1471.1	1497.4	1523.8	1630.7	1740.3	1853.1
355 (433.06)	1332.7	1343.6	1354.5	1365.2	1375.9	1386.6	1397.2	1407.8	1418.3	1444.7	1471.0	1497.3	1523.7	1630.6	1740.2	1853.1
360 (434.40)	1332.4	1343.3	1354.2	1365.0	1375.7	1386.3	1397.0	1407.5	1418.1	1444.5	1470.8	1497.1	1523.5	1630.5	1740.1	1853.0
365 (435.72)	1332.0	1343.0	1353.9	1364.7	1375.4	1386.1	1396.7	1407.3	1417.9	1444.3	1470.6	1497.0	1523.4	1630.4	1740.0	1852.9
370 (437.03)	1331.7	1342.7	1353.6	1364.4	1375.1	1385.8	1396.5	1407.1	1417.7	1444.1	1470.4	1496.8	1523.3	1630.3	1740.0	1852.9
375 (438.32)	1331.4	1342.4	1353.3	1364.1	1374.9	1385.6	1396.2	1406.9	1417.5	1443.9	1470.3	1496.7	1523.1	1630.2	1739.9	1852.8
380 (439.60)	1331.0	1342.0	1353.0	1363.8	1374.6	1385.3	1396.0	1406.6	1417.3	1443.7	1470.1	1496.5	1523.0	1630.0	1739.8	1852.7
385 (440.86)	1330.7	1341.7	1352.7	1363.6	1374.3	1385.1	1395.8	1406.4	1417.0	1443.5	1469.9	1496.4	1522.8	1629.9	1739.7	1852.7
390 (442.12)	1330.3	1341.4	1352.4	1363.3	1374.1	1384.8	1395.5	1406.2	1416.8	1443.3	1469.8	1496.2	1522.7	1629.8	1739.6	1852.6
395 (443.36)	1330.0	1341.1	1352.1	1363.0	1373.8	1384.6	1395.3	1406.0	1416.6	1443.1	1469.6	1496.0	1522.6	1629.7	1739.6	1852.6
400 (444.59)	1329.6	1340.8	1351.8	1362.7	1373.6	1384.3	1395.1	1405.8	1416.4	1442.9	1469.4	1495.9	1522.4	1629.6	1739.5	1852.5
405 (445.81)	1329.3	1340.4	1351.5	1362.4	1373.3	1384.1	1394.8	1405.5	1416.2	1442.8	1469.2	1495.7	1522.3	1629.5	1739.4	1852.4
410 (447.01)	1328.9	1340.1	1351.2	1362.2	1373.0	1383.8	1394.6	1405.3	1416.0	1442.6	1469.1	1495.6	1522.1	1629.4	1739.3	1852.4
415 (448.21)	1328.6	1339.8	1350.9	1361.9	1372.8	1383.6	1394.4	1405.1	1415.8	1442.4	1468.9	1495.4	1522.0	1629.3	1739.2	1852.3
420 (449.39)	1328.3	1339.5	1350.6	1361.6	1372.5	1383.3	1394.1	1404.8	1415.5	1442.2	1468.7	1495.3	1521.9	1629.2	1739.2	1852.2
425 (450.57)	1327.9	1339.2	1350.3	1361.3	1372.2	1383.1	1393.9	1404.6	1415.3	1442.0	1468.6	1495.1	1521.7	1629.1	1739.1	1852.2

TABLE A.2–2 CONT.
SUPERHEATED VAPOR ENTHALPY (Btu/lb_m)

Abs. press. lb_f/sq. in. (sat. temp.)	*Sat. liquid*	*Sat. vapor*	*Temperature—degrees fahrenheit* 460°	470°	480°	490°	500°	520°	540°	560°	580°	600°	620°	640°	660°	680°
430 (451.73)	432.1	1204.6	1211.3	1219.2	1226.7	1234.0	1241.1	1254.8	1267.8	1280.4	1292.6	1304.4	1316.1	1327.6	1338.8	1350.0
435 (452.88)	433.4	1204.6	1210.4	1218.3	1225.9	1233.3	1240.5	1254.2	1267.3	1279.9	1292.1	1304.0	1315.7	1327.2	1338.5	1349.7
440 (454.02)	434.6	1204.6	1209.5	1217.5	1225.2	1232.6	1239.8	1253.6	1266.7	1279.4	1291.7	1303.6	1315.4	1326.9	1338.2	1349.4
445 (455.15)	435.9	1204.6	1208.6	1216.7	1224.4	1231.8	1239.1	1253.0	1266.2	1278.9	1291.2	1303.2	1315.0	1326.5	1337.9	1349.1
450 (456.28)	437.2	1204.6	1207.7	1215.8	1223.6	1231.1	1238.4	1252.3	1265.6	1278.4	1290.8	1302.8	1314.6	1326.2	1337.5	1348.8
455 (457.39)	438.5	1204.6	1206.8	1215.0	1222.8	1230.4	1237.7	1251.7	1265.1	1277.9	1290.3	1302.4	1314.2	1325.8	1337.2	1348.5
460 (458.50)	439.7	1204.6	1205.8	1214.1	1222.0	1229.6	1237.0	1251.1	1264.5	1277.4	1289.9	1302.0	1313.8	1325.4	1336.9	1348.1
465 (459.59)	440.9	1204.6	1204.9	1213.2	1221.2	1228.9	1236.3	1250.5	1264.0	1276.9	1289.4	1301.6	1313.5	1325.1	1336.5	1347.8
470 (460.68)	442.2	1204.6		1212.4	1220.4	1228.1	1235.6	1249.9	1263.4	1276.4	1289.0	1301.2	1313.1	1324.7	1336.2	1347.5
475 (461.75)	443.4	1204.5		1211.5	1219.6	1227.3	1234.9	1249.2	1262.8	1275.9	1288.5	1300.7	1312.7	1324.4	1335.9	1347.2
480 (462.82)	444.6	1204.5		1210.6	1218.8	1226.6	1234.2	1248.6	1262.3	1275.4	1288.0	1300.3	1312.3	1324.0	1335.6	1346.9
485 (463.88)	445.8	1204.5		1209.7	1217.9	1225.8	1233.4	1248.0	1261.7	1274.9	1287.6	1299.9	1311.9	1323.7	1335.2	1346.6
490 (464.93)	447.0	1204.5		1208.8	1217.1	1225.0	1232.7	1247.3	1261.2	1274.4	1287.1	1299.5	1311.5	1323.3	1334.9	1346.3
495 (465.97)	448.2	1204.4		1207.9	1216.3	1224.3	1232.0	1246.7	1260.6	1273.9	1286.7	1299.0	1311.1	1322.9	1334.6	1346.0
500 (467.01)	449.4	1204.4		1207.0	1215.4	1223.5	1231.3	1246.0	1260.0	1273.4	1286.2	1298.6	1310.7	1322.6	1334.2	1345.7
510 (469.05)	451.8	1204.3		1205.2	1213.7	1221.9	1229.8	1244.7	1258.9	1272.4	1285.3	1297.8	1310.0	1321.9	1333.6	1345.0
520 (471.07)	454.1	1204.2			1212.0	1220.3	1228.3	1243.4	1257.7	1271.3	1284.3	1296.9	1309.2	1321.1	1332.9	1344.4
530	456.4	1204.1			1210.3	1218.7	1226.8	1242.1	1256.5	1270.2	1283.4	1296.1	1308.4	1320.4	1332.2	1343.8

Table A.2–2 Cont.
Superheated Vapor Enthalpy (Btu/lb_m)

Abs. press. lb_f/sq. in (sat. temp.)	Temperature—degrees fahrenheit 700°	720°	740°	760°	780°	800°	850°	900°	950°	1000°	1050°	1100°	1200°	1400°	1600°
430 (451.73)	1361.0	1372.0	1382.8	1393.6	1404.4	1415.1	1441.8	1468.4	1495.0	1521.6	1548.3	1575.1	1629.0	1739.0	1852.1
435 (452.88)	1360.7	1371.7	1382.6	1393.4	1404.2	1414.9	1441.6	1468.2	1494.8	1521.4	1548.1	1574.9	1628.9	1738.9	1852.1
440 (454.02)	1360.4	1371.4	1382.3	1393.2	1403.9	1414.7	1441.4	1468.1	1494.7	1521.3	1548.0	1574.8	1628.8	1738.9	1852.0
445 (455.15)	1360.2	1371.2	1382.1	1392.9	1403.7	1414.5	1441.2	1467.9	1494.5	1521.2	1547.9	1574.7	1628.7	1738.8	1851.9
450 (456.28)	1359.9	1370.9	1381.8	1392.7	1403.5	1414.3	1441.0	1467.7	1494.3	1521.0	1547.7	1574.6	1628.6	1738.7	1851.9
455 (457.39)	1359.6	1370.6	1381.6	1392.4	1403.3	1414.0	1440.8	1467.5	1494.2	1520.9	1547.6	1574.5	1628.5	1738.6	1851.8
460 (458.50)	1359.3	1370.3	1381.3	1392.2	1403.0	1413.8	1440.7	1467.4	1494.0	1520.7	1547.5	1574.3	1628.4	1738.5	1851.7
465 (459.59)	1359.0	1370.1	1381.1	1392.0	1402.8	1413.6	1440.5	1467.2	1493.9	1520.6	1547.4	1574.2	1628.3	1738.5	1851.7
470 (460.68)	1358.7	1369.8	1380.8	1391.7	1402.6	1413.4	1440.3	1467.0	1493.7	1520.4	1547.2	1574.1	1628.2	1738.4	1851.6
475 (461.75)	1358.4	1369.5	1380.5	1391.5	1402.3	1413.2	1440.1	1466.8	1493.6	1520.3	1547.1	1574.0	1628.1	1738.3	1851.6
480 (462.82)	1358.2	1369.3	1380.3	1391.2	1402.1	1412.9	1439.9	1466.7	1493.4	1520.2	1547.0	1573.9	1628.0	1738.2	1851.5
485 (463.88)	1357.9	1369.0	1380.0	1391.0	1401.9	1412.7	1439.7	1466.5	1493.3	1520.0	1546.8	1573.7	1627.9	1738.1	1851.4
490 (464.93)	1357.6	1368.7	1379.8	1390.7	1401.7	1412.5	1439.5	1466.3	1493.1	1519.9	1546.7	1573.6	1627.8	1738.1	1851.4
495 (465.97)	1357.3	1368.5	1379.5	1390.5	1401.4	1412.3	1439.3	1466.2	1493.0	1519.7	1546.6	1573.5	1627.7	1738.0	1851.3
500 (467.01)	1357.0	1368.2	1379.3	1390.3	1401.2	1412.1	1439.1	1466.0	1492.8	1519.6	1546.4	1573.4	1627.6	1737.9	1851.3
510 (469.05)	1356.4	1367.6	1378.7	1389.8	1400.7	1411.6	1438.7	1465.6	1492.5	1519.3	1546.2	1573.2	1627.4	1737.7	1851.1
520 (471.07)	1355.8	1367.1	1378.2	1389.3	1400.3	1411.2	1438.3	1465.3	1492.2	1519.0	1545.9	1572.9	1627.2	1737.6	1851.0
530 (473.06)	1355.2	1366.5	1377.7	1388.8	1399.8	1410.8	1438.0	1465.0	1491.9	1518.7	1545.7	1572.7	1627.0	1737.4	1850.9

TABLE A.2–2 CONT.
SUPERHEATED VAPOR ENTHALPY (Btu/lb_m)

Abs. press. lb_f/sq. in. (sat. temp.)	*Sat. liquid*	*Sat. vapor*	*Temperature—degrees fahrenheit* 480°	490°	500°	510°	520°	530°	540°	550°	560°	570°	580°	590°	600°	620°
540 (475.01)	458.6	1204.0	1208.5	1217.1	1225.3	1233.1	1240.8	1248.2	1255.4	1262.3	1269.2	1275.9	1282.4	1288.8	1295.2	1307.6
550 (476.94)	460.8	1203.9	1206.7	1215.4	1223.7	1231.7	1239.4	1246.9	1254.2	1261.2	1268.1	1274.9	1281.5	1287.9	1294.3	1306.8
560 (478.85)	463.0	1203.8	1204.9	1213.7	1222.2	1230.3	1238.1	1245.6	1253.0	1260.1	1267.1	1273.8	1280.5	1287.0	1293.4	1306.0
570 (480.73)	465.2	1203.6		1212.0	1220.6	1228.8	1236.7	1244.3	1251.8	1259.0	1266.0	1272.8	1279.5	1286.1	1292.6	1305.2
580 (482.58)	467.4	1203.5		1210.3	1219.0	1227.3	1235.3	1243.0	1250.5	1257.8	1264.9	1271.8	1278.5	1285.2	1291.7	1304.4
590 (484.41)	469.5	1203.4		1208.5	1217.4	1225.8	1233.9	1241.7	1249.3	1256.6	1263.8	1270.8	1277.6	1284.3	1290.8	1303.6
600 (486.21)	471.6	1203.2		1206.7	1215.7	1224.3	1232.5	1240.4	1248.1	1255.5	1262.7	1269.7	1276.6	1283.3	1289.9	1302.7
610 (487.99)	473.7	1203.1		1205.0	1214.1	1222.8	1231.1	1239.1	1246.8	1254.3	1261.6	1268.6	1275.6	1282.3	1289.0	1301.9
620 (489.75)	475.7	1202.9		1203.1	1212.4	1221.2	1229.6	1237.7	1245.5	1253.1	1260.4	1267.6	1274.6	1281.4	1288.1	1301.1
630 (491.49)	477.8	1202.7			1210.7	1219.6	1228.2	1236.4	1244.3	1251.9	1259.3	1266.5	1273.5	1280.4	1287.1	1300.2
640 (493.21)	479.8	1202.5			1209.0	1218.0	1226.7	1235.0	1243.0	1250.7	1258.2	1265.4	1272.5	1279.4	1286.2	1299.4
650 (494.90)	481.8	1202.3			1207.2	1216.4	1225.2	1233.6	1241.7	1249.5	1257.0	1264.4	1271.5	1278.5	1285.3	1298.6
660 (496.58)	483.8	1202.1			1205.4	1214.8	1223.7	1232.2	1240.4	1248.2	1255.9	1263.3	1270.5	1277.5	1284.4	1297.7
670 (498.24)	485.8	1201.9			1203.6	1213.1	1222.2	1230.8	1239.0	1247.0	1254.7	1262.2	1269.4	1276.5	1283.4	1296.9
680 (499.88)	487.7	1201.7			1201.8	1211.4	1220.6	1229.3	1237.7	1245.7	1253.5	1261.1	1268.4	1275.5	1282.5	1296.0
690 (501.50)	489.6	1201.4				1209.7	1219.0	1227.9	1236.3	1244.5	1252.3	1259.9	1267.3	1274.5	1281.5	1295.2
700 (503.10)	491.5	1201.2				1208.0	1217.5	1226.4	1235.0	1243.2	1251.1	1258.8	1266.3	1273.5	1280.6	1294.3
710	493.4	1201.0				1206.3	1215.9	1224.9	1233.6	1241.9	1249.9	1257.7	1265.2	1272.5	1279.6	1293.4

Table A.2–2 Cont.

Superheated Vapor Enthalpy (Btu/lb_m)

Abs. press. lb_f/sq. in (sat. temp.)	Temperature—degrees fahrenheit															
	640°	*660°*	*680°*	*700°*	*720°*	*740°*	*760°*	*780°*	*800°*	*850°*	*900°*	*950°*	*1000°*	*1200°*	*1400°*	*1600°*
540 (475.01)	1319.7	1331.5	1343.2	1354.6	1366.0	1377.2	1388.3	1399.4	1410.3	1437.6	1464.6	1491.5	1518.5	1626.8	1737.3	1850.7
550 (476.94)	1318.9	1330.8	1342.5	1354.0	1365.4	1376.7	1387.8	1398.9	1409.9	1437.2	1464.3	1491.2	1518.2	1626.6	1737.1	1850.6
560 (478.85)	1318.2	1330.2	1341.9	1353.5	1364.9	1376.1	1387.3	1398.4	1409.4	1436.8	1463.9	1490.9	1517.9	1626.4	1737.0	1850.5
570 (480.73)	1317.5	1329.5	1341.3	1352.9	1364.3	1375.6	1386.8	1398.0	1409.0	1436.4	1463.6	1490.6	1517.6	1626.2	1736.8	1850.4
580 (482.58)	1316.7	1328.8	1340.6	1352.3	1363.7	1375.1	1386.3	1397.5	1408.6	1436.0	1463.2	1490.3	1517.3	1626.0	1736.6	1850.2
590 (484.41)	1316.0	1328.1	1340.0	1351.7	1363.2	1374.6	1385.8	1397.0	1408.1	1435.6	1462.9	1490.0	1517.0	1625.7	1736.5	1850.1
600 (486.21)	1315.2	1327.4	1339.3	1351.1	1362.6	1374.0	1385.3	1396.5	1407.7	1435.2	1462.5	1489.7	1516.7	1625.5	1736.3	1850.0
610 (487.99)	1314.5	1326.7	1338.7	1350.5	1362.1	1373.5	1384.8	1396.1	1407.2	1434.8	1462.2	1489.3	1516.5	1625.3	1736.2	1849.9
620 (489.75)	1313.7	1326.0	1338.0	1349.9	1361.5	1373.0	1384.3	1395.6	1406.8	1434.4	1461.8	1489.0	1516.2	1625.1	1736.0	1849.8
630 (491.49)	1312.9	1325.3	1337.4	1349.2	1360.9	1372.4	1383.8	1395.1	1406.3	1434.0	1461.5	1488.7	1515.9	1624.9	1735.9	1849.6
640 (493.21)	1312.2	1324.6	1336.7	1348.6	1360.4	1371.9	1383.3	1394.7	1405.9	1433.7	1461.1	1488.4	1515.6	1624.7	1735.7	1849.5
650 (494.90)	1311.4	1323.9	1336.1	1348.0	1359.8	1371.4	1382.8	1394.2	1405.4	1433.3	1460.8	1488.1	1515.3	1624.5	1735.5	1849.4
660 (496.58)	1310.6	1323.2	1335.4	1347.4	1359.2	1370.8	1382.3	1393.7	1405.0	1432.9	1460.4	1487.8	1515.0	1624.3	1735.4	1849.3
670 (498.24)	1309.9	1322.4	1334.8	1346.8	1358.6	1370.3	1381.8	1393.2	1404.5	1432.5	1460.0	1487.4	1514.7	1624.1	1735.2	1849.1
680 (499.88)	1309.1	1321.7	1334.1	1346.2	1358.1	1369.8	1381.3	1392.8	1404.1	1432.1	1459.7	1487.1	1514.5	1623.9	1735.1	1849.0
690 (501.50)	1308.3	1321.0	1333.4	1345.6	1357.5	1369.2	1380.8	1392.3	1403.6	1431.7	1459.3	1486.8	1514.2	1623.7	1734.9	1848.9
700 (503.10)	1307.5	1320.3	1332.8	1345.0	1356.9	1368.7	1380.3	1391.8	1403.2	1431.3	1459.0	1486.5	1513.9	1623.5	1734.8	1848.8
710 (504.68)	1306.7	1319.6	1332.1	1344.3	1356.3	1368.1	1379.8	1391.3	1402.7	1430.9	1458.6	1486.2	1513.6	1623.3	1734.6	1848.6

Table A.2–2 Cont.
Superheated Vapor Enthalpy (Btu/lb_m)

Abs. press. lb_f/sq. in. (sat. temp.)	Sat. liquid	Sat. vapor	Temperature—degrees fahrenheit 510°	520°	530°	540°	550°	560°	570°	580°	590°	600°	620°	640°	660°	680°
720 (506.25)	495.3	1200.7	1204.5	1214.2	1223.4	1232.2	1240.6	1248.7	1256.5	1264.1	1271.5	1278.7	1292.5	1305.9	1318.8	1331.4
730 (507.80)	497.2	1200.5	1202.7	1212.6	1221.9	1230.8	1239.3	1247.5	1255.4	1263.0	1270.4	1277.7	1291.7	1305.1	1318.1	1330.7
740 (509.34)	499.0	1200.2	1200.9	1210.9	1220.4	1229.4	1238.0	1246.3	1254.2	1261.9	1269.4	1276.7	1290.8	1304.3	1317.4	1330.0
750 (510.86)	500.8	1200.0		1209.3	1218.8	1227.9	1236.6	1245.0	1253.0	1260.8	1268.4	1275.7	1289.9	1303.5	1316.6	1329.4
775 (514.59)	505.3	1199.3		1205.0	1214.9	1224.3	1233.2	1241.8	1250.0	1258.0	1265.7	1273.2	1287.7	1301.5	1314.8	1327.7
800 (518.23)	509.7	1198.6		1200.5	1210.8	1220.5	1229.8	1238.6	1247.0	1255.2	1263.1	1270.7	1285.4	1299.4	1312.9	1325.9
825 (521.79)	514.0	1197.9			1206.6	1216.7	1226.2	1235.3	1243.9	1252.3	1260.3	1268.1	1283.1	1297.3	1311.0	1324.2
850 (525.26)	518.3	1197.1			1202.3	1212.7	1222.5	1231.9	1240.8	1249.3	1257.5	1265.5	1280.7	1295.2	1309.0	1322.4
875 (528.66)	522.5	1196.3			1197.8	1208.6	1218.8	1228.4	1237.6	1246.3	1254.7	1262.8	1278.4	1293.1	1307.1	1320.6
900 (531.98)	526.6	1195.4				1204.4	1215.0	1224.9	1234.3	1243.2	1251.8	1260.1	1275.9	1290.9	1305.1	1318.8
925 (535.24)	530.6	1194.6				1200.1	1211.0	1221.2	1230.9	1240.1	1248.9	1257.4	1273.5	1288.7	1303.1	1317.0
950 (538.42)	534.6	1193.7				1195.5	1206.9	1217.5	1227.4	1236.9	1245.9	1254.6	1271.0	1286.4	1301.1	1315.2
975 (541.55)	538.5	1192.8					1202.7	1213.6	1223.9	1233.6	1242.9	1251.7	1268.5	1284.2	1299.1	1313.3
1000 (544.61)	542.4	1191.8					1198.3	1209.7	1220.3	1230.3	1239.8	1248.8	1265.9	1281.9	1297.0	1311.4
1050 (550.57)	550.0	1189.9						1201.5	1212.8	1223.4	1233.4	1242.9	1260.7	1277.2	1292.8	1307.6
1100 (556.31)	557.4	1187.8						1192.6	1204.8	1216.1	1226.7	1236.7	1255.3	1272.4	1288.5	1303.7
1150 (561.86)	564.6	1185.6							1196.3	1208.5	1219.7	1230.2	1249.6	1267.5	1284.1	1299.8
1200	571.7	1183.4							1187.2	1200.4	1212.4	1223.5	1243.9	1262.4	1279.6	1295.7

TABLE A.2–2 CONT.
SUPERHEATED VAPOR ENTHALPY (Btu/lb$_m$)

Abs. press. lb$_f$/sq. in. (sat. temp.)	*Temperature—degrees fahrenheit*														
	720°	*740°*	*760°*	*780°*	*800°*	*820°*	*840°*	*860°*	*880°*	*900°*	*1000°*	*1100°*	*1200°*	*1400°*	*1600°*
720 (506.25)	1355.7	1367.6	1379.3	1390.8	1402.3	1413.6	1424.9	1436.1	1447.2	1458.3	1513.3	1568.1	1623.1	1734.4	1848.5
730 (507.80)	1355.2	1367.1	1378.8	1390.4	1401.8	1413.2	1424.5	1435.7	1446.8	1457.9	1513.0	1567.9	1622.9	1734.3	1848.4
740 (509.34)	1354.6	1366.5	1378.3	1389.9	1401.4	1412.7	1424.0	1435.3	1446.4	1457.6	1512.7	1567.6	1622.7	1734.1	1848.3
750 (510.86)	1354.0	1366.0	1377.7	1389.4	1400.9	1412.3	1423.6	1434.9	1446.1	1457.2	1512.4	1567.4	1622.4	1734.0	1848.1
775 (514.59)	1352.5	1364.6	1376.5	1388.2	1399.8	1411.2	1433.9	1422.6	1445.1	1456.3	1511.7	1566.8	1621.9	1733.6	1847.8
800 (518.23)	1351.0	1363.2	1375.2	1387.0	1398.6	1410.1	1421.6	1432.9	1444.2	1455.4	1511.0	1566.2	1621.4	1733.2	1847.5
825 (521.79)	1349.5	1361.8	1373.8	1385.7	1397.4	1409.0	1420.5	1431.9	1443.3	1454.5	1510.3	1565.6	1620.9	1732.8	1847.2
850 (525.26)	1348.0	1360.4	1372.5	1384.5	1396.3	1407.9	1419.5	1430.9	1442.3	1453.6	1509.5	1564.9	1620.4	1732.4	1846.9
875 (528.66)	1346.5	1359.0	1371.2	1383.2	1395.1	1406.8	1418.4	1429.9	1441.4	1452.7	1508.8	1564.3	1619.9	1732.0	1846.6
900 (531.98)	1345.0	1357.5	1369.9	1382.0	1393.9	1405.7	1417.4	1428.9	1440.4	1451.8	1508.1	1563.7	1619.3	1731.6	1846.3
925 (535.24)	1343.4	1356.1	1368.5	1380.7	1392.7	1404.6	1416.3	1427.9	1439.5	1450.9	1507.3	1563.1	1618.8	1731.2	1845.9
950 (538.42)	1341.9	1354.7	1367.2	1379.5	1391.6	1403.5	1415.3	1426.9	1438.5	1450.0	1506.6	1562.5	1618.3	1730.8	1845.6
975 (541.55)	1340.3	1353.2	1365.8	1378.2	1390.4	1402.3	1414.2	1425.9	1437.6	1449.1	1505.9	1561.9	1617.8	1730.4	1845.3
1000 (544.61)	1338.7	1351.7	1364.4	1376.9	1389.2	1401.2	1413.1	1424.9	1436.6	1448.2	1505.1	1561.3	1617.3	1730.0	1845.0
1050 (550.57)	1335.5	1348.8	1361.7	1374.3	1386.7	1398.9	1411.0	1422.9	1434.7	1446.3	1503.7	1560.0	1616.2	1729.2	1844.4
1100 (556.31)	1332.2	1345.8	1358.9	1371.7	1384.3	1396.6	1408.8	1420.8	1432.7	1444.5	1502.2	1558.8	1615.2	1728.4	1843.8
1150 (561.86)	1328.9	1342.7	1356.1	1369.1	1381.8	1394.3	1406.6	1418.8	1430.7	1442.6	1500.7	1557.6	1614.1	1727.7	1843.1
1200 (567.22)	1325.6	1339.6	1353.2	1366.4	1379.3	1392.0	1404.4	1416.7	1428.8	1440.7	1499.2	1556.4	1613.1	1726.9	1842.5

TABLE A.2–2 CONT.
SUPERHEATED VAPOR ENTHALPY (Btu/lb$_m$)

Abs. press. lb$_f$/sq. in. (sat. temp.)	*Sat. liquid*	*Sat. vapor*	*Temperature—degrees fahrenheit* 580°	590°	600°	610°	620°	630°	640°	650°	660°	670°	680°	690°	700°	720°
1250 (572.42)	578.6	1181.0	1191.7	1204.6	1216.5	1227.5	1237.9	1247.8	1257.2	1266.3	1275.0	1283.4	1291.5	1299.5	1307.2	1322.1
1300 (577.46)	585.4	1178.6	1182.4	1196.4	1209.0	1220.7	1231.6	1242.0	1251.8	1261.2	1270.2	1278.9	1287.3	1295.4	1303.4	1318.6
1350 (582.35)	592.1	1176.1		1187.6	1201.2	1213.6	1225.1	1236.0	1246.2	1256.0	1265.3	1274.3	1282.9	1291.3	1299.5	1315.1
1400 (587.10)	598.7	1173.4		1178.1	1193.0	1206.2	1218.4	1229.8	1240.4	1250.6	1260.3	1269.6	1278.5	1287.1	1295.5	1311.5
1450 (591.73)	605.2	1170.7			1184.1	1198.4	1211.3	1223.3	1234.5	1245.0	1255.1	1264.7	1273.9	1282.8	1291.4	1307.8
1500 (596.23)	611.6	1167.9			1174.5	1190.1	1203.9	1216.6	1228.3	1239.3	1249.8	1259.7	1269.3	1278.4	1287.2	1304.1
1600 (604.90)	624.1	1162.1				1171.5	1187.8	1202.2	1215.2	1227.3	1238.7	1249.4	1259.6	1269.3	1278.7	1296.4
1700 (613.15)	636.3	1155.9					1169.2	1186.1	1201.0	1214.4	1226.8	1238.4	1249.3	1259.7	1269.7	1288.4
1800 (621.03)	648.3	1149.4						1167.7	1185.1	1200.3	1214.0	1226.7	1238.5	1249.7	1260.3	1280.1
1900 (628.58)	660.1	1142.4						1145.8	1167.0	1184.7	1200.2	1214.2	1227.1	1239.1	1250.4	1271.4
2000 (635.82)	671.7	1135.1							1145.6	1167.0	1184.9	1200.6	1214.8	1227.8	1240.0	1262.3
2100 (642.77)	683.3	1127.4								1146.3	1167.7	1185.8	1201.6	1215.9	1229.0	1252.9
2200 (649.46)	694.8	1119.2								1121.0	1147.8	1169.1	1187.1	1203.0	1217.4	1243.0
2300 (655.91)	706.5	1110.4									1123.8	1150.0	1171.1	1189.1	1204.9	1232.6
2400 (662.12)	718.4	1101.1										1128.2	1152.9	1173.7	1191.5	1221.6
2500 (668.13)	730.6	1091.1										1099.8	1132.3	1156.6	1176.8	1210.0
2600 (673.94)	743.0	1080.2											1107.1	1137.3	1160.6	1197.5
2700	756.2	1068.3											1072.8	1114.3	1142.5	1184.1

Table A.2–2 Cont.
Superheated Vapor Enthalpy (Btu/lb$_m$)

Abs. press. lb$_f$/sq. in (sat. temp.)	Temperature—degrees fahrenheit 740°	760°	780°	800°	820°	840°	860°	880°	900°	950°	1000°	1050°	1100°	1200°	1400°	1600°
1250 (572.42)	1336.5	1350.3	1363.7	1376.8	1389.6	1402.2	1414.6	1426.8	1438.9	1468.5	1497.7	1526.5	1555.1	1612.0	1726.1	1841.9
1300 (577.46)	1333.3	1347.3	1361.0	1374.3	1387.2	1400.0	1412.5	1424.8	1437.0	1466.9	1496.2	1525.1	1553.9	1611.0	1725.3	1841.3
1350 (582.35)	1330.0	1344.3	1358.2	1371.7	1384.8	1397.7	1410.3	1422.8	1435.1	1465.2	1494.7	1523.8	1552.6	1609.9	1724.5	1840.6
1400 (587.10)	1326.7	1341.3	1355.4	1369.1	1382.4	1395.4	1408.2	1420.8	1433.1	1463.5	1493.2	1522.4	1551.4	1608.9	1723.7	1840.0
1450 (591.73)	1323.4	1338.3	1352.6	1366.5	1380.0	1393.1	1406.0	1418.7	1431.2	1461.8	1491.6	1521.0	1550.1	1607.8	1722.9	1839.4
1500 (596.23)	1320.0	1335.2	1349.7	1363.8	1377.5	1390.8	1403.9	1416.7	1429.3	1460.1	1490.1	1519.6	1548.9	1606.8	1722.1	1838.8
1600 (604.90)	1313.0	1328.8	1343.9	1358.4	1372.5	1386.1	1399.5	1412.5	1425.3	1456.6	1487.0	1516.9	1546.4	1604.6	1720.5	1837.5
1700 (613.15)	1305.8	1322.3	1337.9	1352.9	1367.3	1381.3	1395.0	1408.3	1421.4	1453.2	1484.0	1514.1	1543.8	1602.5	1718.9	1836.2
1800 (621.03)	1298.4	1315.5	1331.8	1347.2	1362.1	1376.5	1390.4	1404.1	1417.4	1449.7	1480.8	1511.3	1541.3	1600.4	1717.3	1835.0
1900 (628.58)	1290.6	1308.6	1325.4	1341.5	1356.8	1371.5	1385.8	1399.7	1413.3	1446.1	1477.7	1508.5	1538.8	1598.2	1715.7	1833.7
2000 (635.82)	1282.6	1301.4	1319.0	1335.5	1351.3	1366.5	1381.2	1395.3	1409.2	1442.5	1474.5	1505.7	1536.2	1596.1	1714.1	1832.5
2100 (642.77)	1274.3	1294.0	1312.3	1329.5	1345.8	1361.4	1376.4	1390.9	1405.0	1438.9	1471.4	1502.9	1533.6	1593.9	1712.5	1831.2
2200 (649.46)	1265.7	1286.3	1305.4	1323.3	1340.1	1356.2	1371.5	1386.4	1400.8	1435.3	1468.2	1500.0	1531.1	1591.8	1710.9	1830.0
2300 (655.91)	1256.7	1278.4	1298.4	1316.9	1334.3	1350.8	1366.6	1381.8	1396.5	1431.6	1464.9	1497.1	1528.5	1589.6	1709.3	1828.7
2400 (662.12)	1247.3	1270.2	1291.1	1310.3	1328.4	1345.4	1361.6	1377.2	1392.2	1427.9	1461.7	1494.2	1525.9	1587.4	1707.7	1827.4
2500 (668.13)	1237.6	1261.8	1283.6	1303.6	1322.3	1339.9	1356.5	1372.5	1387.8	1424.2	1458.4	1491.3	1523.2	1585.3	1706.1	1826.2
2600 (673.94)	1227.3	1252.9	1275.8	1296.8	1316.1	1334.2	1351.4	1367.7	1383.4	1420.4	1455.1	1488.4	1520.6	1583.1	1704.4	1824.9
2700 (679.55)	1216.5	1243.8	1267.9	1289.7	1309.7	1328.5	1346.1	1362.9	1378.9	1416.6	1451.8	1485.5	1518.0	1580.9	1702.8	1823.7

Index

I

J

K

L

M

N

O

P